Managing biodiversity in agricultural landscapes

Conservation, restoration and rewilding

It is widely recognised that agriculture is a significant contributor to global warming and climate change. Agriculture needs to reduce its environmental impact and adapt to current climate change whilst still feeding a growing population, i.e. become more 'climate-smart'. Burleigh Dodds Science Publishing is playing its part in achieving this by bringing together key research on making the production of the world's most important crops and livestock products more sustainable.

Based on extensive research, our publications specifically target the challenge of climate-smart agriculture. In this way we are using 'smart publishing' to help achieve climate-smart agriculture.

Burleigh Dodds Science Publishing is an independent and innovative publisher delivering high quality customer-focused agricultural science content in both print and online formats for the academic and research communities. Our aim is to build a foundation of knowledge on which researchers can build to meet the challenge of climate-smart agriculture.

For more information about Burleigh Dodds Science Publishing simply call us on +44 (0) 1223 839365, email info@bdspublishing.com or alternatively please visit our website at www.bdspublishing.com.

Related titles:

Advances in Conservation Agriculture Volume 1
Print (ISBN 978-1-78676-264-1); Online (ISBN 978-1-78676-266-5, 978-1-78676-267-2)

Advances in Conservation Agriculture Volume 2
Print (ISBN 978-1-78676-268-9); Online (ISBN 978-1-78676-270-2, 978-1-78676-271-9)

Advances in Conservation Agriculture Volume 3
Print (ISBN 978-1-78676-475-1); Online (ISBN 978-1-78676-477-5, 978-1-78676-478-2)

Improving soil health
Print (ISBN 978-1-78676-670-0); Online (ISBN 978-1-78676-672-4, 978-1-78676-673-1)

Promoting pollination and pollinators in farming
Print (ISBN 978-1-80146-098-9); Online (ISBN 978-1-80146-100-9, 978-1-80146-101-6)

Protecting natural capital and biodiversity in the agri-food sector
Print (ISBN 978-1-80146-351-5); Online (ISBN 978-1-80146-352-2, 978-1-80146-353-9)

Reconciling agricultural production with biodiversity conservation
Print (ISBN 978-1-78676-348-8); Online (ISBN 978-1-78676-350-1, 978-1-78676-351-8)

Understanding and fostering soil carbon sequestration
Print (ISBN 978-1-78676-969-5); Online (ISBN 978-1-78676-971-8, 978-1-78676-972-5)

Chapters are available individually from our online bookshop:
https://shop.bdspublishing.com

climate•SMART•publishing
IN AGRICULTURAL SCIENCE

BURLEIGH DODDS SERIES IN AGRICULTURAL SCIENCE
NUMBER 149

Managing biodiversity in agricultural landscapes

Conservation, restoration and rewilding

Edited by Dr Nick Reid and Dr Rhiannon Smith, University of New England, Australia

burleigh dodds
SCIENCE PUBLISHING

Published by Burleigh Dodds Science Publishing Limited
82 High Street, Sawston, Cambridge CB22 3HJ, UK
www.bdspublishing.com

Burleigh Dodds Science Publishing, 1518 Walnut Street, Suite 900, Philadelphia, PA 19102-3406, USA

First published 2024 by Burleigh Dodds Science Publishing Limited

Library of Congress Control Number: 2024933617

British Library Cataloguing in Publication Data
A catalogue record for this book is available from the British Library

ISBN 978-1-80146-454-3 (print)
ISBN 978-1-80146-456-7 (PDF)
ISBN 978-1-80146-455-0 (ePub)
ISSN 2059-6936 (print)
ISSN 2059-6944 (online)

DOI: 10.19103/AS.2024.0134

Typeset by Deanta Global Publishing Services, Dublin, Ireland

Contents

Part 3 The role of government and the private sector in promoting on-farm conservation practices

Part 4 Habitat and animal rewilding

Part 5 Looking ahead

Series list

Title	Series number
Achieving sustainable cultivation of maize - Vol 1 From improved varieties to local applications Edited by: Dr Dave Watson, CGIAR Maize Research Program Manager, CIMMYT, Mexico	001
Achieving sustainable cultivation of maize - Vol 2 Cultivation techniques, pest and disease control Edited by: Dr Dave Watson, CGIAR Maize Research Program Manager, CIMMYT, Mexico	002
Achieving sustainable cultivation of rice - Vol 1 Breeding for higher yield and quality Edited by: Prof. Takuji Sasaki, Tokyo University of Agriculture, Japan	003
Achieving sustainable cultivation of rice - Vol 2 Cultivation, pest and disease management Edited by: Prof. Takuji Sasaki, Tokyo University of Agriculture, Japan	004
Achieving sustainable cultivation of wheat - Vol 1 Breeding, quality traits, pests and diseases Edited by: Prof. Peter Langridge, The University of Adelaide, Australia	005
Achieving sustainable cultivation of wheat - Vol 2 Cultivation techniques Edited by: Prof. Peter Langridge, The University of Adelaide, Australia	006
Achieving sustainable cultivation of tomatoes Edited by: Dr Autar Mattoo, USDA-ARS, USA and Prof. Avtar Handa, Purdue University, USA	007
Achieving sustainable production of milk - Vol 1 Milk composition, genetics and breeding Edited by: Dr Nico van Belzen, International Dairy Federation (IDF), Belgium	008
Achieving sustainable production of milk - Vol 2 Safety, quality and sustainability Edited by: Dr Nico van Belzen, International Dairy Federation (IDF), Belgium	009
Achieving sustainable production of milk - Vol 3 Dairy herd management and welfare Edited by: Prof. John Webster, University of Bristol, UK	010
Ensuring safety and quality in the production of beef - Vol 1 Safety Edited by: Prof. Gary Acuff, Texas A&M University, USA and Prof. James Dickson, Iowa State University, USA	011
Ensuring safety and quality in the production of beef - Vol 2 Quality Edited by: Prof. Michael Dikeman, Kansas State University, USA	012
Achieving sustainable production of poultry meat - Vol 1 Safety, quality and sustainability Edited by: Prof. Steven C. Ricke, University of Arkansas, USA	013
Achieving sustainable production of poultry meat - Vol 2 Breeding and nutrition Edited by: Prof. Todd Applegate, University of Georgia, USA	014

Introduction

Reports from bodies such as the Intergovernmental Science Policy Platform on Biodiversity and Ecosystem Services (IPBES) and the International Union for the Conservation of Nature (IUCN) suggest a 60% reduction in global wildlife populations in the last 40 years with 30,000 species in immediate danger of extinction, and up to 1 million species at risk, primarily due to loss of habitat from agricultural expansion. The magnitude of the global biodiversity crisis from the resulting loss of habitat led signatories to the UN Convention on Biological Diversity to commit to managing 30% of regional lands to conserve biodiversity by 2030. This volume reviews both the challenges as well as key principles, approaches and techniques in meeting this ambitious target.

Part 1 Principles

Chapter 1 sets the scene by reviewing key concepts in biodiversity management in agricultural landscapes and providing a theoretical framework to inform key choices e.g. between 'land sharing' and 'land sparing' approaches to biodiversity conservation, whilst recognising that agricultural land uses are diverse and shape landscapes in different ways.

As the chapter points out, provision of suitable habitat in, and among, farm properties is a fundamental requirement for biota to co-exist with agricultural production and identify key factors (such as issues of scale, context and habitat quality) that influence the value of individual habitat patches and their suitability for different conservation approaches. The chapter also highlights the importance of a landscape perspective for conservation that recognises both the spatial heterogeneity and temporal changes that occur in agricultural landscapes. A landscape perspective offers greater opportunity to manage key landscape properties (such as the overall extent of habitats, composition of land-uses, and landscape configuration) to achieve conservation outcomes. The chapter also highlights ways of managing trade-offs between agricultural production and nature conservation. Finally, the chapter discusses the importance of socio-ecological dimensions of nature conservation and that effective integration of agricultural production and nature conservation must consider the stakeholders involved, including the wellbeing of people and communities who live in agricultural landscapes.

Building on the theoretical framework set up in Chapter 1, Chapter 2 reviews key principles for successful implementation of 'land sharing' approaches where these are appropriate. The chapter examines the principles by which integrated farm management acts to use and cycle energy and matter over time and between different farming activities. Management interventions (e.g.

agronomy or choice of crop) act to regulate channels for the flow of energy and matter in a 'chain of effect' through living organisms and ecological processes to achieve high-level goals such as delivery of ecosystem services for the potential benefit of both society and Nature.

Based on international guidelines, and drawing from historical and current integrated practice, the chapter looks at underlying principles such as scales of interaction influencing agricultural systems and ways of increasing species, functional and habitat diversity. It then outlines key design principles for integrated farm management, looking at stores and fluxes, limits and safe ranges (within which each population or process can operate sustainably) as well as chains of effect that allow delivery of target outcomes. The chapter concludes with a case study on how an integrated farm management process can be successfully implemented in practice.

Building on a theme highlighted in Chapter 1, Chapter 3 outlines engagement frameworks and strategies for understanding and improving the involvement of farmers and their communities in ecological restoration initiatives. It considers approaches for bringing together participants with differing perspectives and interests to address environmental challenges, in ways that account for both biophysical and socio-political contexts. The presentation of two case studies highlights the context, challenges and promise of community engagement, along with overarching principles that support effective community-focused initiatives.

Part 2 Farmland conservation practices

Part 2 focusses on particular conservation practices, starting with Chapter 4 on soil health and ecological restoration. Soil health is central for provisioning and supporting ecosystem services for both agriculture and biodiversity. The chapter begins by reviewing the usefulness of soil health indicators to identify degradation gradients as a basis for designing appropriate restoration programmes. It emphasises the importance of integrating state-of-the-art knowledge of plant-microorganism interactions, the 'engine' driving many ecosystem functions, including the role of nitrogen-fixing microorganisms and mycorrhizal fungi, as well as key ecological concepts of plant-soil feedback and biodiversity-ecosystem function relationships, in developing a restoration strategy. The chapter also discusses how consideration of scale and connectedness in restoration practices is vital to maximise ecosystem multifunctionality. It shows how these principles can be delivered in practice by appropriate manipulation of the plant community, addition of biological agents and changing cultivation practices, including case studies on restoration of species-rich grassland and soil health restoration in semi-arid smallholder agroecosystems.

Heterogeneous agroecosystems with diverse natural areas are more resilient than homogeneous landscapes. As Chapter 5 discusses, one way of achieving this is by creating field margins in order to establish a network of semi-natural vegetation across the landscape. Field margins support biodiversity within agricultural systems and offer a range of ecosystem services. This chapter reviews the ways field margins benefit biodiversity, their design, management and enhancement with seed mixes, as well as the economics of field margins in creating more resilient agricultural landscapes.

Like field margins, farm hedgerows supply diverse benefits to agriculture including pollination, pest control and improved production related to biodiversity. They also provide additional ecosystem services including carbon sequestration, shelter for crops and livestock as well as aesthetic benefits. Hedgerows can be combined with semi-natural habitats and conservation initiatives such as wildflower strips along field margins to maintain and enhance biodiversity of native species in radically transformed agricultural landscapes. Chapter 6 reviews research on key aspects determining their effectiveness such as hedgerow age, height, width, length, opacity, density and species composition. It also discusses key management practices such as less frequent cutting to increase height and width, use of multiple native woody species and ensuring proximity to other semi-natural habitats. As it indicates, financial support and training of farmers in promoting hedgerow biodiversity could reduce farm management costs and increase farm income, assisting with the transition to sustainable agriculture.

Part 3 The role of government and the private sector in promoting on-farm conservation practices

Chapters in Part 3 discuss how governments and the private sector interact with farmers in promoting biodiversity conservation.

Building on themes in Chapter 3, Chapter 7 focuses on the ongoing development of England's new Environmental Land Management (ELM) schemes designed to support a transition to more sustainable and Nature-friendly forms of farming. The co-design process adopted by government in ELM development reflected both a step-change towards participatory policymaking and a commitment to stakeholder engagement and inclusion. This included a programme of tests, trials and pilots (T&Ts) both of individual elements of the ELMs framework as well as longer-term and larger-scale pilot projects to explore change at catchment scale.

By drawing together academic literature, policy updates and research with farmers in the Sustainable Farming Incentive (SFI) National Pilot, this chapter casts a critical eye on successes and challenges in the process of co-design of the SFI National Pilot with farmers. The SFI pilot has offered a promising model for managing biodiversity in agricultural landscapes, emphasising flexibility, co-design and farmer engagement. Farmers' views and experiences have been integrated into scheme design at multiple stages, feeding into the development of the main SFI scheme which continues to evolve and has begun to garner improved levels of interest from the wider farming community.

However, the chapter highlights challenges such as a sometimes uncertain political landscape and stable long-term vision, a short time scale for the pilot, the lack of 'control farms' to measure the environmental effectiveness of measures taken and the lack of a sufficiently representative sample of farms (in terms of farm type and size) involved in the pilot. More fundamentally, experience shows the limitations of co-design if it only seen as one discrete 'step' in policy development, rather than an outlook embedded in the entire process, potentially weakening a shared sense of ownership and responsibility between policymakers and farmers. These issues need to be addressed in future ELM development. Overall, however, the SFI Pilot demonstrates that, even in an imperfect policy development process, co-design can have beneficial results for both farmers and Nature.

Complementing Ch 7, Chapter 8 reviews experience since the 1990s with agri-environmental schemes (AES) in North America involving payment for ecosystem services (PES) as a central component. The chapter reviews research assessing the economic and environmental outcomes of PES programs such as the Conservation Reserve Program, the Environmental Quality Incentives Program and the Wetland Reserve Programs. This research shows that, when well designed, PES can be a major tool to introduce new practices or crops that contribute to more sustainable and biodiverse farmed landscapes. However, more research is needed to assess and improve their design. Future AES improvements may arise also from enhanced monitoring and assessment through advances in information technologies such as remote sensing tools and artificial intelligence, particularly in measuring environmental benefits

Sitting alongside Chapters 7 and 8, Chapter 9 reviews developments in agri-environment schemes (AES) in Australia. Agri-environment schemes have been a part of Australia's environmental policy in some form for more than two decades. In contrast to the broad-ranging and comparatively generic agri-environment schemes employed in Europe and in the USA, schemes in Australia have tended to be smaller in geographic scope and more targeted, both in terms of environmental objectives and in the activities they promote.

More recently, schemes have evolved from community-based initiatives addressing land degradation issues such as soil salinity and poor water quality, and with a focus on extension and education, to the use of market-based mechanisms aimed at the restoration and protection of biodiversity values on farms. Federal and state governments are also increasingly designing schemes which integrate emerging carbon and nature markets in part to defray the cost of biodiversity conservation by attracting private investment. The chapter provides examples of these schemes, including a case study on the federal government's Carbon + Biodiversity Pilot scheme which has trialled providing biodiversity co-payments to carbon projects (e.g. involving measure relating to planting, establishment and maintenance of new crop systems) to generate both carbon and biodiversity benefits.

The chapter discusses the various opportunities and risks associated with this shift in scheme design. While advantages include a more diverse choice of programs to suit farmers' individual needs, challenges include the credibility of still developing carbon and biodiversity offsetting markets, as well as concerns about their actual benefits for both conservation and farmers. The introduction of markets into scheme design warrants more targeted research on barriers and motivations for participation in this evolving policy space. More broadly, there is an ongoing need for rigorous long-term ecological and socio-economic monitoring of schemes. Despite significant public investment in large agri-environment schemes in Australia over the past two decades, there remains a relative paucity of published research into their effectiveness. This includes a continued lack of data on the costs of designing and implementation, at the scheme and individual project level. A lack of knowledge of the cost-effectiveness of alternative strategies for achieving environmental outcomes in agricultural landscapes will hamper the efficiency of markets going forward.

The final missing research component is on optimal governance structures to promote environmental integrity and efficiency. The current system based on a high level of centralised control by government agencies, complemented by a reliance on verification by third-party contractors paid by scheme participants, has limited external scrutiny and not always been effective. There is a greater need to involve external parties (e.g. non-governmental organisations) to help improve outcomes and community confidence in environmental markets. There is also a need to recognise that many farmers already value and protect biodiversity and who may only require relatively modest financial and technical support to build on what they already do.

Part 4 Habitat and animal rewilding

Chapters in Part 4 review the complex issues in balancing 'land sharing' with 'land sparing' and hybrid approaches to achieve successful habitat and animal rewilding outcomes.

Building on themes in Chapters 3 and 7, Chapter 10 looks at lessons from the experience of the UK in how best to implement sustainable land management (SLM) programmes. The chapter explores how government schemes have evolved alongside the principles of SLM and how the role and importance of the landowner has changed from a passive recipient to an active participant in the design process.

However, research suggests mixed evidence on the financial viability of sustainable land use change programmes for different types of farm and local context. This reflect the importance of considering economic, environmental and social capital when attempting to implement change, particularly for more ambitious restoration projects. Successful implementation of SLM requires appropriate targeting of multiple, locally-relevant outcomes, rather than development of broad 'one size fits all' approaches with a single target. It also requires close collaboration and communication between policy makers, researchers, landowners advisors and the wider community at all stages of scheme development and implementation to ensure higher levels of uptake, local community buy-in and better biodiversity outcomes at all scales. In addition, success depends on implementation of verifiable outcome-based performance assessments tailored to the environmental context of each farm business to embed flexibility and adaptive management, whilst giving landowners the confidence and expertise to engage. Finally, schemes need financial remuneration which fairly values of the output and cost of implementation. To illustrate how these elements can be combined, the chapter includes a case study of a successful SLM programme in Wales. Examples like this show that that the recovery of nature and production of food are compatible and achievable objectives if programmes are properly designed and implemented.

Peatlands fulfil essential ecosystem services such as water purification, biodiversity conservation and carbon retention. Restoration of degraded peatlands is therefore key to enhance these services. Chapter 11 describes the hydrological and ecological functioning of undisturbed peatlands, as well as changes resulting from humans activity. It describes the effects of these changes at the landscape level and on peatland structure and function. The cumulative effects of such changes on local site conditions are called land-use legacies, which have significant impacts on restoration prospects of degraded peatlands. The chapter goes on to discuss how to set appropriate restoration targets as well as different choices for landscape planning and management. To illustrate these approaches at work, the chapter includes a case study of restoration of a peatland in Belgium.

As Chapter 12 highlights, natural and semi-natural grasslands and rangelands are disturbance-driven ecosystems that cover 40% of the Earth's land area. Grasslands have often been subjected to agricultural improvement (e.g. sown plant species and nutrient additions), resulting in compromised biodiversity. If the seed bank remains intact and degradation is still in its early stages, rewilding (passive restoration) may restore biodiversity. Otherwise, active restoration may be required e.g. by using 'green hay', biological soil crusts, native forbs, strip seeding and restoration islands. Prescribed burning may be useful to regenerate native vegetation or control invasive species on grasslands and rangelands. A case study from the USA, the Watershed Restoration Initiative, along with a brief discussion of the habitat requirements of the iconic Greater sage-grouse, is provided to illustrate many of these principles in action.

The restoration challenges facing south-eastern Australia's woodlands are not dissimilar to challenges which characterize many agricultural landscapes internationally, and thus form the basis of the case study featured in this chapter, incorporating research conducted since 1998. A key insight from Chapter 13 is that factors operating at multiple scales are critical for informing effective restoration to maximize landscape-scale biodiversity values. The chapter presents a restoration hierarchy to highlight how factors ranging from the individual tree level to farm- and landscape- scales need to be considered in restoration efforts. It then outlines some key lessons for effective translation into on-ground action.

Rewilding is the deliberate (re)introduction of animal species into local or regional areas with the joint aims to assist the conservation of those species and promote the re-establishment of lost or degraded ecological processes. Chapter 14 highlights that ecosystem services obtained by returning wild animals to agricultural landscapes can outweigh disservices (e.g. livestock predation). Balancing of economic and environmental gains in animal rewilding requires effective characterization of wild provisioning services (and buy-in from key stakeholders), coordinating broad spatial heterogeneity in land use, and employing the right mix of coexistence conservation tactics. The chapter uses the examples of two emblematic Australian species, kangaroos and dingoes, to illustrate how these different elements can be combined.

Building on the framework in the previous chapter, Chapter 15 outlines the benefits and the challenges of rewilding in Australia and New Zealand. After reviewing the benefits of rewilding, the chapter outlines key steps in developing rewilding programmes. These include identifying and engaging with key stakeholders, selection of rewilding species and sites as well as appropriate monitoring measures. As the chapter highlights, much successful rewilding occurs in places where invasive species have been removed. Four case studies are presented to illustrate rewilding in practice. The first two describe rewildings on formerly degraded agricultural land, while the third posits a conceptual rewilding program using a continental top predator, the dingo. The fourth case study describes a digitally-enabled survey and monitoring system on a mixed-agriculture farm that provides a step towards rewilding such landscapes in future.

Part 5 Looking ahead

As noted earlier, the magnitude of the global biodiversity crisis from habitat habitat led signatories to the UN Convention on Biological Diversity to commit to managing 30% of regional lands to conserve biodiversity by 2030. This chapter returns to this fundamental challenge. It reviews current bottom-up and top-down approaches to achieving this ambitious goal. It argues that agricultural regions where wild nature is now closer to 1% than 30% by area are unlikely to retain some or perhaps most native species using bottom-up biodiversity-friendly farm management approaches, valuable as these are. The solution in such regions is the restoration of large areas of farmland, perhaps whole farms and clusters of farms, to habitats managed for native biodiversity. This will have to occur either through purchase of large areas of farmland by governments and the non-profit and private sectors and their ecological restoration, or the introduction of schemes that see farmers paid to undertake the restoration. The chapter highlights how this might be achieved through a case study based on the Mount Lofty region of South Australia. It also discusses ongoing challenges in meeting this goal by assessing the impact of over-extraction of water on the ecological health of the Murray–Darling Basin in south-eastern Australia and the effectiveness of attempts to address this problem. The chapter highlights ongoing research needs to achieve biodiversity goals

Part 1

Principles

Chapter 1

Key concepts for conserving biodiversity in agricultural landscapes

Andrew F. Bennett, Angie Haslem*, Frederick W. Rainsford*, Alex C. Maisey* and James Q. Radford*, La Trobe University, Australia*

1 Conserving biodiversity in agricultural landscapes - threats and opportunities

Throughout recorded history, humans have modified the landscapes in which they live to provide for their material needs (Ellis, 2021). Today, agricultural land use to produce food, fibre and biofuels to sustain the human population (~8 billion and increasing) is the most widespread and dominant land use on Earth and has profound consequences for biodiversity (Foley et al., 2005; Ramankutty et al., 2018). The vast scale and extent of agricultural enterprises worldwide means that the future of biodiversity will, in large part, be determined by how effectively species, communities and ecological processes can be sustained in and amongst farmlands (Kremen and Merenlender, 2018; Balmford, 2021). This poses a major challenge because extensive loss of plant and animal species has

* Department of Environment & Genetics and the Research Centre for Future Landscapes, La Trobe University, Melbourne, Victoria 3086, Australia.

http://dx.doi.org/10.19103/AS.2024.0134.01

occurred in agricultural landscapes worldwide and continues to occur under current land-use practices. To counter ongoing loss requires an understanding of the factors that affect the conservation of biota in diverse agricultural landscapes (e.g. Newton, 2017; Fahrig et al., 2015; Kleijn et al., 2011) and applying such knowledge through practical guidance for effective on-ground management (e.g. Breckwoldt, 1983; Norton and Reid, 2013; Macdonald and Feber, 2015; Lindenmayer et al., 2016; Balmford, 2021). It also demands a major commitment, politically and financially, to create and implement measures to retain and enhance biodiversity in agricultural landscapes.

In this chapter we introduce key concepts for the conservation and management of biodiversity in agricultural landscapes (Table 1). First, we outline the threat and the opportunity from a global perspective. Second, we provide an overview of some of the diverse types of agricultural land uses worldwide and the way in which they shape landscapes. Third, we emphasise the provision of suitable habitat as a fundamental requirement for the occurrence of biota in agricultural landscapes and note key factors that influence the value of individual habitat patches. Fourth, we highlight the importance of a landscape perspective for conservation that recognises both the spatial and temporal heterogeneity inherent in agricultural landscapes. Fifth, recognising the inherent trade-off between agricultural production and conservation where land is limited, we summarise the land-sparing–land-sharing paradigm as a means of achieving both objectives. Finally, we note the importance of socio-ecological dimensions of nature conservation and that effective integration of agricultural production and nature conservation must consider the stakeholders involved, including the well-being of people and communities who live in agricultural landscapes.

1.1 Threats of an expanding agricultural footprint to global conservation

Croplands and pasture together occupy at least 34% of the terrestrial, ice-free area of the planet (Ramankutty et al., 2008); additional areas with an agricultural imprint include some semi-arid and arid rangelands, grazed forests and woodlands, and tree plantations (Ramankutty et al., 2008). The demand for land for agriculture is predicted to expand in future decades, driven both by an increasing human population and increasing per capita consumption (Tilman et al., 2001; Green et al., 2005; Laurance et al., 2014). Advances associated with the Green Revolution from the 1970s onwards – such as improvements in plant breeding, use of chemical fertilisers and pesticides, and harnessing of fossil fuel energy sources – have led to increased food production per hectare of land (Ramankutty et al., 2018), but the capacity to feed a human population projected to reach 10 billion or more this century will demand further agricultural lands. Clearly, limiting human population growth is a fundamental

Table 1 Key concepts for the conservation of biodiversity in agricultural landscapes

1. Agricultural landscapes have an essential role in biodiversity conservation.	• The global distribution and extent of agricultural landscapes means that how they are managed has profound implications for future biodiversity conservation. • Agricultural land uses continue to expand, particularly in the tropics. Unless carefully managed, such expansion will foreclose future options for biodiversity conservation.
2. Agricultural land uses have significant impacts on the environment. Improving the compatibility of agricultural production and biodiversity conservation is a critical global issue.	• Agricultural land uses have significant impacts on the environment, including the loss and degradation of native vegetation, alteration to hydrological cycles and the global carbon cycle, spread of exotic species and the use of pesticides. • Agricultural land uses can be managed in ways that support and sustain components of biodiversity on farms – if there is a commitment to do so. • Targeted incentives and co-operative schemes offer ways to enhance nature conservation in agricultural landscapes.
3. Agricultural land uses worldwide are diverse. They shape the structure of agricultural landscapes in different ways.	• Agricultural land uses differ in relation to their scale, the types and amounts of external inputs (e.g. energy, chemicals and mechanisation) and the degree to which they modify natural ecosystems. • Management to enhance biodiversity must be tailored to the agricultural system and the way it influences landscape structure and the socio-ecological context.
4. Availability of suitable habitat for target species is essential for conservation in agricultural landscapes.	Many types of structural components can provide habitat in agricultural landscapes. • Natural elements – e.g. forest patches, wetlands, streams • Modified elements – e.g. hedges, shelterbelts, scattered trees • Anthropogenic elements – e.g. ponds, dams, buildings, ditches, croplands
5. Planning and management for conservation requires a landscape perspective.	A landscape perspective is important for ecological understanding and effective planning and management in agricultural regions: • Species move between different habitats in the landscape, multiple habitats of sufficient total area are needed for viable populations, and the value of habitats is influenced by surrounding land uses; • Many disturbance processes operate at landscape/regional scales; • This scale is relevant to planning by government agencies and community organisations; and • Landholders and local communities in the area must be part of the solution.

(*Continued*)

Table 1 (*Continued*)

6. Agricultural landscapes are mosaics of land uses. Properties of the landscape mosaic can be managed to enhance biodiversity conservation.	Key properties of agricultural mosaics that influence conservation values include: • the total extent of suitable habitat for a species or species guild; • composition of the mosaic, including the heterogeneity of land uses and the occurrence of keystone habitat features; • spatial configuration of habitat elements in the landscape, including the spatial scale and grain of land uses, the degree of aggregation of habitat patches, and the isolation or connectivity of habitats; and • geographic location (and environmental context) of the landscape.
7. Agricultural landscapes change through time. Temporal changes are an important influence on the potential value for biodiversity.	The effects of different kinds of temporal change (seasonal, multi-year, long-term) must be considered in planning for conservation in agricultural landscapes. Key issues include: • the frequency and scale of changes; • interactions between different kinds of change that magnify their effect; • time lags in experiencing the full effects of changes; • the essentially irreversible nature of some changes; and • land-use history as an important driver of present-day and future patterns.
8. Land resources are finite, so trade-offs are required between land allocated to agricultural production and nature conservation.	Land sparing and land sharing represent contrasting ends of a gradient of options for agricultural landscapes. • Land sparing aims for high agricultural yield on farmland committed to production such that other lands can be set aside ('spared') for conservation. • Land sharing involves 'wildlife friendly' farming with production and conservation on the same farmland, accepting that agricultural yield may be reduced. • Empirical evidence supports 'sparing' as the least-worst option for most species. • Most agricultural landscapes will have an intermediate position with areas set aside for nature but also modified habitats amongst productive farmland.
9. To achieve effective integration of agriculture and nature conservation, people and communities who live in the landscape must be part of the solution.	• In contrast with conservation reserves, agricultural landscapes encompass diverse land uses, different land tenures, multiple stakeholders with differing priorities and complex disturbance pressures. • To achieve positive conservation outcomes requires effective engagement with stakeholders, including people and communities who live in the landscape. • Working with multiple stakeholders requires rigorous processes for developing a shared vision, setting objectives, making decisions, allocating resources and responsibilities, resolving conflict and monitoring outcomes.

issue to address, but in the meantime the pressure for new agricultural lands will fall disproportionately on tropical regions, particularly in South America and Sub-Saharan Africa (Laurance et al., 2014; Oakley and Bicknell, 2022). For example, the 200 M ha Cerrado region of tropical savanna in South America, a biodiversity hotspot, is under enormous pressure from ongoing clearing for pasture and to expand agricultural crops (e.g. soybeans and sugar cane), with deforestation rates (2002–11) approximately 2.5 times higher than in the Amazon region (Strassburg et al., 2017).

A second challenge is that the production of foods and other products essential to sustain the human population poses a massive, ongoing cost to the environment and the life-support systems that sustain all of life, including humans (Tilman et al., 2001; Foley et al., 2005). Global consequences of agricultural land uses include alteration to the global carbon cycle; changes to regional and local climates; transformation to hydrologic cycles through alteration of flows, impoundments and extraction of water for irrigation; the use of growing volumes of pesticides; increasing use and release of fertilisers that alter ecosystem function; deliberate or inadvertent introduction of thousands of invasive plant and animal species; and massive loss, transformation or degradation of habitats on which native plant and animal species depend (Tilman et al., 2001; Foley et al., 2005; Ramankutty et al., 2018).

A third challenge is that current trends in land use and the projected expansion of agricultural lands limit or foreclose future options for nature conservation. Disproportionate clearing and agricultural transformation of fertile soils at lower elevations and fragmentation of formerly extensive forests and woodlands into a multitude of smaller patches amongst expanding farmland set in train long-term processes of ecosystem change and decline for plant and animal species (Saunders et al., 1991; Laurance and Bierregaard, 1997). The rate of landscape change compounds the challenge associated with the loss of future options: during the twentieth century there has been an unprecedented expansion of croplands and pastures globally (e.g. croplands increasing from ~8.9 million km^2 to 15 million km^2) (Ramankutty et al., 2018). While this expansion has mostly halted in the developed world, expansion continues in developing countries (Green et al., 2005), particularly in Asia, Africa and South America. Further decline and extinction of species, the endpoint of such sustained pressures, is inevitable unless pro-active measures are taken (Williams et al., 2021).

Conservation reserves have a critical role in the preservation of Earth's biodiversity. The global reserve system continues to expand, but the incomplete representation of ecosystems, the often-small size of reserves and their geographic location and context mean that they cannot, and will not, protect all species. Many species occur in ecosystems poorly represented in reserve systems or now depend on anthropogenic land uses for persistence: examples include the endangered Australasian Bittern *Botaurus poiciloptilus* in rice fields in south-eastern Australia (Herring et al., 2019) and the Corncrake *Crex crex* in Europe (Green et al., 1997), both of which depend heavily on habitats in productive farmland for breeding. Nature conservation in, and among, agricultural lands has a critical and complementary role to the conservation reserve system.

1.2 Opportunities for biodiversity conservation

While recognising the severity and urgency of the global threat to conservation, there are numerous opportunities to maintain and enhance biodiversity in agricultural landscapes. Here, we note three broad areas that encompass a range of opportunities.

First, agricultural production and nature conservation can be managed as complementary activities, such that nature provides a benefit towards production (e.g. Harvey et al., 2008; Karp et al., 2013; Norton and Reid, 2013; Lindenmayer et al., 2022). The 'natural capital' on farms (e.g. healthy soils, wetlands, rivers, scattered remnant trees), including the diversity of plants and animals, provides ecosystem services that benefit farm production, such as pollination of crops and orchards, breakdown of organic material and recycling of nutrients, filtration of water, regulation of insect pests and moderation of the local climate (e.g. Bianchi et al., 2006; Albrecht et al., 2020; Kolkert et al., 2020). For example, in a coffee farming area in Costa Rica, insectivorous birds reduced infestation of a damaging insect pest (the coffee berry borer beetle) by ~50%, substantially reducing crop damage; with more farmland forest cover increasing pest removal (Karp et al., 2013). Nature also provides broader ecosystem services for people living in agricultural landscapes and in the wider community, services such as water filtration, carbon sequestration and storage, aesthetic pleasure from wildlife and attractive places, and opportunities for solitude and spiritual renewal.

The complementarity of farming and nature conservation is emphasised in a range of approaches including those termed eco-agriculture (Scherr and McNeely, 2008), agroforestry (Schroth et al., 2004), organic farming (Hole et al., 2005), natural asset farming (Lindenmayer et al., 2022), working lands conservation (Kremen and Merenlender, 2018), regenerative agriculture (Giller et al., 2021) and land sharing (Phalan et al., 2011). Common elements of these (non-exclusive) approaches include integration of natural or semi-natural vegetation within the production system, low input of chemicals and diverse agricultural products (cf. extensive monoculture), typically creating a heterogenous farm landscape with many edges and ecotones. The first-hand experience of practising farmers is an effective means of sharing, communicating and advocating new approaches and opportunities for combining nature conservation and agriculture (e.g. Campbell, 1994; Doley, 2003; Norton and Reid, 2013; Massy, 2020).

A second broad area is that in which society uses financial incentives or regulation to achieve conservation benefits for the public good in agricultural landscapes. Worldwide, many policies, strategies and incentive schemes are directed towards promoting nature conservation in agricultural landscapes. A prominent example is the range of agri-environment measures implemented in Europe, supported by billions of Euros annually (Batáry et al., 2015). Such measures include, for example, reduction in fertiliser inputs in pastures and crops; restoration of small woods in farmland to enhance habitats for bird species (Bellamy et al., 2022); increasing species richness (plants, invertebrates) in hay meadows (Knop et al., 2006); planting flower strips to support pollinating insects and natural predators (Albrecht et al., 2020); protection of environmentally sensitive areas; and countering farmland abandonment. Knowledge of the biodiversity benefits of such measures is growing (e.g. Kleijn and Sutherland, 2003; Batáry et al., 2015; Redhead et al., 2022), and it is essential that their effectiveness continues to be tested and reviewed. A meta-analysis concluded that agri-environment schemes aimed at enhancing areas *out of production* (e.g. field margins, hedgerows) are more effective at enhancing biodiversity than those targeted at *productive areas* (e.g. grasslands, crops) (Batáry et al., 2015), yet the productive areas are also critical for some species.

A third and growing area for enhancing nature conservation in agricultural landscapes is through the market paying a premium for sustainably produced food and fibre or international trade agreements that specify certain environmental standards for products to be traded. Globally, there is growing attention to initiatives that certify that farms and enterprises comply with environmental and social welfare standards in the production of agricultural products (Blackman and Rivera, 2010). For example, Kilian et al. (2004) reported that the market for sustainable coffee production (e.g. Fair Trade) was growing rapidly, with an annual growth rate greatly exceeding the increase in general worldwide consumption. However, the effectiveness of such schemes in achieving beneficial environmental outcomes on coffee farms is uncertain (e.g. Blackman and Rivera, 2010).

2 Diversity of agricultural land uses and landscapes

Many types of agricultural land uses worldwide provide food, fibre and fuels for the human population (Fig. 1). The specific measures to integrate conservation with agricultural production vary between these land uses. Enhancing biodiversity in a landscape of rice paddies in Taiwan, a silvo-pastoral system in Portugal, extensive cornfields in USA, subsistence agriculture in Papua New Guinea, or cattle grazing on rangelands in Tanzania, each requires specific attention to how important concepts can be most effectively applied. A common theme, however, is to consider the way people and agricultural land uses shape the composition and spatial pattern of the landscape.

To illustrate the diversity of agricultural systems, we have arrayed several broad types in relation to two major dimensions (Fig. 2). A first dimension is the scale of management as indicated by the relative size of properties and their operational units (e.g. fields, paddocks, orchards, paddies). Agricultural systems with smaller operational units often tend to have a greater intensity of human activity (e.g. daily hours worked per area of land) and greater spatial heterogeneity of land uses in a 'fine-grained' landscape (e.g. compare Fig. 1d and e, cropping landscapes). A second dimension is the level of external inputs, such as fertilisers, pesticides, use of machinery and use of fossil fuels for energy. A third potential dimension is the degree of transformation of the natural environment associated with the particular agricultural system (shown by shading, Fig. 2).

The following examples illustrate the variety of agricultural land uses and the ways they interact with, and shape, the landscape. These are illustrative, rather than comprehensive and not mutually exclusive: e.g. cropping and livestock production frequently co-occur on the same farm property.

1 Livestock production based on extensive areas of (semi) natural vegetation

Pastoralism of domestic stock (e.g. cattle, sheep, goats, yaks, alpacas) in rangeland, savanna and alpine landscapes dominated by natural vegetation (e.g. grasslands and shrublands) is widespread in regions such as semi-arid Australia, western USA, Mongolia and southern Africa (e.g. Ludwig et al., 1996; Batpurev et al., 2022) (Fig. 1a). This type of system is characterised by the relatively large areas grazed, the high mobility of stock (and often people), low external inputs of chemicals and energy, sensitivity to overgrazing and degradation of native vegetation, limited availability of water, and often competition with native herbivores. The provision of artificial water points for stock can shape the landscape by concentrating grazing pressure close to water (e.g. 'piosphere' effect) (Andrew, 1988).

2 Livestock production based on enclosed, semi-natural or exotic pastures

Livestock (e.g. sheep, cattle, goats) are grazed on pastures within enclosed fields or paddocks, usually in temperate climatic regions (Fig. 1b). In colder regions, stock may be sheltered in barns during winter months. In some regions, field boundaries may include semi-natural vegetation (e.g. hedges, fencerows, shelterbelts); in others, boundaries comprise fences (e.g. wire, timber, stone). Typically, these grazing landscapes are derived from cleared former forest or

Figure 1 Examples of different kinds of agricultural landscapes: (a) semi-arid rangelands, South Australia; (b) enclosed livestock farms, Devon, UK; (c) intensive feedlot for cattle, Oklahoma, USA; (d) extensive cropland of canola, Wimmera, Australia; (e) fine-scale cropping, Switzerland; (f) rice paddies, Bali; (g) vineyard and olives horticulture, Tuscany, Italy; and (h) smallholder farms, Ecuador. Source: Images adapted from: Andrew Bennett (a, b, c, d, e, g) and Jim Radford (f, h).

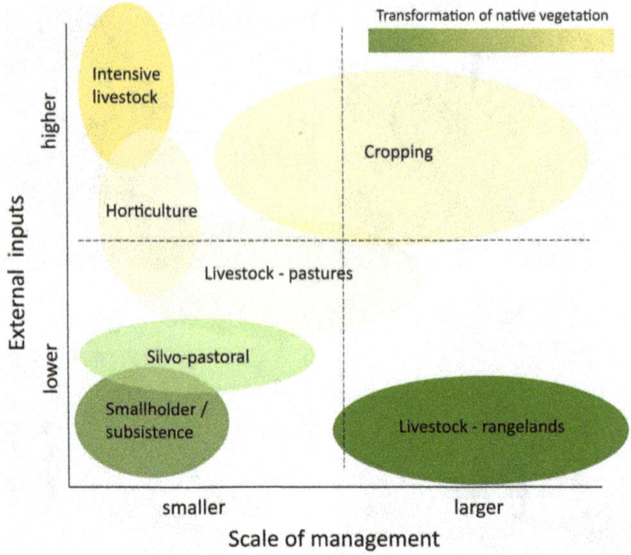

Figure 2 Diagrammatic illustration of different forms of agricultural land uses in relation to two main dimensions: (a) scale of management (i.e. size of production units, farms), and (b) level of external inputs (e.g. fertilisers, pesticides, fossil fuel energy – excluding human labour). Shading is indicative of the degree of transformation of the native vegetation from limited modification (darker green) to total replacement (darker brown).

woodland. They may have scattered trees across fields and small patches of semi-natural forest, woodland or tree plantations.

3 Intensive livestock production

These are farming systems with a high density of domestic stock (e.g. chickens, pigs and cattle) within small enclosures (e.g. sheds, barns and outdoor feedlots), with foodstuffs (e.g. hay and grains) brought in from outside the property (Fig. 1c). Stock and enclosures are concentrated in a relatively small area, giving rise to issues with the management of waste products and nutrient pollution (e.g. nitrogen), such as eutrophication of surface waters and contamination of groundwater (e.g. Hristov et al., 2011). The importation of foodstuffs means this land use also has significant offsite effects, arising from the way these foods are grown on other farms in other agricultural landscapes (e.g. land cleared for cropping, pesticide use, extraction of water for irrigation). Intensive livestock production typically involves the removal of all, or most, native vegetation and requires good road access for transportation.

4 Cropping

This agricultural land use is based on growing crops for food, fibre or biofuels (e.g. wheat, barley, millet, oats, corn, cotton, soybeans, sugar cane, rice), typically as monocultures on arable cultivated land (Fig. 1d and e). It occurs across a wide range of climates from semi-arid (e.g. millet, wheat) to subtropical (e.g. sugar cane) regions. The scale of operation may be large (e.g. wheat crops of hundreds of hectares per field on properties of thousands of hectares) with low spatial heterogeneity across the landscape (Fig. 1d) or smaller with interdigitated crops in areas of high productivity and/or irrigation (Fig. 1e and f). Growing crops for food or fibre commonly involves extensive clearing of native vegetation and marked landscape change; together with substantial external inputs including fertilisers and pesticides and fossil fuels for machinery for cultivating and harvesting. It also results in marked seasonal changes associated with crop life cycles in much of the landscape.

5 Horticultural land uses

Horticultural land uses, such as vineyards, fruit orchards and tea and coffee plantations, typically involve planted monocultures within relatively small areas (e.g. plantation, orchards and vineyards) to produce fruits and berries, seeds or foliage (Fig. 1g). While individual plantings commonly are monocultures, they may collectively form a diverse rural landscape, e.g. vineyards and olive plantations interspersed among other land uses in Mediterranean regions (Fig. 1g). In most cases, horticultural land uses occur where native vegetation has been removed; but in some instances, such as 'shade coffee' grown under a tree canopy (e.g. in Central America), they may also support a range of native vegetation and animal species (Perfecto et al., 1996). Some horticultural crops depend on pollination services from bees (commonly the honeybee *Apis mellifera*), a product of the apiculture industry. Managed pollination can involve complex spatio-temporal movements. In southern Australia, pollination of almond orchards involves transporting thousands of hives from hundreds of kilometres distant (Cowie, 2018). In winter months, many apiarists depend on native vegetation, including nature reserves, to sustain their hives.

6 Smallholder and subsistence agriculture

Smallholder agriculture (i.e. farming an area <2 ha) is estimated to contribute a third of the global food supply (Ricciardi et al., 2018) and represents the livelihood of >2 billion people (Vogel et al., 2023). Smallholder agriculture is widespread in Asia, Africa, South America and Oceania and can take multiple forms, typically involving growing some form of staple crop together with fruit trees and other food plants and supporting some livestock (e.g. chickens, goats, pigs). Use of technology and machinery, and external inputs are usually low, while the input of human labour may be high. Landholdings are small and so a fine-grained and diverse landscape of crops, gardens, dwellings and plantations results (Fig. 1h).

7 Silvo-pastoral farming systems

Silvo-pastoral systems, in which stock are grazed within a structurally open woodland of canopy trees, occur in many parts of the world. Those based on naturally occurring tree species have greater value for biodiversity than systems with non-native tree species, where native vegetation has been removed. A variety of products are obtained from low-density trees (e.g. fuelwood, fruits and nuts, browse for stock). For example, a cultural form of silvo-pastoral system in Spain ('dehesas') and Portugal ('montado') (Ferraz-de-Oliveira et al., 2016) involves a sparse canopy of oak trees (~20–40 trees per hectare) over a natural or cultivated grassy ground layer: products from the oak trees include acorns, cork (from Cork Oak *Quercus suber*) and browse for stock. Incorporating trees into diverse agricultural systems is common in smallholdings in Africa (e.g. Nyong et al., 2020) contributes to a heterogeneous landscape and to the resilience of a farmer's livelihood. Farmer managed natural regeneration (FMNR) is restoring tree cover across millions of hectares of farmland in sub-Saharan Africa (Reij and Garrity, 2016) and other developing countries.

3 Farmland habitats and biodiversity

For species to occur and persist in agricultural landscapes, a fundamental requirement is that suitable habitat is available. What constitutes habitat in farmland depends on the species expected to occur, the types of native vegetation formerly and currently present, and the history of agricultural land use. Throughout most of the world, particularly countries lacking a long history of intensively 'settled' agriculture (e.g. Australia, North America – several hundred years only) and those in which farmlands are currently being carved from extensive natural areas (e.g. South America, Asia, Africa), biodiversity is primarily associated with natural ecosystems. Priority areas for conservation typically are those with natural vegetation, and conservation goals for agricultural landscapes usually involve measures to retain or restore (semi) natural habitats that mimic former ecosystems. Productive farmlands typically have a depleted, modified biota that includes many introduced species.

In contrast, in countries such as those in Europe, a consequence of millennia of settled agricultural land use (e.g. arable lands, pastures and grazing domestic stock) is that many plant and animal species have adapted to farmland environments: indeed, an estimated 50% of species in Europe depend on agricultural habitats (Stoate et al., 2009). Others that did not adapt became locally extinct. Historically, low-intensity management of agricultural lands has delivered a rich assemblage of species, such that agricultural landscapes are viewed as an important part of the conservation estate and many conservation incentives aim to maintain a heterogenous open landscape through farming practices (Sutherland, 2004; Stoate et al., 2009; Batáry et al., 2015). Land abandonment and natural succession towards more-wooded environments are seen as detrimental in many European countries because they result in a loss of plant and animal species.

3.1 Diverse habitats in agricultural landscapes

A wide range of structural components can support biodiversity in agricultural landscapes (Fig. 3). Here, our focus is primarily on habitat components that support native plant and animal species, including providing food and foraging substrates for animals, refuge from predators, shelter from environmental conditions and breeding sites (e.g. nests, dens and burrows). Frequently, many such components are already present in agricultural landscapes but require protection and maintenance to prevent further degradation and loss.

First, natural components such as patches of forest, shrubland or native grassland, wetlands and lakes, and streams are of primary importance, whether within, or interspersed among, farmlands. Many studies, for example, have documented the biodiversity values of patches or fragments of forest in rural landscapes worldwide (e.g. Saunders et al., 1987; Laurance and Bierregaard, 1997; Lindenmayer and Fischer, 2006). These include studies of plant species and communities (e.g. Honnay et al., 1999; Aerts et al., 2006), terrestrial and arboreal mammals (Pardini et al., 2005; Lawton et al., 2021), bats (Law et al., 1999), birds (Newmark, 1991; Bellamy et al., 1996; Boscolo and Metzger, 2009), reptiles (Driscoll, 2004; Lion et al., 2016) and invertebrates (Estrada and Coates-Estrada, 2002; Fuentes-Montemayor et al., 2022; Ouin et al., 2022).

Important insights into farmland habitats can be gained from studies of individual species. For example, in Cambridgeshire, UK, the Tawny Owl *Strix aluco* occurs widely in wooded patches (Redpath, 1995), even in very open farmland. Where woods were highly fragmented (<4 ha in size), owls were able to combine several small woods into a single territory, regularly moving between them. However, such territories were sub-optimal, and the annual turnover of owls was higher than in larger woods (Redpath, 1995). In southern Australia, two arboreal marsupials, the Squirrel Glider *Petaurus norfolcensis* and Brush-tailed Phascogale *Phascogale tapoatafa* were able to persist in a cleared agricultural landscape by using remnant wooded vegetation along roadsides (van der Ree et al., 2001; van der Ree and Bennett, 2003). Important features included the amount and connectivity of roadside networks, and the high-quality habitat provided by many large old trees (e.g. as foraging substrates, tree hollows as den sites).

Second, many modified (semi)natural elements also provide habitat, including hedgerows, fencerows, live fences, flower strips, agroforestry plantations, scattered trees, ditches and drainage lines (Forman and Baudry, 1984; Harvey et al., 2006; Cano et al., 2022). Particularly notable are linear features (Fig. 3b, c and d) which are characteristic ecotonal habitats of agricultural

Figure 3 Examples of habitat components present in farmland. 1. *Natural elements*: (a) wooded remnant amidst cropland (NSW, Australia), (b) stream (Scotland), (c) remnant roadside vegetation (Victoria, Australia). 2. *Semi-natural elements*: (d) hedgerows in farmland (Wales, UK), (e) old trees in farmland (England, UK), (f) wetland/pond (New Brunswick, Canada). 3. *Anthropogenic elements*: (g) farm buildings (Pennsylvania, USA), (h) stone fence (Yorkshire, UK). Source: Images adapted from: Andrew Bennett.

landscapes worldwide – as boundaries between fields, as windbreaks to provide shelter for stock, or along roads and streams. There is a wealth of information concerning hedgerow systems in Europe (e.g. Burel, 1996; Hinsley and Bellamy, 2000), hedges and fencerows in North America (Forman and Baudry, 1984; Sykes and Hannon, 2001; Freemark et al., 2002), live fences in Central America (Harvey et al., 2005), shelterbelts in several continents (Yahner, 1983; Hino, 1985; Kinross, 2004) and roadside vegetation in countries worldwide (Bennett, 1991; Munguira and Thomas, 1992). Not only do such linear features provide habitat for plants and animals in their own right, but they have a wider influence on adjacent farmland by regulating micro-climate (e.g. shade, wind protection) (Baker et al., 2021), facilitating species interactions (e.g. pollination and pest control) (Forman and Baudry, 1984; Garratt et al., 2017) and assisting movements through agricultural landscapes (Bennett et al., 1994; Dániel-Ferreira et al., 2022).

Scattered or isolated trees amidst farmland are another prominent example of an important semi-natural habitat. They have been recognised as 'keystone structures' due to their disproportionate value relative to their abundance/cover (Tews et al., 2004; Manning et al., 2006; Prevedello et al., 2018). Scattered trees serve a range of functions: they can provide foraging habitat for birds, bats, arboreal mammals, reptiles and invertebrates (Harvey and Haber, 1998; Law et al., 2000; Lumsden and Bennett, 2005); they provide perching and nesting sites for birds (Manning et al., 2006); and can serve as stepping stones for movement of individuals through the landscape (Fischer and Lindenmayer, 2002; Doerr et al., 2011). They also play a wider ecosystem role in rural landscapes including as a focal point for restoration, either as a source for natural regeneration of plants or by providing mature-tree habitat within younger plantings. In tropical regions, natural regeneration can occur via seed rain defecated by birds that perch in trees in farmland after flying from a nearby forest (Galindo-González et al., 2000). Such scattered trees also have inherent conservation value in their own right. In Costa Rica, for example, a survey of isolated trees in pastures across 24 dairy farms recorded 5583 trees of 190 species, of which some 57% were primary forest trees (Harvey and Haber, 1998). In wood pastures in Sweden, large old oak trees in farmland are a favoured habitat for a diverse community of saproxylic beetles, including many threatened species (Ranius and Jansson, 2000).

A third type of feature that provides habitat in agricultural landscapes are anthropogenic elements such as irrigation channels, constructed ponds and dams, ditches, buildings and other farm infrastructure (Fig. 3g), and crops and pastures. For example, ponds and dams are habitat for many species of frogs, fish, aquatic insects and wetland plants (Hazell et al., 2001; Biggs et al., 2005; Chen et al., 2019). With appropriate management (e.g. excluding stock, creating emergent vegetation), dams can support frog communities comparable to natural wetlands (Hazell et al., 2004). Buildings are well known as diurnal roost sites for bat species (Voigt et al., 2016); birds such as swallows and martins build nests in farm buildings (Šálek et al., 2018); and a range of species forage in farm pastures, crops and orchards (Fig. 3h). In eastern Europe (Czech Republic, Poland), farmsteads (i.e. farmhouses, associated buildings and gardens) are strongholds for farmland bird species, especially during cold winter months (Rosin et al., 2016; Šálek et al., 2018). These kinds of anthropogenic habitats can each be managed to enhance the quality of the habitat they provide.

3.2 Factors that affect the value of farmland habitats

Numerous factors influence the ability of species to successfully use potential habitats in agricultural landscapes. Here, we summarise several key factors that affect the value of individual habitat components, while in Section 4 we extend the scale to emphasise the importance of a 'landscape perspective' on conservation in agricultural regions.

1 Size (area). This is a particularly important influence on the value of habitats for species, communities and ecological function (Forman, 1995). Individual species show different patterns of occurrence in relation to patch size, typically being more likely to occur as size increases. For example (Fig. 4a), the Hazel Dormouse *Muscardinus avellanarius*, a small arboreal rodent, was recorded in 20% of ancient semi-natural woodland patches of 2–5 ha in Herefordshire, England, ~40% of patches of 11–20 ha, and 100% of woods >100 ha (Bright et al., 1994). Larger woods have greater value and are more likely to sustain a resident population. At the community level, the number of species (species richness) increases with the increasing size (area) of a habitat patch (Fig. 4b). Species richness increases rapidly with size for small patches, then the rate of accumulation of additional species gradually slows (e.g. Loyn, 1987; Fig. 4b). Larger patches of habitat in farmland, whether grassland, shrubland, wetlands, forest patches or hedges will, on average, support more species than smaller patches of the same type. Patch size also influences the composition of communities: those in small species-poor patches often tend to be comprised of more-common and generalist species and typically form a nested subset of the species present in larger, species-rich patches (e.g. Blake, 1991; Wright et al., 1997). Animal species with larger body size or specialist requirements, that depend on natural habitats, are more likely to persist in larger patches and may be absent from smaller areas of natural habitat particularly those that are degraded (e.g. Michalski and Peres, 2007; Peters et al., 2008).

2 Context and isolation. Communities in farmland habitats are dynamic: animal species move to and from patches depending on the availability of food resources, or a resident population may become locally extinct and later recolonise (e.g. Ouin et al., 2004); plant species may colonise new habitat patches over time (Grashof-Bokdam and Geertsema, 1998). A key factor is the context of a habitat patch and its relative isolation from other nearby habitat. This influences how readily organisms can move or disperse their seeds or propagules, access resources, bolster existing populations or recolonise after local extinction. The degree of isolation depends on the distance to move, the scale of movement or dispersal capacity of an organism, and the connectivity of the surrounding environment (Bennett, 2003). Connectivity for an organism is low when the

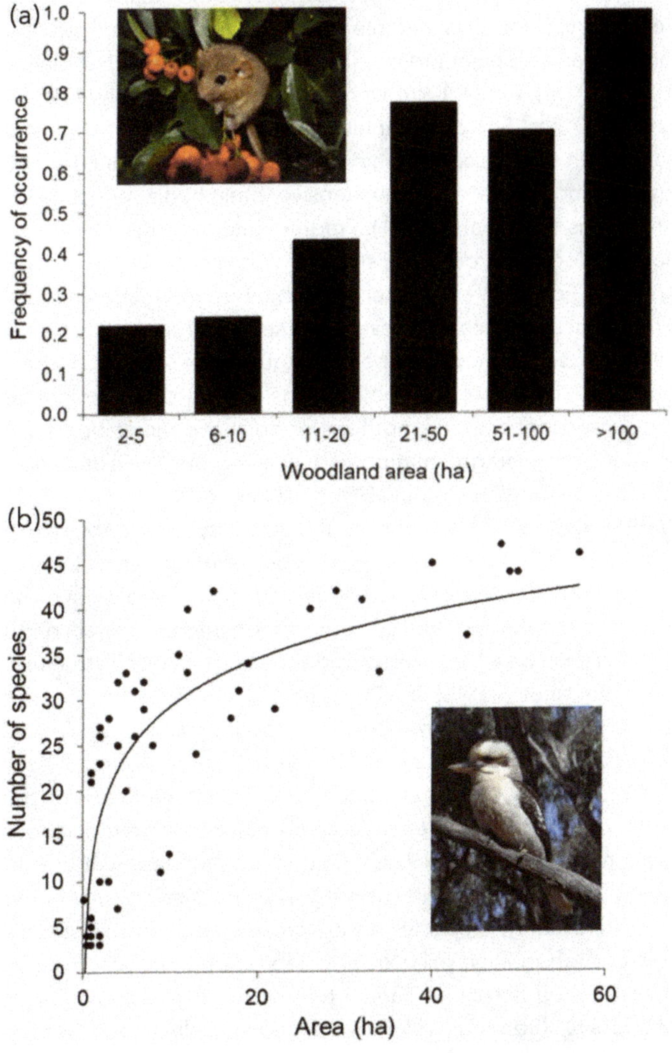

Figure 4 Size (area) is an important influence on the value of habitat patches in agricultural landscapes for species and communities: (a) frequency of occurrence of the Hazel Dormouse *Muscardinus avellanarius* in relation to the size of woodland patches amongst farmland in England (data from Bright et al., 1994); (b) species richness of forest birds in relation to area of forest patches in Gippsland, Australia (data from Loyn 1987). Source: Images adapted from: Danielle Schwarz (a) (Wikimedia) and Andrew Bennett (b).

surroundings impede movement and high when it can readily move through the landscape, whether by continuous 'corridors' or by 'stepping stones' or through land uses that are relatively benign (e.g. scattered trees across farmland).

Spatial context and isolation from other habitat affects species, communities and ecological processes. The Silver-spotted Skipper *Hesperia comma*, a rare butterfly of grazed calcareous grasslands in England, responded to habitat management by recolonising suitable grassland patches (Thomas and Jones, 1993). Recolonisation depended on distance from a source population: more than 75% of patches within 1 km were recolonised, but only ~10% of those isolated by 1–10 km. For communities, nearby extensive habitats serve as source from which species can colonise: in tropical forest remnants in Tanzania, those remnants more isolated from extensive forest had a reduced richness of understorey birds (Newmark, 1991). Fragmentation and isolation of habitats also affects ecological processes, such as pollination and reproduction of flowering plants. For many plant species, habitat fragmentation has a negative effect on reproductive success (measured as seed or fruit production) although the relative impact varies among species (Aguilar et al., 2006). Plants that are self-incompatible (i.e. depend on pollen transfer from another plant) are more susceptible, consistent with a hypothesis of pollination limitation arising from the negative effects of fragmentation and isolation on the relative abundance and behaviour of animal pollinators (Aguilar et al., 2006).

3 Habitat quality. This refers to the capacity of a habitat to provide the resources a species requires (e.g. food, shelter, refuge). In farmland environments, habitats can be altered or degraded in multiple ways. Grazing by domestic stock can alter the floristic composition and structural complexity of vegetation, facilitate weed invasion and degrade surface soil structure (Prober and Thiele, 1995; Yates et al., 2000). The creation of edges between cleared farm pasture and remnant vegetation alters the microclimate (e.g. wind, sunlight and temperature) in remnant habitats and increases invasion of exotic plants and animals (Saunders et al., 1991; Laurance et al., 2002). Wetlands or ponds can be degraded by run-off of nutrients, sediments and chemicals and consequent eutrophication (Chen et al., 2019). Common farm management practices such as tree felling, grass slashing and spraying with chemicals also alter the quality of habitats. Importantly, a species being present does not necessarily mean a habitat is of high quality; poor quality habitat may be occupied but result in unsuccessful breeding, high annual turnover and potentially function as a demographic 'sink' (Redpath, 1995). For example, Holland and Bennett (2010) reported marked variation in population density, age structure, body size, reproduction and recruitment in populations of the Bush Rat *Rattus fuscipes*, a common native rodent, between farmland forest patches that varied in size and habitat quality. The strongest effects were associated with patch size, with some mediated by habitat quality.

4 Interactions between species. These commonly occur in farmland habitats and have both positive and negative effects. Pollination of plants by invertebrates and control of pest species of crops or horticulture can be beneficial for farm productivity. Alternately, interactions detrimental for conservation include increased levels of predation by introduced predators or by open-country predators that move across forest edges (e.g. Andren and Angelstam, 1988); increased competition from farm animals or introduced species; hunting and destruction by humans of wildlife species regarded as pests; and interactions with 'problem' native wildlife (e.g. large predators, herbivores) that move into farmland. In Namibia, for example, conservation of the African Elephant *Loxodonta africana,* is complicated by conflict with farmers arising from crop damage by animals that raid agricultural fields (O'Connell-Rodwell et al., 2000). Similarly, in Sri Lanka, widespread conflict between rural people and the Asian Elephant *Elephas maximus* over crop damage adds complexity to conservation of this endangered species (Fernando et al., 2021). Different regions have other idiosyncratic issues depending on their biota. For example, in North America, the Brown-headed Cowbird *Molothrus ater*, a nest parasite, has a detrimental effect on the populations of numerous migratory bird species where farmland has fragmented forest cover. In Australia, the Noisy Miner *Manorina melanocephala,* a despotic native bird, aggressively excludes small insectivorous species and has altered bird communities across vast areas of fragmented and modified woodlands in agricultural regions (Maron et al., 2013).

5 Agrochemical use. The application of pesticides, herbicides, fertilisers and other chemicals associated with agricultural intensification has complex effects on biodiversity (e.g. Donald et al., 2001). For example, pesticides used to control crop pests have lethal and sub-lethal impacts on a wide-range of non-target invertebrates in terrestrial, aquatic and marine habitats (Pisa et al., 2015; Sánchez-Bayo and Wyckhuys, 2019) and negative consequences for ecosystem functions facilitated by invertebrates such as regulating soil and water quality, pest control, pollination, decomposition, nutrient cycling and soil respiration (Chagnon et al., 2015). There are also impacts on higher trophic levels. Landscape-level declines in insectivorous birds in farmland have been linked to levels of pesticide use in the Netherlands (Hallmann et al., 2014) and more broadly across Europe (Donald et al., 2001).

At the farm-scale, a meta-analysis of 96 studies comparing organic (no input of chemicals) with conventional agricultural practices in temperate regions found that species richness was, on average, 95% greater for plants, 35% greater for field birds, 22% greater for insects and 15% greater for spiders on organic farms than on conventional farms (Stein-Bachinger et al., 2019). The use of chemicals not only affects the biota occurring in fields but also species occurring in field margins and habitat surrounding farmland. In the Netherlands, for example, the abundance of beneficial arthropods in field margins was negatively associated with rates of insecticide application in the surrounding landscape, and pollinator abundance was negatively associated with the frequency of insecticide applications in the adjacent field (Bakker et al., 2022).

4 A landscape perspective for biodiversity conservation

Effective planning for nature conservation in agricultural landscapes requires moving from consideration of individual patches of habitat to a 'landscape perspective'. From a human viewpoint, a landscape typically encompasses an area of kilometres to tens of kilometres across and comprises a mosaic of different land uses (Fig. 5).

4.1 The need for a landscape perspective

Planning and management for conservation requires making decisions about multiple land uses and habitats in a coordinated way. This is important because many species need multiple patches of habitat to meet daily, seasonal or annual requirements for food and breeding and to sustain a viable population. For example, in arable farmland in Britain, individual Wood Mice *Apodemus sylvaticus* occupy a home range that encompasses both hedgerow and adjacent arable crop field, allowing access to resources from both habitats seasonally (Todd et al., 2000). In farmland in Ontario, Canada, the Leopard Frog *Rana pipiens* needs access to three types of habitat in close proximity at different stages of its life cycle: a breeding pond in spring, grassy meadows or fields for foraging in summer, and a stream or lake for overwintering (Pope et al., 2000). Further, single patches of habitat are rarely sufficient to support a viable population of a species, whereas at a landscape scale a species may persist as a metapopulation comprised of multiple local populations. Movements of individuals between local populations allow for gene flow, supplementation of declining populations or recolonisation should extinction occur in a local population; such that the overall metapopulation can persist. The benefit of a metapopulation structure in rural landscapes has been reported for numerous species, as diverse as the Silver-spotted Skipper, a threatened butterfly in Britain (Davies et al., 2005); Orangutan *Pongo* spp. among oil palm landscapes in Malaysian Borneo (Ancrenaz et al., 2021); European Nuthatch *Sitta europaea* in Dutch landscapes (Verboom et al., 1991); and the Eastern Chipmunk *Tamias striatus* in Canadian farmland (Bennett et al., 1994).

A second reason for a landscape perspective is that 'context matters' (Saunders et al., 1991; Wiens, 2002). That is, the function and conservation value of habitats in farmland are influenced by their surroundings. A landscape perspective allows context to be recognised and incorporated in land planning and management. A wetland habitat in farmland may be detrimentally affected by run-off of nutrients from adjacent pastures and crops, or disturbance by domestic stock; or positively affected by having a buffer of fringing vegetation to filter run-off. Context is also important for effective habitat restoration. For example, woodland birds are more likely to use planted restoration sites if they are located close to other nearby wooded vegetation (Haslem et al., 2021). Context is also important when considering complementary resources – different resources must be accessible to individuals, not just present in the landscape.

Finally, a landscape perspective is relevant to the scale at which land-use planning and management occurs, whether by Government agencies, conservation groups or local community groups. Landcare groups in Australia, for example, comprise landholders with farm properties in the same area, often in the same catchment (watershed), who work together at a landscape scale on actions to benefit their land (Campbell, 1994). The landscape is also a scale at which coordinated management can be undertaken in response to processes that influence ecosystems and species, such as management to reduce fire risk,

Figure 5 Examples of spatial heterogeneity and 'grain' in agricultural landscapes: (a) fine-grained heterogeneous landscape in south-east England showing vast, irregular network of hedges and woods amongst farmland; (b) coarse-grained, rectangular grid of cropping paddocks with lines of wooded vegetation along roadsides, Wimmera, Australia; (c) diverse, fine-grained mosaic of crops and orchards, Slovenia; (d) managed mosaic of old and young Heather *Calluna vulgaris* shrubland to provide optimum habitat for Red Grouse *Lagopus lagopus*, Scotland, UK. Source: Images adapted from: Andrew Bennett (a, b, d) and Ian Gordon (c)

integrated control of weeds and pest species across multiple properties, and implementing large-scale restoration.

4.2 Agricultural landscapes are mosaics of different land uses

Agricultural landscapes are a mosaic of different land uses: some contribute to farm production, others are associated with human dwellings and infrastructure and others may include semi-natural or natural elements (Fig. 5). Spatial heterogeneity is evident both *within* a farm property and *between and among* farms in the landscape. Managing land uses *within* a farm property is where the primary opportunity lies for individual landholders to influence conservation in agricultural landscapes. Managing land uses between and among farms will often involve multiple landholders and agencies with responsibility for different land parcels and tenures (e.g. small conservation reserves, roadsides, stream frontages).

A key issue in setting goals and managing agricultural landscapes to increase their value for biodiversity is to identify the characteristics of a landscape that enhance (or limit) its conservation values. That is, which properties of a landscape can be managed to improve conservation values? This has been the focus of a growing body of research (e.g. Villard et al., 1999; Bennett et al., 2006; Mortelliti et al., 2011; Ochoa-Quintero et al., 2015; Zingg et al., 2018), which has shown that measures of biodiversity are indeed strongly related to properties of 'whole' agricultural landscapes. To empirically study such properties requires comparing biodiversity between 'whole' landscapes (as the unit of study) that differ in composition and spatial pattern (McGarigal and Cushman, 2002; Bennett et al., 2006). Here, we outline four broad types of landscape properties that influence biodiversity in agricultural landscapes and summarise hypotheses as to why these key properties are important (Table 2).

First, the *total extent of suitable habitat* refers to the summed area of a particular habitat type in the landscape, regardless of the size, shape or location of individual patches. For example, the total amount of wooded habitat is a strong predictor of the occurrence of *individual* species of forest and woodland birds in rural landscapes in several countries (e.g. Trczinski et al., 1999; Bailey et al., 2002; Radford and Bennett, 2007) and also of the species richness of forest and woodland birds (e.g. Pino et al., 2000; Radford et al., 2005; Cunningham et al., 2014). Similarly, the total amount of cropland habitat is a significant predictor of the richness of cropland birds, such as in rural landscapes in Portugal (Pino et al., 2000). Thus, a critical goal for conservation in rural landscapes is to retain, or restore, as much suitable habitat as possible for particular species or taxa of concern (Table 2). The benefits of restoring habitat was demonstrated for woodland birds in rural landscapes in southern Australia (Bennett et al., 2022): in landscapes dominated by restored (replanted) woodland, the species

Table 2 Properties of agricultural landscapes that influence their value for biodiversity conservation

Landscape property	Influence on landscape-scale biodiversity	Actions to enhance biodiversity
1. Total amount of suitable habitat in the landscape, regardless of size, isolation or location of patches.	The total amount of habitat influences the population size of species in the landscape: the more habitat, the larger the population size, and the less vulnerable to decline and local extinction. Landscapes with a larger total amount of habitat will have more species with populations of sufficient size to persist and hence support greater species richness.	Retain all suitable habitat, maintain and enhance quality. Increase the total amount of suitable habitat through restoration actions (e.g. replanting, natural regeneration).
2. Composition of the landscape mosaic a) Heterogeneity of elements (land uses) b) Occurrence and extent of keystone habitats	a) A greater heterogeneity of elements (land uses) provides habitat for more types of species and also supports species that move between habitats to obtain resources. b) Keystone habitats have a disproportionately positive influence (relative to their area) on the occurrence of particular species in landscapes, or on species richness in a landscape, because of their unique structural properties.	Maintain heterogeneity, particularly of (semi)natural elements, while maximising the overall extent of habitat for priority species Identify, maintain and enhance or replace keystone habitats in the landscape.
3. Spatial configuration of elements in the landscape: a) the subdivision, aggregation and connectivity of a favoured habitat (e.g. forest patches) b) the 'grain' of land parcels (e.g. fields) of diverse elements	Fragmentation per se (i.e. the breaking apart) of a favoured habitat reduces the size and increases the isolation of individual habitat patches and of local populations of species that inhabit them. Increased length of edges with adjacent land can expose patch habitats and populations to increased disturbance. Landscapes with a finer 'grain' (greater structural heterogeneity) are beneficial for organisms that use ecotonal (edge) habitats, or regularly move between different land uses, by reducing the isolation of complementary habitats and increasing opportunity for cross-boundary movements.	While maximising the total amount of suitable habitat (e.g. natural forest, woodland), aim to manage it as a connected system of patches. Use restoration to connect patches, and buffer existing patches. In parts of the agricultural landscape, where possible aim for heterogeneity of complementary land uses at a fine scale.
4. Geographic location and environmental context of a landscape	The geographic location in relation to environmental gradients, ecological source areas for dispersal or areas of disturbance affect the likelihood of species' occurrence in a particular landscape and thereby the richness and composition of the biota in that landscape.	Manage and retain habitats in landscapes that span the full geographic and environmental gradient. Identify and protect source areas for species' dispersal into agricultural landscapes. Locate restoration areas close to sources.

richness of woodland birds increased with increasing extent of restored vegetation in a similar manner to that of landscapes dominated by remnant natural woodlands. Notably, restoration actions by individual landholders that add to the total extent of habitat in the landscape have wider benefits for landscape-scale conservation beyond their own property.

Second, the *composition of the landscape* refers to the types of elements present and their relative proportions. Variation in topography, soils and moisture influence the diversity of natural vegetation communities while anthropogenic elements, such as croplands, pastures, dams and human settlements add further elements. The composition of the land mosaic can influence biodiversity in two main ways (Table 2). First, a greater diversity or heterogeneity of elements provides opportunities for a greater range of habitat types and thus species (Fig. 5). For example, Atauri and de Lucio (2001) found that landscape heterogeneity was the most influential factor driving the species richness of birds, amphibians, reptiles and butterflies in rural mosaics near Madrid, Spain. This relationship with landscape heterogeneity is strongest when the response variable is the richness of a whole assemblage (e.g. all butterflies, Weibull et al., 2000), including species that favour different types of land uses. In southern Australia, the richness of the overall bird community also responded positively to landscape heterogeneity in farmland mosaics (Haslem and Bennett, 2008), but a specific group such as woodland birds responded most strongly to the total amount of woodland habitat. Landscape heterogeneity is also important for anthropogenic elements. A cross-continental comparison of cropland mosaics (Sirami et al., 2019) reported that those with greater cropland heterogeneity (i.e. more crop types, smaller fields) supported greater diversity for a wide range of taxa studied (e.g. plants, bees, butterflies, hoverflies, carabids, spiders and birds).

Another compositional aspect of agricultural landscapes that influences biodiversity is the occurrence of keystone habitats; that is, particular land uses or structural features that have a disproportionate influence relative to their area (Table 2). For example, streamside (riparian) vegetation in landscapes in south-eastern Australia occupies a relatively small area but as agricultural landscapes become depleted of wooded cover, streamside vegetation makes a disproportionate contribution to overall species richness of woodland birds (Bennett et al., 2014). Loss of streamside vegetation has a greater impact on the richness and breeding activity of woodland birds in farmland than loss of wooded roadside strips or stands of scattered trees (Hall et al., 2022). A further keystone habitat in agricultural landscapes is scattered trees in farmland (Tews et al., 2004; Manning et al., 2006), which have disproportionate benefits for biodiversity relative to otherwise-cleared farmland. Identifying, protecting and restoring keystone habitats for plants and animals is an important priority for conservation in agricultural landscapes.

A third landscape property that influences biodiversity in agricultural landscapes is the *spatial configuration* of elements in the mosaic (Table 2). Much attention has been given to changes in forest cover and the relative importance of loss versus fragmentation (i.e. breaking apart) of forest habitats in rural landscapes. While measures of fragmentation (such as subdivision, aggregation, patch shape) do influence the landscape-scale distribution of some species (e.g. Villard et al., 1999; Radford and Bennett, 2007; Mortelliti et al., 2011), overall the evidence is that loss of forest habitat is of much greater influence than its spatial configuration. Forest loss consistently has negative effects on forest-dependent biodiversity, whereas fragmentation per se may have positive or negative effects depending on the species and process (Fahrig, 2003, 2017). Nevertheless, a prudent approach is to avoid fragmentation and isolation and retain and restore as much natural vegetation as possible in the form of a *connected system* of habitats.

The spatial configuration of elements other than wooded habitats is also important and contributes to the 'grain' of the landscape: that is, the relative size of the land parcels that make up the mosaic. In many regions, particularly those dominated by smallholdings, there is a 'fine-grained' landscape comprised of many small elements or fields (e.g. 0.1–2 ha) of different land uses. In contrast, in situations dominated by extensive grain cropping or pastoral grazing, the landscape is 'coarse-grained' with large individual elements (e.g. 100–1000 ha or more in size) (Fig. 5). As a landscape becomes more fine-grained, there is an increasing density of edge habitats between elements, less isolation, and greater opportunity for species that favour ecotones and that move between land uses. Fine-grained landscapes appear beneficial for within-farmland biodiversity. For example, Fahrig et al. (2015) found that across 1 km^2 cropland mosaics in Ontario, Canada, the species richness and abundance of a range of taxa (e.g. birds, plants, butterflies, syrphids, bees, carabids and spiders) was consistently negatively related to the mean size of crop fields. That is, landscapes with smaller crop fields had a greater richness of these taxa. This was attributed to the positive effect of a greater density of field boundaries and closer proximity of crop habitats to semi-natural boundaries (e.g. hedges, fencerows) (Fahrig et al., 2015), which would benefit movements of invertebrate groups involved in pollination or control of natural pests.

Finally, the geographic location and context of an agricultural landscape can influence its value for biodiversity relative to other similar landscapes (e.g. Radford et al., 2005). Landscapes may be located on different parts of an environmental gradient (e.g. rainfall, elevation), or closer to (or away from) an extensive area of habitat that may function as a source area for populations.

4.3 Agricultural landscapes change through time

Agricultural landscapes are dynamic: the composition and configuration of land uses change through time, with diverse consequences for nature conservation (Forman, 1995; Watson et al., 2014; Ridding et al., 2020). Seasonal cycles of crop growth and harvest, flowering and fruiting of trees and shrubs, and flooding or drying of wetlands, for example, result in a changing annual pattern of resources across the landscape and movements of species that track these resources. Other changes occur over longer timescales such as the rotation of crops, growth of tree plantations, introduction of new agricultural products, changes to market forces and disturbance from fire, drought, floods or wind storms (Vesk et al., 2008; Watson et al., 2014; Aspinall et al., 2021). The changes of greatest concern from a conservation perspective are those that most affect the capacity of the land mosaic to support native plant and animal species and maintain ecological processes. These include long-term changes such as the incremental loss and fragmentation of natural vegetation (Fig. 6), simplification and degradation of habitats, loss of cultural habitats to which species have adapted (e.g. coppiced woods, traditional hay meadows), spread of exotic plants and animals, increasing use of chemical pesticides and fertilisers and expansion of human settlements and road systems (e.g. Agger and Brandt, 1988; Bennett, 1990; Donald et al., 2001; Ridding et al., 2020; Aspinall et al., 2021).

It is useful to consider several aspects of the way in which temporal change in rural landscapes affects biodiversity. First, in relation to short- to medium-term changes in land use (e.g. crop rotations, new crop types, restoration measures, set aside), Watson et al. (2014) distinguished key components of the change regime that potentially affect biodiversity: the *frequency* of land-cover changes over a given period of time, the *sequence* of land-cover types, the *time span* over which each land-cover type extends and the *magnitude of difference* between land-cover types. The way in which individual species respond to these different components of change can be related to their ecological traits. For example, Watson et al. (2014) proposed that a higher frequency of land-cover changes is likely to disadvantage many species, with those animal species able to persist being habitat and dietary generalists. Temporal change that involves a marked difference in the types of land cover, such as change from cereal crop to tree-based horticulture (e.g. almonds), is likely to also involve notable differences in the biota that are able use these structurally different habitats.

A second aspect of the way temporal change affects biodiversity is when there are *interactions* between different temporal processes, such that synergistic changes result in even greater impacts than separate effects. For example, drought has severe effects on plant and animal species, but may have even greater effects where there has been long-term loss and fragmentation of habitat. In rural south-eastern Australia, a decade-long drought (~2000–9) had severe consequences for birds (Mac Nally et al., 2009) with many species declining severely in distribution and abundance. Two years post-drought, the relative abundance (reporting rate) of ~50% of species still remained lower than that recorded before or early drought (Bennett et al., 2014). The response of the woodland-dependent bird assemblage during and following drought varied between landscapes and was strongly associated with landscape properties (Haslem et al., 2015; Nimmo et al., 2016). Species richness declined least, and was more stable over time, in landscapes that had a greater area of vegetation in productive (i.e. streamside, floodplain) parts of the landscape (Nimmo et al., 2016). Compositional change was lowest in landscapes with greater overall extent of wooded vegetation, whereas landscapes with less native vegetation (i.e. more clearing for agriculture) had less stable bird assemblages over the duration of the drought (Haslem et al., 2015).

Third, there often are time lags in experiencing the full consequences of temporal processes of change in agricultural landscapes. For example, lagged responses by plant populations (e.g. due to limited dispersal ability) mean that present-day plant communities may still carry the imprint of historic land uses a century or more ago (e.g. Hermy and Verheyen, 2007), and the distribution of some animal species may reflect past, rather than present landscape patterns (Petit and Burel, 1998). The long lifespan of individual organisms, such as trees, means that there may be time lags of many years for the consequences of landscape change to be expressed. For example, scattered old trees in farmland where tree regeneration is absent will take decades to age, senesce and die such that eventually the farmland is treeless (Fig. 7). Such landscapes with lagged responses are said to carry an 'extinction debt' (Tilman et al., 1994). That is, the full consequences of past land uses are still to play out, and further decline and loss of species is expected over time. It is likely that many agricultural landscapes worldwide carry an extinction debt, particularly in countries and regions where farmlands were carved out of natural ecosystems in the last 100 years.

Time lags are also evident in restoration and successional processes in farmland: restored vegetation takes decades for trees and shrubs to grow, develop and mature (e.g. for trees to develop large trunks and limbs, large canopy, hollows; Vesk et al., 2008), and provide new structural habitat for biota. The age of restoration (i.e. time since planting) is a significant influence on the use of such restored sites by forest-dependent bird species (Haslem et al., 2021). Similar time lags in the development of successional habitats, such as secondary forests after farmland abandonment, mean that changes will play out over decades. Such time lags add an urgency to initiate actions to restore farmland habitats now.

Last, there are long-term, directional changes to farmland and (semi) natural habitats that essentially are irreversible or difficult to restore due to

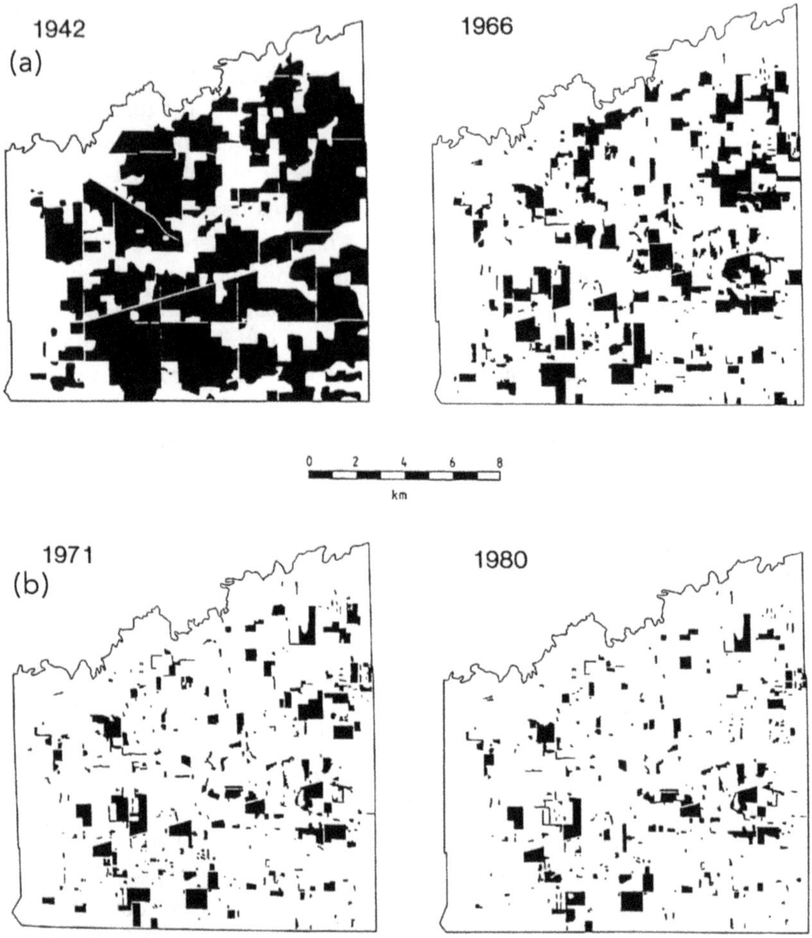

Figure 6 Landscape change and loss and fragmentation of forest cover (black) in an ~200 km^2 study area in south-western Victoria, Australia, from 1942 to 1980 (after Bennett, 1990). Over time there is (a) loss of overall amount of forest cover and (b) fragmentation into smaller and more isolated remnants. This type of pattern of landscape change is played out in agricultural landscapes worldwide.

time, political and social constraints. For example, the spread of towns and cities that converts farmland and forest to roads, housing lots and shopping centres, is a long-term change that also introduces new pressures to remaining lands (e.g. altered drainage, invasive plants, run-off of chemicals and nutrients, recreational pressure). The impoundment and diversion of water from river systems for domestic use and for irrigation agriculture alters the hydrology of landscapes and the ecological function of areas such as floodplains and downstream rivers.

5 Agricultural production and biodiversity conservation: land sparing-land sharing

Producing food and fibre for the growing human population and conserving Earth's biodiversity are two of the most challenging and urgent problems of the twenty-first century. Fundamentally, the common, underlying issue relates to determining priorities for how land is used, often a source of conflict where land is limited relative to the human population size. In the previous sections, we have outlined concepts for the conservation of biodiversity in agricultural lands from an ecological perspective and summarised knowledge on what will best achieve conservation goals. However, the real-world constraint is that both these important goals – feeding the human population and conserving biodiversity – must be achieved within a finite amount of land. Both must be considered simultaneously. Thus, strategies for nature conservation need to be considered and evaluated in relation to competing demands on land to achieve agricultural production (Balmford, 2021).

In seeking solutions, debate has centred on two contrasting approaches, land sparing and land sharing (Green et al., 2005; Phalan et al., 2011; Fischer et al., 2014; Balmford, 2021), which can be likened to the two ends of a continuum (Fig. 8). A land-sparing approach proposes that by maximising the productive yield on lands dedicated to agriculture, the overall requirement for farmland will be smaller such that other areas can be 'spared' from clearing and dedicated for nature conservation. In contrast, a land-sharing approach (or 'wildlife friendly farming', Green et al., 2005) proposes that nature can be retained in farmland by reducing the intensity of farming and incorporating semi-natural elements (e.g. field boundaries, scattered trees, small forest patches) in farmland. Inherent in the land-sparing versus land-sharing debate is the understanding that there are trade-offs between the use of land for conservation and for agricultural production. Both cannot be fully achieved on the same land, and so the underlying question becomes how society can provide for the human population at least cost to the future of Earth's biodiversity.

Empirical testing of sparing versus sharing strategies (e.g. Phalan et al., 2011) has been undertaken by selecting a set of areas that represent differing farming arrangements and also including contrasting areas entirely of natural habitat (with no agriculture). Then for each area, the population density of individual species (a surrogate for conservation) and agricultural yield are quantified. These data are used to generate a 'density-yield function' for each species by plotting population density versus agricultural yield and inspecting the shapes of the responses for each species. 'Winners' (with agriculture) are species that increase in density as agricultural yield in the landscape increases; 'losers' are those that decrease in density with an increase in yield, and an 'intermediate' group are those that peak in density for some mid-level

Figure 7 Many rural landscapes are yet to experience the full consequences of past changes. This landscape has scattered old trees amongst pasture, in various stages of senescence or death, but lacks tree regeneration. In the absence of a regenerative process, it will become a treeless landscape, but the time lag to reach this endpoint may be decades.

of agricultural yield (Fig. 8). Evidence to date from field studies in multiple countries – including Ghana, India, Uganda, Kazakhstan, Brazil, Uruguay and Mexico – and including density-yield functions for >2500 species (birds, butterflies, dung beetles, trees, grasses and daisies) has consistently shown, in every region, that most species decline in density with increasing agricultural yield (i.e. more of the landscape committed to agriculture and producing a greater average yield per ha) and that for most species, the least-worst option is a land-sparing approach (Balmford, 2021). That is, most species will have larger populations in a given landscape if agricultural production is confined to as small an area as possible, while retaining as large an area of native vegetation as possible (Phalan, 2018).

The land-sparing–land-sharing debate highlights the necessity for trade-offs between agriculture and biodiversity conservation and that, for effective long-term conservation in agricultural regions, important choices must be made. In reality, an intermediate position between sparing and sharing will be the most pragmatic outcome in many regions, but with implementation varying geographically and depending on land-use history. In Europe, the long history of fine-grained land use, widespread loss of natural forests and dominance of agricultural land means that a land-sharing approach is the favoured paradigm, supported by billions of euros annually in farm subsidies to enhance within-farm practices (Stoate et al., 2009; Batary et al., 2015). Even so, restoration of strategic tracts of natural vegetation by large-scale replanting or natural regeneration is a complementary option. In contrast, an approach weighted towards land sparing will have the greatest value in the many regions with a limited history of 'settled' agriculture, or where the agricultural frontier is currently expanding into extensive natural areas and few species have experienced, or are associated with, farmland habitats. Retention of large natural areas, or a large total amount of natural vegetation, will be critical to sustaining biodiversity in such regions.

Many questions remain as to how land can be allocated to support both people and nature (e.g. Fischer et al., 2014; Phalan, 2018; Balmford, 2021). Will the land that is spared from high-yielding agriculture actually be retained and used for nature conservation? Will high-quality (for biodiversity) land be set aside for conservation or will it be only marginal, low-productivity land? What is

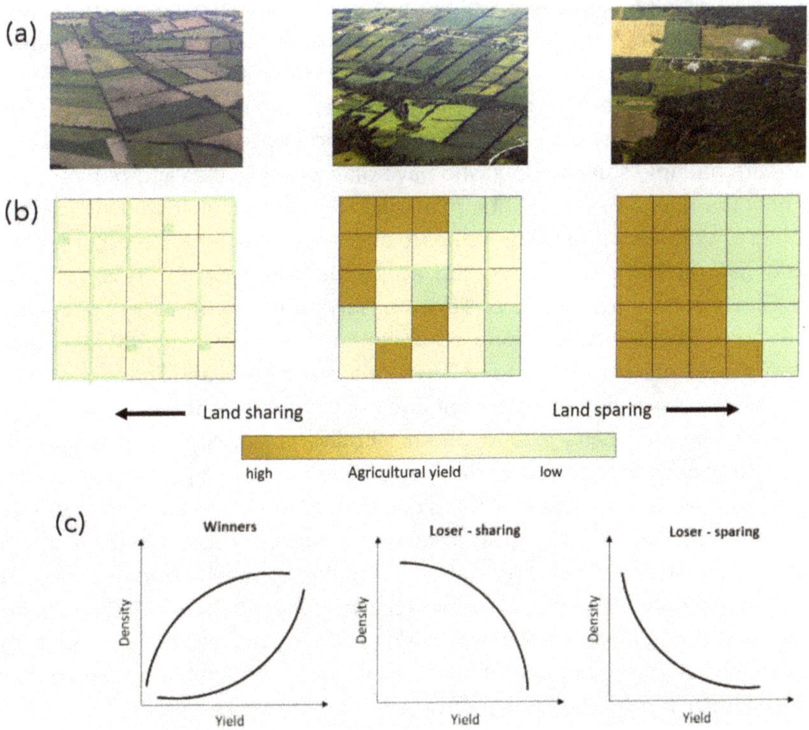

Figure 8 Diagrammatic representation of the land-sharing–land-sparing framework expressed as a gradient of change. (a) Images of rural landscapes. (b) diagrammatic illustration of landscapes with different proportions of the landscape having high, medium or low agricultural yields. Low yield corresponds with natural vegetation. (c) Stylised density-yield curves representing different responses (i.e. change in population density) of species to agricultural yield. 'Winners' increase in density (abundance) as agricultural yield increases; 'losers' decline in density as agricultural yield increases. Source: Adapted from Balmford (2021).

the most appropriate spatial scale for assessing and managing trade-offs and land uses? Which components of biodiversity are best to optimise? How can a land planning approach be carried out in a socially equitable way? These involve political, social and economic considerations along with ecological understanding. It is also evident that retention of sufficient land for conservation (e.g. 30% landscape cover; Smith et al., 2013) is unlikely to be achieved *within* farms alone, emphasising the critical role of other lands within the wider agricultural landscape, such as various forms of conservation reserves, timber reserves, community-managed land or wide streamside reserves on public lands, to complement within-farm measures.

6 Agriculture, nature conservation and people

In contrast with the conservation of biodiversity in national parks and protected areas, integration of conservation in agricultural landscapes poses a different and extraordinarily complex set of challenges. In agricultural landscapes, diverse land uses compete for space, some incompatible with nature conservation; the landscape is comprised of many parcels of land under different land tenures; there are multiple stakeholders who have different priorities and perspectives on how the landscape should be managed; and the agricultural landscape is subject to a different array of pressures to conservation reserves (e.g. continuing loss of natural habitats, urban spread, agricultural chemicals, invasive species).

Given these challenges, positive conservation outcomes in agricultural landscapes require (a) a strong commitment from Governments and their responsible agencies, including both policy commitments and financial support and (b) processes for effective engagement with stakeholders, particularly those who own or manage land, others who live in or visit the landscape and those whose livelihood and well-being are determined by the suite of land uses. This means priorities and decisions about land management cannot be determined solely on ecological or conservation grounds but must take into account the objectives and aspirations of multiple stakeholders – a 'people-centred approach applied at landscape scales' (Sayer et al., 2013). This requires processes for setting objectives, making decisions, allocating resources, governance and monitoring outcomes, that involve the multiple stakeholders. It is a particular challenge in countries where there also are issues of extreme poverty, lack of food security, power imbalances and limited infrastructure for health and education (Garnett et al., 2007; Garcia et al., 2010). A thorough consideration of the social dimensions of integrating nature conservation and agriculture is beyond the scope of this chapter, but we draw attention here to two broad concepts.

First, approaches to engaging stakeholders and local communities to integrate nature conservation and agriculture are diverse – and have had mixed success. Examples include (a) schemes whereby Governmental policy combined with financial incentives influence farmers to change practices to achieve conservation objectives, such as the European agri-environment schemes (Batáry et al., 2015); (b) community-based approaches, sponsored by Government, that involve local groups of landholders in voluntary, participative actions to address local issues (e.g. weeds, pest animals, replanting vegetation), such as the Landcare movement in Australia (Campbell, 1994; Curtis and Lockwood, 2000); and (c) a raft of 'integrated conservation and development programs', supported by international development agencies and philanthropists, that operate in developing countries with the aim of enhancing livelihoods, health, education and sustainable land use (e.g. Garnett et al., 2007).

It is likely that market-based approaches will play an increasingly important role in shaping interventions in agricultural landscapes in future years. This includes, for example, the market for carbon offsets (and/or ecosystem services), with carbon farming having the potential to drive large-scale revegetation or natural regeneration in cleared agricultural lands (Evans et al., 2015). In addition, consumer-led pressure for agricultural products (e.g. meat, wool, cotton) to be produced sustainably and ethically, and verified by certification schemes, is likely to result in changes in land management and greater attention to the 'natural capital' retained on farms. While market-based approaches have the potential to scale up restoration, financial incentives are not the only factor motivating conservation in rural landscapes. Others include a land holder's inherent interest in nature, a commitment to sustainability, aesthetic pleasure and enhanced wellbeing from nature, and cultural and spiritual connections with the land (such as sacred forests, e.g. Ormsby and Bhagwat, 2010).

Second, insights and general principles are emerging for processes likely to enhance engagement with stakeholders and achieve positive outcomes for agriculture and conservation (Garnett et al., 2007; Harvey et al., 2008; Sayer et al., 2013). In a synthesis of the literature and a consensus view of many individuals and organisations, Sayer et al. (2013) listed ten principles for involving people and communities in a landscape approach to management (see Table 3). In reality, working through processes to effectively implement these principles is complex and difficult and requires personal and social skills and strong governance.

7 Conclusion

Agricultural land uses worldwide provide essential food and resources to sustain the human population, but the global extent and scale of the associated changes to the natural environment pose a profound threat to biodiversity conservation and ecosystem function. The global dominance of agriculture means the way agricultural landscapes are managed is pivotal to present and

Table 3 Principles for integrating agricultural production and environmental conservation at the landscape scale through involving people and communities

Principle	Explanation
1. Continual learning and adaptive management	Landscapes are dynamic. It's important to understand change, and to learn from management outcomes to develop improved approaches.
2. Common concern entry point	Solutions require negotiations based on trust. Addressing simpler, short-term objectives as a first step can develop shared learning and build trust. This will build confidence to tackle larger issues, towards a shared vision.
3. Multiple scales	Outcomes at any scale (e.g. landscape) are influenced by processes operating at larger (e.g. regions) or smaller (e.g. local sites) scales. Awareness of these other scales can improve local interventions and actions.
4. Multiple functionality	Landscapes can have multiple uses, valued in different ways by different stakeholders. Trade-offs will usually be required among different uses, recognising stakeholder needs and aspirations.
5. Multiple stakeholders	Different stakeholders have different values and aspirations. Identifying and recognising stakeholder concerns and nurturing communication and respect are important. However, total agreement is difficult and unlikely.
6. Negotiated and transparent change logic	Transparency is the basis of trust, achieved through a mutually understood process of change, and helped by sound governance. A shared vision requires broad consensus on goals, challenges and concerns, options and opportunities.
7. Clarification of rights and responsibilities	Rights and responsibilities of different actors need to be clear to, and accepted by, all stakeholders. When conflict arises there needs to be an accepted, legitimate system for arbitration, justice and reconciliation.
8. Participatory and user-friendly monitoring	To facilitate shared learning and assess progress, monitoring is required, with information widely accessible. All stakeholders should be able to integrate the information they require to interpret activities, progress and threats.
9. Resilience	Disturbance processes affect all landscapes and their social and ecological structures. Actions are needed to address threats, improve the capacity to resist perturbations and support recovery.
10. Strengthened stakeholder capacity	Effective participation in landscape management makes demands of stakeholders. People and groups need the skills and abilities (social, cultural, financial) to participate effectively and to accept roles and responsibilities.

The principles are as given by Sayer et al. (2013), with explanation based on interpretation of their comments.

future conservation efforts. The current extinction crisis, coupled with ongoing expansion of agricultural lands (particularly in the tropics), highlight the urgency of actions to protect, retain and restore biodiversity in agricultural landscapes. A range of options exist, including targeted incentives, co-operative schemes and commercial approaches; but long-term success also requires political and financial commitment from governments and support from the wider community.

Agricultural land-uses are diverse and shape the structure of the landscape in different ways. Nevertheless, common factors that enhance conservation of biodiversity can be recognised. First, availability of suitable habitat is critical for the conservation of plant and animal species. A range of different habitats typically are present in and among farmland, including natural, modified and anthropogenic elements. For a large proportion of species, natural and semi-natural habitats are essential. Second, a landscape perspective is important because many species and ecological processes typically operate at this scale, as does land-use planning by governments and other organisations. A landscape perspective offers greater opportunity to manage key landscape properties – such as the overall extent of habitats, composition of land-uses, and landscape configuration – to achieve conservation outcomes. It also recognises that the structure and function of agricultural landscapes change over time. Third, in all agricultural regions, the outcomes for conservation depend on the inherent trade-offs that arise from the way finite land resources are partitioned among different land uses.

Meeting the challenge of conserving biodiversity in agricultural landscapes requires the participation of multiple stakeholders, including people and communities who live in these landscapes, and governmental and other management agencies. The urgency of this situation is demonstrated by the ongoing loss of species in many agricultural regions. Conversely, measures taken now to protect, retain and restore conservation in agricultural landscapes have potential benefits that will play out over decades to come and help to avoid foreclosing options for the future.

8 References

Aerts, R., Van Overtveld, K., Haile, M., Hermy, M., Deckers, J. and Muys, B. (2006). Species composition and diversity of small Afromontane forest fragments in northern Ethiopia, *Plant Ecol.* 187(1), 127–142.

Agger, P. and Brandt, J. (1988). Dynamics of small biotopes in Danish agricultural landscapes, *Landsc. Ecol.* 1(4), 227–240.

Aguilar, R., Ashworth, L., Galetto, L. and Aizen, M. A. (2006). Plant reproductive susceptibility to habitat fragmentation: review and synthesis through a meta-analysis, *Ecol. Lett.* 9(8), 968–980.

Ancrenaz, M., Oram, F., Nardiyono, N., Silmi, M., Jopony, M. E. M., Voigt, M., Seaman, D. J. I., Sherman, J., Lackman, I., Traeholt, C., Wich, S. A., Santika, T., Struebig, M.

J. and Meijaard, E. (2021). Importance of small forest fragments in agricultural landscapes for maintaining orangutan metapopulations, *Front. Forests Glob. Change* 4, 560944.

Andren, H. and Angelstam, P. (1988). Elevated predation rates as an edge effect in habitat islands: experimental evidence, *Ecology* 69(2), 544–547.

Andrew, M. H. (1988). Grazing impact in relation to livestock watering points, *Trends Ecol. Evol.* 3(12), 336–339.

Aspinall, R., Staiano, M. and Pearson, D. (2021). Emergent properties of land systems: nonlinear dynamics of Scottish farming systems from 1867 to 2020, *Land* 10(11), 1172.

Albrecht, M., Kleijn, D., Williams, N. M., Tschumi, M., Blaauw, B. R., Bommarco, R., Campbell, A. J., Dainese, M., Drummond, F. A., Entling, M. H., Ganser, D., Arjen de Groot, G., Goulson, D., Grab, H., Hamilton, H., Herzog, F., Isaacs, R., Jacot, K., Jeanneret, P., Jonsson, M., Knop, E., Kremen, C., Landis, D. A., Loeb, G. M., Marini, L., McKerchar, M., Morandin, L., Pfister, S. C., Potts, S. G., Rundlöf, M., Sardiñas, H., Sciligo, A., Thies, C., Tscharntke, T., Venturini, E., Veromann, E., Vollhardt, I. M. G., Wäckers, F., Ward, K., Westbury, D. B., Wilby, A., Woltz, M., Wratten, S. and Sutter, L. (2020). The effectiveness of flower strips and hedgerows on pest control, pollination services and crop yield: a quantitative synthesis, *Ecol. Lett.* 23(10), 1488–1498.

Atauri, J. A. and de Lucio, J. V. (2001). The role of landscape structure in species richness distribution of birds, amphibians, reptiles and lepidopterans in Mediterranean landscapes, *Landsc. Ecol.* 16(2), 147–159.

Bailey, S.-A., Haines-Young, R. H. and Watkins, C. (2002). Species presence in fragmented landscapes: modelling of species requirements at the national level, *Biol. Conserv.* 108(3), 307–316.

Baker, T. P., Moroni, M. T., Hunt, M. A., Worledge, D. and Mendham, D. S. (2021). Temporal, environmental and spatial changes in the effect of windbreaks on pasture microclimate, *Agric. Forest Meteorol.* 297, 108265.

Bakker, L., van der Werf, W. and Bianchi, F. J. J. A. (2022). Sweep netting samples, but not sticky trap samples, indicate beneficial arthropod abundance is negatively associated with landscape wide insecticide use, *J. Appl. Ecol.* 59(4), 942–952.

Balmford, A. (2021). Concentrating vs. spreading our footprint: how to meet humanity's needs at least cost to nature, *J. Zool.* 315(2), 79–109.

Batáry, P., Dicks, L. V., Kleijn, D. and Sutherland, W. J. (2015). The role of agri-environment schemes in conservation and environmental management, *Conserv. Biol.* 29(4), 1006–1016.

Batpurev, K., Sinclair, S. J., Avirmed, O., Scroggie, M. P., Olson, K. and White, M. D. (2022). Stakeholders from diverse backgrounds make similar judgments about ecological condition and collapse in Mongolian rangelands, *Conserv. Sci. Pract.* 4(1), e574.

Bellamy, P. E., Charman, E. C., Riddle, N., Kirby, W. B., Broome, A. C., Siriwardena, G. M., Grice, P. V., Peach, W. J. and Gregory, R. D. (2022). Impact of woodland agri-environment management on woodland structure and target bird species, *J. Environ. Manage.* 316, 115221.

Bellamy, P. E., Hinsley, S. A. and Newton, I. (1996). Factors influencing bird species numbers in small woods in south-east England, *J. Appl. Ecol.* 33(2), 249–262.

Bennett, A. F. (1990). Land-use, forest fragmentation and the mammalian fauna at Naringal, south-western Victoria, *Wildl. Res.* 17(4), 325–347.

Bennett, A. F. (1991). Roads, roadsides and wildlife conservation: a review, in: Saunders, D. A. and Hobbs, R. J. (eds), *Nature Conservation 2: The Role of Corridors*. Surrey Beatty & Sons, Chipping Norton, pp. 99–118.

Bennett, A. F. (2003). *Linkages in the Landscape. The Role of Corridors and Connectivity in Wildlife Conservation*. International Union for Conservation of Nature and Natural Resources, Gland Switzerland and Cambridge.

Bennett, A. F., Henein, K. and Merriam, G. (1994). Corridor use and the elements of corridor quality: chipmunks and fencerows in a farmland mosaic, *Biol. Conserv.* 68(2), 155–165.

Bennett, A. F., Holland, G. J., Haslem, A., Stewart, A., Radford, J. Q. and Clarke, R. H. (2022). Restoration promotes recovery of woodland birds in agricultural environments: a comparison of 'revegetation' and 'remnant' landscapes', *J. Appl. Ecol.* 59(5), 1334–1346.

Bennett, A. F., Nimmo, D. G. and Radford, J. Q. (2014). Riparian vegetation has disproportionate benefits for landscape-scale conservation of woodland birds in highly modified environments, *J. Appl. Ecol.* 51(2), 514–523.

Bennett, A. F., Radford, J. Q. and Haslem, A. (2006). Properties of land mosaics: implications for nature conservation in agricultural environments, *Biol. Conserv.* 133(2), 250–264.

Bennett, J. M., Nimmo, D. G., Clarke, R. H., Thomson, J. R., Cheers, G., Horrocks, G. F. B., Hall, M., Radford, J. Q., Bennett, A. F. and Mac Nally, R. (2014). Resistance and resilience: can the abrupt end of extreme drought reverse avifaunal collapse?, *Divers. Distrib.* 20(11), 1321–1332.

Bianchi, F. J., Booij, C. J. and Tscharntke, T. (2006). Sustainable pest regulation in agricultural landscapes: a review on landscape composition, biodiversity and natural pest control, *Proc. Biol. Sci.* 273(1595), 1715–1727.

Biggs, J., Williams, P., Whitfield, M., et al. (2005). 15 years of pond assessment in Britain: results and lessons learned from the work of Pond Conservation, *Aquat. Cons. Mar & Fresh. Ecosyst* 15, 693–714.

Blackman, A. and Rivera, J. E. (2010). The evidence base for environmental and socioeconomic impacts of 'sustainable' certification, *SSRN Journal*, https://doi.org/10.2139/ssrn.1579083.

Blake, J. G. (1991). Nested subsets and the distribution of birds on isolated woodlots, *Conserv. Biol.* 5(1), 58–66.

Boscolo, D. and Metzger, J. P. (2009). Is bird incidence in Atlantic forest fragments influenced by landscape patterns at multiple scales?, *Landsc. Ecol.* 24(7), 907–918.

Breckwoldt, R. (1983). *Wildlife in the Home Paddock*. Angus & Robertson, Sydney.

Bright, P. W., Mitchell, P. and Morris, P. A. (1994). Dormouse distribution: survey techniques, insular ecology and selection of sites for conservation, *J. Appl. Ecol.* 31(2), 329–333.

Burel, F. (1996). Hedgerows and their role in agricultural landscapes, *Crit. Rev. Plant Sci.* 15(2), 169–190.

Campbell, A. (1994). *Landcare. Communities Shaping the Land and the Future*. Allen & Unwin, Sydney.

Cano, D., Martínez-Núñez, C., Pérez, A. J., Salido, T. and Rey, P. J. (2022). Small floral patches are resistant reservoirs of wild floral visitor insects and the pollination service in agricultural landscapes, *Biol. Conserv.* 276, 109789.

Chagnon, M., Kreutzweiser, D., Mitchell, E. A. D., Morrissey, C. A., Noome, D. A. and Van der Sluijs, J. P. (2015). Risks of large-scale use of systemic insecticides to ecosystem functioning and services, *Environ. Sci. Pollut. Res. Int.* 22(1), 119–134.

Chen, W., He, B., Nover, D., Lu, H., Liu, J., Sun, W. and Chen, W. (2019). Farm ponds in southern China: challenges and solutions for conserving a neglected wetland ecosystem, *Sci. Total Environ.* 659, 1322–1334.

Cowie, M. (2018). Beekeepers are renting out their hives to help pollinate almond trees, Available at: https://www.theage.com.au/national/victoria/beekeepers-are-renting-out-their-hives-to-help-pollinate-almond-trees-20180831-p50129.html.

Cunningham, R. B., Lindenmayer, D. B., Crane, M., Michael, D. R., Barton, P. S., Gibbons, P., Okada, S., Ikin, K. and Stein, J. A. R. (2014). The law of diminishing returns: woodland birds respond to native vegetation cover at multiple spatial scales and over time, *Divers. Distrib.* 20(1), 59–71.

Curtis, A. C. M. and Lockwood, M. (2000). Landcare and catchment management in Australia: lessons for state-sponsored community participation, *Soc. Nat. Resour.* 13(1), 61–73.

Dániel-Ferreira, J., Berggren, Å., Wissman, J. and Öckinger, E. (2022). Road verges are corridors and roads barriers for the movement of flower-visiting insects, *Ecography* 2022(2), e05847.

Davies, Z. G., Wilson, R. J., Brereton, T. M. and Thomas, C. D. (2005). The re-expansion and improving status of the silver-spotted skipper butterfly (*Hesperia comma*) in Britain: a metapopulation success story, *Biol. Conserv.* 124(2), 189–198.

Doerr, V. A. J., Doerr, E. D. and Davies, M. J. (2011). Dispersal behaviour of Brown Treecreepers predicts functional connectivity for several other woodland birds, *Emu Austral Ornithol.* 111(1), 71–83.

Doley, A. (2003). "Koobabbie": ecological and economic sustainability, *Pac. Conserv. Biol.* 9(1), 42–48.

Donald, P. F., Green, R. E. and Heath, M. F. (2001). Agricultural intensification and the collapse of Europe's farmland bird populations, *Proc. Biol. Sci.* 268(1462), 25–29.

Driscoll, D. A. (2004). Extinction and outbreaks accompany fragmentation of a reptile community, *Ecol. Appl.* 14(1), 220–240.

Ellis, E. C. (2021). Land use and ecological change: a 12,000-year history, *Annu. Rev. Environ. Resour.* 46(1), 1–33.

Estrada, A. and Coates-Estrada, R. (2002). Dung beetles in continuous forest, forest fragments and in an agricultural mosaic habitat island at los Tuxtlas, Mexico, *Biodivers. Conserv.* 11(11), 1903–1918.

Evans, M. C., Carwardine, J., Fensham, R. J., Butler, D. W., Wilson, K. A., Possingham, H. P. and Martin, T. G. (2015). Carbon farming via assisted natural regeneration as a cost-effective mechanism for restoring biodiversity in agricultural landscapes, *Environ. Sci. Policy* 50, 114–129.

Fahrig, L. (2003). Effects of habitat fragmentation on biodiversity, *Annu. Rev. Ecol. Evol. Syst.* 34(1), 487–515.

Fahrig, L. (2017). Ecological responses to habitat fragmentation per se, *Annu. Rev. Ecol. Evol. Syst.* 48(1), 1–23.

Fahrig, L., Girard, J., Duro, D., Pasher, J., Smith, A., Javorek, S., King, D., Lindsay, K. F., Mitchell, S. and Tischendorf, L. (2015). Farmlands with smaller crop fields have higher within-field biodiversity, *Agric. Ecosyst. Environ.* 200, 219–234.

Fernando, P., De Silva, M. K. C. R., Jayasinghe, L. K. A., Janaka, H. K. and Pastorini, J. (2021). First country-wide survey of the Endangered Asian elephant: towards better conservation and management in Sri Lanka, *Oryx* 55(1), 46–55.

Ferraz-de-Oliveira, M. I., Azeda, C. and Pinto-Correia, T. (2016). Management of Montados and Dehesas for high nature value: an interdisciplinary pathway, *Agrofor. Syst.* 90(1), 1–6.

Fischer, J., Abson, D. J., Butsic, V., Chappell, M. J., Ekroos, J., Hanspach, J., Kuemmerle, T., Smith, H. G. and von Wehrden, H. (2014). Land sparing versus land sharing: moving forward, *Conserv. Lett.* 7(3), 149–157.

Fischer, J. and Lindenmayer, D. B. (2002). The conservation value of paddock trees for birds in a variegated landscape in southern New South Wales. 2. Paddock trees as stepping stones, *Biodivers. Conserv.* 11(5), 833–849.

Foley, J. A., DeFries, R., Asner, G. P., Barford, C., Bonan, G., Carpenter, S. R., Chapin, F. S., Coe, M. T., Daily, G. C., Gibbs, H. K., Helkowski, J. H., Holloway, T., Howard, E. A., Kucharik, C. J., Monfreda, C., Patz, J. A., Prentice, I. C., Ramankutty, N. and Snyder, P. K. (2005). Global consequences of land use, *Science* 309(5734), 570–574.

Forman, R. T. T. (1995). *Land Mosaics: The Ecology of Landscapes and Regions*. Cambridge University Press, Cambridge.

Forman, R. T. T. and Baudry, J. (1984). Hedgerows and hedgerow networks in landscape ecology, *Environ. Manag.* 8(6), 495–510.

Freemark, K. E., Boutin, C. and Keddy, C. J. (2002). Importance of farmland habitats for conservation of plant species, *Conserv. Biol.* 16(2), 399–412.

Fuentes-Montemayor, E., Watts, K., Sansum, P., Scott, W. and Park, K. J. (2022). Moth community responses to woodland creation: the influence of woodland age, patch characteristics and landscape attributes, *Divers. Distrib.* 28(9), 1993–2007.

Galindo-González, J., Guevara, S. and Sosa, V. J. (2000). Bat-and bird-generated seed rains at isolated trees in pastures in a tropical rainforest, *Conserv. Biol.* 14(6), 1693–1703.

Garcia, C. A., Bhagwat, S. A., Ghazoul, J., Nath, C. D., Nanaya, K. M., Kushalappa, C. G., Raghuramulu, Y., Nasi, R. and Vaast, P. (2010). Biodiversity conservation in agricultural landscapes: challenges and opportunities of coffee agroforests in the Western Ghats, India, *Conserv. Biol.* 24(2), 479–488.

Garnett, S. T., Sayer, J. and Du Toit, J. (2007). Improving the effectiveness of interventions to balance conservation and development: a conceptual framework, *Ecol. Soc.* 12(1).

Garratt, M. P. D., Senapathi, D., Coston, D. J., Mortimer, S. R. and Potts, S. G. (2017). The benefits of hedgerows for pollinators and natural enemies depends on hedge quality and landscape context, *Agric. Ecosyst. Environ.* 247, 363–370.

Giller, K. E., Hijbeek, R., Andersson, J. A. and Sumberg, J. (2021). Regenerative agriculture: an agronomic perspective, *Outlook Agric.* 50(1), 13–25.

Grashof-Bokdam, C. and Geertsema, W. (1998). The effect of isolation and history on colonization patterns of plant species in secondary woodland, *J. Biogeog.* 25(5), 837–846.

Green, R. E., Cornell, S. J., Scharlemann, J. P. and Balmford, A. (2005). Farming and the fate of wild nature, *Science* 307(5709), 550–555.

Green, R. E., Rocamora, G. and Schäffer, N. (1997). Populations, ecology and threats to the Corncrake *Crex crex* in Europe, *Vogelwelt-Berl.* 118, 117–134.

Hall, M. A., Nimmo, D. G. and Bennett, A. F. (2022). Birds and insects respond differently to combinations of semi-natural features in farm landscapes, *J. Appl. Ecol.* 59(10), 2654–2665.

Hallmann, C. A., Foppen, R. P. B., van Turnhout, C. A. M., de Kroon, H. and Jongejans, E. (2014). Declines in insectivorous birds are associated with high neonicotinoid concentrations, *Nature* 511(7509), 341–343.

Harvey, C. A. and Haber, W. A. (1998). Remnant trees and the conservation of biodiversity in Costa Rican pastures, *Agrofor. Syst.* 44(1), 37–68.

Harvey, C. A., Komar, O., Chazdon, R., Ferguson, B. G., Finegan, B., Griffith, D. M., Martínez-Ramos, M., Morales, H., Nigh, R., Soto-Pinto, L., Van Breugel, M., Wishnie, M. and Working Group on Biodiversity and Conservation Value of Agricultural Landscapes of Mesoamerica (2008). Integrating agricultural landscapes with biodiversity conservation in the Mesoamerican Hotspot, *Conserv. Biol.* 22(1), 8–15.

Harvey, C. A., Medina, A., Sánchez, D. M., Vílchez, S., Hernández, B., Saenz, J. C., Maes, J. M., Casanoves, F. and Sinclair, F. L. (2006). Patterns of animal diversity in different forms of tree cover in agricultural landscapes, *Ecol. Appl.* 16(5), 1986–1999.

Harvey, C. A., Villanueva, C., Villacís, J., Chacón, M., Muñoz, D., López, M., Ibrahim, M., Gómez, R., Taylor, R., Martinez, J., Navas, A., Saenz, J., Sánchez, D., Medina, A., Vilchez, S., Hernández, B., Perez, A., Ruiz, F., López, F., Lang, I. and Sinclair, F. L. (2005). Contribution of live fences to the ecological integrity of agricultural landscapes, *Agric. Ecosyst. Environ.* 111(1–4), 200–230.

Haslem, A. and Bennett, A. F. (2008). Birds in agricultural mosaics: the influence of landscape pattern and countryside heterogeneity, *Ecol. Appl.* 18(1), 185–196.

Haslem, A., Clarke, R. H., Holland, G. J., Radford, J. Q., Stewart, A. and Bennett, A. F. (2021). Local management or wider context: what determines the value of farm revegetation plantings for birds?, *J. Appl. Ecol.* 58(11), 2552–2565.

Haslem, A., Nimmo, D. G., Radford, J. Q. and Bennett, A. F. (2015). Landscape properties mediate the homogenization of bird assemblages during climatic extremes, *Ecology* 96(12), 3165–3174.

Hazell, D., Cunnningham, R., Lindenmayer, D., Mackey, B. and Osborne, W. (2001). Use of farm dams as frog habitat in an Australian agricultural landscape: factors affecting species richness and distribution, *Biol. Conserv.* 102(2), 155–169.

Hazell, D., Hero, J.-M., Lindenmayer, D. and Cunningham, R. (2004). A comparison of constructed and natural habitat for frog conservation in an Australian agricultural landscape, *Biol. Conserv.* 119(1), 61–71.

Hermy, M. and Verheyen, K. (2007). Legacies of the past in the present-day forest biodiversity: a review of past land-use effects on forest plant species composition and diversity, *Ecol. Res.* 22(3), 361–371.

Herring, M. W., Robinson, W., Zander, K. K. and Garnett, S. T. (2019). Rice fields support the global stronghold for an endangered waterbird, *Agric. Ecosyst. Environ.* 284, 106599.

Hino, T. (1985). Relationships between bird community and habitat structure in shelterbelts of Hokkaido, Japan, *Oecologia*. 65, 442–448.

Hinsley, S. A. and Bellamy, P. E. (2000). The influence of hedge structure, management and landscape context on the value of hedgerows to birds: a review, *J. Environ. Manag.* 60(1), 33–49.

Hole, D. G., Perkins, A. J., Wilson, J. D., Alexander, I. H., Grice, P. V. and Evans, A. D. (2005). Does organic farming benefit biodiversity?, *Biol. Conserv.* 122(1), 113–130.

Holland, G. J. and Bennett, A. F. (2010). Habitat fragmentation disrupts the demography of a widespread native mammal, *Ecography* 33(5), 841–853.

Honnay, O., Hermy, M. and Coppin, P. (1999). Effects of area, age and diversity of forest patches in Belgium on plant species richness, and implications for conservation and reforestation, *Biol. Conserv.* 87(1), 73–84.

Hristov, A. N., Hanigan, M., Cole, A., Todd, R., McAllister, T. A., Ndegwa, P. M. and Rotz, A. (2011). Review: ammonia emissions from dairy farms and beef feedlots, *Can. J. Anim. Sci.* 91(1), 1–35.

Karp, D. S., Mendenhall, C. D., Sandí, R. F., Chaumont, N., Ehrlich, P. R., Hadly, E. A. and Daily, G. C. (2013). Forest bolsters bird abundance, pest control and coffee yield, *Ecol. Lett.* 16(11), 1339–1347.

Kilian, B., Pratt, L., Jones, C. and Villalobos, A. (2004). Can the private sector be competitive and contribute to development through sustainable agricultural business? A case study of coffee in Latin America, *Int. Food & Agrib. Manag. Rev.* 7, 21–45.

Kinross, D. C. (2004). Avian use of farm habitats, including windbreaks, on the New South Wales Tablelands, *Pac. Conserv. Biol.* 10(3), 180–192.

Kleijn, D., Rundlöf, M., Scheper, J., Smith, H. G. and Tscharntke, T. (2011). Does conservation on farmland contribute to halting the biodiversity decline?, *Trends Ecol. Evol.* 26(9), 474–481.

Kleijn, D. and Sutherland, W. J. (2003). How effective are European agri-environment schemes in conserving and promoting biodiversity?, *J. Appl. Ecol.* 40(6), 947–969.

Knop, E., Kleijn, D., Herzog, F. and Schmid, B. (2006). Effectiveness of the Swiss agri-environment scheme in promoting biodiversity, *J. Appl. Ecol.* 43(1), 120–127.

Kolkert, H., Andrew, R., Smith, R., Rader, R. and Reid, N. (2020). Insectivorous bats selectively source moths and eat mostly pest insects on dryland and irrigated cotton farms, *Ecol. Evol.* 10(1), 371–388.

Kremen, C. and Merenlender, A. M. (2018). Landscapes that work for biodiversity and people, *Science* 362(6412), eaau6020.

Laurance, W. F. and Bierregaard, R. O. (1997). *Tropical Forest Remnants: Ecology, Management, and Conservation of Fragmented Communities*. University of Chicago Press, Chicago.

Laurance, W. F., Lovejoy, T. E., Vasconcelos, H. L., Bruna, E. M., Didham, R. K., Stouffer, P. C., Gascon, C., Bierregaard, R. O., Laurance, S. G. and Sampaio, E. (2002). Ecosystem decay of Amazonian forest fragments: a 22-year investigation, *Conserv. Biol.* 16(3), 605–618.

Laurance, W. F., Sayer, J. and Cassman, K. G. (2014). Agricultural expansion and its impacts on tropical nature, *Trends Ecol. Evol.* 29(2), 107–116.

Law, B. S., Anderson, J. and Chidel, M. (1999). Bat communities in a fragmented forest landscape on the south-west slopes of New South Wales, Australia, *Biol. Conserv.* 88(3), 333–345.

Law, B. S., Chidel, M. and Turner, G. (2000). The use by wildlife of paddock trees in farmland, *Pac. Conserv. Biol.* 6(2), 130–143.

Lawton, J. A., Holland, G. J. and Bennett, A. F. (2021). What determines the distribution of a threatened species, the brush-tailed phascogale *Phascogale tapoatafa* (Marsupialia: Dasyuridae), in a highly modified region?, *Austral Ecol.* 46(8), 1404–1417.

Lindenmayer, D. B. and Fischer, J. (2006). *Habitat Fragmentation and Landscape Change: An Ecological and Conservation Synthesis*. Island Press, USA.

Lindenmayer, D. B., Macbeth, S. M., Smith, D. G. and Young, M. L. (2022). *Natural Asset Farming: Creating Productive and Biodiverse Farms*. CSIRO Publishing, Melbourne.

Lindenmayer, D. B., Michael, D., Crane, M., Okada, S., Florance, D., Barton, P. and Ikin, K. (2016). *Wildlife Conservation in Farm Landscapes*. CSIRO Publishing, Melbourne.

Lion, M. B., Garda, A. A., Santana, D. J. and Fonseca, C. R. (2016). The conservation value of small fragments for Atlantic Forest reptiles, *Biotropica* 48(2), 265–275.

Loyn, R. H. (1987). Effects of patch area and habitat on bird abundances, species numbers and tree health in fragmented Victorian forests, in: Saunders, D. A., Arnold, G. W., Burbidge, A. A. and Hopkins, A. J. M. (eds), *Nature Conservation: The Role of Remnants of Native Vegetation*. Surrey Beatty & Sons, Chipping Norton, pp. 65–77.

Ludwig, J., Tongway, D., Noble,, J.. and Hodgkinson, K. (1996). *Landscape Ecology, Function and Management: Principles from Australia's Rangelands*. CSIRO Publishing, Melbourne.

Lumsden, L. F. and Bennett, A. F. (2005). Scattered trees in rural landscapes: foraging habitat for insectivorous bats in south-eastern Australia, *Biol. Conserv.* 122(2), 205–222.

Macdonald, D. W. and Feber, R. (2015). *Wildlife Conservation on Farmland: Managing for Nature on Lowland Farms*. Oxford University Press, Oxford.

Mac Nally, R., Bennett, A. F., Thomson, J. R., Radford, J. Q., Unmack, G., Horrocks, G. and Vesk, P. A. (2009). Collapse of an avifauna: climate change appears to exacerbate habitat loss and degradation, *Divers. Distrib.* 15(4), 720–730.

Manning, A. D., Fischer, J. and Lindenmayer, D. B. (2006). Scattered trees are keystone structures–implications for conservation, *Biol. Conserv.* 132(3), 311–321.

Maron, M., Grey, M. J., Catterall, C. P., Major, R. E., Oliver, D. L., Clarke, M. F., Loyn, R. H., Mac Nally, R., Davidson, I. and Thomson, J. R. (2013). Avifaunal disarray due to a single despotic species, *Divers. Distrib.* 19(12), 1468–1479.

Massy, C. (2020). *Call of the Reed Warbler: A New Agriculture–A New Earth*. University of Queensland Press, Australia.

McGarigal, K. and Cushman, S. A. (2002). Comparative evaluation of experimental approaches to the study of habitat fragmentation effects, *Ecol. Appl.* 12(2), 335–345.

Michalski, F. and Peres, C. A. (2007). Disturbance-mediated mammal persistence and abundance-area relationships in Amazonian forest fragments, *Conserv. Biol.* 21(6), 1626–1640.

Mortelliti, A., Amori, G., Capizzi, D., Cervone, C., Fagiani, S., Pollini, B. and Boitani, L. (2011). Independent effects of habitat loss, habitat fragmentation and structural connectivity on the distribution of two arboreal rodents, *J. Appl. Ecol.* 48(1), 153–162.

Munguira, M. L. and Thomas, J. A. (1992). Use of road verges by butterfly and burnet populations, and the effect of roads on adult dispersal and mortality, *J. Appl. Ecol.* 29(2), 316–329.

Newmark, W. D. (1991). Tropical forest fragmentation and the local extinction of understory birds in the Eastern Usambara Mountains, Tanzania, *Conserv. Biol.* 5(1), 67–78.

Newton, I. (2017). *Farming and Birds*. Harper Collins Publishers, London.

Nimmo, D. G., Haslem, A., Radford, J. Q., Hall, M. and Bennett, A. F. (2016). Riparian tree cover enhances the resistance and stability of woodland bird communities during an extreme climatic event, *J. Appl. Ecol.* 53(2), 449–458.

Norton, D. and Reid, N. (2013). *Nature and Farming: Sustaining Native Biodiversity in Agricultural Landscapes*. CSIRO Publishing, Melbourne.

Nyong, A. P., Ngankam, T. M. and Felicite, T. L. (2020). Enhancement of resilience to climate variability and change through agroforestry practices in smallholder farming systems in Cameroon, *Agrofor. Syst.* 94(3), 687–705.

Oakley, J. L. and Bicknell, J. E. (2022). The impacts of tropical agriculture on biodiversity: a meta-analysis, *J. Appl. Ecol.* 59(12), 3072–3082.

Ochoa-Quintero, J. M., Gardner, T. A., Rosa, I., Ferraz, S. F. and Sutherland, W. J. (2015). Thresholds of species loss in Amazonian deforestation frontier landscapes, *Conserv. Biol.* 29(2), 440–451.

O'Connell-Rodwell, C. E., Rodwell, T., Rice, M. and Hart, L. A. (2000). Living with the modern conservation paradigm: can agricultural communities co-exist with elephants? A five-year case study in East Caprivi, Namibia, *Biol. Conserv.* 93(3), 381–391.

Ormsby, A. A. and Bhagwat, S. A. (2010). Sacred forests of India: a strong tradition of community-based natural resource management, *Environ. Conserv.* 37(3), 320–326.

Ouin, A., Aviron, S., Dover, J. and Burel, F. (2004). Complementation/supplementation of resources for butterflies in agricultural landscapes, *Agric. Ecosyst. Environ.* 103(3), 473–479.

Ouin, A., Holland, G. J., Tessier, M., Clarke, R. H. and Bennett, A. F. (2022). Do butterfly communities benefit from woodland restoration in rural environments? A landscape perspective from south-eastern Australia, *Restor. Ecol.* 30(1), e13478.

Pardini, R., de Souza, S. M., Braga-Neto, R. and Metzger, J. P. (2005). The role of forest structure, fragment size and corridors in maintaining small mammal abundance and diversity in an Atlantic forest landscape, *Biol. Conserv.* 124(2), 253–266.

Perfecto, I., Rice, R. A., Greenberg, R. and van der Voort, M. E. (1996). Shade coffee: a disappearing refuge for biodiversity, *BioScience* 46(8), 598–608.

Peters, M. K., Likare, S. and Kraemer, M. (2008). Effects of habitat fragmentation and degradation on flocks of African ant-following birds, *Ecol. Appl.* 18(4), 847–858.

Petit, S. and Burel, F. (1998). Effects of landscape dynamics on the metapopulation of a ground beetle (Coleoptera, Carabidae) in a hedgerow network, *Agric. Ecosyst. Environ.* 69(3), 243–252.

Phalan, B. T. (2018). What have we learned from the land sparing-sharing model?, *Sustainability* 10(6), 1760.

Phalan, B. T., Onial, M., Balmford, A. and Green, R. E. (2011). Reconciling food production and biodiversity conservation: land sharing and land sparing compared, *Science* 333(6047), 1289–1291.

Pino, J., Rodà, F., Ribas, J. and Pons, X. (2000). Landscape structure and bird species richness: implications for conservation in rural areas between natural parks, *Landsc. Urban Plan.* 49(1–2), 35–48.

Pisa, L. W., Amaral-Rogers, V., Belzunces, L. P., Bonmatin, J. M., Downs, C. A., Goulson, D., Kreutzweiser, D. P., Krupke, C., Liess, M., McField, M., Morrissey, C. A., Noome, D. A., Settele, J., Simon-Delso, N., Stark, J. D., Van der Sluijs, J. P., Van Dyck, H. and Wiemers, M. (2015). Effects of neonicotinoids and fipronil on non-target invertebrates, *Environ. Sci. Pollut. Res. Int.* 22(1), 68–102.

Pope, S. E., Fahrig, L. and Merriam, H. G. (2000). Landscape complementation and metapopulation effects on leopard frog populations, *Ecology* 81(9), 2498–2508.

Prevedello, J. A., Almeida-Gomes, M. and Lindenmayer, D. B. (2018). The importance of scattered trees for biodiversity conservation: a global meta-analysis, *J. Appl. Ecol.* 55(1), 205–214.

Prober, S. and Thiele, K. (1995). Conservation of the grassy white box woodlands: relative contributions of size and disturbance to floristic composition and diversity of remnants, *Aust. J. Bot.* 43(4), 349–366.

Radford, J. Q. and Bennett, A. F. (2007). The relative importance of landscape properties for woodland birds in agricultural environments, *J. Appl. Ecol.* 44(4), 737–747.

Radford, J. Q., Bennett, A. F. and Cheers, G. J. (2005). Landscape-level thresholds of habitat cover for woodland-dependent birds, *Biol. Conserv.* 124(3), 317–337.

Ramankutty, N., Evan, A. T., Monfreda, C. and Foley, J. A. (2008). Farming the planet: 1. Geographic distribution of global agricultural lands in the year 2000, *Glob. Biogeochem. Cycles* 22(1), GB1003.

Ramankutty, N., Mehrabi, Z., Waha, K., Jarvis, L., Kremen, C., Herrero, M. and Rieseberg, L. H. (2018). Trends in global agricultural land use: implications for environmental health and food security, *Annu. Rev. Plant Biol.* 69, 789–815.

Ranius, T. and Jansson, N. (2000). The influence of forest regrowth, original canopy cover and tree size on saproxylic beetles associated with old oaks, *Biol. Conserv.* 95(1), 85–94.

Redhead, J. W., Hinsley, S. A., Botham, M. S., Broughton, R. K., Freeman, S. N., Bellamy, P. E., Siriwardena, G., Randle, Z., Nowakowski, M., Heard, M. S. and Pywell, R. F. (2022). The effects of a decade of agri-environment intervention in a lowland farm landscape on population trends of birds and butterflies, *J. Appl. Ecol.* 59(10), 2486–2496.

Redpath, S. M. (1995). Habitat fragmentation and the individual: tawny owls *Strix aluco* in woodland patches, *J. Anim. Ecol.* 64(5), 652–661.

Reij, C. and Garrity, D. (2016). Scaling up farmer-managed natural regeneration in Africa to restore degraded landscapes, *Biotropica* 48(6), 834–843.

Ricciardi, V., Ramankutty, N., Mehrabi, Z., Jarvis, L. and Chookolingo, B. (2018). How much of the world's food do smallholders produce?, *Glob. Food Sec.* 17, 64–72.

Ridding, L. E., Watson, S. C. L., Newton, A. C., Rowland, C. S. and Bullock, J. M. (2020). Ongoing, but slowing, habitat loss in a rural landscape over 85 years, *Landsc. Ecol.* 35(2), 257–273.

Rosin, Z. M., Skórka, P., Pärt, T., Żmihorski, M., Ekner-Grzyb, A., Kwieciński, Z. and Tryjanowski, P. (2016). Villages and their old farmsteads are hot spots of bird diversity in agricultural landscapes, *J. Appl. Ecol.* 53(5), 1363–1372.

Šálek, M., Bažant, M. and Żmihorski, M. (2018). Active farmsteads are year-round strongholds for farmland birds, *J. Appl. Ecol.* 55(4), 1908–1918.

Sánchez-Bayo, F. and Wyckhuys, K. A. G. (2019). Worldwide decline of the entomofauna: a review of its drivers, *Biol. Conserv.* 232, 8–27.

Saunders, D. A., Arnold, G. W., Burbidge, A. A. and Hopkins, A. J. M. (1987). *Nature Conservation: The Role of Remnants of Native Vegetation*. Surrey Beatty & Sons, Chipping Norton.

Saunders, D. A., Hobbs, R. J. and Margules, C. R. (1991). Biological consequences of ecosystem fragmentation: a review, *Conserv. Biol.* 5(1), 18–32.

Sayer, J., Sunderland, T., Ghazoul, J., Pfund, J. L., Sheil, D., Meijaard, E., Venter, M., Boedhihartono, A. K., Day, M., Garcia, C., van Oosten, C. and Buck, L. E. (2013). Ten principles for a landscape approach to reconciling agriculture, conservation, and other competing land uses, *Proc. Natl Acad. Sci. U.S.A.* 110(21), 8349–8356.

Scherr, S. J. and McNeely, J. A. (2008). Biodiversity conservation and agricultural sustainability: towards a new paradigm of 'ecoagriculture' landscapes, *Phil. Trans. Roy. Soc. B Biol. Sci.* 363(1491), 477–494.

Schroth, G., da Fonseca, G. A., Harvey, C. A., Gascon, C., Vasconcelos, H. L. and Izac, A-M.N. et al. (2004). *Agroforestry and Biodiversity Conservation in Tropical Landscapes*. Island Press.

Sirami, C., Gross, N., Baillod, A. B., Bertrand, C., Carrié, R., Hass, A., Henckel, L., Miguet, P., Vuillot, C., Alignier, A., Girard, J., Batáry, P., Clough, Y., Violle, C., Giralt, D., Bota, G., Badenhausser, I., Lefebvre, G., Gauffre, B., Vialatte, A., Calatayud, F., Gil-Tena, A., Tischendorf, L., Mitchell, S., Lindsay, K., Georges, R., Hilaire, S., Recasens, J., Solé-Senan, X. O., Robleño, I., Bosch, J., Barrientos, J. A., Ricarte, A., Marcos-Garcia, M. Á., Miñano, J., Mathevet, R., Gibon, A., Baudry, J., Balent, G., Poulin, B., Burel, F., Tscharntke, T., Bretagnolle, V., Siriwardena, G., Ouin, A., Brotons, L., Martin, J. L. and Fahrig, L. (2019). Increasing crop heterogeneity enhances multitrophic diversity across agricultural regions, *Proc. Natl Acad. Sci. U.S.A.* 116(33), 16442–16447.

Smith, F. P., Prober, S. M., House, A. P. N. and McIntyre, S. (2013). Maximizing retention of native biodiversity in Australian agricultural landscapes: the 10:20:40:30 guidelines, *Agric. Ecosys. Environ.* 166, 35–45.

Stein-Bachinger, K., Gottwald, F., Haub, A. and Schmidt, E. (2021). To what extent does organic farming promote species richness and abundance in temperate climates? A review, *Org. Agric.* 11(1), 1–12.

Strassburg, B. B. N., Brooks, T., Feltran-Barbieri, R., Iribarrem, A., Crouzeilles, R., Loyola, R., Latawiec, A. E., Oliveira Filho, F. J. B., Scaramuzza, C. A. M., Scarano, F. R., Soares-Filho, B. and Balmford, A. (2017). Moment of truth for the Cerrado hotspot, *Nat. Ecol. Evol.* 1(4), 99.

Stoate, C., Baldi, A., Beja, P., Boatman, N. D., Herzon, I., van Doorn, A., de Snoo, G. R., Rakosy, L. and Ramwell, C. (2009). Ecological impacts of early 21st century agricultural change in Europe: a review, *J. Environ. Manage.* 91(1), 22–46.

Sutherland, W. J. (2004). A blueprint for the countryside, *Ibis* 146(Suppl. 2), 230–238.

Sykes, A. K. and Hannon, S. J. (2001). Fencerows as habitat for birds in an agricultural landscape in central Alberta, Canada, *Ecoscience* 8(4), 441–449.

Tews, J., Brose, U., Grimm, V., Tielbörger, K., Wichmann, M. C., Schwager, M. and Jeltsch, F. (2004). Animal species diversity driven by habitat heterogeneity/diversity: the importance of keystone structures, *J. Biogeog.* 31(1), 79–92.

Thomas, C. and Jones, T. (1993). Partial recovery of a skipper butterfly (*Hesperia comma*) from population refuges: lessons for conservation in a fragmented landscape, *J. Anim. Ecol.* 62, 472–481.

Tilman, D., Fargione, J., Wolff, B., D'Antonio, C., Dobson, A., Howarth, R., Schindler, D., Schlesinger, W. H., Simberloff, D. and Swackhamer, D. (2001). Forecasting agriculturally driven global environmental change, *Science* 292(5515), 281–284.

Tilman, D., May, R. M., Lehman, C. L. and Nowak, M. A. (1994). Habitat destruction and the extinction debt, *Nature* 371(6492), 65–66.

Todd, I. A., Tew, T. E. and Macdonald, D. W. (2000). Arable habitat use by wood mice (*Apodemus sylvaticus*). 1. Macrohabitat, *J. Zool.* 250(3), 299–303.

Trzcinski, M. K., Fahrig, L. and Merriam, G. (1999). Independent effects of forest cover and fragmentation on the distribution of forest breeding birds, *Ecol. Appl.* 9(2), 586–593.

van der Ree, R. and Bennett, A. F. (2003). Home range of the squirrel glider (*Petaurus norfolcensis*) in a network of remnant linear habitats, *J. Zool.* 259(4), 327–336.

van der Ree, R., Soderquist, T. R. and Bennett, A. F. (2001). Home-range use by the brush-tailed phascogale (*Phascogale tapoatafa*) (Marsupialia) in high-quality, spatially limited habitat, *Wildl. Res.* 28(5), 517–525.

Verboom, J., Schotman, A., Opdam, P. and Metz, J. A. J. (1991). European nuthatch metapopulations in a fragmented agricultural landscape, *Oikos* 61(2), 149–156.

Vesk, P. A., Nolan, R., Thomson, J. R., Dorrough, J. W. and Nally, R. M. (2008). Time lags in provision of habitat resources through revegetation, *Biol. Conserv.* 141(1), 174–186.
Villard, M. A., Trzcinski, M. K. and Merriam, G. (1999). Fragmentation effects on forest birds: relative influence of woodland cover and configuration on landscape occupancy, *Conserv. Biol.* 13(4), 774–783.
Vogel, C., Poveda, K., Iverson, A., Boetzl, F. A., Mkandawire, T., Chunga, T. L., Kuestner, G., Keller, A., Bezner Kerr, R. and Steffan-Dewenter, I. (2023). The effects of crop type, landscape composition and agroecological practices on biodiversity and ecosystem services in tropical smallholder farms, *J.Appl. Ecol.* 60, 859–874.
Voigt, C. C., Phelps, K. L., Aguirre, L. F., Corrie Schoeman, M., Vanitharani, J. and Zubaid, A. (2016). Bats and buildings: the conservation of synanthropic bats, In: Voigt, C. C. and Kingston, T. (eds), *Bats in the Anthropocene: Conservation of Bats in a Changing World*. Springer, Cham, pp. 427–462.
Watson, S. J., Luck, G. W., Spooner, P. G. and Watson, D. M. (2014). Land-use change: incorporating the frequency, sequence, time span, and magnitude of changes into ecological research, *Front. Ecol. Environ.* 12(4), 241–249.
Weibull, A. C., Bengtsson, J. and Nohlgren, E. (2000). Diversity of butterflies in the agricultural landscape: the role of farming system and landscape heterogeneity, *Ecography* 23, 743–750.
Wiens, J. A. (2002). Riverine landscapes: taking landscape ecology into the water, *Freshw. Biol.* 47(4), 501–515.
Williams, D. R., Clark, M., Buchanan, G. M., Ficetola, G. F., Rondinini, C. and Tilman, D. (2021). Proactive conservation to prevent habitat losses to agricultural expansion, *Nat. Sustain.* 4(4), 314–322.
Wright, D. H., Patterson, B. D., Mikkelson, G. M., Cutler, A. and Atmar, W. (1997). A comparative analysis of nested subset patterns of species composition, *Oecologia* 113(1), 1–20.
Yahner, R. H. (1983). Seasonal dynamics, habitat relationships, and management of avifauna in farmstead shelterbelts, *J. Wildl. Manage.* 47, 85–104.
Yates, C. J., Norton, D. A. and Hobbs, R. J. (2000). Grazing effects on plant cover, soil and microclimate in fragmented woodlands in south-Western Australia: implications for restoration, *Austral Ecol.* 25(1), 36–47.
Zingg, S., Grenz, J. and Humbert, J.-Y. (2018). Landscape-scale effects of land use intensity on birds and butterflies, *Agric. Ecosyst. Environ.* 267, 119–128.

Chapter 2

Biodiversity for agriculture: the role of integrated farm management in supporting agricultural production through biodiversity

G. R. Squire and C. Hawes, James Hutton Institute, UK

1 Introduction: biodiversity for agriculture and integrated farm management

This chapter examines the links between agricultural management and biodiversity, mainly at the scales of the farm and field. Farms range in size from small holdings to vast tracts of land, but all have similar origins and face similar challenges. Agriculture originated from the conversion of natural vegetation to systems that are repeatedly disturbed to grow crops and grazing. Much original

http://dx.doi.org/10.19103/AS.2024.0134.03

biodiversity was lost in that conversion, but agricultural ecosystems still operate through the same fundamental biophysical processes as natural ecosystems, based on flows of energy and matter (Norman et al., 1984; Vasey, 1992; Sinclair and Weiss, 2010), all mediated by biodiversity (Squire, 2020).

The first part of this chapter (Sections 1–4) sets the context in which farm- and field-scale management can work to regenerate and sustain agricultural ecosystems. A central part of the chapter (Sections 5–9) sets out the main principles governing the design of sustainable and regenerative practices. The final part of the chapter (Sections 10–11) offers some practical approaches, including a case study showing the application of design principles in a particular landscape.

Living organisms mediate every process in agricultural production. The organic matter they make and leave combines with mineral particles to form soil. Without soil, there would be no agriculture and little life on land (Ball, 2022). Organisms form complex trophic webs comprising wild plants, the soil microbiome, detritus-feeders, pollinators, pests and their natural predators, pathogens of crops and livestock and higher organisms, all living within managed crops and grass (Norris and Kogan, 2000; Hawes et al. 2009, 2021). Much of this essential biodiversity is microscopic or just visible to the eye, therefore unseen and easily missed, but it performs key ecosystem services, including building stores of carbon and nutrients, detoxifying pollutants and fixing atmospheric nitrogen in soil. It has been defined as Biodiversity for Agriculture or B for A (Bàrberi et al., 2010; Bàrberi, 2015) to distinguish it from Agriculture for Biodiversity (A for B), where management acts to support iconic or culturally valued plants and animals, some that are now in decline and facing extinction.

The difference between A for B and B for A is not hard or clear-cut, since most organisms affect agriculture in some way (Squire, 2020). However, the distinction is necessary since there is a tendency in discussion on agricultural policy to consider only or mainly A for B. This chapter emphasises the importance of B for A for achieving balanced environmental and economic outcomes for farming.

The term 'Integrated Farm Management' implies that the various activities on a farm – and hence the biodiversity and ecological processes they affect – work together and complement each other (Holland et al., 1994). Individual compartments, such as fields, orchards, woodlots and grazing, continually interact over space and time. Integrated management implies interchange and circularity in the use of resources. While each compartment must be managed to build and conserve soil, carbon stocks and nutrients, the outputs of one activity should be used to benefit another, e.g. when land is mulched or fertilised from straw, hay and livestock manure (Watson et al., 2002; Lemaire et al., 2017; Ball et al., 2018). Integration over time ensures that production in one year is influenced by residual effects from previous years and will, in turn, influence production in future years. Integrated farm management is not a recent concept. Rotation is a long-standing example of temporal integration: one crop might extract nutrients from the soil, while another, such as a legume, regenerates them for the next in sequence. One crop might provide the environment for certain weeds to emerge, while the next might suppress those weeds and stop them from seeding.

Humans nurtured the land and its plants and animals for millennia before formal agriculture began through the domestication of plants around 10 000 years ago (Vasey, 1992; Pearson and Ison, 1997; Cumpston et al., 2022). However, despite examples of sustainable land management that have persisted until today, many examples of poor practice and wrong strategies litter the history of agriculture (Azam Ali, 2021). Failure to appreciate soil, climate and the removal of natural vegetation has been responsible for degrading ancient ecosystems in as little as a few decades (e.g. Bennett, 1935). After the 1950s, the advent of powerful machinery, manufactured fertiliser, pesticides and new crop varieties generated a rise in yield per unit area and allowed profitable crops to be grown more or less continuously (Evans, 1993; Conway, 1997). The rise in agricultural production matched growth in the global human population – a major achievement of science and technology (Kawashima et al., 1997; Keating et al., 2014). But twentieth century intensification led to the demise of many traditional and integrated systems, and concentrated industrial-scale production of a few crop species and varieties (Azam Ali and Squire, 2004; Maxted and Lauvie, 2017; Azam Ali, 2021). The resulting soil erosion, pollution and decline in biodiversity are now a global catastrophe (Doran, 2002; Lal, 2012; Borelli et al., 2013; FAO, 2015).

There is, however, a widespread move to return to spatial and temporal integration (Holland et al., 1994; Leakey, 2012; Ball et al., 2018), but the transition is not simple and straightforward (Syers, 1997). Systems have to be assessed at a range of scales and in relation to present and future stressors whose effects may be highly uncertain (Landis, 2003). Many integrated methods have been lost from memory, while new interventions are needed to restore degraded land and maintain a profitable output without reliance on methods such as intrusive tillage and pesticides.

This chapter continues with a summary of the transitions from natural ecosystems to the various forms of agriculture, the importance of scale and connectedness between local management and global influences, and finally the options for management to modify farmland vegetation. The approach aims to present the principles behind integrated management, using examples from a range of agricultural systems differing in intensity and geographical location.

2 Transitions from natural ecosystems to differing types of farming system

All agricultural systems originated from forest, grassland, wetland and other natural ecosystems that formed on the earth's surface. The various states before, during and after agriculture are shown schematically in Fig. 1. A primary ecosystem (Fig. 1 box 1) can be moved into 'extensive' production, for example, where animals are herded or hunted and plants nurtured and harvested (Fig. 1 box 2). A primary ecosystem can also be moved into a more controlled form of management where the land is repeatedly tilled or grazed. Where such agriculture has survived for hundreds or thousands of years, it is termed here as 'sustainable' (Fig. 1, box 3; examples in Fig. 2). There are many examples of sustainable agriculture remaining today (Vasey, 1992), ranging from small-scale, multi-species 'forest gardens' in the humid tropics (Conway, 1997; McConnell et al., 2003) to large-scale cultivation of major crops such as wetland rice in east Asia (Falvey, 2000).

In many parts of the world, however, agriculture has become more extractive (Fig. 1, boxes 3–4), for example due to over-grazing by livestock and repeated deep and intrusive cultivation (Dlamini et al., 2014; Owuor et al., 2018; Zalidis et al., 2002). Extractive agriculture, if it continues unabated, leads to loss of soil and exhaustion of the land (Fig. 1, box 5), then to either desertification (in which production is not possible) or reversal to a secondary ecosystem. There are also instances in which a transition from a primary ecosystem to extractive

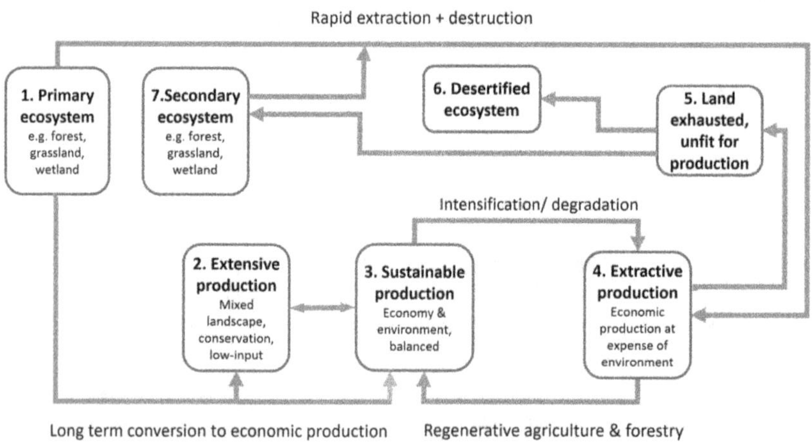

Figure 1 Transitions in land use from a natural ecosystem (1) to systems managed for food and other products (2, 3), which if over-intensified move to a state of extractive production (4). If not reversed (4) leads to land exhausted of resources (5), and then to a desertified system (6) or to a secondary ecosystem if conditions allow regeneration (7). In some forms of agriculture, a primary ecosystem moves rapidly to extractive production (4) and then to (5) and (7), from where the cycle may continue back to (4).

Figure 2 Examples of long-term, sustainable production systems: (top left clockwise) fruit tree orchards in Greece (2013); wetland rice in northern Laos (2008); small grass pasture fields below forest, Romania (2007); mixed rice, fruit, fodder and grazing, south central India (1977). Photographs by G. R. Squire.

production has happened over a very short period without transitioning through a sustainable state (Fig. 1 arrow boxes 1–4). The US Dust Bowl is an example of such rapid conversion (Bennett, 1935). Some instances of shifting cultivation also sit within this category, in which forest or savannah is destroyed by burning or cutting and crops are grown on the nutrients left in the soil (Norman et al., 1984; Anderson, 1993; McConnell et al., 2003). Typically, the nutrients are exhausted after only a few years (Fig. 1, boxes 4–5), after which the process is repeated on another area of primary or secondary ecosystem. Examples of extractive production in Fig. 3 all show large areas of bare or partly vegetated soil, exposed to rain and wind and prone to erosion.

Given the increasing demands on agricultural space by a growing world population (FAO 2021), there is a greater need than ever to return extractive to sustainable production (Hanjra et al., 2017). This is one of the main roles of integrated farm management.

3 Scales of interaction influencing agricultural systems

The spatial focus of integrated farm management is typically the field and farm, the scales at which basic management decisions are made. However, fields and farms are influenced by factors at a range of other scales (Leakey, 2012; Perfecto and Vandermeer, 2017). The concentric rings in Fig. 4 show spatial

Figure 3 Examples of extractive production characterised by large areas of bare or poorly vegetated soil, prone to damage and erosion: (top left clockwise) tilled field, Victoria Australia (2023); shifting cultivation, land cleared, maize planted, Laos (2008); degraded palm grove, Morocco (2014); grazing on sparse vegetation, East Africa (1989). Photographs by G. R. Squire.

scales most relevant to integrated farming. Farms are subject to markets and policy from local to national and global scales. The choices of what to grow and how to manage it are directly influenced by global demand for a product, by governmental factors such as subsidies and regulations, and by the availability of infrastructure such as processing factories, abattoirs and transport networks to connect farmers to differing markets.

In turn, management of the farm aims for gains (or 'goods' or 'services') in terms of products to be consumed or sold, but also regulation of energy and water, and societal and cultural benefits both for those who work the land and the wider population. Achieving such gains inevitably leads to losses, of soil and pollutants, and of crops and livestock not harvested or not sold. Farming needs to repeatedly assess the outputs and value of the gains in relation to the costs of inputs and losses, consider the trade-offs and adjust practice to maximise gains and minimise losses. Part of this adjustment causes feedbacks to higher scales, e.g. to consider the most effective means of gaining grants and subsidies and complying with regulations.

Fields and farms therefore are far from isolated. The global influences in the outer ring apply to both industrialised and subsistence agriculture, since the latter rarely operates in isolation, being influenced by factors such

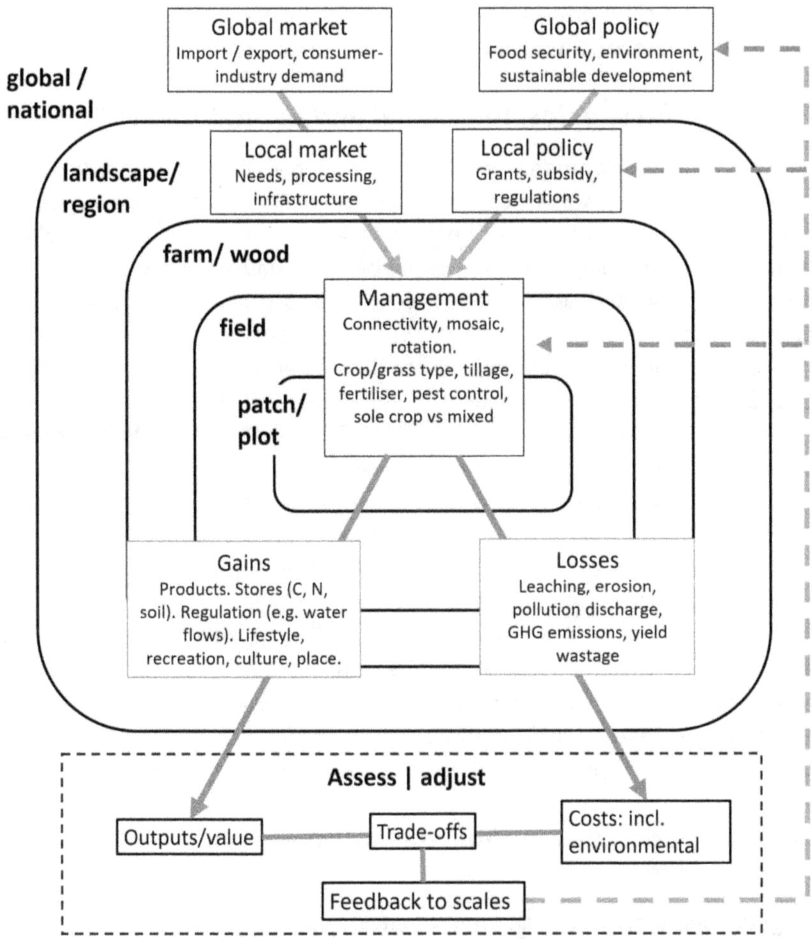

Figure 4 Scales of interaction from global to in-field patch that influence an integrated agricultural system. Diagram adapted from a more comprehensive version by C. Hawes for the Sustainable Crop Systems research programme funded by Scottish Government.

as international aid, global expansion of crops and livestock and the spatial movement of soil and other materials by wind and water.

Whatever other influences there may be, living organisms mediate all management options, gains and losses and the associated processes in the central part of Fig. 4. A 'chain of effect' begins with interventions at field- and farm scales, acts through living organisms which, in turn, modify ecological processes that, in combination, finally determine whether higher-level states are achieved (described at Section 7). Such states include carbon stocks, soil fertility, regulation of water flows, agricultural products, and economic returns.

This part of the chapter ends by considering the options, mainly in choice of species, genotypes and associated agronomy, that initiate a 'chain of effect'.

4 Increasing species, functional and habitat diversity

Technological developments in agriculture in the last century have been concentrated in a few species: the main grain and tuber crops and, to a lesser extent, the main grasses. The greatest effort in international germplasm collections and breeding programmes has been for maize, rice and wheat. These crops were prioritised because they were able to provide the carbohydrate component of the diet for most of the human population. The thousands of other traditionally grown crop plants have been largely neglected (Azam Ali, 2021). The number of genetic varieties of many of the main crop species have diminished (Maxted and Lauvie, 2017). Similarly, of the thousands of tree species used across the world for a multitude of products, only a few have been subjected to major genetic study for use in industrialised production (Leakey, 2012).

A consequence of this concentration of effort has been the spread of resource-demanding species to new regions. Maize, domesticated in the Americas, spread throughout the world, including to resource-poor regions, where the crop is prone to fail if there is insufficient rainfall and no irrigation (Azam Ali and Squire, 2004). A move to more sustainable production will need to reverse this trend, to include more species, a broader choice of functional types in the main crops, and creation of more diverse farmland habitat.

4.1 More species: lessons from traditional and pre-industrial agriculture

As agriculture re-diversifies and regenerates, it can learn from traditional forms of production and gathering. In north-west Thailand, for example, more than 50 cultivated plant species are typically found in and around a field and, in total, thousands of species are used in the area for food, fibre, medicine, dye, construction, firewood, decoration and ceremony (Anderson, 1993). In the home-gardens of Java, many species are grown in small plots for food, spices, medicines and livestock feed, and 'in intricate relationships with each other', occupying a series of canopy heights from those near the ground to tall forms such as coconut and timber trees (Conway, 1997). Similarly, forest gardens in the Kandy region of Sri Lanka each comprise typically around 40 species, including many trees, growing in a tiered arrangement which mimics the natural forest (McConnell et al., 2003; Perera and Rajapakse, 1991).

Botanical diversity in traditional cultures also resides in the variety of end-products from the main species cultivated. Diversity of product is associated with a wide range of phenology and structure possessed by landraces and cultivars. For example, coconut has been grown in semi-wild and in a range of cultivated forms for milk, oil, ointment, vinegar, medicine, livestock 'cake', fibre for matting, palms for thatching, trunk for structure and many other products (Burkhill, 1966). The Australian grass spinifex, in its many forms, has contributed seed-cake, fibre, string, insect repellent, medicinal applications, hafting (joining) metal to handles and shelter (Pitman and Wallis, 2012; Cumpston et al., 2022). The growing and harvesting of a species for such a range of products inevitably raises the number and type of possible interactions that it has with other living organisms.

For integrative or regenerative practice to ameliorate environmental degradation, protect soil and build stores, it will need to displace systems relying on large areas of monoculture with plant assemblages consisting of several species which offer a range of products. As will be described later in the chapter, complex plant systems are better able to regulate the great fluxes of solar radiation and precipitation, use what they need and dissipate the rest.

4.2 Functional diversity in the main crop species

People have relied on maize, rice and wheat for millennia and will continue to need them. However, integrated management should consider reversing the trend in modern technology towards a narrow range of plant form and function (Maxted and Lauvie, 2017). To facilitate harvesting and other agronomic operations, industrial agriculture aims for uniformity among plants in a field. Crop breeding favours determinate and synchronous phenology, mainly in annual forms, traits that ensure most plants in a field mature at about the same time after a well-defined period (Azam Ali & Squire, 2004). With the support of mineral fertiliser and pesticide, management can grow the same species and variety at high plant density year after year. However, systems built around such plants rely on predictable and favourable weather, continued supply of affordable fertiliser, and use of pesticides to prevent proliferation of pests that thrive in such conditions. They may lack resilience and fail if conditions markedly change. A principal mechanism to restore resilience therefore is to increase within-species botanical diversity in terms of plant traits and the configuration in which plants are grown in a field (Azam Ali and Squire, 2004).

Resilience can be enhanced in most crop and grass species by deploying genotypes that are more indeterminate and asynchronous, traits that reduce their susceptibility to abiotic shocks such as bad weather and allow the crop to recover and produce a yield if conditions improve. In addition, species and genotypes may be mixed within a single field, again to reduce susceptibility to unfavourable conditions and to provide more than one product from a field (Willey, 1985). Such mixtures may comprise several genotypes of one crop species (Newton, 2016) or two or more species grown as mixes or intercrops

Table 1 Examples of crop and grass characteristics that may be deployed within fields to facilitate sustainable or regenerative management, arranged in categories of Genotype and Agronomy, adapted from Azam Ali and Squire (2004) with additional input from Hawes et al. (2021)

Characteristic	Typical problems	Examples of regenerative intervention
A Genotype		
Determinate phenology with synchronous reproduction and maturity facilitate management, especially harvesting	High expected returns reduced or lost if reproductive sinks (grain, fruit) fail or harvesting is disrupted in adverse weather	Species/varieties with indeterminate phenology, maturing asynchronously over a long period, allowing plants to continue yielding when adverse conditions are relieved
Annual genotypes give high and regular economic returns	Annuals produce lower above- and below-ground carbon stores, demand regular soil cultivation, and thereby increase soil exposure and erosion	Long-lived or perennial plants in the farming system, e.g. perennial genotypes of annual crops, perennial species such as forages and trees
High mass production the main target of genotypic and agronomic improvement	High fertiliser input and frequent pest control to support high production, both commonly leading to pollution; harvest of grain and fruit with low nutritional value	More efficient (if lower mass-) production using species and genotypes yielding products of high nutritional quality
Reduced diversity among species and genotypes to comply with dominant supply chains	Potential for widespread disruption due to adverse weather affecting many crops simultaneously, and adaptation and spread of pests due to evolved resistance to agrochemicals	Increase the range of phenology, growth habit and product type coupled with agronomic transitions to crop and grass mixtures (see below)
Limited range of harvested products to suit pre-defined major markets	Agriculture locked-in to supply chains and specific technologies, restricting capacity for change	Generate new multiple uses, supply chains and markets for crop products
B Agronomy		
High sowing and stand density to maximise canopy expansion and resource capture	Benefits in favourable conditions not realised when resources are limiting, e.g. soil stores exhausted early, not replenished by rain or irrigation	Reduce stand density to extend period over which a limited resource is used; combine species able to extract resource from different soil depths or at different time in the yearly cycle
Deep and regular soil cultivation facilitates sowing/weed control and encourages rapid crop growth	Repeated disturbance reduces microbial structures, organic carbon and cohesion of soil; exposure to rain, frost and wind raises potential for erosion of soil, together with loss of fertiliser and pesticide.	Minimum tillage to stabilise soil and build carbon; reduce periods of uncovered soil, e.g. undersowing of main crop near or after harvest with a cover to capture unused nutrients

Characteristic	Typical problems	Examples of regenerative intervention
Single species crop stands and limited variability in grass mixtures facilitate planning and agronomic operations	Uniformity across the management unit makes crops prone to adverse weather and pests, weeds and disease	Mixtures of genotypes or species to limit susceptibility to weather and biotic stress, e.g. through intercrops and agroforestry
Simplified rotation allows concentration on the most productive, economic crops	Pests take advantage of continuous cropping with the same genotypes, resulting in rising pesticide usage; high-output crops need high annual inputs of fertiliser.	Re-establish greater crop diversity in rotations, especially through legumes that improve soil and leave N-rich residues for the next crop; introduce sown grass leys (temporary areas of pasture) in arable rotations

(Ehrmann & Ritz, 2014; Suter et al., 2015; Raseduzzaman & Jenson, 2017), including cases where herbaceous grasses or crops are grown with trees in an agroforestry system (Leakey, 2012; Ong et al., 2015). Methods of cultivation and harvesting are necessarily modified during early trialling to cope with the new configurations of genotypes and species. Table 1 summarises some of the genotypic traits and agronomic interventions that have developed from modern technology, common problems arising from their adoption and the phenotypic or agronomic changes that can make fields more resilient. Most interventions in Table 1 directly affect the way in which other living organisms mediate ecosystem processes. A greater diversity in crops means more ecosystem processes can be regulated. Specific examples are given in Sections 5–9 on principles and Section 10–11 on practice.

4.3 Creation of a diverse farmland habitat

While the largest area on farms is usually that of managed fields, integrated farming also aims for regeneration or creation of habitats whose function is not directly for production. While habitats such as field margins, hedgerows, copses, tree lines, wildflower meadows and wetlands are not generally a source of saleable product, they have several important functions that still contribute to economic output (Hawes et al., 2021). One function is to contain resources within the field and farm, e.g. to prevent soil and nutrients moving out of fields in surface water flow and possibly being lost into waterways. Another is to provide habitat for beneficial organisms, such as predators that will move into the managed areas of fields and feed on crop pests. A third is to modify the microclimate of the farm, both for the benefit of crops and to shelter livestock, for example by protecting animals from the effects of high winds or providing shade from intense sunlight.

Habitat is sometimes also created to aid a wider range of biodiversity, under the category A for B. For this to be effective, farms need to connect across the landscape, creating mosaics and corridors that support rare or endangered species (Topping et al., 2003; Perfecto and Vandermeer, 2017). People are also part of this wider biodiversity. Regeneration at all scales from soil to landscape can reconnect people to their source of food and contribute to their culture and sense of 'place' (Ball et al., 2018).

5 Design principles for integrated farm management: stores and fluxes

This section considers the way management interventions can affect living organisms and the way they in turn regulate ecological processes. The emphasis is at the scale of the field and other productive areas. The units of biodiversity range through species, functional types and communities. Biological diversity at smaller scales of the individual or group of individuals interacting in a 'trait space' is the foundation of all higher scales (Squire, 2020) but is generally too fine to be of practical use in farm management.

In natural ecosystems 'stores' (which hold essential chemicals and energy) and 'fluxes' (the movement of these chemicals and energy across space and time) tend to be balanced. The large global fluxes are those of solar radiation, water, air movement and loss of particles from soil and bedrock. Together they allow plant life to turn carbon dioxide in the air to living matter (Fig. 5). The main stores relevant to this chapter are living matter in plants, microorganisms and invertebrates, soil organic matter, mineral nutrients (such as phosphorus, potassium and nitrogen) and water. The main fluxes pass through the soil-plant system, adding to or taking from the stores (Monteith, 1965, 1972).

Incoming short-wave solar energy is either absorbed, reflected back, contained as thermal energy or emitted from plants and soil as long-wave radiation. The energy content of agricultural vegetation is typically around 1% or less of the solar energy it receives (Monteith 1972; Slattery and Ort, 2015). Turning to another kind of flux, precipitation fills soil with water, is taken up by plants and evaporated back to the atmosphere via transpiration (Hoekstra et al., 2011). Excess water runs off the land as surface flow, drains through the soil to the bedrock or evaporates directly from bare surfaces. The water content of plants is a very small fraction of the water they receive and transpire. In producing one unit of dry matter, plants transpire around 1000 times the mass of water (Brendel, 2021; Hoekstra et al., 2011).

Fluxes, therefore, can be very large compared to the stores, and one of the crucial functions of an ecosystem store is to regulate fluxes. In a perennial

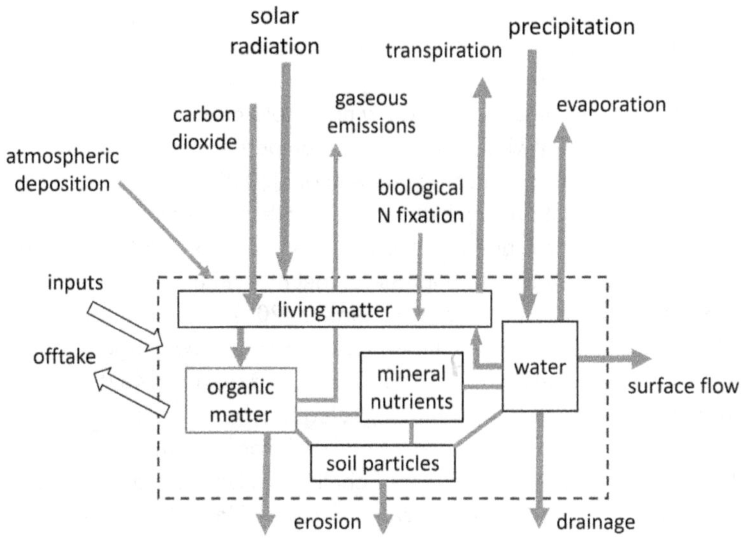

Figure 5 Main stores (in boxes) within the soil-plant system (dashed box), natural fluxes (grey arrows) and indicative human influences (open arrows).

forest or grassland, e.g. high solar radiation is balanced by evaporation of water through vegetation to achieve an equable temperature, and layers of vegetation shield soil from the most intense rain and heat. Stores also allow a system to survive through adversity. In seasonally dry regions, life might wither above ground but, when water returns, the stores re-activate by tapping below-ground resources. If a store is obliterated locally, it can regenerate from neighbouring stores, provided the latter remain in good condition. The capacity of natural systems for survival and regeneration arises through the diversity of their integral life forms. This diversity channels energy and matter through multiple internal pathways and gives the store the capacity to resist and recover from a range of negative external influences.

Agricultural systems are based on the same basic stores and fluxes, but have additional fluxes in the form of human inputs of energy, nutrients and (agri-)toxins, and outputs of harvested crops or livestock. These human-induced fluxes can be large enough to destabilise a system's stores if not carefully regulated. Soil organic matter (SOM), in particular, is crucial to the survival of farming. It influences a soil's capacity to hold water, its penetration by plants and other organisms, and its resistance to erosion. SOM is formed over long periods by micro-organisms processing plant matter and combining the resulting carbon-rich compounds with mineral particles into a complex matrix. However, the burning of vegetation, disturbance by grazing animals, intrusive tillage and application of fertiliser and pesticides reduce the store by detaching carbon compounds from the matrix or encouraging their loss by respiration

(Grandy and Robertson, 2007). Ultimately, a depleting soil store may feed back negatively to reduce the capacity of land to produce crops or livestock and even, to make agriculture impossible.

Among agricultural land uses, growing arable crops with tillage tends to result in loss of carbon while grass and other perennial vegetation result in more carbon being stored in soil (Soussana et al., 2004; Lemaire et al., 2017). Systems of soil cultivation in which seed is drilled directly into soil (e.g. no tillage, minimum tillage, non-inversion tillage) have resulted in higher soil carbon sequestration than when intrusive, mechanical tillage is used to invert and fragment the soil (Grandy and Robertson, 2007; Takoutsing et al., 2016; Rahmati et al., 2020). However, the effects of various low tillage methods on soil carbon dynamics vary with local conditions and do not always increase soil carbon (Valkama et al., 2020). Even where soil carbon is restored, it may not reach the same equilibrium that it attained before agricultural activity started.

Nitrogen (N) poses a unique set of challenges because of its widespread application as an industrially manufactured fertiliser. Natural sources of nitrogen input to soil are by biological fixation through the symbiosis between soil-living bacteria (rhizobia) and the roots of certain plants, mainly legumes (Jensen and Hauggaard-Neilsen, 2003; Kopje and Nemecek, 2010). By the 1990s, however, inputs of manufactured N-fertiliser came to dominate the nitrogen cycle in agriculture (Evans, 1993; Robertson and Vitousek, 2009; Robertson et al., 2013). N-fertiliser now makes a major contribution to greenhouse gas emissions and hence climate change through the use of fossil fuel in the manufacturing process and then through chemical transformations after it has been applied to fields (Smith et al., 2007; IPCC, 2023). The most effective of post-application greenhouse gases are nitrous oxide, much of which originates from conversion of fertiliser nitrogen in soil, and methane arising from livestock and manures. Nitrogen fertiliser also raises emissions of carbon dioxide from agriculture when the organic matter it generates decays or burns. The fluxes of GHG emissions due to nitrogen fertiliser combined with nitrogen losses to water through surface flow and drainage are responsible for much of the pollution from agriculture (Sutton et al., 2011; Norton and Ouyang, 2019; Walling and Vaneeckhaute, 2020). The other two plant macro-nutrients, phosphorus and potassium, present similar but less severe problems because of lower application rates (and lower plant content of these nutrients) and their greater capacity to bind to soil. A major aim of integrated management is to regulate the nitrogen cycle by minimising loss and maximising uptake by plants.

Stabilising and regenerating agro-ecosystems therefore requires knowledge and quantification of the important stores and fluxes that maintain the system and the organisms that mediate the biophysical processes involved. In order to design integrated systems effectively, it is essential to define the limits and 'safe' ranges in which the organisms, processes and states might fluctuate while remaining sustainable for agriculture (Section 6), and then identify and put into practice the 'chains of effect' leading from agronomic interventions through to major outputs (Section 7).

6 Limits and safe ranges

Ideally, a population, process or any biological state can be defined in terms of 'safe limits' within which it can vary without impairing the long-term functioning of the system (Squire, 2017b). In Fig. 6, an attribute varies over time around a mean (the central line) and between two sets of limits. Between X1 and X2, the population or process can operate effectively; in the ranges X2 to Y2 and X1 to Y1, it operates but is impaired in some way; but outside Y1 and Y2 the population declines or the process fails. The asterisks indicate points where the process leaves range X but can be brought back by specific management interventions. The dashed line heading towards and past Y2 represents a point where, because interventions are not applied, the process fails.

Defining the ranges X and Y is straightforward for some but not all factors. For example, the lower limits for biological or chemical contaminants or toxins in feed and food can be defined based on experience and experiment. In contrast, defining limits for many organisms and processes is a major challenge since they depend on multiple interacting environmental and ecological factors. Moreover, the limits for a population or process are rarely absolute, but vary with climate, soil and other aspects of the location, as described below by examples of soil organic carbon and the soil seedbank.

Soil (organic) carbon is one of the major ecosystem stores, produced over long periods by plants sequestering carbon from the atmosphere. It is widely used as an indicator of general soil condition and the capacity of

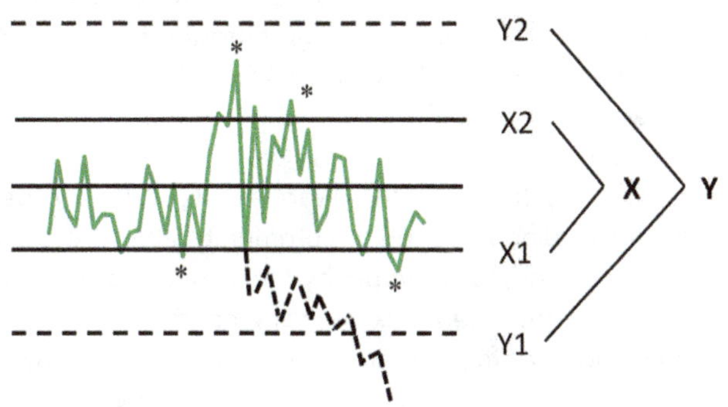

Figure 6 Representation of a population or ecological process varying over time around a mean (central line) and between limits X1 to X2 in which it can operate effectively, and limits Y1 to X1 and X2 to Y2 where it is degrading and needs to be returned to range X by interventions (marked *). Dashed line indicates where a process moves outside X and Y to its permanent damage. Text gives further explanation (adapted from Squire, 2017b).

Published by Burleigh Dodds Science Publishing Limited, 2024.

land to continue absorbing atmospheric carbon. It is a major constituent of soil organic matter (SOM) which forms the habitat for the soil microbiome and associated invertebrates (Jackson et al., 2017; Rees et al., 2005). Soil carbon is usually quantified as the mass of carbon per unit dry mass of soil (e.g. g kg^{-1}), commonly converted to a percentage (typically 0.5% to 10% for agricultural soils), whereas carbon stocks (estimated to a stated soil depth) are usually given as mass per unit land area (e.g. t ha^{-1}). Soil carbon has been estimated over extensive land surfaces by a combination of ground survey, remote sensing and modelling (Luo et al., 2011; Zhang et al., 2015). Data have been combined from many sources into a Global Soil Organic Carbon Map (FAO, 2018), which shows the location of carbon stocks in different soils, climates, and vegetation types. For example, 33% of the global soil carbon to a depth of 0.3 m is in forests, 30% in savannahs and shrublands and 24% in croplands and grasslands.

The main questions for integrated management concern the range of zone X in Fig. 6 for soil carbon and the scope for regeneration if it moves out of X. Soil carbon influences plant growth through several attributes, including soil bulk density and water-holding capacity. It supports the many organisms that live and feed in soil and in the process make nutrients available to crops (Lal, 2012; Valentine et al., 2012). Soil carbon occurs in a range forms or pools, depending on the type of plant that deposits the carbon. The pools break down or transform at different rates and influence plant–soil interactions in different ways (Blair et al., 2006; Grandy and Robertson, 2007; Rahmati et al., 2020). Its safe ranges for agriculture are difficult to generalise but the lower limit X1 has been approximated at around 2% (Loveland and Webb, 2003). The limit at which yield begins to decline depends, however, on other aspects of soil and agronomy. For agricultural soils in the authors' study area in Atlantic zone Europe, a workable range for X was proposed at between 2% and 5% C but, provided crops were supplied with very high nitrogen fertiliser, cereals yield was still near maximum even at and just below 2% C (Squire et al., 2020, 2023).

A second example of defining limits concerns the buried seed, or soil seedbank, that occurs in arable land, grass pasture and regularly disturbed semi-natural ecosystems (Levin, 1990; Warr et al., 1993; Milberg, 1995). Dormant seeds can regenerate a plant community after the above-ground vegetation has been erased, for example by fire, drought or low temperature. In agricultural systems, the seedbank typically contains some species with properties that are positive for the farmed environment, e.g. by supporting food webs (Norris and Kogan, 2000; Marshall et al., 2003), but others that compete aggressively with crops or are poisonous to people and livestock (Squire, 2017b). Poisonous weeds need to be reduced to near zero, and have been in most developed agriculture by targeted weed control and rigorous seed cleaning. For highly competitive species, experiments and models have long been used to define a weed threshold or lower limit of population density at which control is triggered (Cousens, 1987; Walsh et al., 2018).

Limits for the beneficials in arable-grass agriculture have been more difficult to define because of the many plant species and functions they support. For example, soil seedbanks, quantified as number of seeds per unit field surface area to a depth of 0.25 m, in the range 2000–8000 m^{-2} have been proposed as likely to contain a high functional diversity in maritime climates of north-west Europe (Squire, 2017b). Seedbank number itself does not directly equate to emerged weed cover or plant mass, which also depend on field management and competition from the crop. A limit of around 10% ground cover by weeds growing underneath a crop canopy (Smith et al., 2020) is one that would support a diverse above-ground food web without adversely affecting yield. Ground cover has been further quantified in terms of solar radiation flux. A 10% weed cover during the spring-summer period of weed growth intercepted up to 100 MJ m^{-2} (total) solar radiation equivalent to 5% of the 2000 MJ m^{-2} typical for a long-season crop (Squire et al., 2021). This weed cover offered little direct competition with the crop since it resided below the canopy. Therefore the minimum requirements to support beneficial plant species in these climates could be considered to be a seedbank of 2000 m^{-2} coupled with management that regulates emerged species to 10% ground cover during the main phase of growth or 5% of the incoming solar radiation over the life of a crop.

The main challenge is to manage the balance between competitive and beneficial weed species. The practice of *integrated weed management* to regulate this balance, for example through crop rotation, mechanical weed control and highly targeted herbicide application, has been established for some decades (Hatcher and Melander, 2003), but its limited use in high-input cropping has been attributed to farming's lock-in to existing technologies such as chemical control (Swanton and Murphy, 1996; Holst et al., 2007; Baastians et al., 2008).

7 Chains of effect

The crucial role of biodiversity in field-farm management is illustrated by a 'chain of effect'. Its main components are as follows:

1. Management interventions;
2. Living organisms (life forms);
3. Ecological processes; and
4. Higher level outputs or goals.

The definition of the latter can be varied, depending on the nature of the agricultural ecosystem and its role in a society, country or region. For example, goals could be defined as ecosystem services (provisioning, supporting, regulating and cultural services), as in the Millenium Ecosystem Assessment (MEA, 2001), or by the four spheres of sustainability – environmental, economic, societal and political (O'Connor, 2006). Or the goals could align with environmental regulations, governing for example the limit of a pollutant that can be lost from a field or farm.

As an example, a furrowed border is constructed around a field and sown with a plant cover (the intervention). Living microbial matter (life form) and decomposing organic matter hold the furrowed soil in place (process), then the sown plants (life form) emerge and cover the ground, reducing fluctuation in temperature and exposure of soil (process). During heavy rain, the furrows hold water and reduce its rate of lateral flow out of the field (process) and so contribute to reducing soil erosion and loss of fertiliser (supporting and regulating ecosystem services).

In another example, a grain legume is sown in rotation with mainly cereal crops (the intervention). Bacteria in the soil invade the legume's roots to form nodules (life forms). The nodules fix atmospheric nitrogen (process), leading to flows of nitrogen from air to soil (process) and the fixed nitrogen lowers the need to apply mineral nitrogen, thereby supporting crop yield (provisioning) and reducing GHG emissions on the farm (regulating).

While a chain of effect can be used to examine the way to achieve all high-level goals at field and farm scales, it is rarely simple or linear, leading from one intervention to one goal. Rather, a chain branches, one intervention affecting other interventions and on to a range of organisms and processes and in turn several higher goals (Squire and Iannetta, 2018). Examples of more complex chains are given in Fig. 7. Four types of intervention (crop choice, fertiliser, pesticide and tillage) influence life forms (crop, wild plants, invertebrates and soil microorganisms) which, in turn, feed through four sets of processes (resource capture, soil N dynamics, soil organic matter store and soil structural cohesion). The high-level outcomes are expressed in terms of ecosystem services represented by yield (1. provisioning), soil condition (2. supporting), reducing loss and pollution (3. regulating) and wildlife in the form of A for B (4. cultural, aesthetic).

Looking at the chain of effect in Fig. 7, a high-input winter cereal (Fig. 7a) has mainly negative influences on life forms, operating through herbicide (H), insecticide (I) and fungicide (F), fertiliser (especially high nitrogen application to ensure high yield and protein quality) and intrusive tillage. It produces a valuable grain crop but has mainly negative effects on the other processes and services. In contrast, a diverse grass-legume mix for livestock forage or hay (Fig. 7b) has positive influences on the other life forms. Among other benefits, it provides a continuous leaf canopy to protect soil from solar income and rain. Reduced tillage encourages fungal networks to form in soil. The legume

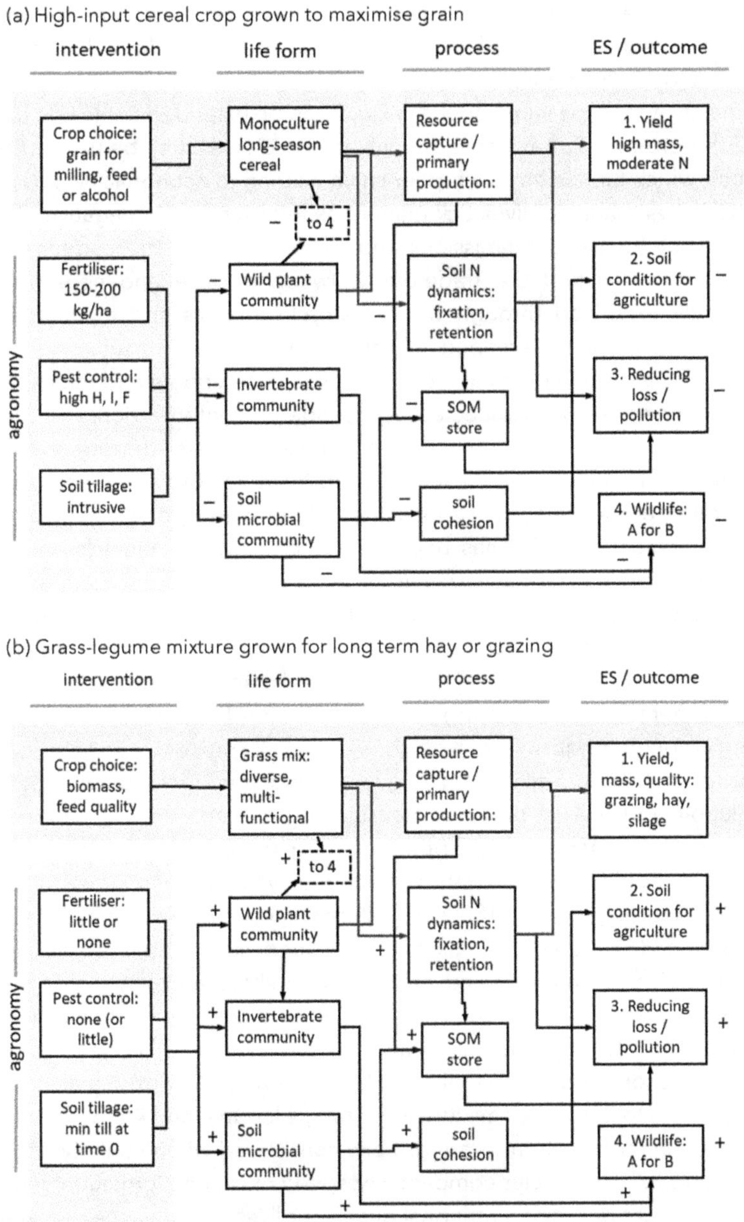

Figure 7 Simplified chain of effect from crop choice and three other interventions to indicative high-level goals (1 = provisioning, 2 = supporting, 3 = regulating, 4 = cultural) for (a) a high-input cereal crop grown to maximise grain yield, and (b) a grass-legume mixture grown for long-term hay or grazing. Plus or minus signs show the direction of effect leading to 2, 3 and 4. Pesticide inputs: H, herbicide; I, insecticide; F, fungicide. The text gives further explanation.

component fixes atmospheric nitrogen, thereby reducing fertiliser input. Together the enhanced, in-field communities also support A for B, for example by providing wildlife habitat.

Chains of the type summarised in Fig. 7 can be extended to include future use of the product. For example, the mixed sward is likely to be used to feed livestock, which then initiates a further chain leading to economic returns from livestock sales, to use of livestock manure as fertiliser and to negative effects through greenhouse gas emissions and water pollution. The complexity of within-farm chains and the trade-offs between positive and negative are increasingly examined through whole-life-cycle analysis and multi-attribute decision support tools (described in Section 10).

The crucial step in defining a chain of effect is to understand the capacity of the life forms, first to be affected by management interventions (functional response), and then to influence an ecological process (functional effect). These types of functions are well known from studies of natural ecosystems but are rarely acknowledged in agriculture. The emphasis on much agricultural science in recent decades has been on the crops, their genetics and to a lesser degree their physiology (Azam Ali, 2021). Knowledge of the functional properties of microbes, other plants and invertebrates is often lacking and hinders regenerative practices, as recognised in FAO's major treatise on global agricultural biodiversity (FAO, 2019). Science and management can in principle work together to provide and apply such information, but appropriate research funding is commonly insufficient.

As described in Section 4, indigenous and traditional knowledge can be valuable today as a guide to transitioning towards diverse, integrative agriculture. Written records of practices long forgotten may also provide guidance to regeneration. An example is offered by the complex 'grass' mixtures constructed throughout the 1800s for different purposes in Europe (Stephens, 1841; Stebler & Schroter,1889; Elliot, 1898). Four of such seed mixtures (Fig. 8) show an increasing number and diversity from short-term hay to long-term grazing. Species of grasses, legumes and other dicotyledonous plants were combined to match their botanical functionality with the local conditions and intended use or products. Number of species ranged from 2 grass and 2 legume in a 1- to 2-year hay mixture to 9 grass, 4 legume and 4 other dicot (17 total) in a mix for long-term grazing. Traits considered when constructing the mixes included inter-species compatibility, speed of canopy formation, ground coverage, height of sward, nutritional quality for livestock, balance of leaf and seed, availability to livestock through the year, ability to penetrate plough-pans and, finally, the capacity to transition from a short-lived to a long-lived perennial state as the sward aged.

In that period, practitioners perfected their seed mixtures by extensive trialling and experience. They did not know, for example, the mechanism by

Published by Burleigh Dodds Science Publishing Limited, 2024.

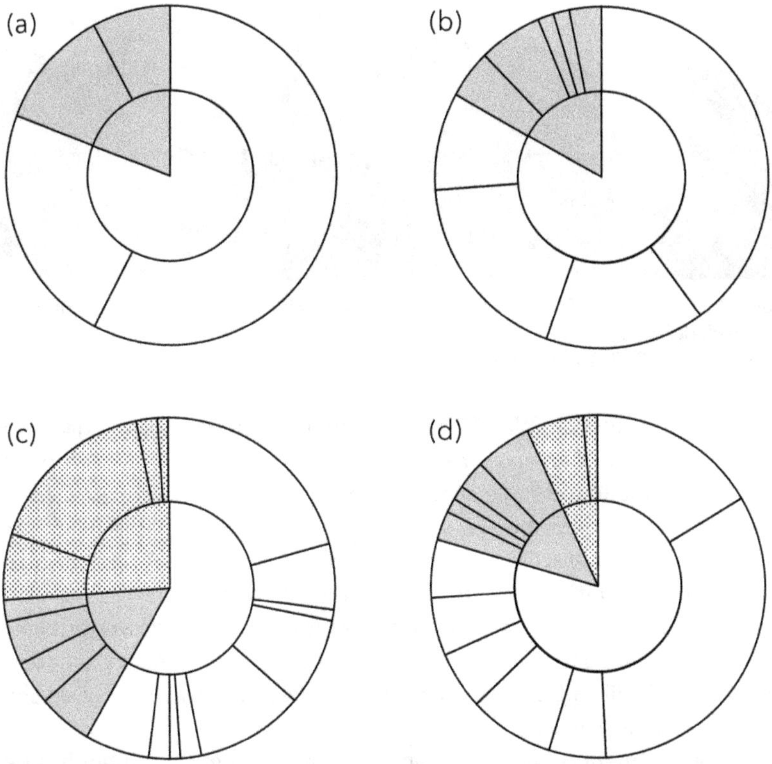

Figure 8 Sown seed mixtures developed in the 1800s in Europe for (a, b) hay or grazing for 2 years, (c, d) long-term grazing, showing in inner circle, proportions by seed weight in the sown mix of grasses (unshaded), legumes (grey), and other dicots (pattern); and in outer circle the proportions of individual species of each type. From data summarised by Findley (1925) based on earlier trials.

which legume species fixed atmospheric nitrogen, but knew from experience that, if grain or forage legumes were grown in a rotation or a mixture, they were nutritionally enriching for the other species. Their approaches and methods would likely have defined a chain of effect as in Fig. 7. Each species or variety opened a new channel or reinforced an existing one. This work provides an example that can be followed today to regenerate degraded land.

The simplified chains in Fig. 7 show that the major aims of field and farm management can rarely, if ever, be achieved by changing just one characteristic of the crop or the agronomy. Rather, a set of interventions needs to be introduced to regulate the various ecological processes that combine to deliver intended outcomes. Given uncertainty in the state of the weather or the presence of beneficial organisms, several interventions may be needed to cover even a single process.

Published by Burleigh Dodds Science Publishing Limited, 2024.

Figure 9 Contrasting fields of tea in Sri Lanka: left, soil eroding, crop failed, land exhausted; right, soil conserved, tiered canopy including N-fixing trees, crop sustainable. Photographs by G. R. Squire, 1989.

A more recent example of applying multiple interventions is given for a perennial crop, tea (*Camellia sinensis*) in Fig. 9. Two fields had been planted on a similar slope in the same climate. The field on the left with bare, eroding soil and poor growth of the tea, has passed through sustainable limits and is transitioning from extractive production (Fig. 1, box 4) to exhaustion (box 5). The field on the right received multiple interventions to keep it in sustainable production (box 3): each row of bushes is contoured to limit downward flow of water, soil is mulched (not visible), and leguminous perennials are planted to fix nitrogen, build soil carbon and form a multi-tiered canopy which gives shade and reduces rainfall impact.

The deployment of multiple changes in crop and management has been termed a 'portfolio approach' to farm management (Robertson and Vitousek, 2009). An example of the approach is now given for the regulation of nitrogen in crop and grass production.

8 Portfolio approaches to crop management: the example of nitrogen

Nitrogen (N) is essential for growth and yield of crops and grass (Cui et al., 2014). Mineral N applied as fertiliser is a major pollutant of water (Ascott et al., 2017; Azevedo et al., 2015) and (as nitrous oxide) makes a major contribution to GHG emissions from cropped land (Norton and Ouyang, 2019; Robertson and Vitousek, 2009). Compared to many other indicators, the processes that determine its stores, flows and losses have been well researched. Management of N therefore provides an instructive example of the potential of regenerative practice based on science (Norton and Ouyang, 2019; Whetton et al., 2022).

N input as fertiliser is primarily determined by economic return based on the N-content of the crop or grass in question. Its limits for environmental safety are more difficult to assess. To reduce waste and environmental damage,

Published by Burleigh Dodds Science Publishing Limited, 2024.

the excess should be zero, i.e. uptake should equal application. Whether an acceptable limit higher than zero can be set (i.e. some waste allowed) then depends on the effect of the nitrogen on biodiversity and habitats, both in and outside the field or farm. In Europe, certain areas or parts of catchments were designated through the EU Nitrates Directive as Nitrate Vulnerable Zones (NVZs) in which farming had to abide by certain rules to limit the application of fertiliser (Van Grinsven et al., 2012).

The designation of NVZs implies that less stringent practice was permissible outside the zones, but there is widespread agreement that agriculture should act to minimise the loss of nitrogen from fields. However, management that relies only on N input being matched by the crop's expected uptake will rarely achieve zero waste. The process of nitrification, which converts nitrogen compounds to nitrate ions, provides plants with a readily accessible source of N, but one which is also highly mobile and readily lost. Losses are especially large during years in which the weather after nitrogen application limits plant growth and hence capacity for N uptake. In a study of the N-balance in European cropping (Leip et al., 2014), the applied nitrogen lost as waste and pollution amounted to around half the nitrogen applied.

To account for uncertainties in the weather and a crop's performance, a 'portfolio approach' to nitrogen management has been recommended to ensure most nitrogen is taken up by crops or retained in soil (Robertson and Vitousek, 2009; Subbarao et al., 2013; Whetton et al., 2022). A portfolio approach consists of a range of interventions applied to a field and any associated livestock, both during a growing season and over longer periods. A portfolio, rather than a single action, is needed because nitrogen uptake, retention and loss are complex processes influenced by many factors (Norton and Ouyang, 2019), whose relative importance in any cropping cycle will be uncertain. Targets of a portfolio approach are shown in Fig. 10 in relation to the main channels for input, offtake, excess, loss and retention of nitrogen. Examples of specific interventions are listed in Table 2, with an indication of the organisms involved (B for A).

Two forms of B for A recur throughout Table 2: the presence of legumes that fix their own N and bring general benefits for soils and food webs (Jensen & Hauggaard-Neilsen, 2003; Kopje & Nemecek, 2010; Iannetta et al., 2016; Stagnari et al., 2017; De Notaris et al., 2021); and the use of mixtures of species and functional types that open and regulate many more ecological processes than can be activated in mono-cultures (Ehrmann & Ritz, 2014; Suter et al., 2015; Kaye and Quemada, 2017; Raseduzzaman and Jenson, 2017).

However, not all interventions in the portfolio are well understood or widely applied. One such is the use of nitrification inhibitors that slow the release of mobile nitrate. Some inhibitors are released, for example as root exudates, by

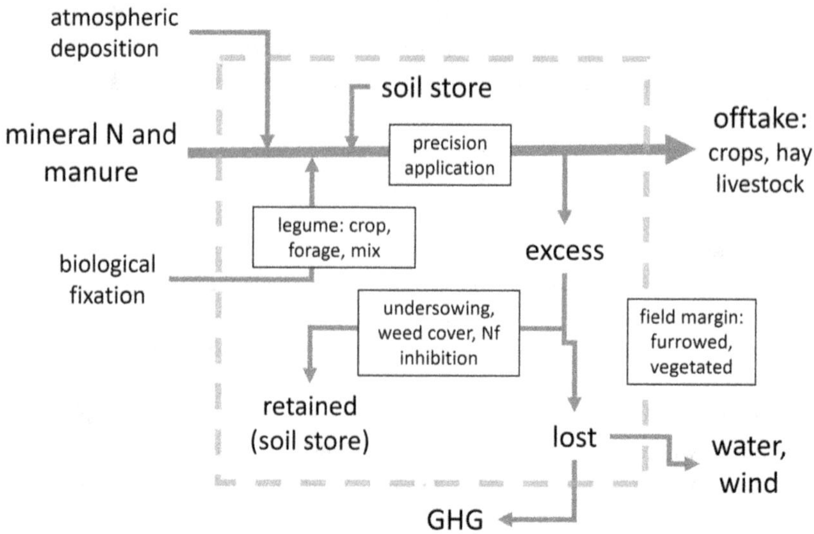

Figure 10 Nitrogen movement and retention in fields, simplified to show main interventions (boxes) in a portfolio approach to minimising N inputs and losses and maximising efficiency of offtake. Text gives information on the interventions. Nf is nitrification.

certain crops and wild plant species (Subbarao et al., 2013; Wang et al, 2021). Others are manufactured chemicals that can be applied to crops or pastures (Moir et al., 2007; Hargreaves et al., 2021). Moreover, the effectiveness of some interventions is uncertain and may vary with context: for example, the application of crop residues to soil may have a range of positive and negative influences (Abalos et al., 2022).

The portfolio approach to the management of nitrogen is a prime example of the need for integration – of crop types, agronomy, sequence, microbiome, food webs – to achieve a result that is simply not possible through any single approach. The use of interventions to achieve multiple aims, to open multiple channels, is commonly more readily adopted by low-input agriculture. Figure 11 shows examples from low-input but highly developed cropping systems that apply legumes, cover crops and mixtures. The latter includes legumes and other species in various configurations, and also perennials for their benefits to soil carbon building and regulating the microclimate.

In contrast, in many regions where intensive and extractive agriculture is the norm, the complexities of low nitrification management coupled with a lock-in to existing technology can result in resistance to change, as shown by the low adoption of legumes in some regions, despite their known benefits

Table 2 Agronomic interventions contributing to a portfolio approach to nitrogen management through the action of living organisms (B for A).

Intervention	B for A
Balancing fertiliser N with crop requirement and measured soil N through precision agriculture	Crop phenology matching soil N availability
Nitrogen-fixing grains and forages that require no or little N application yet increase soil N, available for the next crop.	Bacteria (rhizobia) in soil able to form N-fixing nodules on legume roots
Nitrogen-demanding species or mixed covers and catch crops sown to take up excess N, for example following a cereal	Additional sown species differing from crop in phenology and N-requirement
Use of crop or grass mixtures, including those with legumes, as sown mixed seed, row intercrops or tree-crop systems	Multiple crop species filling more of the trait space and opening more channels
Diverse, naturally regenerated weed understoreys that scavenge excess N in areas and at times where crop N uptake is low and leave high-N organic matter	Non-crop plants, originating in the soil seedbank, enhancing above- and below-ground food webs
Nitrification inhibitors released by certain plants, or applied in chemical form, to limit the rate at which nitrogen compounds are converted to mobile nitrate	Certain crops and wild plant species whose exudates inhibit nitrification in soil
Diverse field margin vegetation designed to trap N in surface water flow, and that can be cut and returned in composted form to the field	Multi-species seed mixes sown outside the cropped area
Low disturbance of soil (no or minimum tillage)	Fungal networks improving soil cohesion, carbon content and nutrient retention
Transfer of livestock manures and crop-grass residues between fields to match need for nutrients	Microbial processes in livestock and soil decomposing plant matter to release nutrients

(Magrini et al., 2016). A widespread transition to legumes and low-nitrification management might well need a redirection of government support and incentives to overcome these barriers to change (Gu et al., 2021).

The outputs from mixed cropping, especially when compared to sole cropping, should not simply be quantified by dry mass per unit field area. Several options exist, including the area of land or period of time that would be needed to reproduce the output of the mixed crops if each component was grown alone (Willey, 1985; Azam Ali and Squire, 2004). Generally, mixed crops use area more efficiently because they interact positively with each other and use different 'fractions' or 'pools' of soil resources. The grass seed mixtures developed in 1800s in Fig. 8 were thought to bring benefit 'because of the interlacing' (Findley, 1925) – the ability of roots of different species to occupy the same soil without competing.

Figure 11 Mixed cropping in Myanmar (Burma): (top left clockwise) groundnut-maize intercrop; rows of sunflower in chickpea field (inset, chickpea pods); multi-species row cropping including two perennials, cotton and pigeon pea (insets, cotton bolls and pea pods); three-species mixed crop, plants withered in dry season, taro, *Colocasia esculenta*, still in soil (inset, tuber in hand), the others already harvested, *Capsicum* sp. (inset, chilli drying) and ginger (not shown). The first three examples contain a legume. Photographs by G. R. Squire, 2014.

9 Considerations of area

The different parts of a farm are linked through the movement of resources between them. Resources such as fertiliser, fuel and seed move into the farm from outside, then agricultural products and waste move out of the farm. The balance of resources into, within and out of the farm have changed through the history of agriculture and differ between its existing forms. Many instances of degradation from Fig. 1 boxes 3–4 and 5 have occurred because the balance is swayed one way or another, as shown in the following examples.

In many early forms of agriculture, and in much current subsistence farming, cultivation and nutrient enrichment were concentrated in small productive areas, supported by a much larger area where nutrients were collected as cut vegetation or manure from livestock grazing there. This internal transfer ensured that at least some harvest was possible, but it relied on the supporting area being rich enough in plant growth and nutrients. Another example is shifting cultivation, described in Section 2, where forest or grassland is burned to return nutrients to the soil, on which crops such as maize and upland rice are then grown (Norman et al., 1984; Anderson, 1993). The harvest may constitute food for the family or contribute to feeding livestock confined to small paddocks. The nutrients in the burnt areas are exhausted after several years, the land is abandoned to revert to a secondary forest and the procedure is repeated in another area of land. In both these cases, limiting nutrients means that adequate provision of food requires large areas of land per unit population. A rising human population density, exacerbated by climatic extremes and wildfires, means less land available for such agricultural practices.

In comparison, the perceived success of high-intensity farming arose from its capacity to increase output per unit area through development of new, high-yielding crop varieties, use of manufactured or mined fertiliser, modern irrigation techniques, mechanised cultivation and agrochemical pest control. In those global regions that could afford the inputs (Azam Ali and Squire, 2004), farming dispensed with the need to scavenge nutrients over a wide area. Between 1960 and 1990, yield per unit field area typically increased 2- to 3-fold. Estimated globally, a five-fold rise in nitrogen fertiliser usage over this period supported a doubling of the human population on a stable agricultural land area (Kawashima et al., 1997). Since 1990, the population has continued to rise while the area of agricultural land has hardly changed, reducing the area of agricultural land per person by 30% to 0.6 ha in 2019 (FAO, 2021). The rise in output per unit area after the 1950s was indeed a major achievement (Keating et al., 2014), but the rising trajectory of yield levelled in many places after the 1990s (Calderini and Slafer, 1998), while 'yield gaps' between actual and potential production remain commonplace (Tittonell and Giller, 2013; Squire, 2017a; Neumann et al., 2022) and farm outflows have continued to cause environmental harm.

A more extreme imbalance in the cycling of resources occurs where one part of the holding supports livestock under cover for all or part of the year. The livestock is fed both from major inflows to the farm, in the form of feed grown and manufactured elsewhere, and cereal grain or hay grown on the farm. Cycling occurs when the livestock manure is spread onto fields to provide nutrients for the next crops. In principle, this cycling can be highly efficient (Lemaire et al., 2017), but excessive flow from livestock enclosures can lead to serious pollution. For example, manure cycled from high-intensity poultry or beef may be so great in quantity that the land is unable to contain it, resulting in excessive outflow and pollution (Jarvie et al., 2003; Post et al., 2020; Rothwell et al., 2020).

The challenge for integrated farm management is to manage land such that internal cycling is maximised, inflows are minimised and outflows tightly regulated. Farming systems can adapt to ensure cycling of energy and matter between years, fields and farms and between crop and livestock enterprises (Pearson and Ison, 1997; Watson et al., 2002; Lemaire et al., 2017). However, the success of integrated management must be gauged by different or at least additional criteria to those such as mass output per unit area which are used to assess high-input farming. Examples include the proportion of land used to grow food, as distinct from other products such as feed and alcohol, metrics on the nutritional content rather than just the mass of food, and the degree to which outflows are causing environmental harm.

The main long-term issues that have to be faced are the declining land area per unit of population and the depletion of soil resources (FAO, 2015, 2021; Hanjra et al., 2017). In principle, soil can be regenerated over long time periods and much of the nitrogen can be provided through biological fixation (and see the case study in Section 11) but, to maintain or improve output where nutrients have been severely depleted, agriculture will have to rely on manufactured and mined plant nutrients, especially phosphate, potash and sulphur.

Integrated management at the scale of the field and farm is unable by itself to meet such future challenges. This is particularly the case in tropical subsistence agriculture where large yield gaps exist between the actual and a defined potential (Squire, 2017a). Tittonell and Giller (2013) in a paper headed 'When yield gaps are poverty traps' have quantified the large gaps in a region of subsistence agriculture between the typical farms and those in the same climate that are well supplied with nutrients. No amount of genetic improvement in crop varieties will close these gaps. It is plant nutrients that are needed. Integrated farm management can contribute, but the global agri-food system has to change drastically to reduce waste of soil, nutrients, crops and food and to recycle on both local and global scales the main plant nutrients before they dissipate to the oceans and atmosphere. The design and implementation of integrated management should therefore be seen as but one practical step towards sustaining future human populations and limiting further environmental damage.

10 Design and implementation of restored farm management systems

This penultimate part of the chapter examines stages in the design and implementation of a system of farm management that both restores biodiversity for agriculture and ensures biodiversity works for the farm.

Restoration occurs in three overlapping phases, beginning with setting the context (Fig. 12 left), followed by quantification of the possibilities given the local topography, climate and infrastructure (Fig. 12 centre) and then implementation of new systems on the farm and the creation of new supply chains (Fig. 12 right).

10.1 Align with global frameworks

The Society for Ecological Restoration (SER) released the second edition of its International Principles and Standards in 2019 (Gann et al., 2019). Several

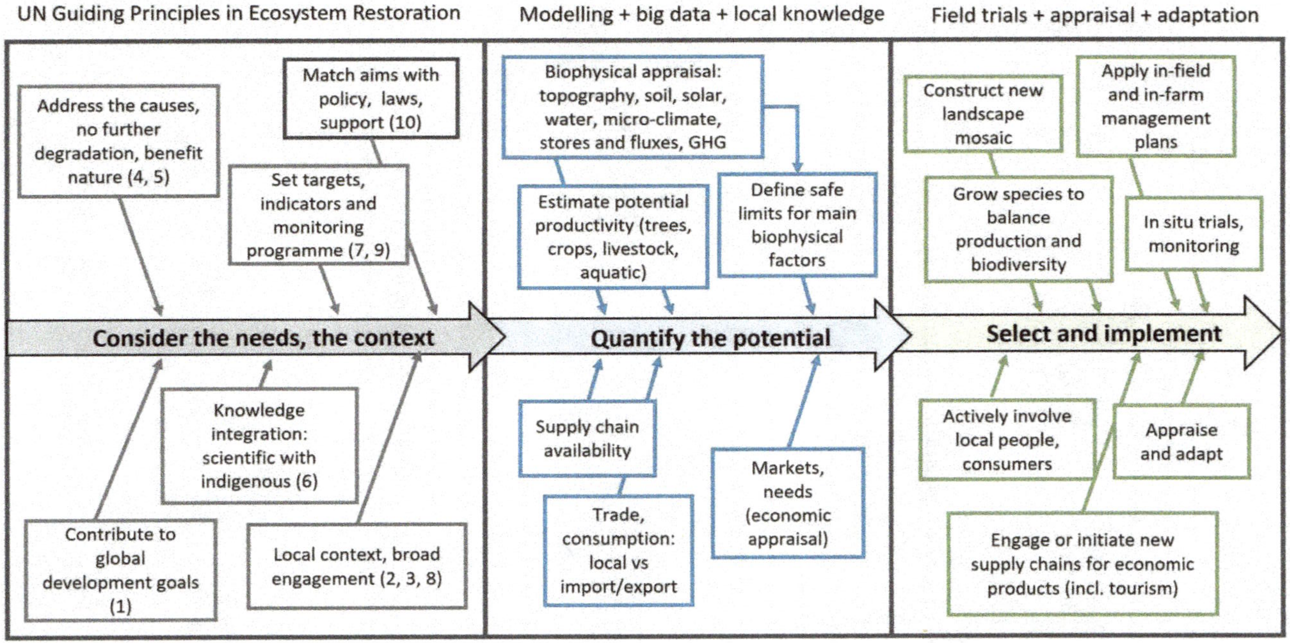

Figure 12 Generalised scheme to guide regeneration of agricultural ecosystems: left-hand panel is based on the UN's Guiding Principles, indicated by numbers in the boxes; central panel compares potential states and options for regeneration; right-hand panel lists activities necessary to put regeneration into practice.

Published by Burleigh Dodds Science Publishing Limited, 2024.

organisations within the UN, working with SER, published the UN's Principles for Ecosystem Restoration in 2021 to coincide with the beginning of the UN Decade of Ecosystem Restoration 2021–2030 (FAO, IUCN, CEM & SER, 2021). Both sets of guidelines are targeted primarily at restoring degraded natural ecosystems, but they can be used to guide regeneration of agricultural systems.

SER summarises its recommendations in eight categories, the UN in ten, but they have similarities. Restoration in any locality should see itself as contributing to global issues through bringing benefits such as carbon capture and reversing biodiversity loss. Planning and implementation should be inclusive of people and combine different types of knowledge, including indigenous knowledge where it exists. Restoration should benefit biodiversity and ecosystem processes but also people and societies. Restoration in any locality should begin by understanding the causes of degradation and preventing them from doing further harm. An implementation plan should include a baseline (the damaged current state) and defined steps in the progression towards a restored state. Monitoring the transition should be aided by measurable biophysical and socio-economic indicators.

The SER approach recommends that 'native reference systems' are identified as an aim. Such systems might be examples of undamaged habitats in the same or similar environment. When the guidelines are applied to agricultural ecosystems, the reference might be a type of farm or production system that has provided sustained output for centuries or millennia (see Section 2). SER is also more specific in recommending that restoration projects work with the capacity of microbes, flora and fauna to revive themselves after damage and that different projects should work together to maximise both local and landscape-scale biodiversity. In an agricultural system, each farm will need to regenerate and sustain its soil microbiome and microfauna, then contribute with other farms to a landscape that regulates water flows, reduces erosion and provides habitat for wildlife (Perfecto and Vandermeer, 2017).

Crucially, 'Policy Integration' is the last of the UN's 10 Principles. Restoration – to be effective – needs to be supported by political will and effective governance that together provide the right incentives. The degradation of farmed ecosystems that has occurred over decades and centuries cannot be reversed in a few years.

In a scheme used here for restoring biodiversity, the UN's 10 Guiding principles are adapted to define the needs and context for regeneration Fig. 12, left panel), including:

- the organisations and people involved in the regeneration scheme;
- the habitats, species and functional life forms that have been degraded;
- the causes of degradation, the means to stop it and possibilities for regeneration;

- targets for regeneration (the reference systems) and a plan for monitoring; and
- scope for alignment with policy and support in respect to land being used for agriculture.

The second phase directs attention to the management units on the farm and what in principle can be achieved (Fig. 12, centre). In many regions of the world, research organisations and farming can work together to quantify this potential in terms of:

- the main stores and flows of energy and matter;
- the different forms of agriculture that the land and climate can support, their potential productivity and required inputs, and the wider biodiversity (A for B) they can support;
- target states and safe ranges for the main indicators;
- the capacity of different plant types to generate multiple outputs (channels);
- supply chains and markets consistent with the regeneration scheme; and
- implications for community and employment.

The third phase (Fig. 12, right) begins the transition to a regenerated system. It tests the organisms that will together regenerate stores and sustain the various outputs. Decisions are made on the biodiversity that sustains agriculture (B for A) and the biodiversity that the transition might aim to support (A for B). This phase can continue for many years as methods are adapted based on the results of trialling and monitoring:

- construction of landscape parcels and connectivity, aiming to balance the internal cycling of materials and minimise external flows;
- trialling new systems, e.g. mixed crops, rotations, agroforestry;
- linking with local communities – food cooperatives, local consumption, short food chains; and
- developing new supply chains.

10.2 Decision tools

The processes involved in moving through the stages in Fig. 12, and the outcomes to the right-hand side, can be facilitated by some form of decision-support system that enables the many possible outcomes to be compared. The main choices have to be made in the central part of Fig. 12. Several tools can be applied to part-quantify potential outcomes, among which multi-attribute decision models (MADM) have become widely used. MADMs or decision trees in various forms are constructed as a hierarchy, the attributes at one level combining to define the attributes at the next higher level, and so on.

Fig. 13 shows an example of some of the attributes whose combined effect is to minimise N losses to water. This branch would be part of a much larger tree.

To the left, boxes with dashed margins show agronomic interventions, whereas boxes with continuous margins show life forms. Each of the latter is the result of previous conditions or choices. The state of the soil microbiome determines whether symbiotic rhizobia exist, and the choice of crop and its performance in the field determines whether symbiosis will happen to allow nitrogen fixation that can replace mineral fertiliser nitrogen. Treatment of the soil and vegetation at the previous harvest combined with activity during the current crop will determine the availability of N in organic residues; then fixation and breakdown will combine to determine how much additional N is needed to satisfy the target crop yield.

Further down, the weed community and field margin vegetation combine with crop growth to give the crop-weed assemblage which, with additional input from tillage, determines the extent of lateral water flow that might move N (and soil and pesticide) out of the field. At this point, as an example, soil organic matter and hence soil cohesion enter the chain from another part of the tree (indicated by an oval rather than a rectangle) to determine surface runoff. Finally, runoff and requirement combine to determine the likelihood of N loss in water moving out of the field. The tree represents a 'chain of effect' linking

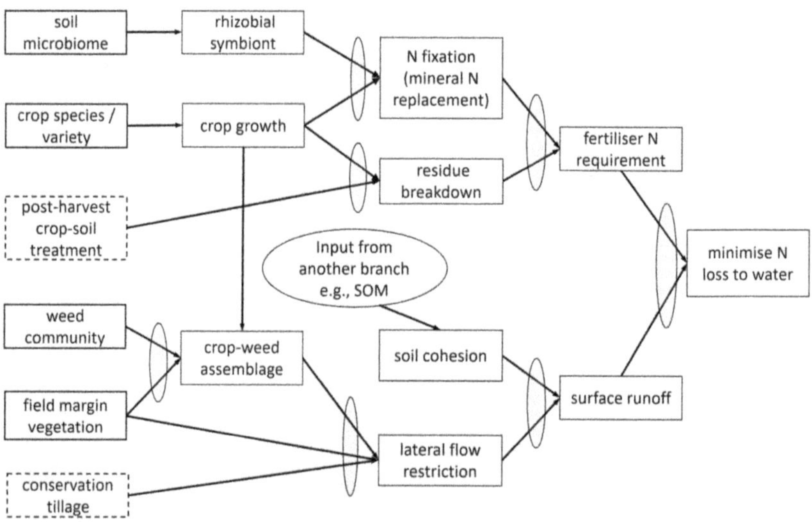

Figure 13 Example of part of a decision tree enabling system planners to evaluate interventions that minimise loss of nitrogen (and other pollutants) to water. The boxes show 'attributes' which can be defined quantitatively or semi-quantitatively, while the vertical ovals show 'utility functions' by which the values of two or more attributes can be combined in DEXi software to determine the value of a dependent attribute. (SOM is soil organic matter, text and Fig. 14 give more information).

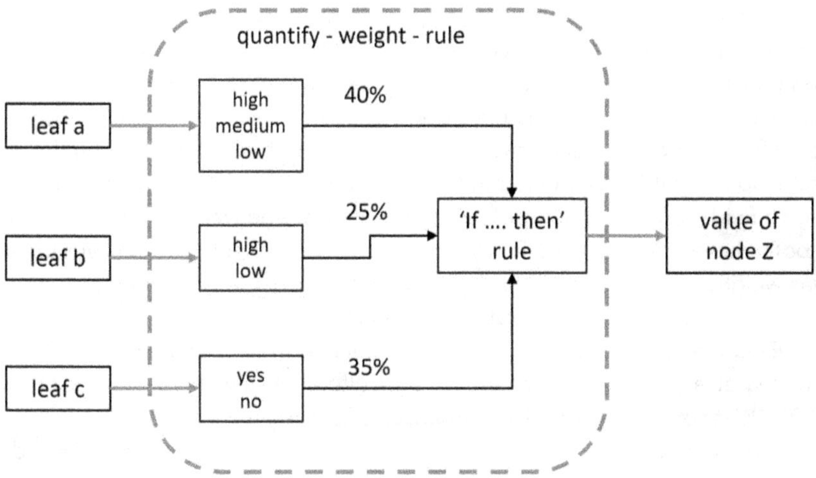

Figure 14 Quantifying attributes in a DEXi model: three attributes (leaves a, b and c) determine the state of a fourth attribute (node, Z) in three steps: the leaf attributes are defined semi-quantitatively (e.g. high/medium/low or yes/no), their contribution to Z is weighted (as %), then conditional (if .... then) rules determine how the values and weightings of a, b and c result in the value of *Z*. Text gives examples.

interventions, organisms, ecosystem processes (states, fluxes), and finally the aim of minimising loss to water. A branch such as in Fig. 13 can be constructed for each of the major high-level outcomes, including GHG emissions targets, food and feed quality and overall soil condition. All branches can then be combined into a single MADM.

A tree of this form is commonly organised using DEXi software (Bohanec, 2014; Bohanec and Znidarsic, 2013). The procedure by which 'lower' (leftward in Fig. 13) attributes are combined to determine the value of 'higher' dependent (rightward) attributes is demonstrated in Fig. 14, where (to continue the analogy of branch and tree) three attributes, named 'leaves' a, b and c, combine to determine the value of a dependent attribute or 'node' *Z*. In an agricultural context, not all attributes are quantified to the same degree: some may be highly quantifiable by measurements or models; others can only be roughly categorised into, for example, high, medium or low states, or bands 1 to 5; and still others may be largely unquantified and regarded as desirable or not (good/bad or yes/no). DEXi models accept such inconsistent classification. First, each leaf is given a semi-quantitative value, then a weighting as to its importance in determining the state of the node; then a set of conditional (if.... then) rules are applied to define the state of *Z* from a, b and c. Functions of this type link all attributes in a tree: if a single leaf value or weighting is changed, there could be ramifications throughout the whole tree.

A simple example of an 'if … then' rule from Fig. 13 would be – if 'rhizobial symbiont present = yes' and 'cereal crop = yes', then 'N fixation = 0'. Greater complexity could be introduced if the crop is a legume, when there would then be a range of 'if …. then' rules depending on the known N-fixing capacity for different legumes and rhizobial types. In another example, lateral flow restriction might be quantified in terms of high, medium or low. It would be high if field margin vegetation and conservation tillage were both present, low if both absent and medium if only one were present. If field margin vegetation was weighted much more than conservation tillage, then field margins alone might be medium but conservation tillage alone would be low.

Examples of MADM are available for soil condition (Bohanec et al., 2007; Griffiths et al., 2010), assessment of agricultural sustainability (Sadok et al., 2009; Pelzer et al., 2012), and a comparison of conventional and agroecological cropping systems (Hawes et al., 2019; Soule et al., 2023). MADMs in DEXi do not rely on prior expertise in modelling. The tree and utility functions can be constructed, trialled and modified as necessary by a range of stakeholders attending a round-table or online conversation. For example, Colomb et al. (2013) described a procedure for evaluating stockless organic farming using DEXi software to bring together the results of numerous consultations with different stakeholders.

11 Case study: Centre for Sustainable Cropping (UK)

A 'portfolio approach' to managing biodiversity for agriculture was established at the Centre for Sustainable Cropping (CSC), a long-term experiment at the James Hutton Institute's research farm near Dundee, UK, based in the lowland, arable-grass agricultural region of Scotland. The experiment began in 2009 with the aim to develop, implement, test and demonstrate a framework for designing integrated arable cropping systems (Hawes et al., 2016, 2019). The CSC contributes to a coordinated effort in regenerative farming that began much earlier and extends to a wide range of agricultural systems.

Agriculture began here over 5000 years ago when land was cleared of post-glacial vegetation to support species of crops and livestock, most of which originated in Europe and western Asia. Little is known of the state of land and its management until the monastic expansion of the 1100s recorded many instances of poor yield and degraded soil. The monasteries introduced regenerative practices such as soil drainage, weed control and rotation. They recommended growing legumes to improve soil fertility, even though their ability to fix atmospheric nitrogen was not discovered scientifically until many centuries later. A phase of further improvement in the late 1700s and the 1800s again reported degradation and methods to restore soil and boost output but, despite many improvements, agriculture in the early twentieth century was unable to provide food security and adequate nutrition for the population (Boyd Orr, 1943).

An Agriculture Expansion Programme, initiated by the government in the 1940s, led to increased output from 1950 to 1990, supported by higher fertiliser and pesticide usage, deeper soil cultivation and improved crop and grass varieties (Squire, 2017b). The lowland landscape remained diverse and output from cereal crops, for example, was high by global standards (Fig. 15, upper). However, the rise in grain output levelled in the 1990s, while fertiliser, pesticide and intrusive tillage continued to damage many fields and cause wider environmental harm, including nitrate pollution (Squire et al., 2022) and soil erosion (Benaud et al., 2020). Over the next three decades, agricultural policy and management maintained yield but failed to achieve substantial reductions in fertiliser usage and greenhouse gas emissions (CCC, 2020), to halt declines in biodiversity, both B for A and A for B (RSPB, 2018) and to provide adequate nutrition for the population (Ritchie, 2017).

Over 5000 years therefore, the agricultural land in the lowland regions of Scotland had moved through repeated cycles of degradation and regeneration, but had not been permanently exhausted or desertified. It was still productive but had reached a point where its negative effects on pollution and biodiversity needed to cease. Non-governmental organisations took the

Figure 15 Agricultural land in lowland east Scotland: (top left clockwise) typical diverse arable-grass land use of the region; high yielding cereal; field where the soil is compacted and waterlogged due to intrusive tillage and heavy traffic; a large area of bare soil prone to erosion.

lead in demonstrating that progress towards regenerating soil and reducing pollution was possible by integrated (Linking Environment and Farming, 2024) or organic (Soil Association, 2024) farm management. Research showed that spatial and temporal integration between grass, crop and livestock in organic systems could be achieved by rotation, including grass leys (temporary pastures) in arable sequences, legumes to fix nitrogen and the cycling of nutrients within the farm in the form of manures and crop residues (Watson et al. 2002). New assessment methodologies were trialled, e.g. to standardise rapid, on-site assessment of soil quality (Ball et al., 2007). A baseline was established through a major regional biophysical appraisal of current farming (central panel of Fig. 12) covering high-input, integrated and organic farming systems. The study confirmed that many attributes of field and farm, including soil condition/health, plant diversity and food webs, were being degraded by high-intensity (mostly arable) management, but could be restored by suitable interventions (Hawes et al., 2010; Valentine et al., 2012; Squire et al., 2020).

Several experimental platforms were established, covering hill farming, mixed livestock and crops. The Centre for Sustainable Cropping (CSC) targeted a high-input system based on cereals for grain. This system was the most damaging to soil in current crop-grass farming and was proving difficult to change. The overall aims of the CSC were:

- to maintain crop yields at comparable levels to those in conventional practice, but with less reliance on non-renewable inputs including fuel, crop protection chemicals and mineral fertilisers;
- to increase the efficiency with which inputs are used, thereby minimising pollution and losses from fields (soil, agrochemicals, greenhouse gas emissions); and
- to achieve these objectives by enhancing in-field and field-margin biodiversity.

The phase of 'select and implement' in the right-hand panel of Fig. 12 began with construction of the CSC platform in which an integrated approach comprising a portfolio of interventions was compared with existing, high-input management. The platform consists of six arable fields covering 42 ha in a rotation of autumn-sown wheat, barley and oilseed rape, and spring-sown field beans, barley and potato. Each field is divided into two halves with the integrated, regenerative system randomly allocated to one half at the outset and then maintained in the same location across multiple crop rotations to detect long-term trends in system properties (Hawes et al., 2019). Fields are monitored annually across a grid of 350 permanent sample locations. The performance of grain legumes was considered the most uncertain. The area sown with peas and beans in the region had declined since the 1850s, reaching a low point in the 1930s from which sown areas had hardly recovered (Squire et al., 2019). It was uncertain

Published by Burleigh Dodds Science Publishing Limited, 2024.

whether rhizobia that combined with legume crops to form the N-fixing symbiosis were still present.

The interventions in the integrated system are listed in the lower four boxes in Fig. 16 under the main components of the strategy:

- organic matter inputs;
- conservation tillage;
- nutrient supply; and
- pest management.

These interventions affect life forms to achieve improvements in soil biology, soil physics, crop health and invertebrate food webs, which combine to achieve goals for the environment, productivity and biodiversity. Progress is monitored by a suite of indicators:

- Soil microbiome and macro-invertebrates;
- Soil carbon and plant nutrient availability;
- Weed seedbank and emerged weed flora;
- Field-margin vegetation;
- Crop phenology and yield;
- Pollinator abundance;
- Predator activity-abundance patterns;
- GHG emissions; and
- Soil water quality.

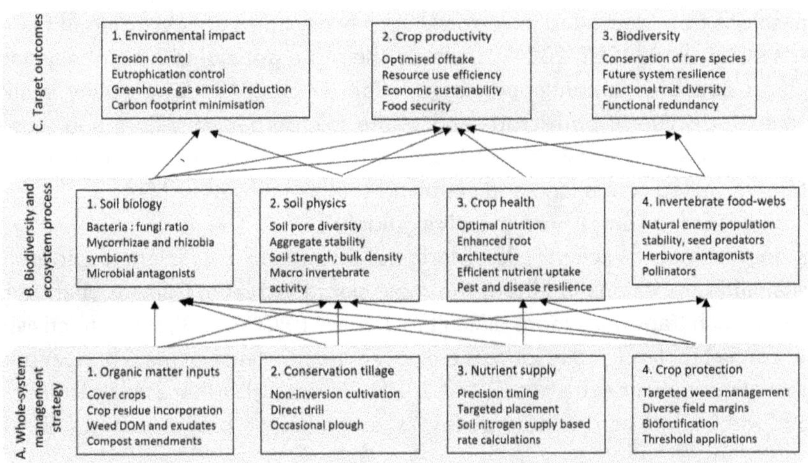

Figure 16 Integrated, outcomes-based approach to designing a regenerative cropping system for biodiversity, environment and food security: case study example from the James Hutton Institute's Centre for Sustainable Cropping long-term platform, Balruddery Farm, Dundee, UK (Hawes et al., 2019; Squire et al., 2022).

Published by Burleigh Dodds Science Publishing Limited, 2024.

The regenerative approach adopted at the CSC applies B for A to open up and regulate chains of effect to achieve high-level targets. Results to date include improved soil biophysical condition, reduced pesticide and fertiliser inputs and increased plant biodiversity. The presence of rhizobia and their capacity to fix atmospheric N in symbiosis with faba bean roots have both been confirmed (Maluk et al., 2022). The grain legume in the rotation coupled with integrated management converted a net N loss from cereals cultivation to a net N gain (Squire et al., 2022).

The portfolio approach was adaptable, in that interventions were modified based on findings. For example, after the first few years, no N fertiliser was applied before winter to reduce losses at times of very high rainfall, while an N-demanding crop was undersown with N-fixing clover to compensate for reduction of mineral N-input.

Activities in the right-hand panel of Fig. 12 are continuing through interactions between science, farming interests, communities and non-governmental organisations. The latter aim to improve all aspects of the food chain (Nourish Scotland, 2024) and promote on-line and in-person debates on the future of land, its management and its products (SEDA Land, 2024). The field research network and the CSC both reside within and contribute to agriculture and food production in an extensive bioregion that aims to achieve food security mainly from local producers (Bioregioning Tayside, 2024).

12 Conclusion

Transitions from degraded to restored agro-ecosystems are occurring in many parts of the world (SER, 2024). However, the more general trend globally and in most countries is continued degradation (FAO, 2019). Biodiversity is still in overall decline in both crops and in the microorganisms, flora and fauna that support agricultural production. Soil is losing structure and eroding in intensively tilled arable land and in upland shifting cultivation. Many ecological transitions fail to comply with the UN's guidelines.

In principle, the scientific knowledge of restoration is sufficient to underpin sustainable transitions in all but the most degraded land. Climate change is making such transitions more difficult to achieve (IPCC, 2023), but extractivism and conflict for land and resources are now a greater cause of mass hunger and famine than climatic extremes (GRFC, 2021). Political will in these circumstances, both within and across boundaries, is essential for progress (FAO, 2019, Chapter 9).

In those regions devoid of serious conflict, regeneration of agricultural land by B for A is entirely feasible. While the functional roles of many organisms contributing B for A are still unknown or uncertain (FAO, 2019) there is sufficient knowledge to guide regeneration. The main factors preventing the right transition are the dominance of global food chains, the reluctance of farming to move away from established practices, and limitations in policy, support and governance. If these obstacles are removed, regenerative practice at the scale of the patch, field and farm can undoubtedly restore the soil microbiome and the flora and fauna of farmland for greater system resilience and overall system sustainability.

13 Where to look for further information

Three sources are suggested for further reading: books, global reports, and organisations directly involved in regeneration.
Books:

- Evans (1993) *Crop evolution, adaptation and yield* is a standard reference.
- Pearson and Ison (1997) *Agronomy of grassland systems* covers grazing and livestock production in many parts of the world.
- Conway (1997) *The Doubly Green Revolution* 'Food for all in the 21st century' deals with ways to tackle the future demands for food and has several chapters relevant to IFM.
- Leakey (2012*) Living with the trees of life* describes the many uses of tree crops and makes the case for multifunctionality in agriculture.
- Findlayson and Warren (2010) *Changing natures* takes a critical look at transitions between hunter-gatherers and early agriculture.
- Tudge (2016) *Six steps back to the land* argues for a greater connection between people and land and a transition to more and smaller holdings.
- Azam Ali (2021) *The Ninth Revolution: 'Transforming food systems for good'* covers past failures (and successes) and presents the case for fundamental change.

Global reports tend to be highly detailed. The best place to begin in each case is the summary, which is sometimes released as a separate document available at the web links in References. Then by perusing the contents, the reader can usually find the topics of most interest.

- United Nations: *Principles for Ecosystem Restoration to guide UN Decade of Ecosystem Restoration 2021–2030* (FAO, IUCN, CEM and SER, 2021).
- The Society for Ecological Restoration: *International Principles and Standards for the practice of ecological restoration* (Gann et al., 2019)
- Food and Agriculture Organisation of the United Nations: *Status of the World's Soil Resources* (FAO, 2015); and *The State of the World's Biodiversity for Food and Agriculture* (FAO, 2019).

Organisations that support and promote sustainable production and agro-ecological transitions:

- The Society for Ecological Restoration (SER): operates globally but points to many local restoration stories, some of which involve production land (SER, 2024).
- The Bioregioning movement: not explicitly about regeneration and farms but the movement is encouraging regions to think and act holistically to provide for their future including their agriculture and biodiversity. An example local to the authors is Bioregioning Tayside (2024).
- Organisations that practice and promote IFM: tend to be local to a country, for example in the UK there are Linking Environment and Farming (LEAF, 2024), Nature Friendly Farming Network (NFFN, 2024) and the Soil Association (2024).

14 Acknowledgements

The authors thank the editors of this book, particularly Nick Reid, and David Paton for their detailed and constructive comments on the early draft, and Francis Dodds for the many positive suggestions on the content, text and layout of this chapter.

15 References

Abalos, D., Recous, S., Butterbach-Bahl, K. et al. (2022) A review and meta-analysis of mitigation measures for nitrous oxide emissions from crop residues. *Sci. Total Environ.*, 828, 154388.

Anderson, A. E. (1993) *Plants and people of the Golden Triangle*, Silkworm Books, Chiang Mai, Thailand.

Ascott, M. J., Gooddy, D. C., Wang, L., Stuart, M. E., Lewis, M. A., Ward, R. S. and Binley, A. M. (2017) Global patterns of nitrate storage in the vadose zone. *Nat. Commun.*, 2017, 8. doi:10.1038/s41467-017-01321-w.

Azam Ali, S. N. (2021) *The ninth revolution: transforming food systems for good*, World Scientific Publishing, Singapore.

Azam Ali, S. N. and Squire, G. R. (2004) *Principles of tropical agronomy*, CAB International, Oxfordshire, UK.

Azevedo, L. B., van Zelm, R., Leuven, R. S. E. W., Hendriks, A. J. and Huijbregts, M. A. J. (2015) Combined ecological risks of nitrogen and phosphorus in European freshwaters. *Environ. Pollut.*, 200, 85–92.

Ball, B. C. (2022) *Healing soil – how soil health will save the planet and us*, Bruce Ball, Roslin, Midlothian, UK, distributed by Kindle Direct Publishing Services, Amazon, Seattle, USA.

Ball, B. C., Batey, T. and Munkholm, L. T. (2007) Field assessment of soil structural quality – a development of the Peerlkamp test. *Soil Use. Manage.*, 23, 329–337.

Ball, B. C., Hargreaves, P. R. and Watson, C. A. (2018) A framework of connections between soil and people can help improve sustainability of the food system and soil functions. *Ambio*, 47, 269–283.

Bàrberi, P., Burgio, G., Dinelli, G, et al. (2010) Functional biodiversity in the agricultural landscape: relationships between weeds and arthropod fauna. *Weed Res.*, 50, 388–401.

Bàrberi, P. (2015) Functional biodiversity in organic systems: the way forward? *Sustain. Agric. Res.*, 4, 26–31.

Bastiaans, L., Paolini, R. and Baumann D. T. (2008) Focus on ecological weed management: what is hindering adoption? *Weed Res.*, 48, 481–491.

Benaud P., Anderson, K., Evans, M. et al. (2020) National-scale geodata describe widespread accelerated soil erosion. *Geoderma,* 371, 114378.

Bennett, H. H. (1935) Facing the erosion problem. *Science,* 81, 321–326.

Bioregioning Tayside (2024) web site https://bioregioningtayside.scot/; accessed 30 April 2024.

Blair, N., Faulkner, R. D., Till, A. R. and Crocker, G. J. (2006) Long-term management impacts on soil C, N and physical fertility: part III: Tamworth crop rotation experiment. *Soil Till. Res.*, 91, 48–56.

Bohanec, M. (2014) *DEXi: Program for multi-attribute decision making user's manual*, Institut Josef Stefan, Slovénie. web site http://kt.ijs.si/MarkoBohanec/dexi.html; accessed 30 April 2024.

Bohanec, M., Cortet, J., Griffiths, B. et al. (2007) A qualitative multi-attribute model for assessing the impact of cropping systems on soil quality. *Pedobiologia*, 51, 239–250.

Bohanec, M. and Znidarsic, M. (2013) DEX methodology: three decades of qualitative multi-attribute modelling. *Informatica,* 37, 49–54.

Borrelli, P., Robinson, D. A., Fleischer, L. R. et al. (2013) An assessment of the global impact of 21st century land use change on soil erosion. *Nat. Commun.*, 2017, 8.

Boyd Orr, J. (1943) *Food and the people*, The Pilot Press Ltd., London.

Brendel, O. (2021) The relationship between plant growth and water consumption. *Ann. For. Sci.,* 78, Article 47.

Burkill, I. H. (1966) *A dictionary of the economic products of the Malay Peninsular*, Second edition, Ministry of Agriculture and Co-operatives, Kuala Lumpur, Malaysia.

CCC (2020) *Reducing emissions in Scotland. Progress report to parliament, and previous reports in this series*, Committee on Climate Change UK. web site https://www.theccc.org.uk/publication/reducing-emissions-in-scotland-2020-progress-report-to-parliament/; accessed 30 April 2024.

Calderini, D. F. and Slafer, G. A. (1998) Changes in yield and yield stability in wheat during the 20th century. *Field Crop. Res.,* 57, 335–347.

Colomb, B., Carof, M., Aveine, A. and Bergez, J. (2013) Stockless organic farming: strengths and weaknesses evidenced by a multicriteria sustainability assessment model. *Agron. Sustain. Dev.*, 33, 593–608.

Conway, G. (1997) *The doubly green revolution*, Penguin Books, London, UK.

Cousens, R. (1987) Theory and reality of weed control thresholds. *Plant Prot. Q.,* 2, 13–20.

Cumpston, Z., Fletcher, M.-S. and Head, L. (2022) *Plants: past, present and future*, Thames and Hudson, Australia.

Cui, Z., Dou, Z., Chen, X., Ju, X. and Zhang, F. (2014) Managing agricultural nutrients for food security in China: past, present and future. *Agron. J.,* 106, 191–198.

De Notaris, C., Mortensen, E. O., Sørensen, P., Olesen, J. E. and Rasmussen, J. (2021) Cover crop mixtures including legumes can self-regulate to optimize N2 fixation while reducing nitrate leaching. *Agr. Ecosyst. Environ.*, 309, 107287.

Dlamini, P., Chivenge, P., Manson, A. and Chaplot, V. (2014) Land degradation impact on soil organic carbon and nitrogen stocks of sub-tropical humid grasslands in South Africa. *Geoderma*, 235–236, 372–381.

Doran, J. W. (2002) Soil health and global sustainability: translating science into practice. *Agr. Ecosyst. Environ.*, 88, 119–127.

Ehrmann, J. and Ritz, K. (2014) Plant:soil interactions in temperate multi- cropping production systems. *Plant Soil*, 376, 1–29.

Elliot, R. H. (1898) *Agricultural changes, later published as The Clifton Park System of Farming* Available online at journeytoforever.org.

Evans, L. T. (1993) *Crop evolution, adaptation and yield*, Cambridge University Press, Cambridge UK.

Falvey, L. (2000) *Thai agriculture – golden cradle of millenia*, Kasetsart University Press, Bangkok, Thailand.

FAO (2015) *Status of the world's soil resources*, Food and Agriculture Organisation of the United Nations, Rome, Technical summary page. https://www.fao.org/documents/card/en/c/i5126e); accessed 30 April 2024.

FAO (2018) *Global soil organic carbon map*, Food and Agriculture Organisation of the United Nations, Rome. https://www.fao.org/soils-portal/data-hub/soil-maps-and-databases/global-soil-organic-carbon-map-gsocmap/en/; accessed 30 April 2024.

FAO (2019) *The state of the world's biodiversity for food and agriculture* (J. Bélanger & D. Pilling, Eds.), FAO Commission on Genetic Resources for Food and Agriculture Assessments, Rome. http://www.fao.org/3/CA3129EN/CA3129EN.pdf; accessed 20 April 2024.

FAO (2021) *Land use statistics and indicators statistics. Global, regional and country trends 1990-2019*, Food and Agriculture Organisation of the United Nations, FAOSTAT Analytical Brief Series No 28, Rome.

FAO, IUCN CEM and SER (2021) *Principles for ecosystem restoration to guide the United Nations Decade of Ecosystem Restoration 2021–2030*, Food and Agriculture Organisation of the United Nations, Rome. https://www.decadeonrestoration.org/publications/principles-ecosystem-restoration-guide-united-nations-decade-2021-2030; accessed 30 April 2024.

Findlay, W. M. (1925) *Grassland in Scotland*. In W. G. R. Paterson (Ed), *Farm crops*, The Gresham Publishing Company, London, pp. 137–177.

Finlayson, B. and Warren, G. M. (2010) *Changing natures – hunter-gatherers, first farmers and the modern world*, Duckworth, London, UK.

Gann, G. D., McDonald, T., Walder, B. et al. (2019) International principles and standards for the practice of ecological restoration. Second edition. *Restoration Ecology*. https://doi.org/10.1111/rec.13035.

GRFC (2021) *Global Report on food crises – 2021*. Global Network Against Food Crises, funded by the EU, AO and WFP. https://www.wfp.org/publications/global-report-food-crises-2021; accessed 30 April 2024.

Grandy, A. S. and Robertson, G. P. (2007) Land-use intensity effects on soil organic carbon accumulation rates and mechanisms. *Ecosystems*, 10, 58–73.

Griffiths, B. S., Ball, B. C., Daniell, T. J. et al. (2010) Integrating soil quality changes to arable agricultural systems following organic matter addition, or adoption of a ley-arable rotation. *Appl. Soil Ecol.*, 46, 43–53.

Gu, B., van Grinsven, H. J., Lam, S. K. et al. (2021) A credit system to solve agricultural nitrogen pollution. *The Innovation*, 2(1), 250.

Hanjra, M. A., Noble, A., Langan, S. and Lautze, J. (2017) Feeding the 10 billion within the sustainable development goals framework. In I. J. Gordon, H. T. Prins and G. R. Squire (Eds.), *Food production and nature conservation: conflicts and solutions*, Routledge, Oxford UK, pp. 15–40.

Hargreaves, P. R., Baker, K. L., Graceson, A. et al. (2021) Use of a nitrification inhibitor reduces nitrous oxide (N2O) emissions from compacted grassland with different soil textures and climatic conditions. *Agr. Ecosyst. Environ.*, 310, 107307.

Hatcher, P. E. and Melander, B. (2003) Combining physical, cultural and biological methods: prospects for integrated non-chemical weed management strategies. *Weed Res.*, 43, 303–322.

Hawes, C., Haughton, A. J., Bohan, D. A. and Squire, G. R. (2009) Functional approaches for assessing plant and invertebrate abundance patterns in arable systems. *Basic Appl. Ecol.*, 10, 34–47.

Hawes, C., Squire, G. R., Hallett, P. D., Watson, C. A. and Young, M. (2010) Arable plant communities as indicators of farming practice. *Agr. Ecosyst. Environ.*, 138, 17–26.

Hawes, C., Begg, G. S., Iannetta, P. P. M., Karley, A. J. and Squire, G. R. (2016) A whole-systems approach for assessing measures to improve arable ecosystem sustainability. *Ecosys. Health Sustain.*, 2(12), e01252. https://doi.org/10.1002/ehs2.125

Hawes, C., Young, M. W., Banks, G., Begg, G. S., Christie, A., Iannetta, P. P. M., Karley, A. J. and Squire, G. R. (2019) Whole-systems analysis of environmental and economic sustainability in arable cropping systems: a case study. *Agronomy*, 9, 438.

Hawes, C., Iannetta, P. P. M. and Squire, G. R. (2021) Agroecological practices for whole-system sustainability. *CAB Reviews*. https://doi.org/10.1079/PAVSNNR202116005

Hoekstra, A. Y., Chapagain, A. K., Aldaya, M. M. and Mekonnen, M. M. (2011) *The water footprint assessment manual: setting the global standard*. Earthscan. ISBN: 978-1-84971-279-8.

Holland, J. M., Frampton, G. K., Cigli, T. and Wratten, S. D. (1994) Arable acronyms analysed - a review of integrated arable farming systems in Western Europe. *Ann. Appl. Biol.*, 125, 399–438.

Holst, N., Rasmussen, I. A. and Bastiaans, L. (2007) Field weed population dynamics: a review of model approaches and applications. *Weed Res.*, 47, 1–14.

Iannetta, P. P. M., Young, M., Bachinger, J. et al. (2016). A comparative nitrogen balance and productivity analysis of legume and non-legume supported cropping systems: the potential role of biological nitrogen fixation. *Front. Plant Sci.*, 21 November 2016. http://dx.doi.org/10.3389/fpls.2016.0170

IPCC (2023) Summary for policymakers. In H. Lee and J. Romero (Eds.), *Climate change 2023: Synthesis report. Contribution of working groups I, II and III to the sixth assessment report of the intergovernmental panel on climate change*, IPCC, Geneva, Switzerland, pp. 1–34. https://doi.org/10.59327/IPCC/AR6-9789291691647.001

Jackson, R. B., Lajtha, K., Crow, S. E. et al. (2017) The ecology of soil carbon: pools, vulnerabilities, and biotic and abiotic controls. *Annu. Rev. Ecol. Evol. S.*, 48, 419–445.

Jarvie, H. P., Neal, A., Withers, P. J. A., Robinson, A. and Salter, N. (2003). Nutrient water quality of the Wye catchment, UK: exploring patterns and fluxes. *Hydrol. Earth Syst. Sc.*, 7, 722–743.

Jensen, E. S. and Hauggaard-Neilsen, H. (2003) How can increased use of biological N2 fixation in agriculture benefit the environment? *Plant Soil*, 252, 177–186.

Kawashima, H., Bazin, M. J. and Lynch, J. M. (1997) A modelling study of world protein supply and nitrogen fertiliser demand in the 21st century. *Environ. Conserv.*, 24, 50–56.

Kaye J. P. and Quemada, M. (2017) Using cover crops to mitigate and adapt to climate change: a review. *Agron. Sustain. Dev.*, 37, 4.

Keating, B. A., Herrero, M., Carberry, P. S., Gardner, J. and Cole, M. B. (2014) Food wedges: framing the global food demand and supply challenge towards 2050. *Global Food Security*, 3, 125–132.

Kopke, U. and Nemecek, T. (2010) Ecological services of faba bean. *Field Crop. Res.115*, 217–233.

Lal, R. (2012) Climate change and soil degradation mitigation by sustainable management of soils and other natural resources. *Agr. Res.*, 1, 199–212.

Landis, W. G. (2003) Twenty years before and hence; ecological risk assessment at multiple scales with multiple stressors and multiple endpoints. *Hum. Ecol. Risk Assess.*, 9(5), 1317–1326.

LEAF (2024) Linking Environment and Farming, web site https://leaf.eco/; accessed 30 April 2024.

Leakey, R. (2012) *Living with the trees of life*, CAB International, Wallingford, UK.

Leip, A., Weiss, F., Lesschen, J. P. and Westhoek, H. (2014) The nitrogen footprint of food products in the European Union. *J. Agric. Sci.* 152(special issue), S20–33.

Lemaire, G., Ryschawy, J., Carvalho, P. C. de F. and Gastal, F. (2017) *Agricultural intensification and diversity for reconciling production and environment*. In I. J. Gordon, H. T. Prins and G. R. Squire (Eds.), *Food production and nature conservation: conflicts and solutions*, Routledge, Oxford UK, pp. 113–132.

Levin, D. A. (1990) The seed bank as a source of genetic novelty in plants. *Am. Nat.*, 135, 563–572.

Loveland, P. and Webb, J. (2003) Is there a critical level of organic matter in the agricultural soils of temperate regions: a review. *Soil Till. Res.* 70, 1–18.

Luo, Z., Wang, E. Sun, O. J., Smith, C. J. and Probert, M. E. (2011) Modeling long-term soil carbon dynamics and sequestration potential in semi-arid agro-ecosystems. *Agr. Forest Meteor.*, 151, 1529–1544.

Magrini, M.-B., Anton, M., Cholez, C. et al. (2016) Why are grain legumes rarely present in cropping systems despite their environmental and nutritional benefits? Analysing lock-in in the French agrifood system. *Ecol. Econ.*, 126, 152–162.

Maluk, M., Ferrando Molina, F., Lopez del Egido, L. et al. (2022) Fields with no recent legume cultivation have sufficient nitrogen fixing rhizobia for crops of faba bean (Vicia faba L.). *Plant Soil*, 472, 345–368.

Marshall, E. J. P., Brown, V. K., Boatman, N. D., Lutman, P. J. W, Squire, G. R. and Ward, L. K. (2003) The role of weeds in supporting biological diversity within cropped fields. *Weed Res.*, 43, 77–89.

Maxted, N. and Lauvie, A. (2017) *Utilising genetic diversity to feed the world*. In I. J. Gordon, H. T. Prins and G. R. Squire (Eds.),*Food production and nature conservation: conflicts and solutions*, Routledge, Oxford UK, pp. 173–203.

McConnell, D. J., Dharmapala, K. A. E. and Attanayake, S. R. (2003) *The forest farms of Kandy*, eBook published 2016, Routledge, London. https://doi.org/10.4324/9781315239637

MEA (2001) *Millenium ecosystem assessment*. https://www.millenniumassessment.org/; accessed 30 April 2024.

Milberg, P. (1995) Soil seed bank after 18 years of succession from grassland to forest. *Oikos*, 72, 3–13.

Moir, J. L., Cameron, K. C. and Di, H. J. (2007) Effects of the nitrification inhibitor dicyandiamide on soil mineral N, pasture yield, nutrient uptake and pasture quality in a grazed pasture system. *Soil Use Manage.*, 23, 111–120.

Monteith, J. L. (1965) Evaporation and environment. *Symposium of the Society for Experimental Biology*, 19, 205–224.

Monteith, J. L. (1972) Solar radiation and productivity in tropical ecosystems. *J. Appl. Ecol.*, 9, 747–766.

Neumann, K., Verburg, P. H., Stehfest, E. and Muller, C. (2022) The yield gap of global grain production: a spatial analysis. *Agr. Syst.*, 103, 316–326.

Newton, A. C. (2016) Exploitation of diversity within crops – the key to disease tolerance? *Front. Plant Sci.*, 7 – 2016. https://doi.org/10.3389/fpls.2016.00665

NFFN (2024) Nature Friendly Farming Network, web site https://www.nffn.org.uk/; accessed 30 April 2024.

Norman, M. J. T., Peason, C. J. and Searl, P. G. E. (1984) *The ecology of tropical food crops*, Cambridge University Press, Cambridge, UK.

Norris, R. F. and Kogan, M. (2000) Interactions between weeds, arthropod pests, and their natural enemies in managed ecosystems. *Weed Sci.*, 48, 94–158.

Norton, J. and Ouyang, Y. (2019) Controls and adaptive management of nitrification in agricultural soils. *Front. Microbiol.*, 10, 3389. https://10.3389/fmicb.2019.01931

Nourish Scotland (2024) web site https://www.nourishscotland.org/; accessed 30 April 2024.

O'Connor, M. (2006) The "four spheres" framework for sustainability. *Ecol. Complex.*, 3, 285–292.

Ong, C. K., Black, C. R. and Wilson, J. (2015) *Tree-crop interactions: agroforestry in a changing climate*, 2nd edition, CAB International, Oxfordshire, UK.

Owuor, S. O., Butterbach-Bahl, K., Guzha, A. C. et al. (2018) Conversion of natural forest results in a significant degradation of soil hydraulic properties in the highlands of Kenya. *Soil Till. Res.*, 176, 36–44.

Pearson, C. J. and Ison, R. L. (1997) *Agronomy of grassland systems*, 2nd edition, Cambridge University Press, Cambridge, UK.

Pelzer, E., Fortino, G., Bockstaller, C. et al. (2012) Assessing innovative cropping systems with DEXiPM, a qualitative multi-criteria assessment tool derived from DEXi. *Ecol. Indic.*, 18, 171–182.

Perera, A. H. and Rajapakse, R. M. N. (1991) A baseline study of Kandyan Forest Gardens of Sri Lanka: structure, composition and utilization. *Forest Ecol. Manag.*, 45, 269–280.

Perfecto, I. and Vandermeer, J. (2017) *A landscape approach to integrating food production and nature conservation*. In I.J. Gordon, H. T. Prins and G. R. Squire (Eds.), *Food production and nature conservation: conflicts and solutions*, Routledge, Oxford UK, pp. 133–152.

Pitman, H. T and Wallis, L. A. (2012) The point of spinifex: aboriginal uses of spinifex grasses in Australia. *Ethnobot. Res. Appl.*, 10, 109–131.

Post, P. M., Hogerwerf, L., Bokkers, E. A. M. et al. (2020) Effects of Dutch livestock production on human health and the environment. *Sci. Total Environ.*, 737, 139702.

Rahmati, M., Eskandari, I., Kouselou, M. et al. (2020) Changes in soil organic carbon fractions and residence time five years after implementing conventional and conservation tillage practices. *Soil Till. Res.*, 200, 104632.

Raseduzzaman, M. and Jenson, E. S. (2017) Does intercropping enhance yield stability in arable crop production? A meta-analysis. *Eur. J. Agron.*, 91, 25–33.

Rees, R. M., Bingham, I. J., Baddeley, J. A. and Watson, C. A. (2005) The role of plants and land management in sequestering soil carbon in temperate arable and grassland ecosystems. *Geoderma*, 128, 130–154.

Ritchie, P. (2017) What would Boyd Orr do? *Nourish Scotland Magazine* 6, 3–4.

Robertson, G. P. and Vitousek, P. M. (2009) Nitrogen in agriculture: balancing the cost of an essential resource. *Annu. Rev. Env. Resour.*, 34, 97–125.

Robertson, G. P., Bruulsema, T. W., Gehl, R. J. et al. (2013) Nitrogen-climate interactions in US agriculture. *Biogeochemistry*, 114, 41–70.

Rothwell, S. A., Doody, D. G., Johnston, C. et al. (2020). Phosphorus stocks and flows in an intensive livestock dominated food system. *Resour. Conserv. Recy.*, 163, 105065.

RSPB (2018) *Balancing act: How farming can support a net-zero emissions target in Scotland*, Royal Society for the Protection of Birds, UK. https://www.scotlink.org/publication/balancing-act-how-farming-can-support-a-net-zero-emission-target-in-scotland/; accessed on 30 April 2024.

Sadok, W., Angevin, F., Bergez, J.-E. et al. (2009) MASC, a qualitative multi-attribute decision model for ex ante assessment of the sustainability of cropping systems. *Agron. Sustain. Dev.*, 29, 447–461.

SEDA Land (2024) *Scottish Ecological Design Association*, web site https://www.seda.uk.net/seda-land; accessed 30 April 2024.

SER (2024) *Society for Ecological Restoration*, web site https://www.ser.org/; accessed 30 April 2024.

Sinclair, T. R and Weiss, A. (2010) *Principles of ecology in plant production*, 2nd edition, CAB International, Oxfordshire, UK.

Slattery, R. A. and Ort, D. R. (2015). Photosynthetic energy conversion efficiency: setting a baseline for gauging future improvements in important food and biofuel crops. *Plant Physiol.*, 168(2), 383–392.

Soil Association (2024) web site https://www.soilassociation.org/; accessed 30 April 2024.

Soule, E., Hawes, C., Young, M. et al. (2023) A predictive indicator assessing effect of cropping system and surrounding landscape on biodiversity. *Ecol. Indic.*, 151, 110289

Soussana J.-F., Loiseau, P., Vuichard, N. et al. (2004) Carbon cycling and sequestration opportunities in temperate grasslands. *Soil Use Manage.*, 20, 219–230.

Smith, B. M., Aebisher, N. J., Ewald, J. et al. (2020) The potential for arable weeds to reverse invertebrate declines and associated ecosystem services in cereal crops. *Front. Sustain. Food Syst.*, 10 January 2020. https://doi.org/10.3389/fsufs.2019.00118

Smith, P., Martino, D., Cai, Z. et al. (2007) *Agriculture*. In B. Metz, O. R. Davidson, P. R. Bosch, R. Dave, and L. A. Meyer (Eds.), *Climate change 2007: Mitigation. Contribution of working group III to the fourth assessment report of the intergovernmental panel on climate change*, Cambridge University Press, Cambridge, United Kingdom and New York, USA.

Squire, G. R. (2017a) Limits to crop production. In I. J. Gordon, H. T. Prins and G. R. Squire (Eds.), *Food production and nature conservation: conflicts and solutions*, Routledge, Oxford UK, pp. 41–67.

Squire, G. R. (2017b) Defining sustainable limits during and after intensification in a maritime agricultural ecosystem. *Ecosys. Health Sustain.*, 3. https://doi.org/10.1080/20964129.2017.1368873

Squire, G. R. (2020) Modelling biodiversity in agriculture. In P. Bárberi and A.-C. Moonen (Eds.), *Reconciling agricultural production with biodiversity conservation*. Burleigh Dodds Science Publishing, Cambridge, UK.

Squire G. R., Quesada N., Begg G. S. and Iannetta P. P. M. (2019) Transitions to a greater legume inclusion in cropland: defining opportunities and estimating benefits for the nitrogen economy. *Food Energy Secur.*, *8*, e00175. https://doi.org/10.1002/fes3.175

Squire, G. R. and Iannetta, P. P. M. (2018) Transition paths to sustainable legume production. *Asp. App. Biol.*, 138, 121–130.

Squire, G. R., Young, M., Ford, L., Banks, G. and Hawes, C. (2020) Defining targets for reversing declines of soil carbon in high-intensity arable cropping. *Agronomy*, 10(7), 973.

Squire, G. R., Young, M. W. and Hawes, C. (2021) Solar radiation flux provides a method of quantifying weed-crop balance in present and future climates. *Plants*, 10(12), 2657. https://doi.org/10.3390/plants10122657

Squire, G. R., Young, M. W. and Hawes, C. (2022) Agroecological management and increased grain legume area needed to meet nitrogen reduction targets for greenhouse gas emissions. *Nitrogen*, 3, 539–554. https://doi.org/10.3390/nitrogen3030035

Squire, G. R., Young, M. W. and Banks, G. (2023) Post-intensification Poaceae cropping: declining soil, unfilled grain potential, time to act. *Plants*, 12(14), 2742. https://doi.org/10.3390/plants12142742

Stagnari, F., Maggio, A., Galiena, A. and Pisante, M. (2017) Multiple benefits of legumes for agricultural sustainability: an overview. *Chem. Biol. Technol. Agric.*, 4(2). https://doi.org/10.1186/s40538-016-0085-1

Stebler, F. G. and Schröter, C. (1889) *The best forage plants*, Translated into English by A. N. McAlpine, 1889. Nutt, London.

Stephens, H. (1841) *The book of the farm*. Blackwood, Edinburgh.

Suter, M., Connolly, J., Finn, J. A. et al. (2015) Nitrogen yield advantage from grass–legume mixtures is robust over a wide range of legume proportions and environmental conditions. *Global Change Biol.*, 21, 2424–2438.

Subbarao, G. V., Rao, I. M., Nakahara, K. et al. (2013) Potential for biological nitrification inhibition to reduce nitrification and N20 emissions in pasture crop-livestock systems. *Animal*, 2013, 7(s2), 322–332. https://doi.org/10.1017/S1751731113000761

Sutton, M. A., Howard, C. M., Erisman, J. W. et al. (2011) *The European nitrogen assessment*, Cambridge University Press, Cambridge, UK.

Swanton, C. J. and Murphy, S. D. (1996) Weed science beyond the weeds: the role of integrated weed management (IWM) on agroecosystem health. *Weed Sci.*, 44, 437–445.

Syers, J. K. (1997) Managing soils for long term productivity. *Philos. T. R. Soc. B*, 352, 1011–1021.

Takoutsing, B., Weber, J., Aynekulu, E. et al. (2016) Assessment of soil health indicators for sustainable production of maize in smallholder farming systems in the highlands of Cameroon. *Geoderma*, 276, 64–73.

Tittonell, P. and Giller, K. E. (2013) When yield gaps are poverty traps: the paradigm of ecological intensification in African smallholder agriculture. *Field Crop. Res.*, 143, 76–90.

Topping, C. J., Hansen, T. S., Jensen, T. S., Jepson, J. U., Nikolajsen, F. and Odderskaer, P. (2003). ALMaSS, an agent-based model for animals in temperate European landscapes. *Ecol. Model.*, 167, 65–82.

Tudge, C. (2016) *Six steps back to the land*, Green Books, Cambridge, UK.

Valentine, T. A., Hallett, P. D., Binnie, K. et al. (2012) Soil strength and macropore volume limit root elongation rates in many UK agricultural soils. *Ann. Bot.,* 110, 259–270.

Valkama, E., Kunypiyaeva, G., Zhapayev, R. et al. (2020) Can conservation agriculture increase soil carbon sequestration? A modelling approach. *Geoderma*, 369, 114298.

Van Grinsven, H. J. M., Ten Berge, H. F. M., Dalgaard, T. et al. (2012) Management, regulation and environmental impacts of nitrogen fertilization in northwestern Europe under the Nitrates Directive; a benchmark study. *Biogeosciences,* 9, 5143–5160.

Vasey, D. E. (1992) *An ecological history of agriculture 10,000 B.C.-A.D. 10,000*, Iowa State University Press, Iowa, USA.

Walling, E. and Vaneeckhaute, C. (2020) Greenhouse gas emissions from inorganic and organic fertiliser production and use: a review of emission factors and their variability. *J. Environ. Manage.*, 15, 276, 111211.

Walsh, M. J., Broster, J. C., Schwartz-Lazaro, L. M. et al. (2018) Opportunities and challenges for harvest weed seed control in global cropping systems. *Pest Manag. Sci.,* 74(10), 2235–2245.

Wang, X., Bai, J., Xie, T. et al. (2021) Effects of biological nitrification inhibitors on nitrogen use efficiency and greenhouse gas emissions in agricultural soils: a review. *Ecotox. Environ. Safe.*, 220, 112338.

Warr, S. J., Thompson, K. and Kent, M. (1993) Seed banks as a neglected area of biogeographic research: a review of literature and sampling techniques. *Prog. Phys. Geogr.,* 17, 329–347.

Watson, C. A., Atkinson, D., Gosling, P., Jackson, L. R. and Rayns F. W. (2002) Managing soil fertility in organic farming systems. *Soil Use Manage.,* 18, 239–247.

Whetton, R. L., Harty, M. A. and Holden, N. M. (2022) Communicating nitrogen loss mechanisms for improving nitrogen use efficiency management, focused on global wheat. *Nitrogen,* 3, 213–246.

Willey, R. W. (1985) Evaluation and presentation of intercropping advantages. *Exp. Agric.,* 21, 119–133.

Zalidis, G., Stamatiadis, S., Takavakoglou, V., Eskridge, K. and Misopolinos, N. (2002) Impacts of agricultural practices on soil and water quality in the Mediterranean region and proposed assessment methodology. *Agr. Ecosyst. Environ.,* 88, 137–146.

Zhang, X., Izaurralde, R. C., Manowitz, D. H. et al. (2015) Regional scale cropland carbon budgets: evaluating a geospatial agricultural modelling system using inventory data. *Environ. Modell. Softw.,* 63, 199–221.

Published by Burleigh Dodds Science Publishing Limited, 2024.

Chapter 3

Engaging local voices: farmers, rural communities, and ecological restoration

Theodore Alter and Kayla Faith Laddin, Pennsylvania State University, USA; and Lauren Hull, Michael Reid and Heidi Kleinert, Department of Energy, Environment and Climate Action, Victoria, Australia

> The practice of regenerative agriculture across vast swaths of the world's farmlands, grasslands, marginal savannahs, and arid regions can provide a major solution to climate change as well as confronting the other key Anthropocene planetary boundary challenges. (Charles Massey, p. 4, Introduction, *The Call of the Reed Warbler: A New Agriculture, A New Earth*)
>
> .... an increasing number of challenges that cannot be solved by government alone (even though government remains an essential resource) require skilled, savvy citizen action if there is to be a resolution. (Harry C. Boyte, pp. ix–x, Preface, *Everyday Politics: Reconnecting Citizens and Public Life*)

1 Introduction

The environmental and sustainability challenges we face today, locally and globally, and for the foreseeable future, are daunting. Water quantity and quality, chemical pollutants, biodiversity loss, climate change, rising sea levels,

http://dx.doi.org/10.19103/AS.2024.0134.05

increasing biosecurity threats, urban air quality, soil contamination, agricultural land degradation, and many other issues are affecting and challenging our livelihoods and quality of life at all scales.

We live in a world characterized by interdependence. As human beings, we impact and are impacted by each other in our social, political, economic, and cultural lives. Given our individual and collective differences, this interdependence is characterized by conflict, politics, and power play (Schmid, 2004). We are also obviously interdependent with nature. Our natural environment in all its forms and beauty affects our lives economically and aesthetically, spiritually and socially, and it provides the fundamental basis for our existence as humans. As we know well, we also impact our natural world through the ways we use and treat it. We live now in what some refer to as the Anthropocene, a period in geologic time where the impact of human activity on climate and natural ecosystems is particularly significant and growing more so (Crutzen, 2006). We are living at the nexus of our human–human and human–nature interdependencies. We cannot escape the existential reality of this nexus of interdependence. We cannot escape understanding nature and we cannot escape engaging and understanding each other if we are to manage those ever-evolving changes in our natural ecosystems upon which our human existence depends.

These complex and interrelated dynamics make achieving coordinated and effective management of environmental and sustainability issues extremely challenging. This is nowhere more evident than in farming regions, where communities are typically dealing with multiple environmental challenges, often in landscapes that have historically been significantly altered to suit the purposes of agricultural production. The voices of farmers and rural communities can often be marginalized in the decision-making processes that determine where and how ecological restoration programs are implemented. Effective engagement of farmers and rural communities in restoration projects involves including them not just in the 'doing' of restoration work but also in planning, designing, monitoring, and evaluating such work. This perspective acknowledges that local knowledge stemming from lived experience is imperative for ecological restoration.

This chapter is about community engagement strategies and tactics for addressing the issues critical to the future overall health of individuals, communities, natural environments, and our planet. The scholarly and practice literature describing the philosophies, processes, politics, and challenges of community engagement and community-led action is voluminous and expanding daily. The challenge of staying current with this burgeoning literature is daunting. Therefore, our goal is two-fold: to provide a review of influential literature and frameworks that are useful to progress our thinking on engaging with multiple actors in diverse situations; and to help frame and guide how best in these situational contexts at various scales to understand and act on issues of ecological restoration at landscape scale.

We examine systems theory; detail two case studies featuring the context, challenges, and promise of broad community-wide engagement; highlight several useful engagement frameworks; and summarize, given our experience and study, key principles associated with effective community engagement and community-led action. We provide selective but not exhaustive listings of additional sources of information to support readers in digging deeper into the many nuances, elements and challenges of community engagement-, and community-led action.

In the following section, we feature the importance of systems and social learning perspectives for understanding our biophysical world, our socio-political world, and the crucial interdependence and intersection of both worlds as the bases for shaping community engagement strategy and action. In the case studies, we focus firstly on community-led action for managing invasive alien species (weeds and rabbits) in Victoria, Australia, and then secondly on engaging farmers and community members in addressing agricultural land-related water pollution problems in the Chesapeake Bay in Pennsylvania and the Mid-Atlantic region of the greater United States. We then highlight frames and insights from community engagement theory and practice, and we summarize, based on our study and experience, key principles necessary for increasing the likelihood of effective community engagement, and community-led action.

2 Perspectives: systems framing and social learning

Systems framing and social learning theories of change provide an important meta-perspective for understanding the roles of community engagement and community-led action in agricultural and ecological restoration. For example, Martin and Reid (2019) highlight the importance of systems thinking and systems strengthening strategies for building human relationships, understanding human differences and individual and collective learning, and co-creating innovative initiatives for addressing invasive vertebrate pest management in Australia. As noted earlier, our existential challenge is working at the nexus of our human-to-human and human-to-nature interdependencies. Pest management and agricultural pollution mitigation are not just issues of biology, chemistry, and soil science. They are also issues that affect the interests of people, and people have different views about the issues and the actions that could and should be taken. This reality causes those issues to be inherently social, cultural, and political.

Understanding the biophysical ecology of whatever issue is at hand is necessary but not sufficient: we must also understand the social, cultural, and political dimensions of the differences, relationships, networks, and power dynamics among people and organizations. In effect, we must understand the socio-political 'wetlands' or ecology of human interdependence (Mathews, 2014).

Learning, innovation, and environmental change occur at the nexus of these two broad ecologies. Addressing the challenges of agricultural land degradation, biodiversity loss, water pollution, invasive species impacts, economic loss to farmers, and myriad other critical environmental and economic sustainability issues requires us to recognize and take on the complex, messy, often politically charged challenge of working at the nexus of our biophysical and socio-political worlds. Engaging each other openly and respectfully in this nexus across our differing values, beliefs, interest, and ways of seeing and knowing our communities and the world around us holds the promise of individual and social learning, creative insight and innovation, democratic practice, and systemic change (Adams et al., 2019; Reid et al., 2021; Johansson, 2006; Bridger and Alter, 2008; Howard and Alter, 2019; Mathews, 2014; Schmid, 2004; Fischer, 2000, 2017; Palmer, 2011; Dzur, 2019).

2.1 Planetary health education

The Planetary Health Education Framework (Redvers et al., 2023; Faerron Guzmán et al., 2021) and the well-known notion of 'wicked problems' (Rittel and Webber, 1973; Batie, 2008) provide powerfully insightful system perspectives for situating the importance and practice of community engagement and community-led action. The Planetary Health Education Framework highlights five interconnected domains central to planetary health knowledge, values, and practice: systems thinking and complexity, equity and social justice, movement building and system change, the Anthropocene and health, and at the center of these four domains, interconnection with nature. These domains are situated within local and global socio-economic, cultural, and environmental conditions and priorities. Redvers et al. (2023) flags a potential limitation to this framework, in that is does not explore how it could operationally inform individual and collective learning and change at multiple scales; in effect, how we can engage each other across the interconnected complexity of the five domains. Nonetheless, this framework provides systems insight into the challenges and promise of community engagement and community-led action.

2.2 Wicked problems

The notion of 'wicked' problems also provides an important frame that can guide our thinking and action on critical issues of biodiversity and ecological restoration. These issues are complex, elusive, conflictual, and often intractable because they involve both biophysical and human considerations. Differing people and interests see and know the world differently and are often in conflict not just concerning what should be done but regarding what is the issue or problem, or even if there is one at all. The wickedness of 'wicked' problems rests not only in their biophysical complexity but also in the complexity of stakeholder perspectives and interests (Rittel and Webber, 1973; Batie, 2008). Moreover, these problems are resistant to resolution, given that in a socio-political context, they evolve over time and are fraught with uncertainty regarding cause and effect and outcomes (Rittel and Webber, 1973; Batie, 2008). Kreuter et al. (2004) note that sorting out the complexities of 'wicked' problems is as much a socio-political process as it is about science.

Batie (2008), drawing on Kreuter et al. (2004), notes the distinction between 'tame' and 'wicked' problems, the former being characterized by relatively low uncertainty and value conflict, while 'wicked' problems are characterized by high uncertainty and value conflict. She argues further that 'wicked' problems are not amendable to conventional science and environmental management but require 'engaged sustainability science' and 'adaptive management'. Engaged sustainability science supporting adaptive management is grounded in ongoing, respectful, democratic dialogue across differing values, beliefs, and ways of knowing and seeing the world amongst issue stakeholders (Batie, 2008). It is in this dialogue, as difficult and conflictual as it might be, that individual and collective social learning, creativity, and innovation occurs. Engagement and learning are central to the application of sustainability science that addresses the evolving complexities and uncertainties of 'wicked' problems.

A note on sustainability science, or as Acosta García et al. (2023) argue, the appropriate term is 'sustainability sciences'. They point out that many scientific disciplines bring their science to issues of sustainability. They also note that in the context of 'wicked' problems, while multiple disciplines have valuable contributions to make, the challenge for scientists is to embrace disciplinary and epistemological pluralism and improve engagement and communication across disciplines. Again, it is at this intersection of difference that learning, creativity, insight, innovation, and change occur. This disciplinary and epistemological pluralism also must embrace local, practical knowledge grounded in the lived experience of stakeholders living and working across the landscape. Jordan et al. (2020) encourage agricultural scientists to become skilled at engaging in a politics of constructive collection action inherent in these pluralisms for biodiversity and environmental challenges to be progressed.

2.3 Social-ecological systems

Approaching wicked problems can be a daunting prospect. They are messy and complex and embedded in particular histories and cultures. Moreover, there can be disagreements between locals, government officials, and scientific experts – because each group uses different concepts and languages, and different scientific disciplines bring differing perspectives as well. This can lead to a myopic understanding of a particular problem, poorly developed interventions, and worse, a sense of hopelessness and the continued deterioration of the natural resources base.

Sustainability and ecological restoration issues, such as water, land degradation, invasive species, and biodiversity conservation, are issues that cross property boundaries and must be tackled at a scale greater than an individual farm. Classic theory posits that no one will change their behavior unless an external authority imposes rules, typically through regulation by the state. However, this is not always the case. Elinor Ostrom's 'social-ecological-system' (SES) framework (Ostrom, 2009) provides a common framework to conceptually and analytically address these issues. Ostrom's work enhances the work of Walker et al. (2002, 2004) which identified resilience, adaptability, and transformability as the three key interrelated attributes of these socio-ecological systems. Ostrom's work adds conceptual and analytical depth to the socio-ecological system notion.

As a 'practical algorithm', Ostrom's framework (2009; McGinnis and Ostrom, 2014) has proven effective locally, nationally, and globally for addressing resource issues through trust, reciprocity, and cooperation over time. It acknowledges that environmental and sustainability challenges are part of wider and complex social and ecological systems. Her SES framework helps us to understand, dissect, and harness the complexity rather than eliminate it. The SES framework contextualizes resource issues in four sub-systems: resource systems, resource units, governance system, and actors. The resource system, for example, could be an agricultural or farming system, a forest landscape, a wetland, or a unique ecological zone. Resource units are the products of the resource system, for example, trees, fish, agricultural products, and environmental goods. Governance systems involve formal rules and cultural norms that govern how people relate to one another and to nature and the resource situation. Actors are the people living with and experiencing the resource situation. Each sub-system has other variables that affect its nature and functioning.

Each sub-system is relatively separable but interacts dynamically with other sub-systems to produce outcomes at the SES level. Central to the framework is the action situation, where the continuous engagement and interactions of individuals are influenced and shaped by the resource and governance sub-systems, affecting the management of resources and outcomes across the wider resource system.

The power of Ostrom's framework lies in re-framing sustainability challenges through a systems perspective, opening a new range of potential possibilities to address these issues. For example, the framework helps identify relevant variables for studying a particular focal point, such as the institutional and governance arrangements for managing rabbits in Australia (Adams et al., 2019), or how local communities organize themselves to sustainably manage their fish stocks without government regulation (Koning et al., 2020).

A fundamental principle supporting collective action is a commitment to deep democracy. Ostrom believed that politics is fundamentally about making decisions with others and that it was likely to be improved with diverse and robust participation (Wall, 2017). She believed that the more power people had over decisions that affected their lives and the more they engaged with diverse perspectives to test and scrutinize ideas, the better the decisions were likely to be. At this intersection, differences intersect, individual and social learning can occur, and creativity and innovation can emerge, holding the promise of new ways of being with each other and the land. Ostrom's insight that humans with interest in collective action situations can often find agency and come to effective collective action through their interaction, their engagement with each other, is the 'magic' that can occur in the 'action situation' of the SES frame.

2.4 Change through engagement

There is a strong sense that community engagement matters for dealing with 'wicked' environmental and natural resource problems at landscape scale. There are also, however, legitimate questions about why community engagement matters and in what ways and to what extent it matters. Eaton et al. (2021) point out that evidence regarding 'how and under what circumstances stakeholder engagement enables desirable changes remains elusive'. They have developed a useful conceptual systems framework for thinking through and systematically examining these questions, titled 'Conceptual framework for change through stakeholder engagement'.

This framework portrays engagement outcomes as a function of engagement processes. Engagement outcomes are depicted in three categories: social learning and community capacity building, behavioral change, and environmental change. The first two outcome categories occur at the individual, group, and 'networked' group levels. The 'networked' group level reflects collective connections across multiple groups and sectors across scales. Environmental change outcomes manifest themselves at three levels or scales – local, regional, and subnational in this framework. Engagement process dimensions include process design features, internal contextual conditions, and stakeholder dialogue, deliberation, and decision-making. Design features are those factors that either help or hinder change: things like when and where are meetings held, who is invited to attend meetings and other events or activities, how is the agenda set and facilitated, and other such matters. Internal contextual conditions refer to pre-existing conditions that shape stakeholder interest and/or willingness to be involved in a meeting or initiative. Stakeholder dialogue, deliberation, and decision-making is the space within which differences in stakeholder values, beliefs, ideologies, understandings of the issue at hand, personal and group interests, and political perspectives intersect and are forged in support or in deterrence of collective action and engagement outcomes.

This framework for change through engagement is dynamic rather than linear and static, capturing the messy, dynamic, and evolutionary nature of community engagement. The dimensions of engagement processes impact and are impacted by each other, and the stakeholder dialogue, deliberation, and decision-making dimension has obvious direct implications for all three engagement outcomes.

Further, the nature of engagement outcomes feedback on each other and, individually and collectively, on the nature of the engagement process dimensions. This continuous looping feedback affords individual and collective or social learning opportunities raising the possibility of change across all dimensions of the framework which then, of course, feed further feedback, learning, and change.

These systems frames for social learning – Planetary Health, 'wicked' problems, socio-ecological systems, and the conceptual framework for change through stakeholder engagement – provide important grounding for thinking about community engagement and community-led action and their relevance and application for issues and situations involving agricultural land degradation, biodiversity loss, and ecological restoration. The *Additional Resources* and *References* sections of this chapter provide references, written materials, and other resources that describe and provide an insightful critique of these approaches. We turn now to two case study examples of community engagement in action which illustrate elements of these systems frames for social learning.

3 Community-led action case studies

This section details two cases studies that illustrate the challenges, processes, principles, and promise of community engagement and community-led action for agricultural and ecological restoration. The first case, community-led pest management in Victoria, Australia, focuses on community engagement strategies and tactics designed to catalyze shared commitment across stakeholders to an approach for managing invasive weeds and rabbits that centers on community.

The second case, engaging local voices in watershed management in eastern United States, focuses on engagement approaches mitigating non-point source pollution stemming from agricultural practices.

3.1 Community-led pest management in Victoria

3.1.1 Introduction

This example concerns a community-focused initiative that provides community groups with greater decision-making powers in the management of four widely established invasive pest species in the state of Victoria, south-eastern Australia. Rabbits, blackberry, gorse, and serrated tussock are particularly problematic pest species in Victoria. For each species, there is an established community-led organization (Community Pest Management Group) that works with stakeholders across the state, particularly in regional and rural areas, to promote community-led management.

3.1.2 Landscape impacts

Australia's most recent State of the Environment report (Murphy and van Leeuwen, 2021) identifies invasive species as the most prevalent threat to Australian fauna and a primary cause of extinction. Rabbits, gorse, blackberry, and serrated tussock are all considered established, widespread, and beyond eradication from the whole of Victoria.

The European rabbit (*Oryctolagus cuniculus*) is one of the most widely distributed pest animals in Victoria, posing a threat to biodiversity in a wide range of habitats, including subalpine regions, arid landscapes, grasslands, and coastal plains (Woolnough et al., 2020). Wild rabbits destroy crops, cause soil erosion, threaten native plant populations by grazing and browsing (Finlayson et al., 2022), compete with native fauna, and breed prolifically (https://weedsandrabbitsstory.com/).

Blackberry (*Rubus fruticosus*) is among the most devastating weeds in southern Australia because of its invasiveness and its impact on productive land and natural ecosystems (NSW Department of Primary Industries, 2009). Infestations decrease the productivity of land, alter water flow, cause erosion, harbor pest animals, and displace native species (https://weedsandrabbitsstory.com/).

Gorse (also known as furze; *Ulex europaeus*) is one of Victoria's most destructive weeds. Highly invasive and persistent, it establishes quickly particularly on poor soils and disturbed areas, harbors pest animals, clogs waterways, and damages natural environments (https://vicgorsetaskforce.com.au/faqs/).

Serrated tussock (*Nassella trichotoma*) has become widespread in south-east Australia, readily invading pastures, native grasslands and open scrub. Similar in appearance to many native tussocks, it can spread unnoticed for many years until significant infestations have developed. It is particularly problematic in native grasslands, where it outcompetes native species and reduces habitat for native fauna (http://serratedtussock.com/).

All four species cause damage across millions of hectares, and they readily re-infest areas where they have been removed. This puts agricultural enterprises, community and cultural assets, and landscapes at risk – whether it is agricultural or residential land, culturally significant sites of Traditional Owners, or publicly managed parks and reserves. The movement of these pests and weeds across property boundaries is a significant source of tension between farmers, communities, and public land managers.

For each of the four species, there are a range of effective control options available for different sites and situations. However, the cross-boundary nature of weed and rabbit infestations makes them a 'common property' problem, requiring coordinated approaches for effective management. Martin and Smith-Heron (2019) highlight the wide range of institutional and social challenges to achieving effective, coordinated action on invasive species, emphasizing the need for a greater focus on the human elements of the issue if we are to achieve sustained and effective involvement of citizens and coordinated action.

3.1.3 Community–government partnerships

Victoria's four Community Pest Management Groups (CPMGs) have been established to manage the challenges posed by these four invasive species: the Victorian Rabbit Action Network (VRAN), the Victorian Serrated Tussock Working Party (VSTWP), the Victorian Gorse Taskforce (VGT), and the Victorian Blackberry Taskforce (VBT).

Each of the CPMGs was established separately, since the 1990s, in response to a range of issues, but significantly the re-direction of government investment away from established pest management and a shift toward 'shared responsibility' between government and land managers (Martin and Smith-Heron, 2019). Each CPMG has approached their role differently over time, but with common principles of coordination across land tenures and stakeholders in promoting best practice control methods. The CPMGs are each comprised of representatives from communities and government agencies involved in pest management, with a strong focus on rural and regional community-based leadership. They work with community organizations, businesses, and government agencies, including community Landcare groups, Traditional

Owner corporations, local government, public land management agencies, and research institutions.

3.1.4 Wicked problem

The management of established pest species is well-recognized for its complexity. Martin (2019) points to the significant social challenge of invasive species management, highlighting the need to secure and sustain the involvement of a diverse range of people and expertise over time.

In 2016, a four-year project was funded through the Australian Government's Agricultural White Paper (Commonwealth of Australia, 2015). It aimed to improve the management of established pest species in Victoria by deepening community participation and engagement. The project re-framed the challenge of established species as a wicked problem (Batie, 2008; Head and Alford, 2015; Rittel and Webber, 1973). This framing recognized that although there are many technical tools and options for managing rabbits, blackberry, gorse, and serrated tussock, the success of control efforts is determined largely by human behavioral and social capacity elements (Woolnough et al., 2020). Sustained behavioral change and capacity building (Australian Public Service Commission, 2012) is central to effective management and control.

The project was managed by Victoria's state agriculture department (Agriculture Victoria). It centered on the premise that building the capability and capacity of people who lead, coordinate, and support invasive species management activities would make weed and rabbit management more effective and sustainable.

With the CPMGs in a position to act as 'facilitating institutions' between government agencies and local, on-ground action (Martin and Reid, 2019), the project focused on empowering and strengthening community-led management through the CPMG structures and changing Victoria's existing management systems.

3.1.5 System approach

The project took a system strengthening approach to re-frame the complexity of issues from multiple perspectives, foster agreement on the challenges, and identify interventions. System thinking processes and mapping can act as tools for individuals to learn together and develop a collective strategy (Martin and Reid, 2019). A system approach also acknowledges that better relationships across system 'sectors' (e.g. organizations and citizens) is critical to 'strengthen' the system supporting co-ordination and co-operation. The approach was modeled on an initiative that led to the establishment of the Victorian Rabbit Action Network. This earlier work was undertaken on the premise that widespread ownership of a collective action strategy would most likely arise when parties with fundamental and differing interests in the problem participated in discussions about their differences and possible solutions (Adams et al., 2019; Reid et al., 2021).

At the start of the project, Agriculture Victoria facilitated a system mapping exercise, conducting interviews with 80 system participants representing a diverse range of perspectives and expertise in the pest management space. Interviews focused on understanding how people were managing weeds and rabbits, the challenges they were facing and who they were working with. The system maps were discussed at a workshop involving 114 participants, where people were invited to consider the 'system' in its entirety and generate interventions to better support community-led action on weeds and rabbits. An example of one of the system maps is available at https://embed.kumu.io/21b3609cc7d900909c2a7f634ef3f08a#generic-systems-map.

The exercise defined the social, political, and ecological elements of the system and the interaction between them, as a basis for identifying potential interventions that could be applied to improve outcomes. It brought people together to consider different ideas and perspectives and helped facilitate shared understanding.

Farmers, individuals working closely with the farming sector in operational and advisory roles, and citizens involved in community-based environmental groups (e.g. Landcare) were key participants in interviews and the workshop. This ensured that local and on-ground experience was front and center in both the intelligence gathering and the subsequent discussions about interventions. As a result, the interventions identified were focused on improving how people and their institutions and communities can work more effectively together. This included a community grants program, research, and leadership programs. A detailed account of the systems approach is provided in Martin and Reid (2019). The use of a system approach for feral pig management in the Australian state of Queensland is also described by Martin and Reid (2019), with the authors detailing the value of the mapping process in building shared awareness and commitment to collective action across different stakeholders.

3.1.6 Collective governance and decision-making

At the conclusion of the system exercise, a co-operative governance structure was formed, bringing together community leaders from the four CPMGs and Agriculture Victoria in the Delivery Leadership Group (DLG) to guide the project's investment and decision-making. This marked a significant shift from the historically favored government-led decision-making on management of established pest species, in which rural and regional stakeholders managing pest infestations day-to-day are marginalized from governance of public programs. In contrast, the DLG placed community leaders at the decision-making table, ensuring that community-based knowledge of the problems and solutions guided public investments and improved coordination and relationships between government and community. Rather than focusing on linear extension of technical information or enforcement action, the project has sought to build individual and collective community capacity for pest management.

The cooperative governance model ensures that the people most affected by the weed and rabbit problems are central in the process of defining the problems, co-creating strategies to address these problems, and sharing in investment decisions.

3.1.7 Impacts of the approach

The metrics of success for the project were deliberately focused on empowering community and improving invasive species management systems – that is, the first two engagement outcomes in the Eaton et al. (2021) framework for change – rather than numbers of rabbits killed or hectares of weeds treated. This focus acknowledges the often-temporary nature of weed and rabbit suppression (given their mobility across land boundaries). Each CPMG also reported a range of short term, local environmental change outcomes – the third engagement outcome category in the Eaton et al. framework. Longer term on-ground outcomes, for example increased stocking capacity or regeneration of native species, were not able to be assessed over the four-year project duration.

An independent evaluation (Gilmour et al., 2021) of the partnership and the investments stemming from the DLG between 2016 and 2020 found that the approach has

- raised awareness amongst landholders of weed and rabbit management issues;
- supported individuals to work together in collaboration across the landscape;
- extended CPMGs' reach across Victoria, connecting them and their resources with more landholders;
- raised the standard of weed and rabbit control across the state by hosting more intensive field days, support programs, and training events;
- increased the capabilities of community members by delivering training in engagement, best-practice control techniques, and awareness of how to work in culturally sensitive landscapes; and
- increased community members' confidence so they can share their new knowledge and skills with others in their community.

For example, the Victorian Rabbit Action Network (VRAN) has focused on capacity building and network development, based on the hypothesis that there is no 'quick-fix' for the complex problem of rabbit management. VRAN's central program is a training and mentoring program to boost rabbit control expertise in the community and public land management sectors. Led by experts from community and government, it has equipped participants with rabbit management knowledge which they take back to their rural communities and their land manager colleagues. A key focus is on using integrated and proven methods that are applied in the right order: assessing rabbit numbers, baiting, destroying burrows, follow-up maintenance, and monitoring. The training program also recognizes that 'best-practice' is not possible in all environments, and teaches adapted techniques for sensitive sites, for example in peri-urban areas and where there are Aboriginal ancestral remains.

Training participants continue to meet as a 'community of practice', sharing ideas and using the collective knowledge of the group to solve problems (see Dickson et al., 2019, p. 63, for more information). The approach was recognized by the United Nations in 2019 (https://publicadministration.un.org/unpsa/database/Winners/2019-

Winners/Victorian-Rabbit-Action-Network) for its efforts to 'deliver more inclusive and equitable services', with an independent evaluation indicating a fundamental shift in invasive species management toward a community-led approach with the exchanging of knowledge and expertise (ACIL Allen, 2017).

The Victorian Blackberry Taskforce extended their reach to community groups operating at the urban–rural interface, where management of blackberry can be particularly challenging due to landowners' differing motivations for land management and conservation. Drawing on the knowledge of local leaders, engaging local landowners in project planning, and advocating for a range of different control methods enabled the group to adapt their advice and approach to suit the local context. The VBT's work has been critical in supporting better collaboration between neighboring landowners, for example through the formation of local-level Blackberry Action Groups, where farmers and other community members come together with government land managers to tackle blackberry across land boundaries. These partnerships aim to provide community members with greater ownership of blackberry management and promote locally appropriate solutions. They are built on the principles of inclusive decision-making and representation, and focused on partnership and relationship development that promotes community capacity building (https://vicblackberrytaskforce.com.au/community-partnership/).

At a more local level, each of the CPMGs has reported significant short-term on-ground impacts. For example, during one of the Victorian Serrated Tussock Working Party's 2019–20 local extension programs, 143 properties were visited over an area of 8700 ha in central Victoria; of the 71 ha of serrated tussock infestation discovered, 65 ha were treated in the first year (Victorian Serrated Tussock Working Party, 2020). The Victorian Gorse Taskforce's community grants program has enabled small groups of neighboring landholders to undertake gorse control programs in their local areas, with 296 ha of gorse controlled through 13 local projects in 2019–20 (Victorian Gorse Taskforce, 2020).

The results from the partnership between Agriculture Victoria and CPMGs highlight that rural leaders and community groups need to be involved not just in 'doing' established invasive species control, but also in the design and delivery of management programs. This approach is a major change in both government and community thinking about the management of pests and weeds. When programs are built on the community's knowledge and experience – and participants can see their concerns and local context reflected in program design – there is a much stronger sense of shared ownership and responsibility across stakeholders. The increased connections and capacity at a community or local level make programs more sustainable, long-lasting and effective.

Following the conclusion of the White Paper investment in 2021, Agriculture Victoria has facilitated the formation of a new Victorian Invasive Pest Advisory Board. It will provide an ongoing forum for community and government stakeholders to collectively govern investment and management of established invasive pests in Victoria. The Board will provide strategic advice on the design and delivery of Victorian Government community programs that support community-led action and share the responsibility in the management of established invasive pests. In addition, recognizing the integral role the CPMGs provide across the state of Victoria, the Victorian Government is also continuing to support the four groups to plan and deliver community-based initiatives, with a focus on building community knowledge, community leadership and engagement skills, and establishing partnerships with agencies.

The approaches used by each of the CPMGs recognize that there is no 'quick fix' for established pest management: the biological and social systems make management highly complex. By focusing on the 'human' factor of the problem, the partnership has empowered people to take collective action. The strength of the networks created, and capacity built, prepares these groups and the communities to deal with other community issues which will arise in the future.

3.1.8 Lessons learned

The case study highlights important practices and approaches for addressing some of the key socio-political challenges of building and sustaining community and farmer engagement in complex ecological management projects:

- Spending time seeking to understand and account for the human dimensions of a complex problem is critical for making a difference with intractable problems.
- Engaging in system thinking can help us appreciate the complexity in the system. It can illustrate the interdependency and connectedness of both biophysical and socio-political dimensions of problems.
- Farmers, land managers, industry representatives, and government officers can achieve more together through knowledge and perspective sharing rather than alone.
- Community-led, government-supported approaches are not only feasible but can lead to transformational change.
- Engaging community and government together at the inception is important for understanding the system and re-framing the intractable, complex issue. This generates shared ownership and a greater commitment to collective action.
- Managing established invasive species is more than providing technical information and regulating behaviors of landholders. Rather, it is about creating the space for ongoing co-learning, and acknowledging both the technical aspects and the socio-political challenges of achieving sustained behavioral change.
- Shifting the focus from temporary measures of success (e.g. number of pests killed) to human-centered metrics (e.g. increased capability) can support longer term, more sustainable system-level change.
- The partnership and governance structure enabled a sharing of responsibility through facilitating a model of distributive leadership, in which community representatives can play an active role.

3.2 Engaging local voices in watershed management

3.2.1 Background

The Chesapeake Bay watershed is the largest estuary in the United States extending across six states in the Mid-Atlantic region (Penn State University, no date). The bay stretches through Delaware, Maryland, New York, Pennsylvania, Virginia, and West Virginia, impacting a population of 17 million (Penn State University, no date). The Chesapeake Bay watershed is rich in ecological biodiversity while also being densely populated, causing ongoing challenges with ecological preservation. The watershed has been facing significant pollution from runoff produced by agricultural practices. Nitrogen, phosphorus, and sediment are the major non-point source pollutants damaging local water quality and ecosystems in the bay (Pennsylvania Department of Environmental Protection (DEP), no date). A large portion of the Chesapeake Bay watershed is in Pennsylvania, and an estimated 34 000 Pennsylvania farms are in the watershed (Penn State University, no date). In 2010, Pennsylvania contributed to approximately 44% of the nitrogen pollution and 24% of the phosphorus pollution in the Bay (Pennsylvania Water, 2022). Environmental management in the Chesapeake Bay watershed, like similar situations, is characterized by complex and interdependent environmental, economic, social, and political factors.

Production agriculture in Pennsylvania impacts the livelihoods of thousands and feeds millions and is one of the state's largest economic sectors. Certain agricultural practices exacerbate water pollution in the watershed. Most farmers lack the financial, educational, and technical resources to tackle these issues (Penn State University, no date). Entire communities are affected as well, not just farmers. According to *Pennsylvania's Community Clean Water Implementation Guide,* the watershed provides millions of people with drinking water, supplies water for industries such as agriculture and tourism, supports recreation, fishing, and electricity generation, and is home to wildlife and native plant species. The waters of the Chesapeake Bay watershed have enormous value (DEP, 2019).

Various initiatives spearheaded by organizations such as the Pennsylvania Department of Environmental Protection, Pennsylvania's Chesapeake Bay Program, the United States Department of Agriculture's (USDA) Natural Resources Conservation Service initiative, the Environmental Defense Fund, Penn State University Extension, and others are addressing watershed issues. These initiatives resulted in an increase of no-till agriculture by 52% and a decrease in conventional tillage by 61% from 2006 to 2016 (USDA, 2022). In those same 10 years, 'edge-of-field practices' helped reduce nutrient and sediment runoff like riparian buffers and field borders nearly tripled (USDA, 2022). Cover crop implementation in crop rotations also increased from 3% to 33% in this period.

This considerable progress in recent years is due to many efforts from organizations, communities, and most importantly farmers. Using a community-based approach to this issue has been essential. This approach ensures local farmers remain at the center of the conversation and creates practical and sustainable solutions.

3.2.2 Programs

Pennsylvania has invested heavily in water quality and environmental restoration initiatives for its numerous waterways for decades. Under the terms of the Federal Clean Water Act, the U.S. Environmental Protection Agency (EPA) can establish a total maximum daily load (TMDL), which indicates the limits to the amount of pollution the body of water can withstand (DEP, 2021). In 2010, these pollution limits were released for the Chesapeake Bay with pollution reduction goals to be met by the deadline of 2025, focusing on sediment reduction.

Emerging from these new targets outlined by the EPA was the Chesapeake Bay Watershed Implementation Program (WIP). WIP is in partnership with the DEP, EPA, and different counties in Pennsylvania to assist in crafting solutions to the various problems impacting the watershed (DEP, 2019).

WIP has already accomplished the goals outlined in Phases 1 and 2 of the programs and has made immense progress in restoring Pennsylvania's main waterways. Phase 1 was established in 2011 to identify pollutant sources and find reduction solutions (DEP, 2022a). Phase 2 was started in 2012 and relied heavily on community input and the inclusion of adaptive management principles in the plan (DEP, 2022a). From these efforts, there has been an annual reduction of over 16 million pounds [7260 tons] of nitrogen and nearly 2.5 million pounds [1135 tons] of phosphorus (almost 50%) compared to 1985 (DEP, 2019). The current Phase 3 WIP sets new goals for the watershed and involves county-wide action plans, requiring participation and collaboration with the community. New targets, to be completed by 2025, include reducing nitrogen by 34 million pounds [15 420 tons] per year, phosphorus by 0.7 million pounds [320 tons] per year, and sediment by 531 million pounds [240 860 tons] per year (DEP, 2019). These goals and targets cannot be reached without collaborating with communities and farmers affected by these issues. Phase 3 WIP is working with volunteers and farmers at the county level to foster local participation in restoration and conservation projects across the bay, on the basis that the direct input of locals will achieve local water quality improvements that work best for that specific community (DEP, 2019).

A recent separate initiative complementing the Chesapeake Bay WIP is a project called Water for Agriculture (Water for Ag). The project is funded by the USDA and the National Institute of Food and Agriculture (NIFA) in partnership with Penn State University, Arizona State University, and the University of Nebraska Lincoln. Like the WIP, Water for Ag is focused on working with key stakeholders to better implement natural resource solutions instead of taking an expert-led approach.

This initiative was undertaken in four pilot counties to assist five communities across Nebraska, Arizona, and Pennsylvania (Ishler et al., 2021). Many farmers lack access to resources to handle these problems on their own, government officials and scientists usually have limited local knowledge, and there is typically a lack of substantive collaboration among these parties. A community-led approach fosters a collaborative relationship between these key stakeholders, equipping them with the proper resources and knowledge. Effectively empowering stakeholders allows them to lead efforts and design solutions to the most pertinent challenges (Water4Ag, no date). This holistic way of thinking establishes a network of resources and allows community members, especially farmers, to stay included in the process.

3.2.3 Community engagement strategies

The WIP and Water for Ag projects demonstrate that the process is equally as important as the outcomes and, in fact, can lead to meaningful outcomes for community-led projects. There must be intentional engagement strategies and guiding principles to successfully implement such programs locally at the landscape scale.

Phase 3 of the WIP has developed a comprehensive and effective engagement strategy, grounded in program experience and centered on farmers and other community stakeholders (DEP, 2022b). The strategy focuses on

- widespread collaboration with many partners to develop, write, and implement Phase 3;
- strategic inclusion and engagement with various sectors and members of the community to ensure all concerns and needs are addressed; and
- a strategic communication effort reaching out to key stakeholders and the public to provide support and understanding throughout the project.

Similarly, Water for Ag has created a step-by-step engagement process that guides community leaders on how to implement an action plan (Water4Ag, no date):

- Initiation: setting the stage, interview findings, celebrating success to date, building relationships, exploring expectations, and assessing commitment;
- Getting organized: collaboratively establishing goals, intentions, and expectations;
- Where are we now: current programs, successes, opportunities, research, perspectives, policies, involvement, and future forces;
- Where do we want to go: vision, needs, priorities, collaboration, engagement, information, research, involvement, education, etc.;
- How do we get there: resources, partnerships, strategies, capacity, involvement, information, implementation, and evaluation planning;
- Making it happen: organization, resources (human, financial, political, etc.) engagement structure; and
- Evaluation and assessment: project and site assessment; adjustment, adaptation, and sustainability.

3.2.4 Engagement strategies in practice

In practice, community engagement is quite complex, but with clear planning and input that reflects the perspectives of a broad array of local community members along the way, stakeholder-owned strategies can be developed and implemented. A 'bottom-up, county-based clean water planning approach' has been declared for the work done with Phase 3 WIP (DEP, 2019). As part of the collaboration process in Phase 3, an active 20-member steering committee and seven workgroups discuss the environmental pollution issues and gain insights from different perspectives (Pennsylvania, 2022). These committees consist of members from the public, non-profit, and private sectors including state legislative representatives (DEP, 2022b). This plan fosters a well-rounded and holistic approach as many fields of study were taken into consideration, which is fundamental when approaching complex environmental resource management issues. Natural resource management issues like this can be categorized as wicked problems, as their interconnected nature with the environment, communities, the economy, and a range of other factors make solutions difficult to find and require multidisciplinary efforts.

Phase 3 WIP has also created the County-wide Action Plan (CAP) initiative which is an effort to create county-specific solutions to improve water quality in the Chesapeake Bay. This was in continuing partnership with the DEP and a Local Area Goals Workgroup. From this collaborative process with four pilot counties in Pennsylvania including Adams, York, Franklin, and Lancaster, a Communities Clean Water Planning Guide was created. The planning guide included 'priority goals and initiatives, action steps, the identification of responsible parties, and available and needed technical and financial resources' (DEP, 2022a). It is a tool to be used by other counties that lays out the framework of how to engage the local community and its leaders to improve water quality, economic development, and increase community knowledge about the issues (DEP, 2019). This part of the project is county specific as each county is facing slightly different circumstances. This highlights the complex nature of these wicked problems as there is no one-size-fits-all solution and the solutions are context-dependent. This collaborative process helps the implementation process as each county will have a unique action plan which best suits the specific needs of the community, farmers, and other stakeholders across the Bay watershed.

As part of the WIP strategic inclusion process, there have been numerous deliberative discussions, focus groups, roundtables, and listening sessions focused on gaining community input and working together to create solutions. For example, in 2018 around 200 participants assembled for a day-long workshop to listen, learn, and discuss local planning and expand local engagement (DEP, 2022c). The goals were to increase stakeholder knowledge, answer questions, and provide an opportunity for members to develop plans for a Community Clean Water Toolbox used to help develop the Countywide Action Plans (DEP, 2022c). From this session, the toolbox included in the guide was finalized and the necessary feedback was applied to the work being done (DEP, 2022c).

Effective communication is essential when working with communities and different networks. Phase 3 WIP has developed accurate and accessible outreach materials for anyone interested or wanting to become involved. These informative publications include summaries of the work being done, planning guides for communities and stakeholders, and various templates which can be used for reference. The Pennsylvania Clean Water Academy is an online educational tool with an extensive course catalogue and updated news and information. Resources like the Water Academy provide communities with the most up-to-date information and enable local engagement. Also, 40 representatives from various backgrounds commented on the Phase 3 WIP plan (DEP, 2022d). Ongoing feedback and commentary are necessary checkpoints for the leaders to make sure the work being done encapsulates the feelings and goals of the communities they are working with.

Mifflin County is part of the Chesapeake Bay Watershed in central Pennsylvania. Mifflin County is a key research site in the Water for Ag project and has used an engagement framework and other similar approaches to facilitate conversations with local committees and community members to help prioritize their concerns. The Local Leadership Committee Interim Summary Report for 2019–20 explained how the first year of the project entailed laying the groundwork for relationships and gathering information to fully understand the scope of the problem. The leadership committee is composed of 16 representatives of farmers, agricultural organizations, local government, local and state agencies, and technical service and education providers (Mifflin County Local Leadership Committee (MCLLC), 2021). This includes various backgrounds and perspectives for a well-rounded perspective of the issues. The committee focused on understanding the impacts of the proposed programs, bridging any in data and information, and appreciating previous efforts to address these issues. The committee was then able to create a list of goals in order of priority based on these deliberative discussions and identify which areas were most critical to stakeholders.

The committee encouraged public participation through 43 face-to-face interviews with local farmers, local government representatives, farmers, civic organizations, and technical and educational service providers (MCLLC, 2021). These provided them with detailed perspectives and made sure farmers knew their voices were heard and opinions were valued. Additionally, surveys directed at non-farmers in the local community were distributed simultaneously to further understand public perception of the issues. Generating ongoing feedback and views from the community is very important in the community-based leadership process. The results of these surveys showed that 82% think that it is their personal responsibility to help protect water quality and 51% would be willing to change their land management practices improve water quality and quantity (MCLLC, 2021).

The survey results have allowed the committee to conclude that most farmers and non-farmers in this community believe preserving and enhancing water quality is important (MCLLC, 2021). This information will help the committee and other leadership organizations determine the next steps and how to best implement solutions.

3.2.5 Impacts of the programs

The Chesapeake Bay WIP has been instrumental in reducing nutrient runoff by working with farmers and local communities. Farmers have reduced the amount of phosphorus pollution going downstream in the watershed by over one-third and cut the amount of nitrogen pollution by about one-sixth (DEP, no date). Many publications, reports, and guides have been produced from the collaborative efforts of this program, allowing anyone to access the resources for similar projects. These have been used as part of the communication strategy as outreach to the entire community and for anyone who wants to be involved.

Additionally, from the Water for Ag Mifflin County project site, many insights were gained that will be valuable moving forward with the continuation of this program and for any future community engagement initiatives. Again, these resources are openly available, supporting future community-led initiatives and creating a foundation for any future environmental and natural resource issues which the community will be equipped to handle.

3.2.6 Lessons learned

The Chesapeake Bay WIP and Water for Ag projects demonstrate how effective engagement with the community leads to sustainable solutions. They have used the following key approaches to guide their work and accomplish their goals.

Water for Ag (Ishler et al., 2021):

- Recognizing that everyone has valuable expertise and perspectives to contribute, and building relationships is critical to successful collaborative efforts.
- The collaborative nature of bringing different perspectives to the table can provide important opportunities to identify ideas and develop solutions that no individual or organization could achieve on its own.
- Building bridges of mutual respect and open communications with members of the community around water issues is essential.
- Structured and well-facilitated processes can be useful in allowing communities and groups to efficiently focus on what matters most – creating solutions to the issues they have identified.

- Taking the time for deliberative and in-depth discussions and looking at an issue from multiple perspectives are important ingredients in learning from one another and creating long-term and successful strategies to address the issues that matter most to communities.
- A cooperative effort between the farm and non-farm communities is essential to addressing water quality challenges reflective of everyone's concerns.
- Building momentum and creating an environment of continued community engagement related to recreation and stewardship of the county's rivers and streams among all residents is essential.

Phase 3 WIP (DEP, 2022b):

- Developing and providing timely, mainstream, and relatable messaging, and avoiding government and academic jargon, helps make the information accessible to all.
- Increasing efforts to garner positive mainstream media coverage improves the spread of information.
- Enlisting and leveraging supportive advocates such as farmers, hunters and other outdoor sports participants, business owners, sports figures, and others who are not conventional environmental advocates helps showcase support for clean water to their audiences through their own channels.
- Publicly recognizing progress and highlighting success stories through social media, blogs, newsletters, and press events helps motivate further positive actions.

4 Engaging farmers and rural communities in agricultural and ecological restoration projects

Although both cases highlight a range of regionally specific challenges, invasive species and water pollution issues are widespread globally. Invasive alien species are one of the biggest causes of biodiversity loss and species extinctions and are also a global threat to food security (IUCN, 2021). On a global scale, nutrient pollution from agricultural runoff is considered one of the most widespread water quality challenges and the single-largest contributor of non-point source pollution (du Plessis, 2022).

Wherever they occur, these and other such environmental and ecological issues are community- and regional-level common problems and require collective action. As illustrated by these case studies, agricultural and ecological restoration at the landscape scale requires people to develop productive relationships, despite differences of view, and work together to address critical issues. To reiterate a point made earlier in this chapter, while these issues are biophysical issues, they are also fundamentally human issues, given the reality of differing values, beliefs, and ideas about how best to proceed across our human differences. Effective sustained action on our local, regional, and global agricultural and ecological restoration challenges requires people to develop shared visions and commitment to one another, collaborative decision making and, critically, action. Sustained action, in turn, requires that all community members and stakeholders feel and believe that they are respected for their knowledge and experience, their voices are genuinely heard, and they have the power to affect actions taken and to act themselves. This goal is aspirational and difficult, if not impossible, to achieve, but the engagement challenge is that people must feel respected, regardless of role or status, and in productive relationships with one another and the issue at hand. They must be genuinely engaged (Gunderson et al., 1995; Olsson et al., 2006; Bridger and Alter, 2006, 2008; Brennan et al., 2013; Bridger et al., 2013, 2019; Howard and Alter, 2019; Frumento et al., 2019).

What does genuine engagement look like? Arnstein's (1969) 'ladder of participation' was developed in the context of community planning projects. Participation or engagement in this framework ranges from non-participation to informing and consulting to co-creating and delegation of power. This work has been instrumental in shaping much of the thinking about community engagement processes over the years. The International Association for Public Participation (IAP2, 2018) defines five levels of public participation: inform, consult, involve, collaborate, and empower. These goals range from minimal to more extensive and intensive public participation. Note that these goals are not necessarily mutually exclusive and that certain goals may be appropriate for some project initiatives and not others.

The phrase 'community-led' is sometimes used when describing approaches to public participation and community engagement across the spectrum from consultation to variations of empowerment (Attygale, 2020). Attygale notes, however, that 'community-led' has specific meaning and should be used to help us nuance our understanding of approaches to community engagement generally. Citing Wessells (2018), she describes 'community-led' approaches as 'those that are led not by an organization or other outsiders but by a collective community process'. Along a continuum ranging from more community-led to more organization-led approaches, Attygale outlines a spectrum of community-led approaches to change. This spectrum includes community-owned, community-driven, community-shaped, and community-informed approaches to engagement, with the spectrum as listed moving from more community-led to more-organization-led. In this framing, 'community-led' is about inclusive involvement of community members and shifting power and decision making to community, reflecting 'empowerment' in the IA2P framework.

The Spectrum of Community Engagement to Ownership (Gonzalez, 2019) is another useful framework that extends and nuances the IA2P model and complements Attygale's framing of the relationships between community engagement and community-led action. This framework is grounded in the philosophy that deep, inclusive, democratic public participation is critical for effectively addressing the socio-political, economic, environmental, and ecological issues facing us locally and globally. González (2019) extends the IA2P framework by noting that a form of 'engagement' is to ignore rather than even inform the community. She notes that the impact of this 'stance towards the community' is marginalization of the voices of particular interests. Also following the IA2P model, Gonzales speaks of 'deferring to' the community. The impact of this stance, she argues, is foster public participation through community-driven decision making. This 'defer to' language reflects the 'empower' language in the IA2P model but is more suggestive about whose power should count. The impact of this stance, Gonzales argues, is 'community ownership' of decisions and actions. The other dimensions of the IA2P framework – inform, consult, involve, and collaborate – complete the Spectrum of Community Engagement to Ownership. Gonzales characterizes the impact of these stances towards community as placation, tokenism, voice, and delegated power, respectively.

Nabatchi and Leighninger (2015) note that civic participation includes various forms of engagement with differing processes, levels of commitment, and impact. They argue that the public participation and engagement space is confusing for community members and public officials alike. Leighninger's Types of Engagement Model (Organizing Engagement, no date) is designed to bring clarity to this space. This model focuses on what he terms thick, thin, and conventional forms of engagement. Thick engagement, comprised of actions deemed 'intensive, informed, and deliberative', has been delineated by Leighninger as the most effective form of engagement. Contrasted with the broad and shallow outreach efforts of thin engagement, thick engagement allows community members autonomy over typically unreachable democratic processes. Deeper levels of community participation in democratic processes tend to result in a more engaged, dedicated, and caring community. Thick engagement typically occurs through focused and extended discussions and deliberations where participants can interact with diverse opinions and work through disagreements (Carcasson, 2020).

While Leighninger highlights thick engagement as the ideal form of civic participation, thin and conventional forms of engagement also provide necessary democratic power to community members. Thin engagement comprises the less demanding forms of civic participation, for both community members and public officials – easily accessible and convenient, these strategies allow individuals to politically align themselves with their beliefs and opinions at a low level of commitment. Surveys, polls, and petitions are some of the ways thin civic engagement can be enacted. Similarly, conventional engagement provides low-commitment opportunities for community members to take part in democratic processes. Unlike thick and thin engagement, conventional engagement is built into our democratic processes to ensure power is not abused. Typical local government meetings are examples of bureaucratic processes that conventional engagement provides. While still providing for interaction among community members and public decision-makers, most forms of conventional engagement fail to provide the discourse and deliberation necessary for meaningful community participation. While decision-makers may feel that a conventional level of community engagement is enough, citizens are often left under-represented and disillusioned by these processes. To empower citizens and increase civic participation, a broad application of thick, thin, and conventional engagement in conjunction is necessary. Carcasson (2020) provides an insightful discussion of the tensions inherent along this continuum of thick and thin engagement.

Frumento et al. (2019) highlight and discuss the importance for practitioners and those managing engagement processes of five key engagement practices and having an engagement roadmap. The key engagement practices are particularly salient for more intensive or 'thick' engagement. These practices are described as: reflection and continuous learning, building relationships and trust, sharing power with stakeholders and community members, collaboratively choosing metrics for program or project success, and actively conducting action research in real time (Frumento et al., 2019). These practices are grounded in the importance of respectfully recognizing and learning from differences among stakeholders and community members. They recognize differing ways of knowing and understanding the world, both expert and practical local knowledge (Fischer, 2000), and help level organizational and epistemic politics and power relationships among conservation policy makers, practitioners, and community members. These are practices, not step-by-step recipes, and they can be confronting and challenging for all involved. Personal and professional reflexivity, intentionality, and practical engaged field experiences are all practices that themselves build individual and collective capacity to engage more effectively.

Effective community engagement for all issues, including biodiversity conservation and ecological restoration, does not just happen. It requires intentional thoughtful planning, organizing, and managing of the engagement process. Frumento et al. (2019) also outline an engagement roadmap. The main components include strategy development, identifying stakeholders and building capacity, defining the issue, communicating with stakeholders, facilitating dialogue, maintaining momentum, and evaluating engagement efforts.

An engagement roadmap is useful in organizing our thinking about the community engagement processes generally. It is critical, however, to remember that the roadmap journey is not linear and that the steps noted by Frumento et al. (2019) are not mutually exclusive. The reality is that the steps overlap, they often occur simultaneously, and they circle back on one another. For example, identifying stakeholders and defining the issue are obviously important to think about in planning and organizing an engagement initiative but initial thinking about these matters will shift and become more nuanced as communication with stakeholders and community members occurs and dialogue is facilitated. Further, the human interaction or engagement associated with each step and with the process of steps recursively looping back on one another are contexts for new insights to surface, individual and collective learning, creativity, co-creation, and innovation, occurring at the intersection of difference.

Engagement processes are replete with politics. Working through the differences of perspective and view endemic to critical wicked issues, such as agricultural land and ecological restoration, biodiversity conservation, and sustainability, is about dealing with and managing conflict which is inherently political. It is paramount to recognize, too, that engagement itself is politics. One way to reflect on this political aspect is considering how inclusive engagement shifts the composition of voices at the table by making space for voices previously muted, marginalized, or not heard, by centering local practical knowledge in addition to 'expert' ways of knowing. This shift in epistemic politics reflects a shift in power, which can be threatening to vested interests. It is in this sense that engagement itself is political. Frumento et al. (2019) provide a series of scenarios illustrating the nature of each dimension of this engagement roadmap framework and the engagement process, including this idea of inclusive, deep, and thick engagement itself as political action.

Acknowledging the inherently political process of engagement, being intentional about our practices and the recursive nature of engagement strategy can be difficult territory to navigate. *Community Engagement for Collective Action* (Alter et al., 2017) is a practitioner-oriented handbook and serves as a valuable individual and group capacity-building complement to the engagement practices and engagement roadmap components detailed by Frumento et al. (2019). This handbook integrates community engagement theory and practice to support practitioners in strengthening their engagement practices, planning, and actions. An important feature of this handbook is its emphasis on reflective and reflexive learning (Boyd and Fales, 1983; Schon, 1984) for community engagement practitioners. A series of 'practice tips' to assist practitioners to strengthen these skills are included.

The literature on community engagement and engagement processes is vast, and we detail only a small subset of that literature. The frameworks presented in this chapter are helpful in thinking about what has or has not been done in a specific issue or local context and how these general frameworks might be adapted to the situation at hand. There is no 'one-size-fits-all' prescription for doing community engagement. Policy makers, practitioners, and stakeholders must work at co-creating the frameworks, activities, and actions most appropriate for their situation (Frumento et al., 2019).

5 Overarching engagement principles

Drawing on the case studies and the theory and framework presented earlier, several overarching principles emerge. The following summary outlines the behavioral, process, and socio-political considerations that we view as pertinent to effective community engagement and community-led action.

Draw on different knowledge sets and interdisciplinary approaches

- Respect local voices, experiences, expertise, rights, and norms through deep engagement and listening across difference.
- Exchange knowledge and work with experts in different disciplines including those outside of technical and scientific fields.
- Stay open-minded to different practices and methods.
- Clearly define the problem and the historical context using local knowledge.

Spend time on the process and connections, understand relationship dynamics

- Process is almost always as important as the output.
- Acknowledge tensions, power imbalances, and socio-political challenges upfront – seek to bring in excluded voices and perspectives, note and be aware of in future discussions.
- Encourage ongoing, iterative feedback and evaluations.
- Actively listen before proposing ideas.
- Foster a relationship with the community that values transparency and trust.
- Engage stakeholders in complex problem-solving through dialogue and deliberation.
- Take the time to build a strong connection with the community and allow them to be the driving force.

Embrace and hold complexity, diversity, and humility

- Acknowledge 'wickedness' of the issues and the connectedness of the different systems that influence people and landscapes.

- Don't look for simple solutions: there is rarely a 'silver bullet' or a 'one-size-fits-all' answer.
- Look beyond the 'golden hammer': tools which worked in the past may not work in the future.
- Encourage diversity and inclusion of community members and perspectives.
- Acknowledge and engage with tension between stakeholders to understand differences, negotiate on conflicts and find commonality.
- Recognize and acknowledge the new information, creativity, and innovation that comes from bringing together the experiences and perspectives of diverse stakeholders.
- Listen to and respect everyone. Embrace humility.

Empower local people and communities to lead

- Provide participants with information, resources and connections that support them to have meaningful contributions to the project.
- Formally recognize participants' contributions.
- Empower community members to make change – through learning, capacity building, and decision-making power, giving them the opportunity to prioritize which issues are important and how they're addressed.
- Be aware of engagement fatigue and impacts on communities of short-term engagement – look to ways to embed change at community level rather than in institutions.
- Support and partner with the institutions that already engage directly with farmers and rural communities.
- Ensure you are creating ongoing future opportunities for participants to engage with each other.

6 Conclusion: engagement, uncertainty, and ecological restoration

This chapter outlines systems perspectives and engagement frameworks, strategies, and principles for understanding and improving the involvement of farmers and their communities in biodiversity conservation, agricultural land use issues, and ecological restoration. It considers approaches for bringing together participants with differing perspectives and interests to address both the biophysical and socio-political ecologies of these challenges. Biodiversity conservation, agricultural land use, and ecological restoration are critical to planetary health, socio-economic growth and development, and our human survival. Our agricultural lands, wild areas, watersheds, rivers and streams, and animal and plant biodiversity, soil health, and other such issues are the lifeblood of our farms, families, communities, nations, and global planetary existence.

Addressing these matters and specific issues associated with them effectively and democratically, regardless of scale, requires that we engage each other at the intersection of our biophysical and socio-political worlds and our human differences.

The reality is that we need each other to deal with the environmental and ecological challenges facing us today and the issues we do not know about yet that will no doubt emerge in the future. The central overarching challenge we face locally and globally in our groups, organizations, and communities is strengthening our individual and collective capacity for working and learning together across our differences, including our ways of understanding and knowing the world, in the face of these ecological and other challenges. The intersection of those differences holds great promise for emergent collective insights, creativity, and innovations (Johansson, 2006) regarding how to address these dynamically changing ecological challenges. This is the challenge of engagement, the challenge of community, given our shared interests in local as well as planetary outcomes. One might argue that community engagement, ideally an ongoing, intensive, and inclusive socio-political process, is at the heart of ecological sustainability. Going further, one might argue that our capacity to engage each other effectively and democratically is the heart of ecological and socio-political sustainability. The two are intimately intertwined.

Uncertainty characterizes the world within which we live (Knight, 1921; Taleb, 2007, 2012; Kay and King, 2020), including our understanding of biodiversity conservation and ecological restoration. We have differing views on what the future holds regarding possible environmental and ecological changes and their consequences. Then, there are the ecological futures which we cannot even imagine even despite our experiences and scenario forecasting. As Batie (2008) notes, dealing with the 'wickedness' of this reality, along with the high levels of uncertainty and differences in human values and beliefs, requires 'engaged' sustainability sciences and adaptive management strategies. Human engagement at all levels and scales, intensive engagement that supports open and respectful sharing of expertise, experience, and imagination across our human differences, is central to progressing biodiversity conservation and ecological restoration in an uncertain world.

7 Acknowledgments

Alter's work on this chapter was supported by the USDA National Institute of Food and Agriculture and Multistate/Regional Research Appropriations under Project #PEN04802 and Accession #7003365.

8 Where to look for further information

8.1 Books and monographs

- Jane Goodall Institute (2022). *Local Voices, Local Choices: The TACARE Approach to Community-Led Conservation*. Redlands, CA: Esri Press.
- Martin, P., Low Choy, D., LeGal, E. and Lingard, K. (2016). *Effective Citizen Action on Invasive Species: The Institutional Challenge*. https://doi.org/10.13140/RG.2.1.3927.7045.

8.2 Guides and reports

- Bassler, A., Brasier, K., Fogle, N. and Taverno, R. (2008). Guiding principles of effective community engagement, Guiding Principles of Effective Community Engagement. Penn State College of Agricultural Sciences: Department of Agricultural Economics, Sociology, and Education. https://aese.psu.edu/research/centers/cecd/engagement-toolbox/engagement/guiding-principles-of-effective-community-engagement.
- Biennial strategy review system Chesapeake Bay program (2023). Logic and action plan. Chesapeake Bay Program.https://d18lev1ok5leia.cloudfront.net/chesapeakeprogress/logic-action- plans/Final_2025_wip_outcome_srs_logic_and_action_plan_v01.11.2023.pdf.
- Dale, A., Curnow, J., Campbell, A. and Seigel, M. (Eds.) (2022). Building global sustainability through local self-reliance: lessons from Landcare. *Australian Centre for International Agricultural Research*.
- Friedman, S. (2018). Water quality guide for dairy sustainability in the Chesapeake Bay an open-source road map for cooperatives and processors, Chesapeake Bay water quality dairy sustainability guide. Environmental Defense Fund. https://www.edf.org/sites/default/files/documents/hesapeake-bay-water-quality-dairy- sustainability-guide.pdf.
- Howard, T., Thompson, L., Alter, T. and Frumento, P. (2016). Community action for wild dog management. PestSmart publication, Invasive Animals Cooperative Research Centre, Canberra, Australia. https://www.researchgate.net/publication/312469019_Community_action_for_wild_dog_management.
- Karp, M. and Lundy-Wagner, V. (2016). *Collective Impact: Theory versus Reality*. Corridors of College Success Series. CCRC Research Brief. Number 61, Community College Research Center, Teachers College, Columbia. University. https://files.eric.ed.gov/fulltext/ED564894.pdf.
- Torjman, S. and Makhoul, A. (2012). Community-led development. The Caledon institute of social policy. https://maytree.com/wp-content/uploads/978ENG.pdf.

- Weaver, L. (2017). Turf, trust, co-creation, and collective impact. Tamarack Institute. https://www.tamarackcommunity.ca/library/turf-trust-co-creation-collective-impact.

8.3 Popular articles

- DEP Newsroom (2022). DEP Submits Final Pennsylvania Chesapeake Bay Watershed Plan with New State Funds Supporting Partners' Progress. Department of Environmental Protection. https://www.dep.pa.gov/Citizens/Pages/DEP-Newsroom-.aspx.
-

8.4 Refereed journal articles

- Ban, N. C., Frid, A., Reid, M., Edgar, B., Shaw, D. and Siwallace, P. (2018). Incorporate Indigenous perspectives for impactful research and effective management. *Nature Ecology and Evolution,* 2, 1680–1683. https://doi.org/10.1038/s41559-018-0706-0.
- Castleden, H., Morgan, V. S. and Lamb, C. (2012). "I spent the first year drinking tea": Exploring Canadian university researchers' perspectives on community-based participatory research involving Indigenous peoples. *The Canadian Geographer/Le Géographe Canadien*, 56, 160–179. https://doi.org/10.1111/j.1541-0064.2012.00432.x.
- Douwe van der Ploeg, J. (2021). The political economy of agroecology. *The Journal of Peasant Studies,* 48(2), 274–297, https://doi.org/10.1080/03066150.2020.1725489.
- Margavio Striley, K. and Lawson, S. (2014). Theorizing communication orientations of privilege: how white discourses (de) construct Australian aboriginals. *Journal of International and Intercultural Communication*, 7(2), 170–191. https://doi.org/10.1080/17513057.2014.898368.
- Ridges, M., Kelly, M., Simpson, G., Leys, J., Booth, S., Friedel, M. and Country, N. (2020). Understanding how aboriginal culture can contribute to the resilient future of rangelands – the importance of aboriginal core values. *The Rangeland Journal*, 42, 247–251. https://doi.org/10.1071/RJ20031.
- Tennet, R. and Lockie, S. (2013). Vale landcare: the rise and decline of community-based natural resource management in Australia. *Journal of Environmental Planning and Management*, 56(4), 572–587. https://doi.org/10.1080/09640568.2012.689617.
- Vanclay, F. (1994). Hegemonic landcare: further reflections from the national landcare conference. *Rural Society*, 4(3–4), 45–47. https://doi.org/10.1080/10371656.1994.11005135.

8.5 *Webinars*

- Muth, A. (2022). Working at the intersection of people and forests. January 26, 2022. https://aese.psu.edu/research/centers/cecd/stories-from-the-field/recordings/working-at-the-intersection-of-people-and-forests.
- Marshall, D. (2022). People & pigs: community engagement in invasive species management. November 17, 2022. https://aese.psu.edu/research/centers/cecd/stories-from-the-field/recordings/people-pigs-community-engagement-in-invasive-species-management.
- Reid, M. (2023). The Australian rabbit problem and a people solution. March 15, 2023. https://aese.psu.edu/research/centers/cecd/stories-from-the-field/recordings/the-australian-rabbit-problem-and-a-people-solution.
- Royer, M. (2017). Water quality protection practices in Pennsylvania agriculture. Penn State Extension. https://extension.psu.edu/water-quality-protection-practices-in-pennsylvania-agriculture.
- Whitmer, W. (2022). The Water for Agriculture Project: building the evidence base for engagement research and practice. February 24, 2022. https://aese.psu.edu/research/centers/cecd/stories-from-the-field/recordings/water-4-ag-evidence-base-engagement-research-practice.

8.6 *Websites and interactive story maps*

- Community engagement: Tamarack Institute. https://www.tamarackcommunity.ca/community-engagement.
- Engagement toolbox, Pennsylvania State College of Agricultural Sciences, Department of Agricultural Economics, Sociology, and Education. https://aese.psu.edu/research/centers/cecd/engagement-toolbox.
- Healthy Waters, Healthy Communities. Pennsylvania Department of Environmental Protection, Pennsylvania Department of Conservation and Natural Resources. https://gis.dep.pa.gov/chesapeakebaywatershed/index.html.

9 References

ACIL Allen Consulting (2017). Consulting. *Victorian Rabbit Action Network Impact Analysis. Report to the Department of Economic Development, Jobs, Transport and Resources*. Available at: https://vran.com.au/wp-content/uploads/2020/10/VRAN-Impact-Analysis.pdf (accessed 29 March 2023).

Acosta García, N., Boyd, E., Gillette, M. B., Rask, N., Saldert, H. and Esmond Singleton, B. E. (2023). Disagreeing well in an unparadigmatic field: a response to Bodin (2021). *Sustainability Science* 18(2), 1049–1052. https://doi.org/10.1007/s11625-022-01228-6.

Adams, L. B., Alter, T. R., Parkes, M. W., Reid, M. and Woolnough, A. P. (2019). Political economics, collective action and wicked socio-ecological problems: A practice story from the field. *Gateways: International Journal of Community Research and Engagement* 12(1), 1, Article ID 6342. https://doi.org/10.5130/ijcre.v12i1.6496.

Alter, T., Driver, A., Frumento, P., Howard, T., Shufstall, B. and Whitmer, W. (2017). Community engagement for collective action: a handbook for practitioners. Invasive Animals CRC, Australia. Available at: https://community.invasives.com.au/wp-content/uploads/sites/4/2020/07/UNE-Invasive_Species_Handbook_2017_w.pdf.

Arnstein, S. R. (1969). A ladder of citizen participation. *Journal of the American Institute of Planners* 35(4), 216–224. https://doi.org/10.1080/01944366908977225.

Attygale, L. (2020). Understanding community led approaches to community change. Tamarack Institute. Available at: https://www.tamarackcommunity.ca/hubfs/Resources/Publications/2020%20PAPER%20%7C%20Understanding%20Community-Led%20Approaches.pdf (accessed March 8, 2023).

Batie, S. S. (2008). Wicked problems and applied economics. *American Journal of Agricultural Economics* 90(5), 1176–1191. https://doi.org/10.1111/j.1467-8276.2008.01202.x.

Boyd, E. M. and Fales, A. W. (1983). Reflective learning: key to learning from experience. *Journal of Humanistic Psychology* 23(2), 99–117.

Brennan, M. A., Bridger, J. C. and Alter, T. R. (Eds.) (2013). *Theory, Practice, and Community Development*. London: Routledge, Taylor and Francis Group.

Bridger, J. C. and Alter, T. R. (2008). An interactional approach to place-based rural development. *Community Development* 39(1), 99–111. https://doi.org/10.1080/15575330809489744.

Bridger, J. C. and Alter, T. R. (2006). The engaged university, community development, and public scholarship. *Journal of Higher Education Outreach and Engagement* 11(1), 163–178.

Bridger, J. C., Frumento, P. Z., Alter, T. R. and Brennan, M. A. (2013). A framework for thinking and acting critically in community. In: Brennan, M. A., Bridger, J. C. and Alter, T. R. (Eds). *Theory, Practice, and Community Development*. London: Routledge, Taylor and Francis Group, pp. 180–190.

Bridger, J., Alter, T., Frumento, P., Howard, T. and Adams, L. (2019). Community-engagement theory for a new natural resource paradigm. In: Martin, P., Alter, T., Hine, D. and Howard, T. (Eds). *Community-Based Control of Invasive Species*. Australia: CSIRO Publishing, pp. 84–96.

Carcasson, M. (2020). Between thick and thin: improving public engagement through a wicked problems lens. *National Civic Review* 109(3), 6–14.

Commonwealth of Australia (2015). *Agricultural Competitiveness White Paper Stronger Farmers Stronger Economy*. Canberra: Commonwealth of Australia.

Crutzen, P. J. (2006). The "Anthropocene". In: Ehlers, E. and Krafft, T. (Eds). *Earth System Science in the Anthropocene*. Berlin, Heidelberg: Springer.

Department of Environmental Protection (n.d.). Agriculture – Pennsylvania's Chesapeake Bay Program office. Available at: https://www.dep.pa.gov/Business/Water/Pennsylvania%E2%80%99s%20Chesapeake%20Bay%20Program%20Office/agriculture/Pages/Agriculture.aspx (accessed February 21, 2023).

Dickson, K., Bartel, R., Roux, D. and Alter, T. (2019). Achieving organisational learning in natural resource management. In: Martin, P., Alter, T., Hine, D. and Howard, T. (Eds).

Community-Based Control of Invasive Species. Australia: CSIRO Publishing, pp. 48–73.

du Plessis, A. (2022). Persistent degradation: global water quality challenges and required actions. *One Earth* 5(2), 129–131. https://doi.org/10.1016/j.oneear.2022.01.005.

Dzur, A. (2019). *Democracy Inside: Participatory Innovation in Unlikely Places*. Oxford: Oxford University Press.

Eaton, W. M., Brasier, K. J., Burbach, M. E., Whitmer, W., Engle, E. W., Burnham, M., Quimby, B., Kumar Chaudhary, A., Whitley, H., Delozier, J., Fowler, L. B., Wutich, A., Bausch, J. C., Beresford, M., Hinrichs, C. C., Burkhart-Kriesel, C., Preisendanz, H. E., Williams, C., Watson, J. and Weigle, J. (2021). A conceptual framework for social, behavioral, and environmental change through stakeholder engagement in water resource management. *Society and Natural Resources* 34(8), 1111–1132, https://doi.org/10.1080/08941920.2021.1936717.

Faerron Guzman, C. A. and Potter, T. (2021). The planetary health education framework. Available at: https://www.planetaryhealthalliance.org/education-framework-pdf (accessed March 3, 2022).

Faerron Guzmán, C. A., Aguirre, A. A., Astle, B., et al. (2021). A framework to guide planetary health education. *Lancet Planet Health* 2021(5), e253–255.

Finlayson, G., Taggart, P. and Cooke, B. (2022). Recovering Australia's arid-zone ecosystems: learning from continental-scale rabbit control experiments. *Restoration Ecology* 30(4), e13552, https://doi.org/10.1111/rec.13552.

Fischer, F. (2000). *Citizens, Experts, and the Environment: The Politics of Local Knowledge*. Durham and London: Duke University Press.

Fischer, F. (2017). *Climate Crisis and the Democratic Prospect: Participatory Governance in Sustainable Communities*. Oxford: Oxford University Press.

Frumento, P., Whitmer, W., Alter, T., et al. (2019). Strategy and practice for community engagement. In: Martin, P., Alter, T., Hine, D. and Howard, T. (Eds). *Community-Based Control of Invasive Species*. Australia: CSIRO Publishing, pp. 97–121.

Gilmour, P., Walker, L. and Smith, J. (2021). Impact Publications Analysis of the Weeds and Rabbits Project. Available at: https://weedsandrabbitsstory.com/wpcontent/uploads/2021/09/AGVicWeedsRabbits_ImpactAnalysis_Report-Final-10092021.pdf.

González, R. (2019). The spectrum of community engagement to ownership. Facilitating Power and Movement Strategy Center. Available at: https://movementstrategy.org/wp-content/uploads/2021/08/The-Spectrum-of-Community-Engagement-to-Ownership.pdf.

Gunderson, L. H., Holling, C. S. and Light, S. S. (Eds.) (1995). *Barriers and Bridges to the Renewal of Ecosystems and Institutions*. New York: Columbia University Press.

Head, B. W. and Alford, J. (2015). Wicked problems: implications for policy and management. *Administration and Society* 47(6), 711–739, https://doi.org/10.1177/0095399713481601.

Howard, T. and Alter, T. (2019). Framing community engagement. In: Martin, P., Alter, T., Hine, D. and Howard, T. (Eds). *Community-Based Control of Invasive Species*. Melbourne: CSIRO Publishing, pp. 74–83.

IA2P International Federation (2018). IAP2 spectrum of public participation. Brisbane: IA2P Federation. Available at: https://iap2.org.au/wp-content/uploads/2020/01/2018_IAP2_Spectrum.pdf (accessed 26 March 2023).

IUCN (2021). Issues brief: invasive alien species and climate change. Switzerland: International Union for Conservation of Nature. Available at: https://www.iucn.org/sites/default/files/2022-04/ias_and_climate_change_issues_brief_2021.pdf (accessed 4 June 2023).

Ishler, V., Whitmer, W., Laughner, J. and Santangelo, N. (2021). Water for Agriculture local leadership committees share initial outcomes, Water for Ag, Water for Agriculture's project publications to date. Penn State College of Agricultural Sciences Department of Agriculture. Available at: https://bpb-use1.wpmucdn.com/sites.psu.edu/dist/4/97528/files/2021/06/DAS-page-February.pdf (accessed 21 February 2023).

Johansson, F. (2006). *The Medici Effect: What Elephants and Epidemics Can Teach Us About Innovation*. Massachusetts: Harvard Business School Press.

Jordan, N., Gutknecht, J., Bybee-Finley, K. A., et al. (2020). To meet grand challenges, agricultural scientists must engage in the politics of constructive collective action. *Crop Science* 61, 1–8. https://doi.org/10.1002/csc2.20318.

Kay, J. and King, M. (2020). *Radical Uncertainty: Decision-Making Beyond the Numbers*. New York: W. W. Norton & Company. Inc.

Knight, F. (1921). *Risk, Uncertainty, and Profit*. New York: Houghton Mifflin.

Koning, A. A., Martin Perales, K., Fluet-Chouinard, E. and McIntyre, P. B. (2020). A network of grassroots reserves protects tropical river fish diversity. *Nature* 588, 631–635. https://doi.org/10.1002/fee.2293.

Kreuter, M. W., De Rosa, C., Howze, E. H. and Baldwin, G. T. (2004). Understanding wicked problems: a key to advancing environmental health promotion. *Health Education and Behavior* 31(4), 441–454.

McGinnis, M. D. and Ostrom, E. (2014). Social-ecological system framework: initial changes and continuing challenges. *Ecology and Society* 19(2), 30. http://doi.org/10.5751/ES-06387-190230.

Martin, P., Alter, T., Hine, D. and Howard, T. (Eds.) (2019). *Community-Based Control of Invasive Species*. Canberra: CSIRO Publishing.

Martin, P. (2019). Preface: supporting citizen stewardship. In: Martin, P., Alter, T., Hine, D. and Howard, T. (Eds). *Community-Based Control of Invasive Species*. Canberra: CSIRO Publishing, p. xiv–xx.

Martin, P. and Smith-Heron, A. (2019). A focus on citizen-led action. In: Martin, P., Alter, T., Hine, D. and Howard, T. (Eds). *Community-Based Control of Invasive Species*. Canberra: CSIRO Publishing, pp. 1–27.

Martin, P. and Reid, M. (2019). The socio-ecological systems approach. In: Martin, P., Alter, T., Hine, D. and Howard, T. (Eds). *Community-Based Control of Invasive Species*. Canberra: CSIRO Publishing, pp. 28–47.

Mathews, D. (2014). *Ships Passing in the Night? Cousins Research Group Report on Higher Education in Democracy*. Dayton, OH: The Kettering Foundation Press.

Mathews, D. (2014). *The Ecology of Democracy: Finding Ways to Have a Stronger Hand in Shaping Our Future*. Dayton, OH: The Kettering Foundation Press.

Matson, L., Ng, C., Dockry, M., et al. (2020). Transforming research and relationships through collaborative tribal-university partnerships on Manoomin (wild rice). *Environmental Science and Policy* 115, 108–115. https://doi.org/10.1016/j.envsci.2020.10.010.

Mifflin County Local Leadership Committee Interim Summary Report (2021). Summary report Jan 2021. Water for agriculture. Available at: https://bpb-us-e1.wpmucdn

.com/sites.psu.edu/dist/4/97528/files/2021/07/Summary-Report-Jan21-2.pdf (accessed 21 February 2023).

Murphy, H. T. and van Leeuwen, S. (2021). Biodiversity: invasive species, problematic native species, and diseases. In: *Australia State of the Environment 2021*. Canberra: Australian Government Department of Agriculture, Water and the Environment. Available at: https://soe.dcceew.gov.au/biodiversity/pressures/invasive-species-problematic-native-species-and-diseases; https://doi.org/10.26194/ren9-3639.

Nabatchi, T. and Leighninger, M. (2015). *Public Participation for 21st Century Democracy*. Hoboken, NJ: John Wiley & Sons.

NSW Department of Primary Industries Weed Management Unit (2009). *Blackberry Control Manual: Management and Control Options for Blackberry (Rubus spp.) in Australia*. Victoria: Department of Primary Industries.

Olsson, P., Gunderson, L. H., Carpenter, S. R., Ryan, P., Lebel, L., Folke, C. and Holling, C. S. (2006). Shooting the rapids: navigating transitions to adaptive governance of social-ecological systems. *Ecology and Society* 11(1), 18. [online]. Available at: http://www.ecologyandsociety.org/vol11/iss1/art18/ I.

Organizing Engagement (n.d.). *Types of Engagement*. Available at: https://organizingengagement.org/models/types-of-engagement-thick-thin-and-conventional/ (accessed 9 March, 2023).

Ostrom, E. (2009). A general framework for analyzing sustainability of social-ecological systems. *Science* 325(5939), 419–422.

Palmer, P. (2011). *Healing the Heart of Democracy: the Courage to Create a Politics Worthy of the Human Spirit*. San Francisco, CA: Jossey-Bass.

Pennsylvania Department of Environmental Protection (2019). Pennsylvania's community clean water planning. Available at: https://files.dep.state.pa.us/Water/ChesapeakeBayOffice/WIPIII/FinalPlan/Community%20Clean%20Water%20Planning%20Guide.pdf (accessed 21February 2023).

Pennsylvania Department of Environmental Protection (2021). Pennsylvania's Community clean water implementation guide. Available at: https://files.dep.state.pa.us/Water/ChesapeakeBayOffice/WIPIII/2021/PA_Community_Clean_Water_Implementation_Guide-2021_Update.pdf (accessed February 21, 2023).

Pennsylvania Department of Environmental Protection DEP (2022a). Pennsylvania Phase 3 Chesapeake Bay watershed implementation plan final amended PA Phase 3 WIP. Available at: https://files.dep.state.pa.us/Water/ChesapeakeBayOffice/WIPIII/FinalPlan/FINAL_AMENDED_PA_PHASE_3_WIP.pdf (accessed 21 February 2023).

Pennsylvania Department of Environmental Protection (2022b). Final amended PA phase 3 WIP executive summary. Available at: https://files.dep.state.pa.us/Water/ChesapeakeBayOffice/WIPIII/FinalPlan/FINAL_AMENDED_PA_PHASE_3_WIP_EXECUTIVE_SUMMARY.pdf (accessed 21February 2023).

Pennsylvania Department of Environmental Protection (2022c). Appendix 2. Summary of local engagement. Available at: https://files.dep.state.pa.us/Water/ChesapeakeBayOffice/WIPIII/FinalPlan/Appendix%202%20Summary%20of%20Local%20Engagement.pdf (accessed 21 February 2023).

Pennsylvania Department of Environmental Protection (2022d). Public comment response document for the draft Phase 3 WIP. Available at: https://files.dep.state.pa.us/Water/ChesapeakeBayOffice/WIPIII/FinalPlan/Appendix%204%20Public%20Comment%20Response%20Document.pdf (accessed 21 February 2023).

Pennsylvania Department of Environmental Protection (n.d.). Pennsylvania's phase 3 watershed implementation plan: healthy waters, healthy communities. Available at: https://www.dep.pa.gov/Business/Water/Watershed-Restoration/Chesapeake-Bay-Watershed-Restoration/WIP3/Pages/Phase-III-WIP-(Watershed-Implementation-Plans).aspx (accessed 20 February 2023).

Pennsylvania Water Science Center (2022, February 15). Pennsylvania and the Chesapeake Bay watershed. *Pennsylvania and the Chesapeake Bay Watershed*. United States Geological Survey. Available at: https://www.usgs.gov/centers/pennsylvania-water-science-center/science/pennsylvania-and-chesapeake-bay-watershed.

Penn State University (n.d.). *Saving the Chesapeake Penn State University Extension*. Available at: https://www.psu.edu/impact/story/saving-the-chesapeake/ (accessed 21 February 2023).

Redvers, N., Faerron Guzman, C. A. and Parkes, M. W. (2023). Towards an educational praxis for planetary health: a call for transformative, inclusive, and integrative approaches for learning and relearning in the Anthropocene. *Lancet. Planetary Health* 7(1), e77–e85. https://doi.org/10.1016/S2542-5196(22)00332-1.

Reid, M. J., Hull, L. A., Alter, T. R., Adams, L. B., Kleinert, H. M. and Woolnough, A. P. (2021). New development: public sector responses to complex socio-ecological issues–no silver bullets for rabbits. *Public Money and Management* 41(4), 351–355, https://doi.org/10.1080/09540962.2019.1685168.

Rittel, H. W. J. and Webber, M. M. (1973). Dilemmas in a general theory of planning. *Policy Sciences* 4(2), 155–169.

Schmid, A. A. (2004). *Conflict and Cooperation: Institutional and Behavioral Economics*. Oxford: Blackwell Publishing, pp. 262–266.

Schon, D. (1984). *The Reflective Practitioner: How Professionals Think in Action*. New York: Basic Books.

Taleb, N. (2007). *Black Swan: the Impact of the Highly Improbable*. New York: Random House.

Taleb, N. (2012). *Antifragile: Things That Gain from Disorder*. New York: Random House.

USDA (2022). USDA announces initiative invests $22.5 million in water quality improvements in Chesapeake Bay. United States Department of Agriculture. Available at: https://www.fsa.usda.gov/news-room/news-releases/2022/usda-announces-initiative-invests-225-million-in-water-quality-improvements-in-chesapeake-bay (accessed 20 February 2023).

Victorian Gorse Taskforce (2020). *Victorian Gorse Taskforce Progress Report 2019/20*. Available at: https://vicgorsetaskforce.com.au/wp-content/uploads/2021/10/SDD200960-VGT-Progress-Report-2019-20-web.pdf (accessed 4 June 2023).

Victorian Serrated Tussock Working Party (2020). *Victorian Serrated Tussock Working Party Progress Report 2019/20*. Available at: http://serratedtussock.com/wp-content/uploads/files/VSTWP-Progress-Report-2019-2020.pdf (accessed 5 June 2023).

Walker, B., Carpenter, S. R., Anderies, J. M., Abel, N., Cumming, G., Janssen, M. A., Lebel, L., Norberg, J., Peterson, G. D. and Pritchard, R. (2002). Resilience management in social-ecological systems: a working hypothesis for a participatory approach. *Conservation Ecology* 6(1), 14. Available at: http://www.consecol.org/vol6/iss1/art14/.

Walker, B., Holling, C. S., Carpenter, S. R. and Kinzig, A. P. (2004). Resilience, adaptability and transformability in social–ecological systems. *Ecology and Society* 9(2), 5. Available at: http://www.ecologyandsociety.org/vol9/iss2/art5/.

Wall, D. (2017). Elinor Ostrom's rules for radicals:Cooperative alternatives beyond markets and states. London: Pluto Press.

Water4Ag (n.d.). *Water for Agriculture: Engagement: An Engaged Approach to Water and Agricultural Issues*. Available at: https://water4ag.psu.edu/about/approach-to-engagement/ (accessed 21 February 2023).

Wessells, M. G. (2018). *A Guide for Supporting Community-Led Child Protection Processes*. New York: Child Resilience Alliance.

Woolnough, A. P., Hull, L. J., Reid, M. J., Adams, L. B. and Kleinert, H. M. (2020). Community-led approaches can lead to better outcomes for management of European Rabbits *Oryctolagus cuniculus* and other invasive species. *The Victorian Naturalist* 137(6), 210–219.

Chapter 4

Implementing sustainable land use change programmes

Liz Lewis-Reddy, ADAS Policy and Economics, UK

1 Introduction

To meet the needs of a growing global population, reverse the decline in global biodiversity and adapt to the challenges of climate change, we are going to have to change the way we farm. One of the ways Governments across the UK have sought to address these challenges is through the creation of publicly funded agri-environment schemes (AES). Originally designed to have a singular focus on the protection and enhancement of on-farm biodiversity, modern iterations of these schemes have evolved to reflect the growing awareness of the need to promote a multifunctional landscape. This broader perspective allows for the recognition of the true value of these sustainably managed landscapes through the delivery of a diversity of outcomes; from food, to wildlife habitat, to carbon sequestration. This chapter explores how these schemes have evolved alongside the principles of sustainable land management (SLM), how they are shaped by the culture of the UK nations within which they develop, and how the role and importance of the landowner has changed from a passive recipient of instruction to an active participant in the design process. The chapter will also

http://dx.doi.org/10.19103/AS.2024.0134.18

explore a case study scheme whose ambition of economic viability achieved through ecologically resilient land management enables us to explore the challenges and opportunities associated with achieving SLM.

2 Sustainable land (SLM) management in a British context

To sustainably meet the needs of an estimated 9.1 billion people in 2050[1] (United Nations, 2017), global agricultural productivity (as measured by total factor productivity) would need to increase from the 2010–2018 average annual growth rate of 1.51% to a growth rate of 1.75% per annum (Steensland and Zeigler, 2021). Given limited natural resources and the continued degradation of resources already in use, increases in agricultural productivity will need to occur through gains in efficiency, intensification (e.g. increasing yield per hectare) or technological change. One example of such technological change is the widespread adoption of techniques associated with SLM.

The United Nations defines SLM as 'the use of land resources, including soils, water, animals and plants, for the production of goods to meet changing human needs, while simultaneously ensuring the long-term productive potential of these resources and the maintenance of their environmental functions'.[2] To deliver SLM, Sayer et al. (2012) proposed 'ten principles for a landscape approach to reconciling agriculture, conservation, and other competing land uses', a key principle of which is multifunctionality. The objective of multifunctionality can be achieved through the delivery of a range of outputs (environmental, economic, and social) from each hectare of land use, with each output delivered in such a way that it does not undermine the potential delivery of another. Agricultural production that delivers benefits for biodiversity while maintaining sustainable quantities of food produced without an over reliance on antimicrobial products (i.e. antibiotics and anthelmintics) would be one example of how a multifunctional approach could be applied to land management. The potential for multifunctionality underpins recent natural resource legislation enacted across the UK, with subtle variations reflecting the context (socioeconomic, environmental, and political) of each devolved nation.

In Northern Ireland (NI), a multifunctional approach is reflected in the launch of 'A Sustainable Agricultural Land Management Strategy for Northern Ireland'. In this example, there is a specific focus on working with the landowning community to enable behaviour change. Through better data and upskilling, the objective is to ensure that agricultural productivity is considered in parallel with environmental performance. Scotland has a slightly different focus as

1 https://www.unfpa.org/press/new-population-projections-underline-need-voluntary-family-planning-programmes.
2 https://www.fao.org/land-water/land/sustainable-land-management/en/.

illustrated through the various iterations of their 'Land Use Strategy'.[3] These strategies have a focus on transitioning to a low carbon economy through a greater understanding of the 'value' of Scotland's natural assets whilst enabling a more positive and constructive engagement between landowners and communities (as advocated through the Land Rights and Responsibility Statement (LRRS)).[4] In England, the 25 Year Environment Plan approach to SLM has a focus on improving the economic health of the agricultural sector through improvements in economic, ecological, and animal health resilience. The mechanisms that underpin this plan, such as the Sustainable Farming Incentive,[5] aim to reward farmers and land managers for the delivery of public goods.

The approach to implementing SLM in Wales embeds elements of all these approaches. This nation has considerable environmental assets (estimated to generate in excess of £8.8billion in goods and services annually (9% of GDP)) (National Trust, 2001), with most (88%)[6] managed under an agricultural system based in less favoured areas (LFA).[7] SLM has, understandably, been a focus of key legislation (Well-being of Future Generations Act, Environment (Wales) Act, and the Planning (Wales) Act). The Environment (Wales) Act, in particular, sets out nine principles or ways of working towards the objective of the sustainable management of natural resources. These principles include a focus on the following:

- Adaptive management: To ensure that the evidence gathered following the implementation of an action is reviewed and that any lessons learnt will be used to improve future actions and outputs;
- Spatial scale: Whereby any actions to repair and prevent serious damage to ecosystems are undertaken at an ecosystem scale;
- Collaboration and cooperation: Across and between all stakeholders within an ecosystem;
- Public participation and wider stakeholder engagement: To ensure buy-in to the concept of the value of the natural environment;
- Benefits and intrinsic value of ecosystems (species, habitats, and ecosystem services): These are identified and valued within decision making;
- Preventing damage to ecosystems: Through the recognition that the long-term sustainability of ecosystem service delivery relies on the restoration of healthy, diverse, and robust ecosystems; and

3 https://www.gov.scot/publications/scotlands-third-land-use-strategy-2021-2026-getting-best-land/.
4 https://www.gov.scot/publications/scottish-land-rights-responsibilities-statement/.
5 https://www.gov.uk/government/publications/sfi-handbook-for-the-sfi-2023-offer.
6 https://gov.wales/sites/default/files/publications/2019-06/agriculture-in-wales-evidence.pdf.
7 LFA refers to land located and included in the list of less favoured areas adopted by Article 2 of European Council Directive No.75/268EEC. In the UK, this includes land generally suitable for extensive livestock production and for the growing of crops for livestock feed, but agricultural production is restricted by soil, relief, aspect or climate conditions.

- Taking account of the short-, medium-, and long-term consequences of actions (including farming practices) to support the 'resilience of ecosystems'.

These principles underpin the development of Wales' post-Brexit agricultural support system, the Sustainable Farming Scheme. The same can be said for each of the aforementioned policies targeted at the delivery of SLM across the UK. Providing public funds to deliver land management focussed policy priorities is not a new approach. It is the principle that public funds could (and should) be used to pay for the environmental outcomes which arise from changes to farming practices that represent a shift in focus. Seventy years ago, with food rationing looming large in the collective British consciousness, the primacy of food production resulted in large swathes of land being given over to intensive single-use livestock or crop production. This approach became enshrined in various domestic (upland drainage policies) and EU policies (early version of the Common Agricultural Policy (CAP)), targeted at stimulating primary production. Over time, the narrow focus of these mechanisms and the negative impact they had on the natural environment, began to shift public opinion. Pressure began to mount on the UK Government to find a way for public funds to expand beyond subsidising food production and towards stewardship (Buckwell et al., 1997; Carey et al., 2002; Walker et al., 2004; Wynn, 2002). It was this change in approach that led to the development of the UK AES (Carey et al., 2002).

3 Evolution of the concept of SLM: from from landscape protection to ecosystem service delivery

In 1986, Countryside Stewardship initiatives, or AES, in the UK were formalised in law via the Agriculture Act (Buckwell et al., 1997). This led to the development of the Environmental Stewardship Act (ESA) (Carey et al., 2002; Coates, 1997) in 1987. The premise of the ESA was to 'protect the landscape, wildlife, and historic interest of specific areas within the UK that are of national environmental significance' (DEFRA, 2004).

In 1991, the ESA was expanded to incorporate the wider countryside through the creation of the Countryside Stewardship Scheme (CSS) (Carey et al., 2002; Coates, 1997). The CSS was taken over by the Ministry of Agriculture, Fisheries and Food (MAFF) in 1996, and had its remit expanded further in 2001 when MAFF became the Department for Environment, Fisheries and Rural Affairs (DEFRA). The multiple objectives of the new scheme came to include protection of all features of the rural landscape (DEFRA, 2004). These included biodiversity, wildlife habitats, archaeological sites, and historic features. Such changes in remit were enabled through reforms of the CAP (MacSharry Reform,

1992; Agenda 2000 reform; Luxembourg and Fischler Reform, 2003) into two pillars of payment. Pillar I, or *direct payments,* retained much of the original focus on public subsidy (i.e. were made on eligible land, conditional on land owners maintaining their land under good agricultural and environmental condition). Pillar II payments had a broader remit and were targeted at the *rural development* objectives of the individual members states and included novel agri-environmental payments and investment support.

This broadening of the remit reflected growing recognition (and an increased volume of evidence) of the importance of working beyond the boundaries of individual farms when attempting to address landscape-scale objectives. The Environment Stewardship Scheme in England 'at scale' introduced financial incentives or 'bonuses' to support coordinated action across landscapes (e.g. measures to improve water quality in a catchment such as fencing off watercourses). By affecting change at a landscape scale, the delivery of ecological outcomes became important to not only the public sector but also the private sector. With farmers working together to implement land management practices that benefited water quality, flood water management and beyond, those industries for whom these outcomes were a fundamental part of their business model began to take interest. Private sector initiatives often fell under the banner of Payment for Ecosystem Service (PES) schemes. PES are voluntary agreements between at least one 'seller' and one 'buyer' conditional on the supply of a well-defined ecosystem service, or a land use presumed to produce that service (Wunder, 2005).

Rather than being based on compensating landowners for enhancing the ecological value of farmland, these business-to-business schemes were based on the principle of paying for a defined environmental deliverable. These novel mechanisms represented a step change away from the early AES of the CAP. The environmental deliverable associated with a land management intervention was no longer a by-product of 'not farming properly'. The environmental outcome was now a primary commodity. Under a PES scheme, the economic resilience of the farm business was now linked to ecologically SLM.

4 Challenges and lessons from implementing SLM programmes

As these novel mechanisms became more established, they became a testing ground for more commercial approaches to addressing key challenges associated with publicly funded schemes (e.g. demonstration of value for money). One such approach was to set the payment rates for capital works and the delivery of environmental outcomes via 'Reverse Auctions'. Through this approach, farmers opt into a bidding process, offering to manage their land under prescribed management options at a price that more closely reflects the actual cost of doing so. To determine whether these approaches resulted in a more cost-efficient investment, Elliott et al. (2015) considered aspects of auction design and farmer responses to auctions through a combination of laboratory experiments, simulation modelling, and farmer workshops. The analysis demonstrated that over time, although the environmental outcomes of the schemes did not diminish, collusion between participants impacted the price at which interventions were set. This weakness in the reverse auction approach highlighted an opportunity for future scheme implementation; even within commercial environments, landowners can and will work together to achieve the best outcome for their collective businesses. The emphasis on the cost-effectiveness of schemes, however, remains a consideration for investors (public and private). For the public sector, demonstrating value for money is essential when associated with a limited pot of funding. For private sector investors who are participating in, or developing, schemes that are designed to achieve specific (and reportable) objectives, demonstrating delivery is key. The issue of 'deadweight' for both investors is therefore a significant problem.

Deadweight in an AES is defined as 'outcomes which would have occurred without intervention' (HM Treasury, 2003). Deadweight is important when estimating the net impact of AES on environmental delivery and wider behaviour change. However, the perception that it is a significant feature of publicly funded schemes may be overstated. An evaluation of an earlier Welsh AES, Tir Gofal (Agra Ceas, 2006), found a very small percentage (e.g. 5%) of deadweight within the scheme. Only 5% of those enrolled in the scheme would have undertaken the activity irrespective of funding. However, even for these 5%, the additional funding significantly increased the amount of work undertaken (i.e. less work would have been done and without support it may have been deferred to the future). A similar outcome was observed from an evaluation of an early Higher Level Stewardship (HLS) scheme in England. CCRI (2010) determined that in the absence of HLS scheme payments, 79% of the environmentally focussed management work would not have been undertaken, equating to only 21% deadweight.

Modelling work (Uthes et al., 2010; Kuhn et al., 2019) has suggested that the low levels of deadweight observed may reflect the environmental context within which participating farms are operating. Farms participating in AES were more likely to be able to make money from participation than other agricultural businesses. This could be because the management practices already underway on farm were similar to those dictated by the AES. The outcome of this would be on-farm costs which were well below the scheme payment rate (Uthes et al., 2010). In essence, farms which select interventions that match their environmental context and capability do not represent deadweight but are instead illustrating the opportunity (economic and environmental) associated with appropriate targeting of land management interventions.

Post Brexit, the agricultural support framework across the UK has shifted away from the two pillar segregation of agricultural production from environmental deliverables. The new approaches, as outlined earlier in this chapter, are targeted at supporting SLM, and have ambitious objectives, including the delivery of sustainable food production; to provide a response to the climate emergency; and to halt, and ultimately reverse, the loss of biodiversity. These objectives will require working at scale across many land holdings. Successful implementation of SLM across landscapes will require high levels of uptake and collaboration from across the landowning community. The main factors that will impact uptake are linked to the three overarching pillars of sustainability: the economics of the farm business, the environmental constraints and opportunities within which these businesses operate, and the capacity and capability of the farmer, and the wider landowning community.

4.1 Economic context

Agricultural production is subject to a wide variety of external variables (e.g. price fluctuations and climate stresses). Diversification (agricultural and non-agricultural) of the farming business to include a variety of income streams has often been promoted as a means by which can farms can reduce their reliance on, at times, an unreliable source of income. Previous AES have sometimes been viewed as one such means of diversification but research on the economic outcomes of AES (the best proxy for understanding the potential profitability of some aspects of SLM) is mixed. Sauer et al. (2012) noted that farms participating in voluntary AES tended to become less specialised and more diversified with respect to their production structure. However, the 'costs plus income foregone' model for the setting of AES payment rates is predicated on the presumption that participating farmers will not profit from the implementation of AES measures. This objective is borne out by the fact that there is a general consensus within the farming community that implementing measures to benefit the environment or biodiversity incurs a direct economic cost through a reduction in primary production capability. For the implementation of sustainable management schemes, this perspective will have to be overcome with robust economic evidence. For example, Harkness et al. (2021) determined that for some farming systems (e.g. dairy, general cropping, and mixed farms), participation in AES (as an indication of system and business flexibility) can have a demonstrably positive impact upon long term financial stability. The authors attributed the increased stability not only to the value of the payment received but also the potential of these farming practices to increase natural processes on farm alongside an associated reductions in input cost (Harkness et al., 2021).

By contrast, Udagawa et al. (2014) observed no evidence that entry into AES (Entry Level Stewardship (ELS)) in England significantly increased levels of farm income and found that there was a modest level of cost for participation, at least initially. The authors suggested that earlier studies which suggested otherwise (Harrison and Jones, 2010) were too focussed on a subset of farms and therefore not representative of the wider sector. A causal relationship between AES payment level and financial stability was also disputed by Zhu and Oude Lansink (2012). The authors identified an inverse relationship between the share of total subsidies in total farm revenues and business efficiency, termed an 'income and insurance effect'. They demonstrated that farm business models which obtain the majority of their income from subsidies or public programmes have less motivation to improve efficiency.

It is important, however, to understand what is driving this variation in response to subsidy across farm types and environmental contexts. The outcome of these studies suggests that a degree of self-selection is taking place whereby the profitability of participation in AES is not solely a reflection of payment rates but also of the environmental attributes of the participating farms.

4.2 Environmental constraints and opportunities

By contrast with the findings from dairy, general cropping and mixed farms, Harkness et al. (2021) found that AES payments on LFA farms reduced the stability of income, despite these farms receiving more money from agri-environment schemes per hectare, on average, than other farm types. This counterintuitive outcome was linked to the fact that LFA farms are based in agriculturally challenging environments. As such, there would be little room for improvements in productivity through the implementation of sustainable management. Soil management-focussed AES options may also generate improvements in yield whereas the same options applied to upland soils are unlikely to yield a similar productivity enhancement. Upland areas are, however, capable of delivering significant environmental outcomes (e.g. water quality and quantity management, and carbon sequestration). The outcome of the Harkness et al. (2021) study could be a reflection of the misalignment between outcomes and payment rates. In essence, by only valuing the outcomes related to improvements in soil structure the additional environmental delivery achieved by these LFA farms was not being accounted for in the payment rate.

By contrast with LFA farmers, Harrison and Jones (2010) found that farms with an abundance of features (e.g. hedgerows, small field parcels) were able to profit from participation in the English entry-level scheme (ELS) for Countryside Stewardship through careful selection of measures. By selecting, intentionally or sub-consciously, options which would have limited income-foregone impacts, these farmers were able to maximise profitability. Farm businesses based in more featureless landscapes had to instead resort to more costly interventions which limited the overall profitability of each measure. It is worth noting that this profitability was viewed solely from the perspective of the payment rate and did not take account of any increase in natural processes arising from the intervention (as highlighted in Harkness et al., 2021). It could be argued that (as per the previous LFA farm examples) overall scheme design was flawed. If payment rate had been targeted at environmental output rather than capital delivery, the environmental output on both farm types would have been appropriately remunerated. The feature-rich farms would not have had to incur significant capital costs as they were already delivering value (albeit at lower rate of change), whereas increasing feature richness on the feature poor farms would have resulted in significant additional environmental value. This challenge of only 'paying for change' links back to the concerns over 'deadweight'. The aforementioned evaluation work undertaken by Agra Ceas (2006) and CCRI (2010), however, has demonstrated that deadweight is more perception than reality. There is always more than can be achieved, especially when landowners collaborate.

Lobley and Butler (2010), Morris et al. (2017) and Uthes et al. (2010) have flagged that multifunctional objectives also require multifunctional policy frameworks. Changes to agricultural support mechanisms to focus on SLM will need to be prefaced by in-depth policy analysis (and the delivery of a range of targeted interventions to increase capacity and capability of the landowning community) if contradictory effects are to avoided. Otherwise less agriculturally productive regions with their deminished financial capital, may struggle to respond to policy and be left further behind (Severini and Tantari, 2013).

4.3 Human capital

Work undertaken by Wynne-Jones (2013) highlighted that most farmers see their primary role as being producers of food, and few understand the concept of being producers of ecosystem services or other environmental outcomes. Those who participate in agri-environment schemes generally do so as a way to support an approach to farming that makes less demand of their time and resources, or as a risk management solution to smooth incomes during uncertain economic times (Ingram et al., 2013). This 'business as usual' option selection often reflects a farmer's understanding of what works best for their business but not necessarily an understanding of what the scheme is trying to achieve (Turner et al., 2013) and how it would benefit their business in the long term.

5 Building capacity and capability within the farm business

The ability of the farming community to respond to change is not simply a reflection of the current current farmer but is also dependent on the next generation. The presence of a successor has long been identified as a key driver of innovation and expansion of farm businesses (Barnes et al., 2014; Chiswell, 2014; Potter and Lobley, 1992). Farms with an identified successor and an established plan for the handover of responsibility tend to be motivated, as well as increasingly disposed to adaptation, investment, and expansion (Chiswell, 2014). This greater receptiveness to innovation was highlighted by Vecchio et al. (2022) in their work on the adoption of precision agriculture (PA) by Italian farmers. The authors identified three groups of farmers: adopters, non-adopters, and planners. Adopters were typically young, university educated farmers who frequently informed themselves on new innovations by reading relevant literature.

Busse et al. (2014) and Läpple and Hennessy et al. (2015) also highlighted the importance of intergenerational factors influencing adoption. Both publications identified that younger, highly qualified farmers, who had been given management responsibility were more likely to be open minded to innovation. Failing to embed collaboration at all stages of innovation and scheme design can result in missed opportunities to engage farmers who could be classed as 'early adopters'. Work undertaken by Läpple and Hennessy (2015), Toma et al. (2018), and Farrell et al. (2022) identified that public policy interventions which promoted structural change and earlier inter-generational transfer may facilitate adoption. In their study on the Irish organic farming sector as part of the Irish EIP-Agri funding programme, Läpple et al. (2015) found that targeted policy interventions that attract the next generation back to farming have the potential to enhance innovation acceptance (embracing novel opportunities).

The diversity of perspectives within a farm business in the wider farming community will consequently require a diverse range of mechanisms to support change. Knowledge exchange service providers associated with the agricultural community have responded to this by not only providing a range of different methods for promoting knowledge exchange activities, but also bundling methods (e.g. small group learning and coaching) to best suit different types of farmers and or different types of problems. Research undertaken by Nettle et al. (2022) reported that the best methods (based on extent, reach, and time to change) were small group learning and one-to-one advice or coaching. They noted that a focus on farmer needs, and the 'journey of change' were key to having a greater impact. Supporting farmers on this journey of change towards more SLM will also require improved links between the farming community and the development of the evidence base for change. As these developments often take place through institutions that seem remote from on-farm practicalities, this can represent a challenge to gaining uptake across the farming sector.

6 Building capacity and capability around the farm business

In their paper exploring the on-farm experimentation (OFE) process, Lacoste et al. (2022) reiterated the importance of restructuring the farmer-researcher relationship. The OFE process refers to a collaborative experimental research process that brings agricultural stakeholders together to develop and implement mutually beneficial experiments. OFE is based on three principles: (1) experimentation occurs on-farm; (2) interests and insight from all participants are explicitly acknowledged at the outset, and (3) experimentation is a deliberative, iterative process of joint exploration. This helps to build trust not only between the stakeholders taking part in the OFE but in the outcome of the research.

The importance of building trust is reinforced by work undertaken by Ketterings (2014). Their research suggested that a successful knowledge transfer programme is one where the actual message might not be as important as the trust-based partnership and understanding of farmer reality that leads to asking the right questions. Based on the nutrient management experiences of a dairy farm in New York State, Ketterings (2014) listed the key ingredients for farm level adoption as: (1) understanding the concerns and recognition that change is necessary among all involved; (2) identification of win-win situations first; (3) posing of relevant questions, and generation of believable results based on reliable data (replicated trials where feasible); (4) farmer involvement and accountability in the process (farmers as drivers of the adaptive management process), and (5) development and maintenance of a trust-based relationship between farmer, farm advisor, and university researcher. Ketterings (2014) identified that an innovative, outcome-focussed adaptive management approach that includes research, extension, and a focus on human dimensions (a people-based approach) was the most effective way to obtain long term on-farm change.

Their analysis of dairy farm experiences included a review of three different adaptive management approaches, implemented at field, whole farm, and regional/state levels. In the study, the participant farmers, in collaboration with the research institutions (Cornell University), played a fundamental role in the development, design, and implementation of on-farm trials. In addition, the outcome of the trials was evaluated through the production of summary reports and discussed at crop management team meetings. This enabled context-specific conclusions and impact to be reported and acted upon. The lessons learned from these on-farm research partnerships were then explored with farmers across the wider catchment, the outcome of which was significant improvements in field, farm, and regional/state level balances for nitrogen and phosphorus in New York State.

Research undertaken in Germany by Busse et al. (2014), which focussed on the German Agricultural Innovation System, reinforced the importance of the involvement of farmers at the design stage of innovation. This study illustrated that the role of farmers in generating innovations is often not recognised by researchers. The outcome of this lack of recognition is the inability of the farmers to function not only as adopters but also co-designers or co-developers. A review of 11 case studies from the international RETHINK research programme by Šūmane et al. (2018) identified the importance of recognising all stakeholders as equal co-authors of knowledge generation and facilitators of knowledge exchange. They also recognised that national level ambitions are best delivered by supporting local-level initiatives; including those linked to regulation.

This latter point was demonstrated two case studies, one involving a group of organic Austrian farmers, and another a group of smallholders in Latvia (Šūmane et al., 2018). The former regarded inflexibility around dates for spreading manure and slurry as not enabling farmers to take into consideration regional weather. The latter was constrained by national regulations related to the administration of vaccinations. In both cases, by not being part of the decision-making process, the farmers felt that their expertise and experience-based knowledge and skills were being ignored. The impact of their exclusion ultimately undermined the sustainability of their agricultural practices and their ability to drive innovation within their own systems. The paper by Nettle et al. (2022) previously referred to, outlined the importance of locally designed innovations, developed through trusted networks of farmers, scientists, and commercial firms, with specific reference to the Australian dairy sector. The flexibility to iterate and adapt at a local level proved to be the most successful strategy for increasing uptake across the sector.

The process of a farmer adopting a new technology or practice has been conceptualised as a 'dynamic learning process', which is influenced not only by the characteristics of the farmer (Hill, 2007) and the innovation, but also by the broader social environment (Prager et al., 2017). A farmer's social and information networks are important in becoming aware of a problem, evaluating known information, and shaping the decision whether (or not) to adopt. It is influenced by farmer-to-farmer interaction, and exchange of information between peers and with specialists (Prager et al., 2017). Whether production or sustainability focussed, the research exploring the influence of scheme design on farmers' adoption of new practices or technologies/innovations identifies similar influencing factors. These include the role of the farmer in idea development and implementation and the establishment of supportive structures (policy and institutions, including centres of excellence) which facilitate collaboration and innovation development at a local level (Läpple et al., 2015).

6.1 Role of Institutions

Reflecting on the implementation and institutionalisation of Innovation Platforms and climate smart agriculture in sub-Saharan Africa, Schut et al. (2016) and Makate (2019) demonstrate how adaptation of these interventions from their 'old' institutional way of working, towards approaches where stakeholders have collective agency, may increase impact. Schut et al. (2016) identified that the efficacy (and impact) of the Innovation Platform is as much a result of the quality of the multi-stakeholder process (jointly identifying constraints, sharing information, and learning) as it is of the specific innovation it develops.

Schut et al. (2016), however, recognised that many Innovation Platforms are structured along disciplinary lines, have a long history of 'top down' technology development and transfer, and have limited expertise with regards to institutional innovation, facilitation of interactive multi-stakeholder processes or how to address structural power inequalities between actors in the supply chain. Effective dissemination (adoption, use, uptake or commercialisation of existing knowledge) of alternative approaches to land management requires an understanding of local farming systems and contexts, which can only be achieved through strong linkages and active participation of a range of actors. Makate (2019) reinforced this observation by highlighting the value of resource pooling (capital, knowledge, technology, land, etc.) and that interventions work better when multiple stakeholders are harnessed as a resource and brought to the table.

In the Welsh context, the development of robust institutions supporting Innovation Platforms is essential to supporting change. Statistics gathered by Defra in 2015[8] highlight that the agricultural workforce in Wales comprises approximately 2% of the Welsh population. Despite their responsibility for the management of over 70% of the Welsh land area, this workforce is not required to achieve any educational training beyond secondary education. Only 20% of the workforce has a qualification which equates to level 4 (equivalent to a Foundation Degree or Higher National Diploma) or above, compared to 40% in other sectors of the economy (Jones, 2015). It could be argued that this lack of education explains the reliance of the Welsh agricultural sector on anecdotal approaches to farming and ecological awareness.

8 https://www.gov.uk/government/statistical-data-sets/structure-of-the-agricultural-industry-in-england-and-the-uk-at-june

The above insight from the literature reflects the importance of considering economic, environmental and social capital when attempting to implement change at all levels, from farm business to sector wide. Successful implementation of SLM will therefore require:

- appropriate targeting of locally relevant multifunctional options, rather than development of broad 'one size fits all' approaches, to ensure the best outcomes are achieved for both the farm business, the investor, and wider society;
- close collaboration and communication between policy makers, researchers, landowners advisors and the wider community at all stages of scheme development and implementation to ensure higher levels of uptake, local community buy-in and better biodiversity outcomes at all scales;
- implementation of verifiable outcome-based performance assessments tailored to the environmental context of each farm business to embed flexibility and adaptive management, whilst giving landowners the confidence and expertise to engage; and
- financial remuneration which fairly values of the output and cost of implementation.

7 Case study: Pumlumon Project

The Pumlumon Project (PP) is a place-based project named after the Pumlumon Mountain (Fig. 1). It covers a watershed area of 40000 ha, contains the highest part of the Cambrian Mountain range and straddles the counties of Powys

Figure 1 Landscape in the Pumlumon project area.

and Ceredigion. The whole project area is home to 15000 people, spread across 11 local communities. There are 250 farms in the project area with farming, forestry, and tourism the main economic activities. It is also the largest watershed in Wales with the reservoirs and streams that drain into the Wye, Severn, Rheidol, Dyfi and Leri river catchments supplying water to four million people in England. Encompassing over 9000 ha of key habitats including river valleys, semi-natural woodland, species-rich grassland, heather moorland, and blanket bog, at the heart of the project area sits the 5000 ha Pumlumon Site of Special Scientific Interest (SSSI), currently in unfavourable ecological condition and declining.[9]

For over 10 years, the Montgomeryshire Wildlife Trust (MWT) has been working with local communities, land managers, statutory agencies, and both local and national businesses, to restore and enhance the resilience of the ecosystem within the project area; piloting an integrated approach whereby the ecosystem services (i.e. water quality, flood risk reduction, carbon safeguarding) can be better delivered via the mechanism of SLM.

7.1 Economic

Early on in project development, MWT recognised the importance of exploring novel and innovative economic models that could help support the long-term sustainability of the rural economy. The degree of scientific uncertainty around verifiable delivery of ecosystem services represented an early challenge to secure investment. However, this was overcome in the short term by working closely with the research community to develop robust methods for quantifying change. The project team also undertook extensive research with the local and national business community to better align the supply of ecosystems services from the project area with private sector demand.

In 2014, substantial capital funding was provided from the National Lottery for a major ecotourism infrastructure build within the project area, the Dyfi Osprey Project. This development had a significant impact on the local community (valued at £612500/year)[10] and highlighted the economic opportunities associated with nature-based tourism.

7.2 Environmental

In collaboration with the landowning community, the project team undertook a thorough review of environmental deliverables from the project area (i.e. what can, or could be, produced in the Pumlumon project area and could contribute

9 http://wyndrushwild.co.uk/wp-content/uploads/2012/11/MSE_Survey-of-Summit-Vegetation-on-Pumlumon-SSSI-contract-science-report.pdf
10 Dyfi Osprey Project Economic Impact Summary (2010–2013)

to the area's future income and sustainability (carbon sequestered, sustainable food, water quality, and quantity management). This included a spatial mapping element and quantitative element. The quantitative element, where reasonable, used economic methods to estimate potential monetary value associated with each deliverable.

The mapping of key environmental deliverables enabled a detailed assessment of the spatial variation in the supply. This 'opportunity mapping' (Fig. 2) also highlighted where interventions could be most appropriately targeted to deliver maximum outcomes.

7.3 Social

From the project's inception through to implementation, a farmer/landowner working group was established to co-produce each of the project. Early opportunities were explored with 'early adopters' from the group who became project champions within the wider community. The trusted relationships that developed because of this approach continue today and have enabled the exploration of new ideas from PES to more locally based collaborative economies.

Since its inception in 2008, the PP continues to grow. The farmer/landowner working group remains active and continues to collaborate on

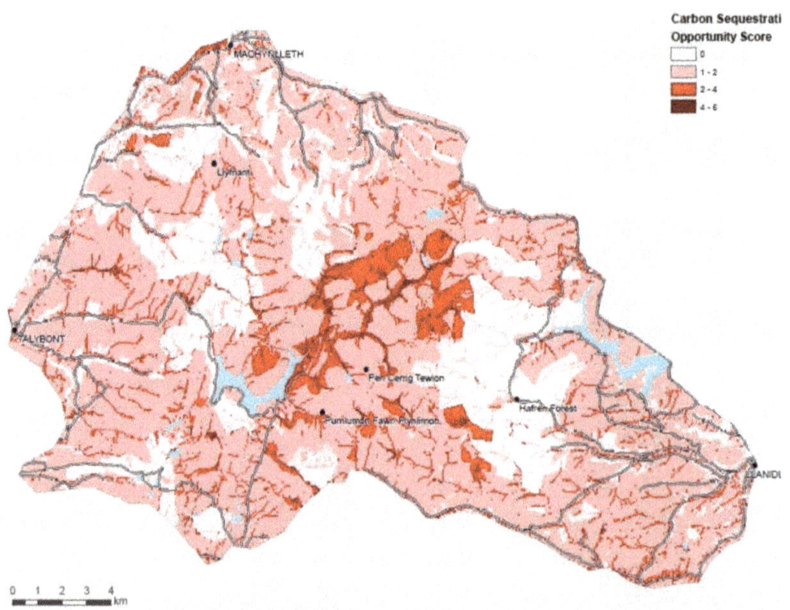

Figure 2 Opportunity mapping of peatland restoration opportunities in the Pumlumon project area. Darker areas represent highest restoration priority.

emerging opportunities with MWT (Fig. 3). A number of the lessons learnt via the project (e.g. the importance of collaboration and co-production with local communities, the ability to embed private sector investment mechanisms into farm business models and, most importantly, the capacity to produce a range of viable ecosystems services through the application of SLM (Fig. 4a and b)) helped to underpin the SLM objectives of the current Welsh legislative framework.

8 What the future could look like

SLM models such as those explored in the Pumlumon Project demonstrate that a future where the decline in UK biodiversity is halted, recovery is well underway, and the direct links between the health of the natural environment and the health and well-being of society are valued and understood are possible (Fig. 5). To achieve this, the UK requires an economically viable landowning community working with the environment to ensure that it can provide both for the needs of the natural environment and society.

The post-Second World War decoupling of ecosystem processes from agricultural practice damaged the natural environment and degraded its ability to deliver ecosystem services (e.g. water quality and quantity). Valuing agricultural fields solely by crop yield meant that the collateral damage to the wider environment became irrelevant. Value should instead incorporate the multiple societal benefits accrued from SLM (e.g. clean water, reduced flood

Figure 3 Pumlumon Project knowledge transfer event with the International Federation of Agricultural Journalists and Landowner working group.

Figure 4 (a) Degraded peat bog before restoration of water table. (b) Degraded peat bog after restoration of water table.

Figure 5 Restored blanket bog within the Pumlumon project area. The image is reproduced with the permission of the Montgomeryshire Wildlife Trust.

risk, reduced greenhouse gas emissions) as well as the true 'costs' associated with production. If SLM is not pursued, unsustainable practices which negatively impact upon the delivery of the associated benefits will ultimately reduce the overall value of what is being produced, shrinking the profit margin and resulting in economic instability. It is damaging to silo one area to production and another to environmental deliverables.

Deploying a single scale of merit (e.g. number of plant species within a given area) should also be avoided. Biodiversity across a range of farm types and elevations (e.g. lowlands and uplands) is different, but equally valuable. The implementation of SLM needs to avoid being reductionist, and to exercise caution in the use of proxies for simple indices of quality or change. Effort needs to be expended in understanding local environments and tailoring solutions and monitoring programmes. The on-going reduction of expert input, poor monitoring, lack of standardised and accepted metrics, delivery of one-size-fits-all solutions, and single outcome approaches have all contributed to poor outcomes. The objectives of the integrated natural resource policies proposed across the UK suggest that the recovery of nature and production of food is compatible, and multiple outcomes are possible. To be successful in the long term, however, they will need to embed the following key principles:

- Create multiple outcomes – ecosystem services produced everywhere and anywhere can make a contribution to biodiversity;
- Use societal funds for societal good – public funds should be used to provide equitable access to health and well-being benefits, reductions in flood risk, improvements in water and air quality, and safeguarding of vulnerable habitats and species;
- Foster greater public understanding of the value of the natural environment and role of consumer choice; and
- Support institutions capable of upskilling and professionalisation of those who work with the natural environment.

If these principles are adhered to, the recovery of biodiversity will be secured and the economically and ecologically sustainable management of the UK environment into the future will be assured.

9 Conclusion and future trends

Significant work is underway across the devolved nations of the UK, and the wider world, to better quantify and value the ecosystem services arising from SLM. Further research is required to ensure these metrics are robust enough to give the private sector the confidence to invest, sensitive enough to detect change (even over short timescales) to warrant public investment, and resilient enough to help guide decision making into the future. ADAS is working closely with the devolved UK nations on each of the SLM programmes. We are helping them to develop structures and tools (earth observation, artificial intelligence, natural capital accounting and governance protocols) to help address these needs, but there is always more work to be done.

10 Where to look for further information

UK Sustainable Land Management Policy documents can be found here:

Northern Ireland: SLM strategy

- https://www.daera-ni.gov.uk/sites/default/files/publications/daera/16.17.079%20Sustainable%20Land%20Management%20Strategy%20final%20amended.PDF.

Scotland: Land Use Strategy

- https://www.gov.scot/publications/scotlands-third-land-use-strategy-2021-2026-getting-best-land/.

Scotland: Land Rights and Responsibilities

- https://www.gov.scot/publications/scottish-land-rights-responsibilities-statement/.

England: Environmental Improvement Plan

- https://www.gov.uk/government/publications/environmental-improvement-plan.

Wales: Sustainable Management of Natural Resources

- https://naturalresources.wales/media/678063/introducing-smnr-booklet-english-final.pdf.

11 References

Agra CEAS Consulting (2006). Socio-economic Evaluation of TIR GOFAL. Final Report for Countryside Council for Wales, Fforwm Tirlun and the Welsh Development Agency, p. 103.

Barnes, A., Toma, L., Mathews, K., Sutherland, L. and Thomson, S. (2014). Intensify, diversify, opt-out: testing farmer stated intentions to past and future CAP reform scenarios. Contributed Paper prepared for presentation at the 88th Annual Conference of the Agricultural Economics Society, AgroParisTech, Paris, France.

Buckwell, A., Blom, J., Commins, P., Hervieu, B., Hoferither, M., Meyer, H., von Rabinowicz, E., Sotte, E. and Sumpsi Vina, J. (1997). *Towards a Common Agricultural and Rural Policy for Europe.* European Commission, Brussels.

Busse, M., Doernberg, A., Siebert, R., Kuntosch, A., Schwerdtner, W., König, B. and Bokelmann, W. (2014). Innovation mechanisms in German precision farming. *Precision Agriculture* 15(4), 403–426. https://doi.org/10.1007/s11119-013-9337-2.

Carey, P. D., Barnett, C. L., Greenslade, P. D., Hulmes, S., Garbutt, R. A., Warman, E. A., Myhill, D., Scott, R. J., Smart, S. M., Manchester, S. J., Robinson, J., Walker, K. J., Howard, D. C. and Firbank, L. G. (2002). A comparison of the ecological quality of land between an English agri-environment scheme and the countryside as a whole. *Biological Conservation* 108(2), 183–197.

Chiswell, H. M. (2014). The importance of next generation farmers: a conceptual framework to bring the potential successor into focus. *Geography Compass* 8(5), 300–312.

Coates, D. (1997). UK policy for the ESAs. Grassland management in the 'environmentally sensitive areas'. In: Occasional Symposium No. 32, British Grassland Society, Lancaster.

DEFRA (2004, 26/08/2006). History.. Available at: http://www.defra.gov.uk/erdp/docs/css_esas_report/history.htm.

Elliott, J., Day, B., Jones, G., Binner, A. R., Smith, G., Skirvin, D., Boatman, N. D. and Tweedie, F. (2015). Scoping the strengths and weaknesses of different auction and PES mechanisms for Countryside Stewardship. Defra Project LM0105. Final report.

Farrell, M., Murtagh, A., Weir, L., Conway, S. F., McDonagh, J. and Mahon, M. (2022). Irish organics, innovation and farm collaboration: a pathway to farm viability and generational renewal. *Sustainability* 14(1). https://doi.org/10.3390/su14010093.

Harkness, C., Francisco, A., Semenov, M., Senapati, N., Shield, I. and Bishop, J. (2021). Stability of farm income: the role of agricultural diversity and agri-environment scheme payments. *Agricultural Systems* 187, 1-13.

Harrison, G. and Jones, J. (2010). A farm level assessment of the profitability of Entry Level Scheme participation in the Lincolnshire Wolds. 84th Annual Conference, Agricultural Economics Society, March 29–31.

Hill, B. (2007). *Business Competence*. Department of the Environment, Food and Rural Affairs.

Ingram, J., Gaskell, P., Mills, J. and Short, C. (2013). Incorporating agri-environment schemes into farm development pathways: a temporal analysis of farmer motivations. *Land Use Policy* 31, 267–279.

Jones, W. (2015). Independent Review of Learning delivered by FE colleges and the relevance of that delivery supporting Farm businesses in Wales.

Ketterings, Q. M. (2014). Extension and knowledge transfer: adaptive management approaches for timely impact. *The Journal of Agricultural Science* 152(S1), 57–64. https://doi.org/10.1017/S002185961300066X.

Kuhn, T., Schäfer, D., Holm-Müller, K. and Britz, W. (2019). On-farm compliance costs with the EU-nitrates directive: a modelling approach for specialized livestock production in northwest Germany. *Agricultural Systems* 173, 233–243.

Lacoste, M., Cook, S., McNee, M., Gale, D., Ingram, J., Bellon-Maurel, V., MacMillan, T., Sylvester-Bradley, R., Kindred, D., Bramley, R., Tremblay, N., Longchamps, L., Thompson, L., Ruiz, J., García, F. O., Maxwell, B., Griffin, T., Oberthür, T., Huyghe, C., Zhang, W., McNamara, J. and Hall, A. (2022). On-farm experimentation to transform global agriculture. *Nature Food* 3(1), 11–18. https://doi.org/10.1038/s43016-021-00424-4.

Läpple, D. and Hennessy, T. (2015). Exploring the role of incentives in agricultural extension programs. *Applied Economic Perspectives and Policy* 37(3), 403–417. https://doi.org/10.1093/aepp/ppu037.

Läpple, D., Renwick, A. and Thorne, F. (2015). Measuring and understanding the drivers of agricultural innovation: evidence from Ireland. *Food Policy* 51, 1–8. https://doi.org/10.1016/J.FOODPOL.2014.11.003.

Lobley, M. and Butler, A. (2010). The impact of CAP reform on farmers' plans for the future: some evidence from South West England. *Food Policy* 35(4), 341–348.

Makate, C. (2019). Effective scaling of climate smart agriculture innovations in African smallholder agriculture: a review of approaches, policy and institutional strategy needs. *Environmental Science and Policy* 96, 37–51. https://doi.org/10.1016/J.ENVSCI.2019.01.014.

Mills, J., Courtney, P., Gaskell, P., Reed, M. and Ingram, J. (2010). Estimating the incidental socio-economic benefits of Environmental Stewardship Schemes. Final Report by CCRI.

Morris, W., Henley, A. and Dowell, D. (2017). Farm diversification, entrepreneurship and technology adoption: analysis of upland farmers in Wales. *Journal of Rural Studies* 53, 132–143.

National Trust (2001). Valuing our environment: economic impact of the environment of Wales. *The Valuing Our Environment Partnership*, July 2001.

Nettle, R., Major, J., Turner, L. and Harris, J. (2022). Selecting methods of agricultural extension to support diverse adoption pathways: a review and case studies. *Animal Production Science*. https://doi.org/10.1071/AN22329.

Potter, C. and Lobley, M. (1992). Aging and succession on family farms: the impact on decision making and land use. *Sociologia Ruralis* 32(2/3), 317–331.

Prager, K., Creaney, R. and Lorenzo-Arribas, A. (2017). Criteria for a system level evaluation of farm advisory services. *Land Use Policy* 61, 86–98. https://doi.org/10.1016/J.LANDUSEPOL.2016.11.003.

Sauer, J., Walsh, J. and Zilberman, D. (2012). Producer behaviour and agri-environmental policies: a directional distance based matching approach. Agricultural and Applied Economics Association Conferences. Annual Meeting, Seattle, Washington, DC.

Sayer, J., Sunderland, T. and Gazoul, J. (2012). Ten principles for a landscape approach to reconciling agriculture, conservation, and other competing land uses. *Proceedings of the National Academy of Sciences of the United States of America* 110(21), 8345–8348.

Schut, M., Klerkx, L., Sartas, M., Lamers, D., Campbell, M. M., Ogbonna, I., Kaushik, P., Atta-krah, K. and Leeuwis, C. (2016). Innovation platforms: experiences with their institutional embedding in agricultural research for development. *Experimental Agriculture* 52(4), 537–561. https://doi.org/10.1017/S001447971500023X.

Severini, S. and Tantari, A. (2013). The effect of the EU farm payments policy and its recent reform on farm income inequality. *Journal of Policy Modeling* 35(2), 212–227.

Steensland, A. and Zeigler, M. (2021). Productivity in agriculture for a sustainable future. In: Campos, H. (Ed.). *The Innovation Revolution in Agriculture*. Springer, Cham. https://doi.org/10.1007/978-3-030-50991-0_2.

Šūmane, S., Kunda, I., Knickel, K., Strauss, A., Tisenkopfs, T., Rios, Id. I., Rivera, M., Chebach, T. and Ashkenazy, A. (2018). 'Local and farmers' knowledge matters! How integrating informal and formal knowledge enhances sustainable and resilient agriculture. *Journal of Rural Studies* 59, 232–241. https://doi.org/10.1016/J.JRURSTUD.2017.01.020.

Toma, L., Barnes, A. P., Sutherland, L.-A., Thomson, S., Burnett, F. and Mathews, K. (2018). Impact of information transfer on farmers' uptake of innovative crop technologies: a structural equation model applied to survey data. *The Journal of Technology Transfer* 43(4), 864–881. https://doi.org/10.1007/s10961-016-9520-5.

Turner, A., Wilson, L., Edgington, P., Nias, I. and Drake, B. (2013). Final report on project 23768 resource protection monitoring of uptake and management of ES options to address DWPA. Report for Natural England.

Udagawa, C., Hodge, I. and Reader, M. (2014). Farm level costs of agri-environment measures: the impact of entry level stewardship on cereal farm incomes. *Journal of Agricultural Economics* 65(1), 212–233.

Uthes, S., Sattler, C., Zander, P., Piorr, A., Matzdorf, B., Damgaard, M., Sahrbacher, A., Schuler, J., Kjeldsen, C., Heinrich, U. and Fischer, H. (2010). Modeling a farm population to estimate on-farm compliance costs and environmental effects of a grassland extensification scheme at the regional scale. *Agricultural Systems* 103(5), 282–293.

Vecchio, Y., De Rosa, M., Pauselli, G., Masi, M. and Adinolfi, F. (2022). The leading role of perception: the FACOPA model to comprehend innovation adoption. *Agricultural and Food Economics* 10(1), 5. https://doi.org/10.1186/s40100-022-00211-0.

Walker, K. J., Stevens, P. A., Stevems, D. P., Mountford, J. O., Manchester, S. J. and Pywell, R. F. (2004). The restoration and re-creation of species-rich lowland grassland on land formerly managed for intensive agriculture in the UK. *Biological Conservation* 119(1), 1–18.

Wunder, S. (2005). *Payments for Environmental Services: Some Nuts and Bolts*. Centre for International Forestry Research, Indonesia.

Wynn, G. (2002). The cost-effectiveness of biodiversity management: a comparison of farm types in extensively farmed areas of Scotland. *Journal of Environmental Planning and Management* 45(6), 827–840.

Wynne-Jones, S. (2013). Ecosystem service delivery in Wales: evaluating farmers' engagement and willingness to participate. *Journal of Environmental Policy and Planning* 15(4), 493–511.

Zhu, X. and Oude Lansink, A. (2012). Impact of CAP subsidies on technical efficiency of crop farms in Germany, the Netherlands and Sweden. *Journal of Agricultural Economics* 61, 545–564.

Part 2

Farmland and conservation practices

Chapter 5

Soil health and ecological restoration

Alice Day, The University of Manchester, UK; Ezekiel M. Njeru, Kenyatta University, Kenya; and David Johnson, The University of Manchester, UK

1 Introduction

Soils underpin the functioning of terrestrial ecosystems, and both directly and indirectly contribute to the provision of multiple ecosystem services, including the production of food and fibre, water quality, carbon capture, and biodiversity (Smith et al., 2021). For decades, the value of soils for humanity has been known and has led to well-known popularist works and statements, including the famous phrase by Franklin D. Roosevelt in 1937 that 'a nation that destroys its soil destroys itself'. Yet, many ecosystems and the soils on which they depend are still being degraded at an alarming rate, which threatens the ability to feed growing populations and mitigate the effects of global climate and environmental change. With this degradation comes the opportunity to restore land back either to productive, sustainable agriculture or to provide other valuable ecosystem services.

Many large-scale intensive agricultural systems that currently drive global food markets use yield-focused approaches with high chemical (fertiliser and pesticide) and physical (tillage) interventions. While the short-term (decadal) benefits of these high-input systems have in many cases been positive, there are three major concerns. First, many innovations are simply maintaining

http://dx.doi.org/10.19103/AS.2024.0134.06

the status quo with limited net progress, such as the evolution of resistance to pesticides by pests in a process akin to the 'Red Queen hypothesis' in evolutionary biology (the idea that species must evolve and adapt simply to compete against other evolving species; van Valen, 1973; Denison et al., 2003). Second, over longer time periods, the cost of intensification in terms of impacts on soil health and ecosystem function, such as losses in soil organic matter and key functional groups of soil microorganisms, outweighs the benefits, leading to loss of yield (Kopittke et al., 2019). Third, there are combined challenges facing agriculture from global climate and environmental change, land sparing and sharing, and the requirements for sustainability. The damage caused by sustained use of high-input systems combined with climatic perturbations leaves land unsuitable for agriculture and in need of restoration.

The concept of 'healthy soil' is seen as key in restoration efforts (Bender et al., 2016). Soil health is notoriously difficult to define in a meaningful way, so here we use the definition 'the continued capacity of soil to function as a vital living ecosystem that sustains plants, animals, and humans' (USDA, 2019) and apply the term as a guiding concept rather than an explicit quantifiable measure. In this chapter, we focus on the biological elements of soil health (Rashid et al., 2016), particularly on interactions among soil biodiversity and plants as regulators of ecosystem services. However, soil is part of a larger ecosystem that plays a critical role in connecting Earth's spheres, and this connectedness must be considered when restoring abandoned agricultural lands back to resilient systems. Moreover, spatial scale becomes highly relevant for implementation strategies. Soil properties vary over small distances, and it is likely that even at a scale relevant to an individual (i.e. farmer), different factors need to be considered when restoring land. The variability of soil properties partly explains why a universal and practical guide on effective soil health indicators and restoration practices remains elusive. Despite this issue, there are several guiding principles that can be used for soil health and ecosystem restoration, not least the consideration of ideas derived mainly from the ecological literature.

Increasingly, farmers and land managers are turning to concepts derived from ecology to enhance soil health and ecosystem services. Some of these concepts contribute to 'regenerative' agricultural practices (a loose term, but defined here as practices that restore ecosystem health and impart resilience) often using 'nature-based solutions' where features of natural ecosystems (e.g. soil biodiversity) are used to promote services and ameliorate effects of disservices (Mrunalini et al., 2022). In the last decade, significant advances have been made in our understanding of ecological processes, and here we discuss those which we think are most readily translatable to soil health and ecological restoration. For example, there is strong evidence for 'plant-soil feedbacks' (PSFs; van der Putten et al., 2013): the idea that plants shape the biotic and abiotic properties of soils which can have either positive (positive feedback) or negative impacts (negative feedback) on the growth of subsequent generations of plants. Some of these PSF effects are driven by interaction with beneficial and pathogenic soil microorganisms, such as soil-dwelling mycorrhizal fungi that have coevolved to form symbioses with 90% of all land plant species. The ubiquity of these relationships has opened up a promising line of research into using these fungi to enhance plant productivity and ecosystem services.

Finally, there is an abundance of evidence showing that biodiversity often positively influences individual ecosystem functions (Loreau et al., 2001) and increasingly, ecosystem multifunctionality in natural systems (Manning et al., 2018). These so-called biodiversity-ecosystem function (BEF) relationships could be embraced in restoration practices and can be applied at different temporal, spatial, and taxonomic scales (i.e. genetic/intraspecific, species/interspecific, functional group, and trophic level). Understanding and using plant-soil interactions could provide a low-input, 'natural' solution to restore degraded lands globally.

Here, we outline (i) the major degradation gradients and challenges in soil health and ecosystem restoration facing agricultural landscapes, (ii) the key attributes of healthy soil and ecosystems, and (iii) how recent progress in the ecological literature may help address restoration challenges and increase ecological resilience against climate and environmental change. Specifically, we focus on the potential roles of mutualistic mycorrhizal fungi and bacteria, PSF, and the relationship between biodiversity and ecosystem function to 'bioengineer' soils and ecosystems (Bender et al., 2016; Coban et al., 2022). We also describe two contrasting case studies where these approaches have been tested in restoration experiments in Kenya and UK grassland.

2 Major degradation gradients and challenges in soil health and ecosystem restoration

The Food and Agriculture Organization of the United Nations estimates that 33% of the world's soils are already degraded, but this figure is set to rise substantially in the next 30 years as the combined drivers of climate change, land-use change, and population growth increase (FAO and ITPS, 2015). However, there is a wide variation in the databases used to derive estimates and a large spatial variation in degradation, and the availability of data for some regions is poor (Gibbs and Salmon, 2015). Soil degradation is complex: the broad impacts of the degradation of soils act on three axes (physical, chemical, and biological), which themselves encompass a vast range of specific effects and span a range of magnitudes (Fig. 1). The impacts and global significance have been well described elsewhere, and we refer readers to key papers on this topic (Borelli et al., 2017; Gardi et al., 2013; Guerra et al., 2020; Arcurus, 2017).

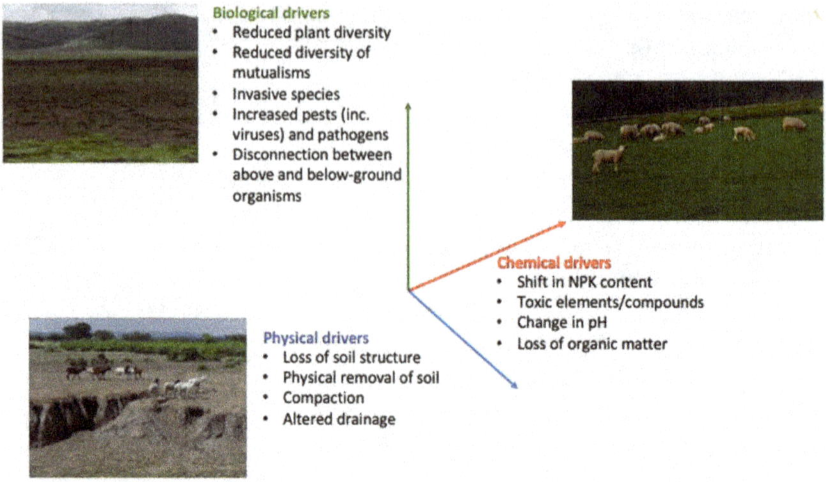

Figure 1 The three primary axes of degradation of soil health with examples of some of the key specific drivers. Drivers may act on all axes simultaneously. The photographs illustrate examples of degradation caused mainly by the primary axis: the biological axis is represented by a grassland on the Tibetan plateau that has been over-grazed (photo credit: Richard Bardgett); the chemical axis is represented by a UK grassland that has received sustained inputs of artificial fertilisers, leading to dominance of fast-growing grass species such as *Lolium perenne* (photo credit: David Johnson); the physical axis is represented by a grassland in Kenya that has succumbed to extreme physical erosion (photo credit: Richard Bardgett).

A major challenge is that these axes of degradation are not mutually exclusive and can interact with each other, sometimes in opposing directions, and are spatially and temporally variable. Also, there are still many unknowns: for example, it has only been possible to measure the species richness of soil microorganisms since the molecular revolution and massive gaps in data remain. Finally, some of the drivers of degradation may be initially counter-intuitive. Planting trees, for example, has received huge press globally recently, which has led to numerous national-level policy changes to dramatically increase tree cover. Yet, it is essential to plant the right tree in the right place (Bateman et al., 2023); there are many examples where the lack of consideration of PSF or environmental context in tree planting schemes or following tree expansion has serious negative effects on some key ecosystem services, including carbon storage. This can occur particularly in early stages of tree planting because trees can allocate substantial amounts of labile carbon into soils through roots and mycorrhizal fungi, which can catalyse the breakdown of more recalcitrant forms of carbon in a process known as the 'priming effect' (Dijkstra and Cheng, 2007). The three primary axes of degradation (Fig. 1) also are acted on by a suite of climate, environmental, and anthropogenic drivers. These are serious challenges that make the implementation of restoration efforts and their outcomes difficult to predict.

Restoration practices need to take a holistic approach. While the focus of this chapter is on soil health, this concept is inextricably linked to the functioning of the other major 'spheres' to which soil connects either physically or functionally. For example, soils influence the hydrosphere and are crucial for storing and filtering water, but if managed poorly they can also contaminate it (water companies spend huge sums removing 'soil' usually in the form of particulate organic matter from drinking water supplies); soils are a key component of the biosphere and interact with plants and supply them with growth-limiting nutrients and water, but soil life is also fuelled by plant production; soils are globally massive sources of greenhouse gases (notably CO_2, CH_4, and N_2O) to the atmosphere, but they can also be managed to mitigate these emissions. Thus, a key take-home message of this chapter is to consider soil health from the perspective of landscape scale ecosystem multifunctionality; such consideration may require trade-offs in some functions and some components of the landscape.

3 Key attributes of healthy soil and ecosystems

A wealth of literature has attempted to define soil health (e.g. Lehmann et al., 2020), which is a useful catch-all that brings together myriad properties. Soils are critical for the functioning of ecosystems and delivery of ecosystem services, and a tight definition potentially allows for the development of diagnostic indicators that can be rolled out to non-experts. We hope this endeavour continues, but in the meantime, a single diagnostic tool is not available. We also think that a detailed discussion about optimal diagnostic measures is unhelpful. Therefore, what is important is adherence to the principle of soil health in order to inform decisions about land management. This principle is also flexible enough to consider recent findings or application of existing ideas from different disciplines to improve ecosystem multifunctionality ('ecosystem health': Lu et al., 2014).

Soil biodiversity is a central component of soil health, regardless of how soil health is defined and measured. The myriad biotic interactions in soil have profound effects on regulating and provisioning services, and in influencing plant performance, which is a fundamental ecosystem function. This is because plant photosynthesis (primary production) drives virtually all soil biological activity and itself generates food, fuel, and fibre essential for humankind.

Our view (Fig. 2) of healthy soils therefore places plant-microorganism interactions at the centre. Implementing techniques to enhance the efficacy of these interactions is likely to generate cascades to influence other important attributes. We suggest that the most important plant-microorganism interactions are those where there is a direct influence of each partner on each

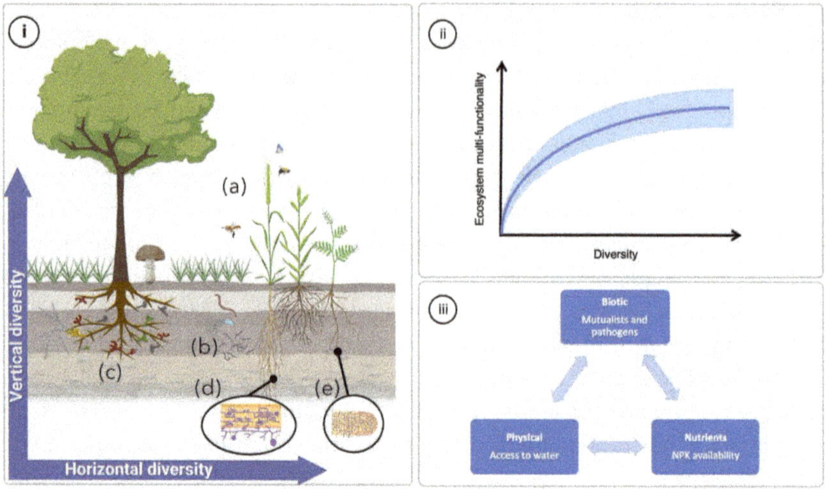

Figure 2 Schematic representation of the three processes discussed in this chapter that could enhance soil health and enhance ecosystem services: (i) increasing vertical (i.e. bridging trophic levels) and horizontal (within a trophic level but also across spatial scales) diversity through manipulation of (a) plant communities and maintenance of soil cover; (b) minimising soil disturbance and increasing soil organic matter to promote diverse soil food webs; utilising plants that associate with ectomycorrhizal (c) and (d) arbuscular mycorrhizal fungi and (e) nitrogen-fixing bacteria on their root systems. The abundance and diversity of mutualistic microbes is a crucial component of soil health and a critical aspect of plant-soil feedback; (ii) idealised relationship between biodiversity and ecosystem multifunctionality as a result of positive species interactions, selection effect and niche partitioning for resources; and (iii) plant-soil feedback connects above- and below-ground sub-systems, and plants act on soil biotic, chemical, and physical properties to influence the performance of subsequent generations. Source: Created with BioRender.com.

other, i.e. mutualistic and pathogenic symbioses (we use the term 'symbiosis' in the strict sense to consider all forms of direct interaction). These are the 'engine' of soil functioning, and so the rationale for this bias is based on the evidence that reciprocal exchange of limiting resources (carbon, nitrogen, phosphorus, and water) is fundamental to ecosystem function and in driving a change in ecosystem functions (such as in restoration efforts). These effects lead to rapid cascades of energy and matter through the ecosystem, to enable other important groups of free-living organisms to thrive and to generate knock-on effects on soil properties (such as soil structure and organic matter content).

3.1 Trade-offs are (almost) inevitable

We also highlight that the generation of healthy soils may lead to trade-offs, as will a narrow focus on one aspect of ecosystem health. For example, the recent policy has seen a shift towards net zero; indeed, excellent initiatives such as the 4 per mil campaign (www.4p1000.org) have put a much-needed spotlight on increasing the carbon content of soils. However, it is possible that the benefits of a sole focus on carbon sequestration could lead to negative consequences on other ecosystem functions under some circumstances. For example, traits that promote carbon sequestration may negatively affect other traits such as litter quality that is vital for nutrient cycling (Giweta, 2020). Such trade-offs are widespread in evolution and can in fact be used positively to shape new agricultural practices (Denison, 2012). Trade-offs therefore need to be understood and embraced but minimised, as part of the journey towards ecosystem restoration.

4 Addressing restoration challenges using ecological concepts

In this section, we provide important information on plant-microorganism interactions, with a focus on mutualisms, PSF, and BEF relationships, emphasising how these interactions influence ecosystem services relevant to agricultural landscapes. We then discuss how these aspects of ecology and ecosystem function either have been or can be applied to aid restoration.

4.1 Plant-microorganism interactions

Typically, an individual plant is a vast reservoir of biodiversity. Plants harbour invertebrates (many of which cause them damage) and their tissues contain numerous viruses and microbial endophytes (the function of which is often poorly understood). The root systems are colonised either actively (i.e. by organisms that have coevolved with plants, such as mycorrhizal fungi and nitrogen-fixing bacteria) or passively (i.e. microorganisms that associate with the surface of plant roots) by fungi and bacteria. The collective term for communities of microorganisms is the 'microbiome', and it is increasingly recognised that the plant microbiome offers both opportunities and threats to plant health and broader ecosystem processes (Sergaki et al., 2018). Some of the associations between plants and other organisms may be distinct, or even specific, and so for most ecosystems it is the functions of the 'community microbiome' that is most important. Here, we provide an overview of some of the features of the biology of microorganisms associating directly and indirectly with plants that are most likely to have an important role in affecting soil health and restoration of degraded lands. We then provide evidence of where these organisms have been implicated in successful restoration strategies and offer some practical advice for their inclusion in land management.

4.1.1 Nitrogen-fixing microorganisms

The evolution of symbiotic nitrogen fixation was a major adaptation that enabled plants to overcome growth limitation by nitrogen (Poole et al., 2018). The atmosphere is a vast source of nitrogen which diffuses into soil pores, enabling nitrogen-fixing microbes to access this source of nutrition. Plants that fix nitrogen are generally found in the family Fabaceae and include major crop plants (e.g. beans and peas) as well as a diverse range of other species (shrubs, trees, and herbs). Incorporation of nitrogen-fixing plants into agricultural and restoration systems is common and should be encouraged. One major consequence of doing this is that soils become gradually enriched in nitrogen because the litter (dead roots and shoots) of the nitrogen-fixing plants generally has high concentrations of nitrogen relative to other species and can be incorporated back into the soil as a natural and sustainable fertiliser (Kebede, 2021). One popular approach is to consider nitrogen-fixing trees in silvopastoral or alley-cropping systems. The production of nitrogen-rich litter by perennial nitrogen-fixing trees provides a long-term input of nutrients (Danso et al., 1992; Sierra and Nygren, 2006). Moreover, nitrogen-fixing trees also associate with arbuscular mycorrhizal (AM) fungi, so there are likely to be added benefits (see below) (Atangana et al., 2013). The use of nitrogen-fixing plants may not provide such a quick return on soil fertility as artificial fertiliser inputs but clearly bring many advantages including cost and the wider benefits associated with mycorrhizal plants and fungi.

4.1.2 Mycorrhizal fungi

The successful colonisation of land (the so-called 'greening of the Earth') and subsequent diversification of plants that took place around 450 million years ago was only possible because of the evolution of the mycorrhizal symbiosis. Thus, mycorrhizal plants and fungi have coevolved to cope with the challenges of life on land; now, the majority of land plants form symbioses with mycorrhizal fungi. There are several types of mycorrhizal symbiosis, which differ according mainly to the host plant and fungal identity but which have similar functional attributes. These are arbuscular mycorrhiza, ectomycorrhiza, ericoid mycorrhiza, and orchid mycorrhiza, with the first two types being the most relevant to most areas of land requiring restoration (but there are certain circumstances where ericoid and orchid mycorrhiza may need to be considered from a restoration and soil health perspective).

The basis of all mycorrhizal types is an exchange of resources between plant and fungal partners: plants have the capacity through photosynthesis to fix carbon and thus allocate some of this carbon to the fungi. By contrast, the fungi have access to mineral nutrients in the soil, notably nitrogen and phosphorus, and allocate some of these resources to the host plant, acting as natural 'bio-fertilisers'. In many cases (hundreds of experiments), inoculation of a plant with a mycorrhizal fungus leads to bigger plants and higher yields, mediated by greater acquisition of growth-limiting nutrients. This has led to the notion, often in the popular scientific press, that mycorrhizal fungi could be a saviour for agriculture and restoration and an essential part of soil health. While we agree there is certainly potential in many circumstances, such a simplistic view of symbiosis ignores some very important effects that must be considered in the context of restoration.

First, not all fungi have the same magnitude of effect on host plant performance, and there can be considerable variation in host plant growth. Second, some fungi have a *negative* effect on host plant performance. Third, only a tiny proportion of the global diversity of mycorrhizal plants and fungi have been tested together. There are about 250 000 species of plant species that form mycorrhizas, while the estimate for compatible fungi is much less certain (Brundrett, 2009). There are estimated to be only a few hundred to a thousand species of AM fungi and about 20 000 species of ectomycorrhizal fungi. In addition, we know little about the importance of intraspecific diversity of the fungi; other than that there can be a significant variation in how genotypes of mycorrhizal plants and fungi respond to each other (Johnson et al., 2012). Fourth, the local environment (e.g. light, water, temperature, and soil properties) can influence the outcomes of plant growth experiments, and, in most cases, experiments testing the response of plants to mycorrhizal fungi have been conducted in highly controlled and artificial conditions. Thus, some of the variations in responses in host response likely arises from how mycorrhizal plants and fungi interact with the environment they are being tested in. This variation has led to the idea of a 'mycorrhizal phenotype' based on plant genotype × fungal genotype × environment interactions (Johnson et al., 1997). Fifth, mycorrhizal fungi and plants exist in communities, and the response of a plant to a community of mycorrhizal fungi may not be the same as the response to an individual (see later section on biodiversity × ecosystem function relationships). These five issues require practitioners to be wary of the idea that mycorrhizal fungi are a 'silver bullet' to plant production, restoration, and soil health. Ultimately, a clear idea of the objectives of restoration and a knowledge of the system in question is important to help determine when mycorrhizal fungi do have an important role.

Despite the important caveats listed above, the encouragement of mycorrhizal fungi in restoration strategies may generate important benefits (Table 1). For example, most mycorrhizal fungi produce extensive networks of extraradical mycelium in soil (Leake et al., 2004), comprising tiny hyphae (typically 1–20 μm diameter), which has a huge absorptive area. This network enables the fungi to grow well beyond the root zone and penetrate tiny pores in the soil, thus opening up new volumes of soil for the exploitation of

Table 1 Potential beneficial effects on ecosystem properties and services of promoting mycorrhizal fungi in restoration of agricultural land

Type of effect	Potential impact on properties	Link to ecosystem services
Biological	Increased inter-plant signalling	Plants can respond to pests, diseases, and climatic perturbations more rapidly, reducing reliance on pesticides and increasing food production
	Promotion of above-ground biodiversity	Enhances pest reduction, nutrient cycling, and resilience to stress due to the high genetic diversity
	Promotion of below-ground biodiversity	Fungi, bacteria, and fauna interact to form a complex web that maintains processes such as nutrient cycling, decomposition, and soil turnover
Chemical	Increased nutrition and biomass of plants	Higher crop yields
	Increased nutrient and mineral liberation and uptake	Greater quality crops that are more resilient to stress with benefits for food, fuel, and fibre production
	Greater soil organic matter formation	Increased water retention and infiltration to prevent flooding; promotes nutrient cycling and reduces leaching
	Greater carbon sequestration and storage	Reduces greenhouse gas emissions helping to combat global change
	Bioremediation	Degrades toxic chemicals creating a healthier system and healthier food
Physical	Hyphal networks stabilise the soil to reduce soil erosion	Increased nutrient retention (reduced leaching) and promotes water retention
	Rebuilds and maintains soil structure by soil aggregation	Allows for greater water and dissolved oxygen retention within the soil pore spaces Protects soil carbon pools

water and nutrients. In addition, some fungi, especially ectomycorrhizal and ericoid fungi, have an arsenal of degradative capabilities that enable the fungi to acquire nutrients from more complex sources. Indeed, there is very good evidence that simple organic forms of nitrogen can even be taken up by the fungi intact (Nasholm and Persson, 2001). This ability circumvents the need for microorganisms to mineralise organic forms of nitrogen into mineral forms (e.g. nitrate and ammonium) prior to uptake by plants. Thus, ectomycorrhizal and ericoid mycorrhizal fungi effectively 'short-circuit' the traditional view of nutrient cycling. Encouragement of mycorrhizal fungi may thus provide plants with opportunities to more effectively acquire growth-limiting resources in the soil and reduce or eliminate the need to add artificial fertilisers. The catch, as outlined previously, is that not all fungi have the same effect on plants and the outcome is context dependent. Moreover, symbiosis has fundamentally evolved under nutrient-limiting conditions, whereas inputs of fertilisers create nutrient environments that are essentially unnatural; the result of the latter is often very good plant growth and almost certainly greater growth than could be achieved by relying on symbioses. But the key point here is that mycorrhizal symbioses provide slow and steady, sustainable growth and provide a suite of other benefits that artificial fertilisers fail to provide (Table 1). Practitioners simply need to be comfortable with the likely trade-off between 'slow and steady' versus 'short and fast' responses from fertiliser inputs.

A further suite of benefits arises from the fact that mycorrhizal fungi receive substantial amounts of carbon from their host plant. The amount of carbon allocated to the fungi varies depending on species (of host plant and fungus) and the environmental context (including time of year). Nonetheless, plants would be expected to allocate typically 5–20% of their photosynthesised carbon to mycorrhizal fungal biomass (Leake et al., 2004). An important consequence of this carbon supply is the cascade of energy through soil food webs. Many fungi form extensive networks of mycelium that are large pools of carbon: ultimately, this mycelium dies and decomposes, but it contributes to the organic matter content of soils, which is a key attribute linked to ecosystem health.

Mycorrhizal hyphae are relatively easy to degrade compared to thicker, woodier roots (Wilkinson et al., 2011; Drigo et al., 2012), and so the fungi provide a conduit for the rapid transfer of plant-derived carbon that becomes available to other organisms. In addition, some soil fauna, such as collembola and nematodes, consume hyphae (Kanters et al., 2015). Thus, carbon flow via mycorrhizal fungi is a crucial way in which soil food webs proliferate and function, thus contributing to healthy soils. The contribution of mycorrhizal fungi to long-term stable pools of carbon in soils is difficult to assess, but there is some evidence for this process. In relatively undisturbed boreal forest systems, roots and production of mycorrhizal mycelium are thought to have made a substantial contribution (~70%) to the large stores of carbon in these soils. Boreal systems have cold temperatures and low pH and are characterised by ectomycorrhizal and ericoid mycorrhizal fungi. Whether ectomycorrhizal and ericoid mycorrhizal plants and fungi have such an impact on carbon storage in other ecosystems remains unclear, although analyses of patterns globally suggest this may be the case (Averill et al., 2014). Encouraging the growth of species that form these mycorrhizal associations in degraded agricultural ecosystems is likely to be a good strategy from a carbon storage perspective.

AM fungi are also implicated in promoting carbon storage. Grasslands and forests dominated by AM plants tend to store carbon as mineral-associated organic matter, whereas ectomycorrhizal and ericoid mycorrhizal systems store carbon in particulate organic matter (Cotrufo et al., 2019). However, the specific contribution of AM fungi for carbon storage remains unclear. For example, under climate change scenarios, notably elevated atmospheric CO_2 concentrations, there is evidence that AM fungi may accelerate carbon losses from soils (Cheng et al., 2012). These findings have been contested, including emphasising that the symbiosis is multifunctional (Verbruggen et al., 2013) and emphasising the need to consider the collective benefit of promotion of mycorrhizal fungi in ecosystems. Despite the findings from Cheng et al. (2012), from a restoration perspective, AM fungi may well be crucial for a range of services in addition to carbon storage. For example, in heavily degraded grasslands, organic matter content is often significantly reduced and plant cover can be patchy (Fig. 1). Promotion of AM plants and fungi will be critical in returning such systems to health and in so doing accumulating organic matter. In other words, a key principle in restoration is to consider the starting point of the degraded system in question and evaluate the collective benefits of a restoration strategy.

A further suite of potential effects of mycorrhizal fungi is on the physical properties of soil, which link with other attributes mentioned above (Lehmann et al., 2017). The production of hyphae helps provide structure through physical enmeshment (Tisdall et al., 1997), and in the case of AM fungi, in particular, through production of glycol proteins (Agnihotri et al., 2022) or other molecules. Such structural effects are important because of the well-established role of soil structure in providing more favourable growth conditions for plants, better protection of soil carbon pools, more effective retention of water and resistance to wind and water erosion.

4.1.2.1 Common mycorrhizal networks

One fascinating feature of mycorrhizal fungi is their ability to form symbioses with more than one host plant simultaneously and produce so-called 'common mycorrhizal networks' (CMNs; Selosse et al., 2006). This capacity immediately raises important questions about the functioning of plant-mycorrhizal symbioses, notably: how are soil-borne resources distributed through these networks and transferred to host plants? Is nutrient allocation equal or controlled in part by plant hosts? How do these interactions shape plant competition and growth? There is no space here to discuss all these questions, but from a restoration and soil health perspective, the following summary is useful.

First, there are two important caveats to be made about the significance of CMNs. Although there have been some elegant experiments measuring aspects of CMN extent (Beiler et al., 2010; Grelet et al., 2010), there is currently very little information on the extent and ubiquity of CMNs across the major ecosystems on Earth (Alaux et al., 2021). The fact that many mycorrhizal fungi have the potential to form CMNs does not mean they do so in nature. In addition, it is unlikely that CMNs formed by plants are functionally very large, although this remains to be tested rigorously. The idea, often put forward in the popular scientific press, that most plants (especially trees) are interconnected by enormous CMNs connecting individual plants over distances of 10s or 100s of metres is not supported by any firm evidence.

Despite the points above, CMNs have been shown to influence several important aspects of ecosystem functioning. For example, the ability of new plant recruits (seedlings) to become colonised quickly is vital for the fitness of that plant and thus the development of healthy ecosystems. Seedlings are rapidly colonised by hyphae associated with neighbouring plants and integrated into a CMN. Integration into a CMN gives seedlings access to an established absorptive structure otherwise unavailable to them, but what benefit do plants get from integration into such a network? Currently, there is evidence for a variety of effects; for example, some studies, particularly with AM plants, show that integration into CMNs can have negative effects on seedling growth (Liang et al., 2020; Dassen et al., 2021). Indeed, the findings from Dassen et al. (2021) may indicate that severance of networks by physical actions may promote the growth of seedlings in grassland restoration. On the other hand, many studies have found that plants that do rapidly become colonised often are larger than those that do not (Nara, 2006), and the use of mesh cores to control the ability of plants to become integrated into CMNs also show that seedling survival and growth improves (Liang et al., 2020, 2021). The emerging view is that plants forming ectomycorrhiza are more likely to gain benefit from integration into CMNs than those that form arbuscular mycorrhiza, but clearly the specific context matters.

Plants may also gain other benefits from being integrated into a CMN. Much work has focused on the movement of carbon (and other materials) from plant to plant via CMNs. This is a much-debated topic, but it certainly can and does occur in some circumstances to such an extent that some plants (mycoheterotrophs) have completely lost their ability to photosynthesise and are either partially or entirely dependent on carbon moving from a neighbouring green plant via CMNs (Leake, 1994). Many mycoheterotrophs are of high conservation value, and so efforts to restore systems to promote these species must consider not just the fungi they associate with but also the 'third-party' green plant species that is supplying carbon. The ecological significance of the movement of carbon from green plant to green plant remains to be fully realised, however. There may be some circumstances where this is nutritionally important (e.g. when seedlings are in shade or when the relatedness of seedlings to other plants in the CMN is high; Pickles et al., 2017).

Recent laboratory work also provides evidence that formation of CMNs can act as conduits for signalling processes related to plant defence against herbivores, such as aphids (Babikova et al., 2014). This mechanism has many important and fascinating implications (Johnson and Gilbert, 2015; Gilbert and Johnson, 2015, 2017; Babikova et al., 2013), but these effects have yet to be realised in field settings.

From a restoration and soil health perspective, promotion of CMNs should be encouraged simply because the likely benefits outweigh the negative effects, especially in forests containing ectomycorrhizal trees. Importantly, while there are methods to promote mycorrhizas generally, there are currently no interventions that can promote CMNs specifically (Alaux et al., 2021). Thus, the formation of CMNs is a potential, perhaps inevitable, outcome of having healthy soils containing a rich diversity and abundance of mycorrhizal fungi, which bring a suite of benefits and increase ecosystem multifunctionality.

4.2 Plant-soil feedbacks

PSF may be defined as 'changes in the properties and functions of soils from one generation of plants that affects the performance of subsequent generations of plants'. This idea therefore has important implications for the way communities develop through time and the way they influence soil health. Although this concept is relatively recent in ecology (van der Putten et al., 2013), the basic idea has been well known to agronomists for many decades using crop rotations, although this application belies the extent and detail of PSF processes. One of the main reasons for introducing rotations is to limit the accumulation of pathogens in the soil. Historically, disastrous pathogen outbreaks such as potato blight can decimate production and hugely impact societies. Even in modern agriculture, with all the technological innovations available, build-up of pathogens such as the fungus *Gaeumannomyces graminis* var. *tritici* that causes 'take-all' in wheat has substantial effects on yield acting through processes that fall under the banner of PSFs. Interestingly, association of wheat with AM fungi may have a role in suppressing the effect of take-all, thus also highlighting the importance of mutualists in combatting negative effects of pathogens (Siasou et al., 2009). Nonetheless, a major component of PSF that is relevant to agriculture and restoration is the accumulation of pathogens. But such-negative effects of PSFs encompass far more than root pathogens. PSFs may also promote soil fauna that have negative effects on plant growth, such as root herbivores.

Negative PSF is counteracted by positive PSF, and so it is the net effect that matters for ecosystem functioning. The recruitment of beneficial bacteria, such as nitrogen-fixers, and mycorrhizal fungi can have dramatic positive effects on nutrient capture, protection from pathogens and pests and plant growth (see above). Indeed, elements of BEF theory also come into effect because the diversity, as well as the identity, of beneficial and pathogenic microorganisms also matters (Semchenko et al., 2018). Finally, PSFs may be expressed through abiotic factors. For example, plants may modify the chemical environment of the soils in which they are growing that may lead to positive or negative effects on subsequent generations. Such effects may also be seen, albeit less commonly, through physical effects (e.g. modifying soil structural properties).

The process of measuring the direction and strength of PSF in experiments is relatively easy and involves the use of simple pot-based experiments. These comprise an initial 'conditioning' phase, whereby plants modify their environment, and then a 'feedback phase', where all plant species in question are grown on their own soil (conspecific or 'home' soil) or from soil conditioned by other species (heterospecific or 'away' soil). The biomass of plants in the home-and-away comparisons enables the calculation of a simple index reflecting the extent to which positive or negative PSF effects are seen.

The experimental approaches used in PSF could quite easily be incorporated by enthusiastic land managers to determine the precise PSF effects of target species on specific soil under restoration and indeed whether PSF is a process worth considering. For example, if PSF effects are weak in shaping community development relative to other factors (such as the extent of degradation of land), there may be no reason to think about their incorporation into restoration strategies (Yelenik and Levine, 2011).

On-site analysis of PSF is by far the best way of deciding which species to use in a particular restoration context (Kulmatiski et al., 2006; Bauer et al., 2017), even though it requires a modest amount of time (1-2 growing seasons) and effort to complete. More recent experimental tests have begun to implement PSF methodologies in the field, which is more challenging from an experimentalist's perspective but provides more realistic outcomes that are likely of greater benefit to restoration practitioners (De Long et al., 2019). There is no reason why such an experimental approach could not be applied as part of an active restoration project enabling poorly performing plant communities (negative PSF) to be identified and replaced by those that generate positive PSF. Without on-site measurements, extrapolation of findings from the growing numbers of studies may be applied, but the obvious issue is that, by definition, PSF is highly dependent on the specific plant species × environment interaction. We therefore advocate the development of a global database of PSF effects that integrate metadata on biological, chemical, and physical properties of soils, other key environmental data, and plant species and genotype identities, which ultimately could become a valuable resource to restoration practitioners.

4.3 Biodiversity and ecosystem function

A huge research effort has been undertaken over the last decades attempting to characterise the relationship between BEF. Much early work concentrated on plant species richness as the 'biodiversity' component and plant biomass as the corresponding response variable representing 'ecosystem function'. Now, more aspects of these two axes have been explored, including consideration of intraspecific (i.e. genotypic; Johnson et al., 2012), species, functional group (categorisation of organisms according to their traits or other characteristics), trophic levels and interaction network topology as components of biodiversity, in addition to a suite of broader functions as response variables, such as carbon capture, greenhouse gas emissions, and decomposition rates. Biodiversity can also have positive impacts on the stability of ecosystems by making them more resistant and resilient to perturbations, such as drought. As ever, there is much variation in outcomes depending on the specific context, but an overarching theme is that biodiversity in whatever form often has some positive effect on 'ecosystem multifunctionality', that is, the collective response of several individual ecosystem functions (Manning et al., 2018).

A key mechanism behind BEF is niche partitioning, whereby growth-limiting resources are more effectively used by communities that are more diverse than those with reduced diversity, and this reduces competitive interactions. For example, different plant species preferentially acquire certain chemical forms of nutrients such as nitrogen, so the net effect is overall greater resource capture than what could be achieved if all species used the same form equally (von Felten et al., 2009). This idea extends across trophic levels and covers many different niches including light, water, nutrients, symbiotic partners, temperature, and so on. Crucially, these processes also apply to all biodiversity groups, including soil microorganisms. Thus, the principles of BEF can be used to enhance soil health. For example, manipulation of species richness of ectomycorrhizal fungi can generate positive effects on host plant diversity (Hazard et al., 2017a,b; Hazard and Johnson, 2018), and similarly, variation of ectomycorrhizal strains may have similar effects (Wilkinson et al., 2010, 2012; Hazard et al., 2017b). Furthermore, the fact that biodiversity groups interact (and sometimes very strongly in the case of plant-microorganism interactions; Johnson et al., 2012) means that BEF relationships generate complex interactive effects across trophic levels, leading to the idea of vertical diversity (Duffy et al., 2007; Mursinoff and Tack, 2017). Thus, it is possible to manage plant diversity to enhance the diversity and function of soil.

The concepts underpinning BEF relationships offer opportunities to enhance the restoration potential of nature-based solutions. Yet, there are few restoration schemes reported in the peer-reviewed literature that have actively manipulated biodiversity (Hughes et al., 2018). This may reflect the fact that established BEF relationships are too abstract and not understood by restoration workers or that there is inadequate explanation of how to implement these ideas in practice. Some mechanisms behind BEF are certainly challenging to comprehend for non-specialists, such as the issue around 'selection' (or 'sampling effect'; Huston, 1997). This is when particularly productive members of a community dominate the response seen in mixtures: often, such species have traits that promote productivity, such as being particularly responsive to colonisation by mycorrhizal fungi or forming symbioses with N-fixing microbes. The selection effect offers an opportunity for restoration in that such high-performing species could be actively included in mixtures from the outset.

A further factor to consider is spatial scale: biodiversity effects are inherently related to the area in which the species occur. Small-scale variation in soil properties is such that consideration needs to be given to intervention strategies at local scales, such as an individual farm scale. For example, species choice in biodiversity manipulations needs to reflect the localised conditions. Indeed, even at large scales such as a region, the collective level of diversity can make a difference. Recent work shows that regions were more resilient against climate extremes when they had greater crop diversity compared to regions with fewer crop species (Renard and Tilman, 2019), while greater heterogeneity in crops promoted multitrophic diversity (Sirami et al., 2019). This raises the opportunity for collaborative measures between land managers and owners to modify individual practices and biodiversity at large scales 'for the common good'. Likewise, restoration of very large areas offer scope to implement plant strategies that actively increase the diversity or heterogeneity of species, even in habitats that have similar soil properties when a 'common' set of species may have originally been considered.

5 How to promote ecological processes in restoration strategies

There are three main ways to enhance soil health via the processes outlined previously (Fig. 2), namely: (a) manipulation of plant community, (b) changing cultivation practices of land, and (c) active addition of biological agents. Often, a combination of all three approaches will increase the likelihood of positive outcomes because they will have varying effects on mitigating the three main axes of soil degradation (Fig. 1). Furthermore, soil properties can also be enhanced through the addition of organic residues and biochar, but these are outside the scope of this review (see Bolan et al., 2022; Sessions et al., 2019).

5.1 Manipulation of plant community

Managing plant communities is probably the most effective way of influencing soil biological communities. As noted previously, beneficial mycorrhizal fungi are directly dependent on plants for their energy, and plants also are the main conduit of energy (carbon) into soil food webs. Therefore, when restoration strategies are being considered, care should be taken to ensure target plant species and communities are compatible with mycorrhizal fungi of interest and also offer greatest potential to influence the functioning of soil food webs. The classic 'checklist' of mycorrhizal plants remains a valuable source of information for UK flora (Harley and Harley, 1987), while more recent online databases have been developed with a global perspective (Soudzilovskaia et al., 2020). Additional databases of plant traits provide information on a suite of characteristics relevant to restoration (Kattge et al., 2020).

The principles of BEF relationships should also be considered where possible. This may require the establishment of more diverse communities than hitherto thought. In addition, exploiting the selection effect can also generate benefits greater than otherwise possible. For example, using the ideas of the selection effect may lead to inclusion of nitrogen fixing species in mixtures, or species that are known to perform favourably in the situation of interest.

Grazing intensity is also critical in shaping plant community diversity and function and below-ground properties (Byrnes et al., 2018; Zhou et al., 2017). Most grassland species have co-evolved alongside herbivory, and a modest level of grazing may be required not-only to limit successional processes but also to ensure plant-root symbioses are 'balanced'. Too much grazing is a common feature of degraded lands, however, and so stocking density needs to be critically evaluated. Similarly, the type of grazing can also have mixed outcomes: 'mob grazing' can for example negatively affect soil structure (Roberts and Johnson, 2021). In addition, grazing pressure from wild species also needs to be factored into any assessment of grazing pressure.

Fire and fire frequency can cause large-scale shifts in plant community diversity and functions with subsequent impacts on soil health. Fires enhance ecological heterogeneity, creating niches that can influence plant diversity (Pausas and Ribeiro, 2017). Frequent fires, however, can have negative effects on plant community diversity and function, and plants that rely on specific seasonal intervals are at greater risk of extinction (Gallagher et al., 2021). The impact of fires on soil health, particularly soil microbial communities, is more complex and less certain. Some studies indicate large variability in the response of microbial communities to fire events, but the impacts appear to persist over long time periods with subsequent consequences for soil function and health (Barreiro and Díaz-Raviña, 2021).

5.2 Addition of biological agents

The addition of microbial inocula to aid restoration offers considerable promise. This can be achieved using 'natural' inocula, for example root fragments and soil, which contain diverse microbial communities. The use of soil inocula has, for example, been used to steer the direction of PSF effects in restoration trials (Wubs et al., 2019), and techniques have been developed to generate biocrusts (Velasco Ayuso et al., 2017).

The availability of commercial mycorrhizal fungal inocula has increased markedly in recent years, no doubt reflecting demand for 'sustainable' ways of increasing productivity. The principle of these products is laudable: simply spread the inocula on soil or, in some cases, inoculate seedling roots directly (possibly more likely for high-value horticultural crops). Independent scientific field trials using verified inocula have generally found positive results for grain yield (e.g. Zhang et al., 2019) and restoration (Koziol and Bever, 2017, 2019). However, there are major concerns that potential practitioners should be aware of. First, there is no industry standard for inocula. Therefore, the quality varies considerably, with the potential for some inocula to be absent of any viable propagules. Recent studies have rigorously tested different inocula and found striking differences in colonisation of roots, including some plants that remained uncolonised (Yao et al., 2018; Perkins and Bennett, 2018). Thus, it is vital that a reputable supplier is sought and that practitioners obtain evidence that the product contains what is claimed. Second, some products may contain 'carrier' material that is in effect a fertiliser: thus, positive growth responses may simply be a standard fertiliser response. Third, it is vital that the correct type of mycorrhizal fungus is used for the target plants. On a broad level, this means that inoculum for plants that form arbuscular mycorrhiza must contain AM fungi. This sounds obvious, but sometimes products do not specify the type of fungus used. As noted previously, there can be substantial plant genotype × fungal genotype interactions. This means that some inocula, regardless of its quality, may generate plant responses on a continuum of negative to positive. Finally, the environmental context is also important. Mycorrhizal fungi must compete with the plethora of soil biodiversity (including any native mycorrhizal fungi) and also be able to thrive in particular abiotic and edaphic conditions. These points above highlight that, as ever, testing for specific contexts under controlled conditions is important before large amounts of money are spent on inocula. Suggestions on the practicalities of such testing have recently been described (Salomon et al., 2022).

The introduction of other soil microorganisms (e.g. nitrogen fixers and phosphate-solubilising bacteria) also offers potential, but the development of commercial microbial inocula has been limited to date. Part of the issue is how to scale up production in a way that is suitable for field application. Seed inoculation is an obvious way, and this approach has been developed for research trials, but there is still a gap for commercially available material (O'Callaghan, 2016). New ways of producing 'biocrusts' have been developed to aid the restoration of dryland systems (Velasco Ayuso et al., 2017), and the addition of native whole soil from pristine areas to degraded systems offers promise in accelerating restoration success (Duell et al., 2023).

5.3 Changing cultivation practices

Minimising physical disturbance of soils is rapidly becoming a key strategy to restore soil health and ecosystem functioning, using techniques such as minimum tillage or no-till (Roger-Estrade et al., 2010). The extensive extraradical mycelium produced by some species of mycorrhizal fungi is particularly sensitive to physical disturbance from ploughing and also requires a substantial investment of carbon from host plants. As a result, ploughing can significantly reduce the function and diversity of mycorrhizal fungi (Helgason et al., 1998; Schnoor et al., 2011). Thus, ensuring land has a suitable cover of compatible plant species goes hand in hand with soil management strategies to minimise the disruption of fungal networks. The success of reduced or no-till approaches for enhancing soil health likely depends on the soil properties in question, but overall, reducing tillage has positive effects on key parameters such as aggregate stability (Nunes et al., 2020).

6 Case studies

6.1 Restoration of species-rich grassland: the Colt Park case study

6.1.1 Problem and context

Colt Park is situated in the Yorkshire Dales in northern England and historically comprised traditionally managed species-rich hay meadows, developed on limestone bedrock. Since the 1940s, the management of these meadows became more intense (through ploughing, reseeding, and fertilisation) and led to a reduction in plant biodiversity and the composition of soil microbial communities, reflected by a decrease in soil fungal: bacterial ratio. The aim over the last 23 years has been to find the most effective land management regimes to restore plant diversity and increase the soil fungal:bacterial ratios in these agriculturally improved grassland areas to the original species-rich state (Smith et al., 2008). Subsequent research focused on how the most effective land management regimes for biodiversity also impacted carbon sequestration, soil nutrient retention, and ecosystem resilience to climate change (Cole et al., 2019).

6.1.2 Approach

In collaboration with Natural England (a non-departmental public body who advises the UK government on the natural environment in England), a field experiment was initiated that tested the effectiveness of various treatments to increase plant biodiversity and restore soil microbial communities. These included a combination of reseeding, mineral fertilisers, and farmyard manure applications, addition of the hemi-parasite *Rhinanthus minor*, and changing hay cut dates (Smith et al., 2008). *Rhinanthus minor* is important because its hemi-parasitic habit means that it preferentially suppresses the growth of fast-growing grasses contributing to more successful restoration of species-rich grasslands (Fry et al., 2017). Hay cut dates heavily determine plant species composition.

6.1.3 Ecosystem services

A combination of a late July hay cut date, reseeding, mineral fertiliser reductions, and farmyard manure applications was found to be the most beneficial land management regime to restore the hay meadows (Smith et al., 2003, 2008). Plant biodiversity and soil microbial communities responded positively to this regime, including the soil fungal:bacterial ratio that benefits multiple ecosystem services (Smith et al., 2008). A key ecological process relevant to the successful restoration of Colt Park meadows is the 'priority effect'. This process reflects the importance of the order of plant species arrival and their impact on the soil environment. The functional traits of these initial species determine the success of the target species establishment (Fry et al., 2017).

Increasing grassland biodiversity, especially the promotion of the legume, *Trifolium pratense*, enhanced carbon capture and carbon transfer to soil microbial communities and reduced loss of CO_2 from ecosystem respiration (De Deyn et al., 2011, 2011; De Long et al., 2019). The increased fungal:bacterial ratio promoted soil nutrient retention, another major ecosystem service, ensuring the growth of healthier, nutrient-rich plants.

6.1.4 Resilience

Regenerating biodiversity creates a resilient system that is less sensitive to perturbations (Craven et al., 2018). Resilience is becoming more important in the face of climate change, which is generating more frequent and extreme weather events. At Colt Park, rain shelters were used to test the impacts of grassland biodiversity restoration on resilience to drought (Cole et al., 2019). Grassland biodiversity restoration increased the resilience of carbon fluxes but with subsequent trade-offs on plant productivity, which must be considered in agricultural systems. Increasing soil fungal biomass, in line with the proposed land management regime, also aided resilience to drought and promoted the retention of soil carbon and nitrogen (Cole et al., 2019).

6.1.5 Conclusion

The Colt Park case study indicates the importance of combination treatments for grassland restoration and highlights the need to act swiftly because of the time taken for substantive effects on ecosystem services to be detected and the need to stay ahead of the climate curve. Typically, grassland management schemes cover 5- 10-year periods: Colt Park has been under land management regimes for over 15 years, but target species composition and diversity have still not been fully restored (Smith et al., 2008). The regime used at Colt Park has proved hugely successful in accurately restoring species-rich grasslands and provides a reliable example of the importance of combination treatments in land management to restore above- and below-ground diversity.

6.2 Soil health restoration in semi-arid smallholder agroecosystems using soil microorganisms: the Eastern Kenya case study

6.2.1 The problem

Agricultural intensification has led to over-dependence on agrochemical inputs that are often too costly for the majority of smallholder farmers in Kenya. Plant-growth-promoting microorganisms provide one of the most viable alternatives for restoration of soil health, improvement of crop production, and resilience to climate change perturbations (Koskey et al., 2021; Njeru and Koskey, 2021). However, until now, the understanding and use of beneficial microorganisms among smallholder farmers in developing countries is limited. During the last decade, native microorganisms have been isolated, characterised, and screened for their ability to enhance growth and tolerance of crop genotypes to biotic and abiotic stresses using greenhouse bioassays and field experiments within smallholder farms (e.g. Musyoka et al., 2020; Mburu et al., 2020; Njeru et al., 2020; Koskey et al., 2017).

6.2.2 Methods

With support from the Royal Society UK under the Future Leaders–African Independent Research (FLAIR) Fellowship programme, extensive laboratory, greenhouse, and field experiments were conducted in Embu, Kitui, and Tharaka Nithi counties of Eastern Kenya. High ecological value microorganisms including nitrogen-fixing bacteria, phosphate-solubilising bacteria, and AM fungi were targeted, which have great potential for sustainably improving crop production, soil health, and tolerance to drought in smallholder farming systems. Extensive bioprospecting of these microorganisms was done followed by isolation, purification, identification, and determination of the microbial isolate potential in improving crop production under greenhouse and field conditions.

6.2.3 Results

Remarkably, a consortium of three native Rhizobia isolates identified as *Rhizobium tropici* clone H53, *Mesorhizobium* sp. WSM3874, and *Rhizobium pusense* strain Nak353 improved cowpea grain yield under field conditions by up to 29%

(Nyaga and Njeru, 2020). Measurements made after the field trials in the same experimental fields showed enhanced production of the subsequent maize crop, demonstrating soil health improvement following the growth of the inoculated cowpea. Moreover, retrieved beta-rhizobia (*Paraburkholderia phenoliruptrix* BR3459a) and *Rhizobium mesosinicum* strain CCBAU 25010 demonstrated higher symbiotic efficiency in greenhouse studies compared to the commercially available isolate, *Bradyrhizobium* sp. USDA 3456 (Muindi et al., 2021).

Native phosphate-solubilising bacteria and azotobacter were also tested for their ability to improve cowpea and maize production. Some of the isolates obtained demonstrated significantly high plant growth-promotion efficiency in comparison to a phosphate-treated control under greenhouse conditions (Kirui et al., 2022; Wakarera et al., 2022). Notably, inoculation of maize with Rhizatech (a commercial AM fungal inoculant) and a consortium of AM fungal isolates (*Rhizophagus irregularis* and *Funneliformis mosseae*) significantly enhanced grain yield by 19% and 14%, respectively, besides increasing shoot phosphorus content, shoot moisture content, and chlorophyll readings.

6.2.4 Conclusion

These findings demonstrate the existence of superior native isolates within smallholder semi-arid agroecosystems with potential to be further developed as low-cost, microbial inoculants for soil health restoration. Further field trials will inform and promote the development of affordable and sustainable crop production strategies to improve the performance, quality, sustainability, and resilience of crops cultivated by smallholders in the semi-arid tropics of Kenya and beyond.

7 Conclusion and future trends

The effective management and treatment of soils and associated plant communities is vital for the implementation of effective restoration strategies. We highlight the need to consider key ecological concepts and processes in restoration strategies, notably BEF relationships, the consideration of all levels of diversity from genotypes to functional groups, and PSF. A critical component of these concepts is the interaction between plants and soil organisms, especially mutualistic fungi (mycorrhizas) and nitrogen-fixing bacteria. New technologies are emerging to facilitate the incorporation of mycorrhizal fungi into ecosystems, but the effectiveness of these products needs to be carefully evaluated on a case-by-case basis. An overarching requirement to enhance soil health is to consider soils as part of an ecosystem, and the vital interactions with plants in shaping ecosystem processes and services. Mycorrhizal fungi have great potential to aid restoration and play a part in sustainable intensification of agriculture, but further work is urgently required to gain a predictive understanding of plant-mycorrhizal-fungal interactions for specific contexts. Our understanding of the functioning of plant-soil interactions continues to grow rapidly driven by 'fundamental' research; the challenge now is to translate this understanding into practical solutions, which requires renewed collaborative efforts between academics, businesses, and practitioners.

8 Acknowledgements

EMN acknowledges the FLAIR Fellowship program administered by the African Academy of Sciences and The Royal Society, UK, and funded through the UK Government's Global Challenges Research Fund (grant no. FLR/ R1/190944). DJ and EMN also acknowledge the Royal Society for funding through a FLAIR Collaboration grant (FCG\R1\201026).

9 Where to look for further information

There is a wealth of scientific literature on the topic of soil health (key papers of which are reviewed and discussed in this chapter), alongside a number of new research programmes that tackle this topic, such as the European Joint Programme Soils (www.ejpsoil.eu). In the UK, some of the principles discussed in this chapter are supported by a number of organisations, such as the Sustainable Soils Alliance, the Soil Association, UKSoils, and Championing the Farmed Environment, and we encourage practitioners to consult their guidance.

10 References

Agnihotri, R., Sharma, M. P., Prakash, A., Ramesh, A., Bhattacharjya, S., Patra, A. K., Manna, M. C., Kurganova, I. and Kuzyakov, Y. (2022). Glycoproteins of arbuscular mycorrhiza for soil carbon sequestration: Review of mechanisms and controls. *Science of the Total Environment* 806(2), 150571.

Alaux, P., Zhang, Y., Gilbert, L. and Johnson, D. (2021). Can common mycorrhizal fungal networks be managed to enhance ecosystem functionality? *Plants, People, Planet* 3(5), 433–444.

Arcurus, C. (2017). Safeguarding our soils. *Nature Communications* 8.

Atangana, A., Khasa, D., Chang, S. and Degrande, A. (2013). Biological nitrogen fixation and mycorrhizal associations in agroforestry. *Tropical Agroforestry*, 173-202.

Averill, C., Turner, B. L. and Finzi, A. C. (2014). Mycorrhiza-mediated competition between plants and decomposers drives soil carbon storage. *Nature* 505(7484), 543–545.

Babikova, Z., Gilbert, L., Bruce, T. J. A., Birkett, M., Caulfield, J. C., Woodcock, C., Pickett, J. A. and Johnson, D. (2013). Underground signals carried through common mycelial networks warn neighbouring plants of aphid attack. *Ecology Letters* 16(7), 835–843.

Babikova, Z., Johnson, D., Bruce, T., Pickett, J. and Gilbert, L. (2014). Underground allies: How and why do mycelial networks help plants defend themselves?: What are the fitness, regulatory, and practical implications of defence-related signaling between plants via common mycelial networks? *BioEssays: News and Reviews in Molecular, Cellular and Developmental Biology* 36(1), 21–26.

Barreiro, A. and Díaz-Raviña, M. (2021). Fire impacts on soil microorganisms: Mass, activity, and diversity. *Current Opinion in Environmental Science & Health* 22, 100264.

Bateman, I. J., Anderson, K., Argles, A., Belcher, C., Betts, R. A., Binner, A., Brazier, R. E., Cho, F. H. T., Collins, R. M., Day, B. H., Duran-Rojas, C., Eisenbarth, S., Gannon, K., Gatis, N., Groom, B., Hails, R., Harper, A. B., Harwood, A., Hastings, A., Heard, M. S., Hill, T. C., Inman, A., Lee, C. F., Luscombe, D. J., MacKenzie, A. R., Mancini, M. C., Morison, J. I. L., Morris, A., Quine, C. P., Snowdon, P., Tyler, C. R., Vanguelova, E. I., Wilkinson, M., Williamson, D. and Xenakis, G. (2023). A review of planting principles to identify the right place for the right tree for 'net zero plus' woodlands: Applying a place-based natural capital framework for sustainable, efficient and equitable (SEE) decisions. *People and Nature* 5(2), 271–301. doi: 10.1002/pan3.10331.

Bauer, J. T., Blumenthal, N., Miller, A. J., Ferguson, J. K. and Reynolds, H. L. (2017). Effects of between-site variation in soil microbial communities and plant-soil feedbacks on the productivity and composition of plant communities. *Journal of Applied Ecology* 54(4), 1028–1039.

Beiler, K. J., Durall, D. M., Simard, S. W., Maxwell, S. A. and Kretzer, A. M. (2010). Architecture of the wood-wide web: Rhizopogon spp. genets link multiple Douglas-fir cohorts. *New Phytologist* 185(2), 543–553.

Bender, S. F., Wagg, C. and van der Heijden, M. G. A. (2016). An underground revolution: Biodiversity and soil ecological engineering for agricultural sustainability. *Trends in Ecology and Evolution* 31(6), 440–452.

Bolan, N., Hoang, S. A., Beiyuan, J., Gupta, S., Hou, D., Karakoti, A., Joseph, S., Jung, S., Kim, K., Kirkham, M. B., Kua, H. W., Kumar, M., Kwon, E. E., Ok, Y. S., Perera, V., Rinklebe, J., Shaheen, S. M., Sarkar, B., Sarmah, A. K., Singh, B. P., Singh, G., Tsang, D. C. W., Vikrant, K., Vithanage, M., Vinu, A., Wang, H., Wijesekara, H., Yan, Y., Younis, S. A. and Van Zwieten, L. (2022). Multifunctional applications of biochar beyond carbon storage. *International Materials Reviews* 67(2), 150–200.

Borrelli, P., Robinson, D. A., Fleischer, L. R., Lugato, E., Ballabio, C., Alewell, C., Meusburger, K., Modugno, S., Schütt, B., Ferro, V., Bagarello, V., Oost, K. V., Montanarella, L. and Panagos, P. (2017). An assessment of the global impact of 21st century land use change on soil erosion. *Nature Communications* 8(1), 2013.

Brundrett, M. C. (2009). Mycorrhizal associations and other means of nutrition of vascular plants: Understanding the global diversity of host plants by resolving conflicting information and developing reliable means of diagnosis. *Plant and Soil* 320(1-2), 37–77.

Byrnes, R. C., Eastburn, D. J., Tate, K. W. and Roche, L. M. (2018). A global meta-analysis of grazing impacts on soil health indicators. *Journal of Environmental Quality* 47(4), 758–765.

Cheng, L., Booker, F. L., Tu, C., Burkey, K. O., Zhou, L., Shew, H. D., Rufty, T. W. and Hu, S. (2012). Arbuscular mycorrhizal fungi increase organic carbon decomposition under elevated CO_2. *Science* 337(6098), 1084–1087.

Coban, O., De Deyn, G. B. and van der Ploeg, M. (2022). Soil microbiota as game-changers in restoration of degraded lands. *Science* 375(6584), abe0725.

Cole, A. J., Griffiths, R. I., Ward, S. E., Whitaker, J., Ostle, N. J. and Bardgett, R. D. (2019). Grassland biodiversity restoration increases resistance of carbon fluxes to drought. *Journal of Applied Ecology* 56(7), 1806–1816.

Cotrufo, M. F., Ranalli, M. G., Haddix, M. L., Six, J. and Lugato, E. (2019). Soil carbon storage informed by particulate and mineral-associated organic matter. *Nature Geoscience* 12(12), 989–994.

Craven, D., Eisenhauer, N., Pearse, W. D., Hautier, Y., Isbell, F., Roscher, C., Bahn, M., Beierkuhnlein, C., Bönisch, G., Buchmann, N., Byun, C., Catford, J. A., Cerabolini, B. E. L., Cornelissen, J. H. C., Craine, J. M., De Luca, E., Ebeling, A., Griffin, J. N., Hector, A., Hines, J., Jentsch, A., Kattge, J., Kreyling, J., Lanta, V., Lemoine, N., Meyer, S. T., Minden, V., Onipchenko, V., Polley, H. W., Reich, P. B., van Ruijven, J., Schamp, B., Smith, M. D., Soudzilovskaia, N. A., Tilman, D., Weigelt, A., Wilsey, B. and Manning, P. (2018). Multiple facets of biodiversity drive the diversity-stability relationship. *Nature Ecology and Evolution* 2(10), 1579–1587.

Danso, S. K. A., Bowen, G. D. and Sanginga, N. (1992). Biological nitrogen fixation in trees in agro-ecosystems. *Plant and Soil* 141(1–2), 177–196.

Dassen, S., Putten, W. H. and De Deyn, G. B. (2021). Severance of arbuscular mycorrhizal fungal mycelial networks in restoration grasslands enhances seedling biomass. *New Phytologist* 232(2), 753–761.

De Deyn, G. B., Quirk, H., Oakley, S., Ostle, N. and Bardgett, R. D. (2011). Rapid transfer of photosynthetic carbon through the plant-soil system in differently managed species-rich grasslands. *Biogeosciences* 8(5), 1131–1139.

De Deyn, G. B., Shiel, R. S., Ostle, N. J., McNamara, N. P., Oakley, S., Young, I., Freeman, C., Fenner, N., Quirk, H. and Bardgett, R. D. (2011). Additional carbon sequestration benefits of grassland diversity restoration. *Journal of Applied Ecology* 48(3), 600–608.

De Long, J. R., Heinen, R., Steinauer, K., Hannula, S. E., Huberty, M., Jongen, R., Vandenbrande, S., Wang, M., Zhu, F. and Bezemer, T. M. (2019). Taking plant-soil feedbacks to the field in a temperate grassland. *Basic and Applied Ecology* 40, 30–42.

Denison, R. F. (2012). *Darwinian Agriculture: How Understanding Evolution Can Improve Agriculture*. Princeton, NJ: Princeton Press, p. 272.

Denison, R. F., Kiers, E. T. and West, S. A. (2003). Darwinian agriculture: When can humans find solutions beyond the reach of natural selection? *Quarterly Review of Biology* 78(2), 145–168.

Dijkstra, F. A. and Cheng, W. X. (2007). Interactions between soil and tree roots accelerate long-term soil carbon decomposition. *Ecology Letters* 10(11), 1046–1053.

Drigo, B., Anderson, I. C., Kannangara, G. S. K., Cairney, J. W. G. and Johnson, D. (2012). Rapid incorporation of carbon from ectomycorrhizal mycelial necromass into soil fungal communities. *Soil Biology and Biochemistry* 49, 4–10.

Duell, E. B., O'Hare, A. and Wilson, G. W. T. (2023). Inoculation with native soil improves seedling survival and reduces non-native reinvasion in a grassland restoration. *Restoration Ecology* 31(1), e13685.

Duffy, J. E., Cardinale, B. J., France, K. E., McIntyre, P. B., Thébault, E. and Loreau, M. (2007). The functional role of biodiversity in ecosystems: Incorporating trophic complexity. *Ecology Letters* 10(6), 522–538.

FAO and ITPS. (2015). *Status of the World's Soil Resources (SWSR)–Main Report*. Rome, Italy: Food and Agriculture Organisation of the United Nations and Intergovernmental Technical Panel on Soils. Available at: http://www.fao.org/3/a-i5199e.pdf.

Fry, E. L., Pilgrim, E. S., Tallowin, J. R. B., Smith, R. S., Mortimer, S. R., Beaumont, D. A., Simkin, J., Harris, S. J., Shiel, R. S., Quirk, H., Harrison, K. A., Lawson, C. S., Hobbs, P.

J. and Bardgett, R. D. (2017). Plant, soil and microbial controls on grassland diversity restoration: A long-term, multi-site mesocosm experiment. *Journal of Applied Ecology* 54(5), 1320–1330.

Gallagher, R. V., Allen, S., Mackenzie, B. D. E., Yates, C. J., Gosper, C. R., Keith, D. A., Merow, C., White, M. D., Wenk, E., Maitner, B. S., He, K., Adams, V. M. and Auld, T. D. (2021). High fire frequency and the impact of the 2019-2020 megafires on Australian plant diversity. *Diversity and Distributions* 27(7), 1166–1179.

Gardi, C., Jeffery, S. and Saltelli, A. (2013). An estimate of potential threats levels to soil biodiversity in EU. *Global Change Biology* 19(5), 1538–1548.

Gibbs, H. K. and Salmon, J. M. (2015). Mapping the world's degraded lands. *Applied Geography* 57, 12-21.

Gilbert, L. and Johnson, D. (2015). Plant-mediated 'apparent effects' between mycorrhiza and insect herbivores. *Current Opinion in Plant Biology* 26, 100–105.

Gilbert, L. and Johnson, D. (2017). Plant-plant communication through common mycorrhizal networks. In: Becard, G. (Ed.), *How Plants Communicate With Their Biotic Environment: Advances in Botanical Research* 82, 83–97.

Giweta, M. (2020). Role of litter production and its decomposition, and factors affecting the processes in a tropical forest ecosystem: A review. *Journal of Ecology and Environment* 44(1), 11.

Grelet, G. A., Johnson, D., Vrålstad, T., Alexander, I. J. and Anderson, I. C. (2010). New insights into the mycorrhizal Rhizoscyphus ericae aggregate: Spatial structure and co-colonization of ectomycorrhizal and ericoid roots. *New Phytologist* 188(1), 210–222.

Guerra, C. A., Rosa, I. M. D., Valentini, E., Wolf, F., Filipponi, F., Karger, D. N., Xuan, A. N., Mathieu, J., Lavelle, P. and Eisenhauer, N. (2020). Global vulnerability of soil ecosystems to erosion. *Landscape Ecology* 35, 823–842.

Harley, J. L. and Harley, E. L. (1987). A check-list of mycorrhiza in the British flora. *New Phytologist* 105(S1), 1–102.

Hazard, C. and Johnson, D. (2018). Does genotypic and species diversity of mycorrhizal plants and fungi affect ecosystem function? *New Phytologist* 220(4), 1122–1128.

Hazard, C., Kruitbos, L., Davidson, H., Mbow, F. T., Taylor, A. F. S. and Johnson, D. (2017a). Strain identity of the ectomycorrhizal fungus Laccaria bicolor is more important than richness in regulating plant and fungal performance under nutrient rich conditions. *Frontiers in Microbiology* 8, 1874.

Hazard, C., Kruitbos, L., Davidson, H., Taylor, A. F. and Johnson, D. (2017b). Contrasting effects of intra- and interspecific identity and richness of ectomycorrhizal fungi on host plants, nutrient retention and multifunctionality. *New Phytologist* 213(2), 852–863.

Helgason, T., Daniell, T. J., Husband, R., Fitter, A. H. and Young, J. P. (1998). Ploughing up the wood-wide web? *Nature* 394(6692), 431–431.

Hughes, A. R., Grabowski, J. H., Leslie, H. M., Scyphers, S. and Williams, S. L. (2018). Inclusion of biodiversity in habitat restoration policy to facilitate ecosystem recovery. *Conservation Letters* 11(3), 1–8.

Huston, M. A. (1997). Hidden treatments in ecological experiments: Re-evaluating the ecosystem function of biodiversity. *Oecologia* 110(4), 449–460.

Johnson, D. and Gilbert, L. (2015). Interplant signalling through hyphal networks. *New Phytologist* 205(4), 1448–1453.

Johnson, D., Martin, F., Cairney, J. W. G. and Anderson, I. C. (2012). The importance of individuals: Intraspecific diversity of mycorrhizal plants and fungi in ecosystems. *New Phytologist* 194(3), 614–628.

Johnson, N. C., Graham, J. H. and Smith, F. A. (1997). Functioning of mycorrhizal associations along the mutualism-parasitism continuum. *New Phytologist* 135(4), 575-585.

Kanters, C., Anderson, I. and Johnson, D. (2015). Chewing up the wood-wide web: Selective grazing on ectomycorrhizal fungi by Collembola. *Forests* 6(12), 2560-2570.

Kattge, J., Bönisch, G., Díaz, S., Lavorel, S., Prentice, I. C., Leadley, P., Tautenhahn, S., Werner, G. D. A., Aakala, T., Abedi, M., Acosta, A. T. R., Adamidis, G. C., Adamson, K., Aiba, M., Albert, C. H., Alcántara, J. M., Alcázar C, C., Aleixo, I., Ali, H., Amiaud, B., Ammer, C., Amoroso, M. M., Anand, M., Anderson, C., Anten, N., Antos, J., Apgaua, D. M. G., Ashman, T. L., Asmara, D. H., Asner, G. P., Aspinwall, M., Atkin, O., Aubin, I., Baastrup-Spohr, L., Bahalkeh, K., Bahn, M., Baker, T., Baker, W. J., Bakker, J. P., Baldocchi, D., Baltzer, J., Banerjee, A., Baranger, A., Barlow, J., Barneche, D. R., Baruch, Z., Bastianelli, D., Battles, J., Bauerle, W., Bauters, M., Bazzato, E., Beckmann, M., Beeckman, H., Beierkuhnlein, C., Bekker, R., Belfry, G., Belluau, M., Beloiu, M., Benavides, R., Benomar, L., Berdugo-Lattke, M. L., Berenguer, E., Bergamin, R., Bergmann, J., Bergmann Carlucci, M., Berner, L., Bernhardt-Römermann, M., Bigler, C., Bjorkman, A. D., Blackman, C., Blanco, C., Blonder, B., Blumenthal, D., Bocanegra-González, K. T., Boeckx, P., Bohlman, S., Böhning-Gaese, K., Boisvert-Marsh, L., Bond, W., Bond-Lamberty, B., Boom, A., Boonman, C. C. F., Bordin, K., Boughton, E. H., Boukili, V., Bowman, D. M. J. S., Bravo, S., Brendel, M. R., Broadley, M. R., Brown, K. A., Bruelheide, H., Brumnich, F., Bruun, H. H., Bruy, D., Buchanan, S. W., Bucher, S. F., Buchmann, N., Buitenwerf, R., Bunker, D. E., Bürger, J., Burrascano, S., Burslem, D. F. R. P., Butterfield, B. J., Byun, C., Marques, M., Scalon, M. C., Caccianiga, M., Cadotte, M., Cailleret, M., Camac, J., Camarero, J. J., Campany, C., Campetella, G., Campos, J. A., Cano-Arboleda, L., Canullo, R., Carbognani, M., Carvalho, F., Casanoves, F., Castagneyrol, B., Catford, J. A., Cavender-Bares, J., Cerabolini, B. E. L., Cervellini, M., Chacón-Madrigal, E., Chapin, K., Chapin, F. S., Chelli, S., Chen, S. C., Chen, A., Cherubini, P., Chianucci, F., Choat, B., Chung, K. S., Chytrý, M., Ciccarelli, D., Coll, L., Collins, C. G., Conti, L., Coomes, D., Cornelissen, J. H. C., Cornwell, W. K., Corona, P., Coyea, M., Craine, J., Craven, D., Cromsigt, J. P. G. M., Csecserits, A., Cufar, K., Cuntz, M., da Silva, A. C., Dahlin, K. M., Dainese, M., Dalke, I., Dalle Fratte, M., Dang-Le, A. T., Danihelka, J., Dannoura, M., Dawson, S., de Beer, A. J., De Frutos, A., De Long, J. R., Dechant, B., Delagrange, S., Delpierre, N., Derroire, G., Dias, A. S., Diaz-Toribio, M. H., Dimitrakopoulos, P. G., Dobrowolski, M., Doktor, D., Dřevojan, P., Dong, N., Dransfield, J., Dressler, S., Duarte, L., Ducouret, E., Dullinger, S., Durka, W., Duursma, R., Dymova, O., E-Vojtkó, A., Eckstein, R. L., Ejtehadi, H., Elser, J., Emilio, T., Engemann, K., Erfanian, M. B., Erfmeier, A., Esquivel-Muelbert, A., Esser, G., Estiarte, M., Domingues, T. F., Fagan, W. F., Fagúndez, J., Falster, D. S., Fan, Y., Fang, J., Farris, E., Fazlioglu, F., Feng, Y., Fernandez-Mendez, F., Ferrara, C., Ferreira, J., Fidelis, A., Finegan, B., Firn, J., Flowers, T. J., Flynn, D. F. B., Fontana, V., Forey, E., Forgiarini, C., François, L., Frangipani, M., Frank, D., Frenette-Dussault, C., Freschet, G. T., Fry, E. L., Fyllas, N. M., Mazzochini, G. G., Gachet, S., Gallagher, R., Ganade, G., Ganga, F., García-Palacios, P., Gargaglione, V., Garnier, E., Garrido, J. L., de Gasper, A. L., Gea-Izquierdo, G., Gibson, D., Gillison, A. N., Giroldo, A., Glasenhardt, M. C., Gleason, S., Gliesch, M., Goldberg, E., Göldel, B., Gonzalez-Akre, E., Gonzalez-Andujar, J. L., González-Melo, A., González-Robles, A., Graae, B. J., Granda, E., Graves, S., Green, W. A., Gregor, T., Gross, N., Guerin, G. R., Günther, A., Gutiérrez, A. G., Haddock, L., Haines, A., Hall, J., Hambuckers, A., Han, W., Harrison, S. P., Hattingh, W., Hawes, J. E., He, T., He, P., Heberling, J. M., Helm, A., Hempel, S., Hentschel, J.,

Hérault, B., Hereş, A. M., Herz, K., Heuertz, M., Hickler, T., Hietz, P., Higuchi, P., Hipp, A. L., Hirons, A., Hock, M., Hogan, J. A., Holl, K., Honnay, O., Hornstein, D., Hou, E., Hough-Snee, N., Hovstad, K. A., Ichie, T., Igić, B., Illa, E., Isaac, M., Ishihara, M., Ivanov, L., Ivanova, L., Iversen, C. M., Izquierdo, J., Jackson, R. B., Jackson, B., Jactel, H., Jagodzinski, A. M., Jandt, U., Jansen, S., Jenkins, T., Jentsch, A., Jespersen, J. R. P., Jiang, G. F., Johansen, J. L., Johnson, D., Jokela, E. J., Joly, C. A., Jordan, G. J., Joseph, G. S., Junaedi, D., Junker, R. R., Justes, E., Kabzems, R., Kane, J., Kaplan, Z., Kattenborn, T., Kavelenova, L., Kearsley, E., Kempel, A., Kenzo, T., Kerkhoff, A., Khalil, M. I., Kinlock, N. L., Kissling, W. D., Kitajima, K., Kitzberger, T., Kjøller, R., Klein, T., Kleyer, M., Klimešová, J., Klipel, J., Kloeppel, B., Klotz, S., Knops, J. M. H., Kohyama, T., Koike, F., Kollmann, J., Komac, B., Komatsu, K., König, C., Kraft, N. J. B., Kramer, K., Kreft, H., Kühn, I., Kumarathunge, D., Kuppler, J., Kurokawa, H., Kurosawa, Y., Kuyah, S., Laclau, J. P., Lafleur, B., Lallai, E., Lamb, E., Lamprecht, A., Larkin, D. J., Laughlin, D., Le Bagousse-Pinguet, Y., le Maire, G., le Roux, P. C., le Roux, E., Lee, T., Lens, F., Lewis, S. L., Lhotsky, B., Li, Y., Li, X., Lichstein, J. W., Liebergesell, M., Lim, J. Y., Lin, Y. S., Linares, J. C., Liu, C., Liu, D., Liu, U., Livingstone, S., Llusià, J., Lohbeck, M., López-García, Á., Lopez-Gonzalez, G., Lososová, Z., Louault, F., Lukács, B. A., Lukeš, P., Luo, Y., Lussu, M., Ma, S., Maciel Rabelo Pereira, C., Mack, M., Maire, V., Mäkelä, A., Mäkinen, H., Malhado, A. C. M., Mallik, A., Manning, P., Manzoni, S., Marchetti, Z., Marchino, L., Marcilio-Silva, V., Marcon, E., Marignani, M., Markesteijn, L., Martin, A., Martínez-Garza, C., Martínez-Vilalta, J., Mašková, T., Mason, K., Mason, N., Massad, T. J., Masse, J., Mayrose, I., McCarthy, J., McCormack, M. L., McCulloh, K., McFadden, I. R., McGill, B. J., McPartland, M. Y., Medeiros, J. S., Medlyn, B., Meerts, P., Mehrabi, Z., Meir, P., Melo, F. P. L., Mencuccini, M., Meredieu, C., Messier, J., Mészáros, I., Metsaranta, J., Michaletz, S. T., Michelaki, C., Migalina, S., Milla, R., Miller, J. E. D., Minden, V., Ming, R., Mokany, K., Moles, A. T., Molnár, A., Molofsky, J., Molz, M., Montgomery, R. A., Monty, A., Moravcová, L., Moreno-Martínez, A., Moretti, M., Mori, A. S., Mori, S., Morris, D., Morrison, J., Mucina, L., Mueller, S., Muir, C. D., Müller, S. C., Munoz, F., Myers-Smith, I. H., Myster, R. W., Nagano, M., Naidu, S., Narayanan, A., Natesan, B., Negoita, L., Nelson, A. S., Neuschulz, E. L., Ni, J., Niedrist, G., Nieto, J., Niinemets, Ü., Nolan, R., Nottebrock, H., Nouvellon, Y., Novakovskiy, A., Nutrient Network, Nystuen, K. O., O'Grady, A., O'Hara, K., O'Reilly-Nugent, A., Oakley, S., Oberhuber, W., Ohtsuka, T., Oliveira, R., Öllerer, K., Olson, M. E., Onipchenko, V., Onoda, Y., Onstein, R. E., Ordonez, J. C., Osada, N., Ostonen, I., Ottaviani, G., Otto, S., Overbeck, G. E., Ozinga, W. A., Pahl, A. T., Paine, C. E. T., Pakeman, R. J., Papageorgiou, A. C., Parfionova, E., Pärtel, M., Patacca, M., Paula, S., Paule, J., Pauli, H., Pausas, J. G., Peco, B., Penuelas, J., Perea, A., Peri, P. L., Petisco-Souza, A. C., Petraglia, A., Petritan, A. M., Phillips, O. L., Pierce, S., Pillar, V. D., Pisek, J., Pomogaybin, A., Poorter, H., Portsmuth, A., Poschlod, P., Potvin, C., Pounds, D., Powell, A. S., Power, S. A., Prinzing, A., Puglielli, G., Pyšek, P., Raevel, V., Rammig, A., Ransijn, J., Ray, C. A., Reich, P. B., Reichstein, M., Reid, D. E. B., Réjou-Méchain, M., de Dios, V. R., Ribeiro, S., Richardson, S., Riibak, K., Rillig, M. C., Riviera, F., Robert, E. M. R., Roberts, S., Robroek, B., Roddy, A., Rodrigues, A. V., Rogers, A., Rollinson, E., Rolo, V., Römermann, C., Ronzhina, D., Roscher, C., Rosell, J. A., Rosenfield, M. F., Rossi, C., Roy, D. B., Royer-Tardif, S., Rüger, N., Ruiz-Peinado, R., Rumpf, S. B., Rusch, G. M., Ryo, M., Sack, L., Saldaña, A., Salgado-Negret, B., Salguero-Gomez, R., Santa-Regina, I., Santacruz-García, A. C., Santos, J., Sardans, J., Schamp, B., Scherer-Lorenzen, M., Schleuning, M., Schmid, B., Schmidt, M., Schmitt, S., Schneider, J. V., Schowanek, S. D., Schrader, J., Schrodt, F., Schuldt, B., Schurr, F., Selaya Garvizu, G., Semchenko, M., Seymour, C., Sfair, J. C., Sharpe, J. M., Sheppard, C. S., Sheremetiev,

S., Shiodera, S., Shipley, B., Shovon, T. A., Siebenkäs, A., Sierra, C., Silva, V., Silva, M., Sitzia, T., Sjöman, H., Slot, M., Smith, N. G., Sodhi, D., Soltis, P., Soltis, D., Somers, B., Sonnier, G., Sørensen, M. V., Sosinski, E. E., Soudzilovskaia, N. A., Souza, A. F., Spasojevic, M., Sperandii, M. G., Stan, A. B., Stegen, J., Steinbauer, K., Stephan, J. G., Sterck, F., Stojanovic, D. B., Strydom, T., Suarez, M. L., Svenning, J. C., Svitková, I., Svitok, M., Svoboda, M., Swaine, E., Swenson, N., Tabarelli, M., Takagi, K., Tappeiner, U., Tarifa, R., Tauugourdeau, S., Tavsanoglu, C., Te Beest, M., Tedersoo, L., Thiffault, N., Thom, D., Thomas, E., Thompson, K., Thornton, P. E., Thuiller, W., Tichý, L., Tissue, D., Tjoelker, M. G., Tng, D. Y. P., Tobias, J., Török, P., Tarin, T., Torres-Ruiz, J. M., Tóthmérész, B., Treurnicht, M., Trivellone, V., Trolliet, F., Trotsiuk, V., Tsakalos, J. L., Tsiripidis, I., Tysklind, N., Umehara, T., Usoltsev, V., Vadeboncoeur, M., Vaezi, J., Valladares, F., Vamosi, J., van Bodegom, P. M., van Breugel, M., Van Cleemput, E., van de Weg, M., van der Merwe, S., van der Plas, F., van der Sande, M. T., van Kleunen, M., Van Meerbeek, K., Vanderwel, M., Vanselow, K. A., Vårhammar, A., Varone, L., Vasquez Valderrama, M. Y., Vassilev, K., Vellend, M., Veneklaas, E. J., Verbeeck, H., Verheyen, K., Vibrans, A., Vieira, I., Villacís, J., Violle, C., Vivek, P., Wagner, K., Waldram, M., Waldron, A., Walker, A. P., Waller, M., Walther, G., Wang, H., Wang, F., Wang, W., Watkins, H., Watkins, J., Weber, U., Weedon, J. T., Wei, L., Weigelt, P., Weiher, E., Wells, A. W., Wellstein, C., Wenk, E., Westoby, M., Westwood, A., White, P. J., Whitten, M., Williams, M., Winkler, D. E., Winter, K., Womack, C., Wright, I. J., Wright, S. J., Wright, J., Pinho, B. X., Ximenes, F., Yamada, T., Yamaji, K., Yanai, R., Yankov, N., Yguel, B., Zanini, K. J., Zanne, A. E., Zelený, D., Zhao, Y. P., Zheng, J., Zheng, J., Ziemińska, K., Zirbel, C. R., Zizka, G., Zo-Bi, I. C., Zotz, G. and Wirth, C. (2020). TRY plant trait database - Enhanced coverage and open access. *Global Change Biology* 26(1), 119–188.

Kebede, E. (2021). Contribution, utilization, and improvement of legumes-driven biological nitrogen fixation in agricultural systems. *Frontiers in Sustainable Food Systems* 5, 1–18.

Kirui, C. K., Njeru, E. M. and Runo, S. (2022). Diversity and phosphate solubilization efficiency of phosphate solubilizing bacteria isolated from semi-arid agroecosystems of eastern Kenya. *Microbiology Insights* 15, 11786361221088991. doi: 10.1177/11786361221088991.

Kopittke, P. M., Menzies, N. W., Wang, P., McKenna, B. A. and Lombi, E. (2019). Soil and the intensification of agriculture for global food security. *Environment International* 132, 105078.

Koskey, G., Mburu, S. W., Awino, R., Njeru, E. M. and Maingi, J. M. (2021). Potential use of beneficial microorganisms for soil amelioration, phytopathogen biocontrol, and sustainable crop production in smallholder agroecosystems. *Frontiers in Sustainable Food Systems* 5, 130.

Koskey, G., Mburu, S. W., Njeru, E. M., Kimiti, J. M., Ombori, O. and Maingi, J. M. (2017). Potential of native Rhizobia in enhancing nitrogen fixation and yields of climbing beans (*Phaseolus vulgaris* L.) in contrasting environments of eastern Kenya. In: Drakou, E. G., Orgiazzi, A., Hedlund, K. and Ritz, K. (Eds.), *Optimizing the Delivery of Multiple Ecosystem Goods and Services in Agricultural Systems*. Lausanne: Frontiers Media. doi: 10.3389/978-2-88945-296-5.

Koziol, L. and Bever, J. D. (2017). The missing link in grassland restoration: Arbuscular mycorrhizal fungi inoculation increases plant diversity and accelerates succession. *Journal of Applied Ecology* 54(5), 1301–1309.

Koziol, L. and Bever, J. D. (2019). Mycorrhizal feedbacks generate positive frequency dependence accelerating grassland succession. *Journal of Ecology* 107(2), 622–632.

Kulmatiski, A., Beard, K. H. and Stark, J. M. (2006). Soil history as a primary control on plant invasion in abandoned agricultural fields. *Journal of Applied Ecology* 43(5), 868–876.

Leake, J. R. (1994). The biology of myco-heterotrophic ('saprophytic') plants. *New Phytologist* 127(2), 171–216.

Leake, J. R., Johnson, D., Donnelly, D., Muckle, G., Boddy, L. and Read, D. (2004). Networks of power and influence: The role of mycorrhizal mycelium in controlling plant communities and agroecosystem functioning. *Canadian Journal of Botany* 82(8), 1016–1045.

Lehmann, A., Zheng, W. and Rillig, M. C. (2017). Soil biota contributions to soil aggregation. *Nature Ecology and Evolution* 1(12), 1828–1835.

Lehmann, J., Bossio, D. A., Kögel-Knabner, I. and Rillig, M. C. (2020). The concept and future prospects of soil health. *Nature Reviews. Earth and Environment* 1(10), 544–553.

Liang, M., Johnson, D., Burslem, D. F. R. P., Yu, S., Fang, M., Taylor, J. D., Taylor, A. F. S., Helgason, T. and Liu, X. (2020). Soil fungal networks maintain local dominance of ectomycorrhizal trees. *Nature Communications* 11(1), 2636.

Liang, M., Shi, L., Burslem, D. F. R. P., Johnson, D., Fang, M., Zhang, X. and Yu, S. (2021). Soil fungal networks moderate density-dependent survival and growth of seedlings. *New Phytologist* 230(5), 2061–2071.

Loreau, M., Naeem, S., Inchausti, P., Bengtsson, J., Grime, J. P., Hector, A., Hooper, D. U., Huston, M. A., Raffaelli, D., Schmid, B., Tilman, D. and Wardle, D. A. (2001). Biodiversity and ecosystem functioning: Current knowledge and future challenges. *Science* 294(5543), 804–808.

Lu, Y., Wang, R., Zhang, Y., Su, H., Wang, P., Jenkins, A., Ferrier, R. C., Bailey, M. and Squire, G. (2015). Ecosystem health towards sustainability. *Ecosystem Health and Sustainability* 1(1), 1–15.

Manning, P., van der Plas, F., Soliveres, S., Allan, E., Maestre, F. T., Mace, G., Whittingham, M. J. and Fischer, M. (2018). Redefining ecosystem multifunctionality. *Nature Ecology and Evolution* 2(3), 427–436.

Mburu, S. W., Koskey, G., Njeru, E. M., Ombori, O., Maingi, J. M. and Kimiti, J. M. (2020). Differential response of promiscuous soybean to local diversity of indigenous and commercial Bradyrhizobium inoculation under contrasting agroclimatic zones. *International Journal of Plant Production* 14(4), 571–582.

Mrunalini, K., Behera, B., Jayaraman, S., Abhilash, P. C., Dubey, P. K., Swamy, G. N., Prasad, J. V. N. S., Rao, K. V., Krishnan, P., Pratibha, G. and Srinivasa Rao, C. (2022). Nature-based solutions in soil restoration for improving agricultural productivity. *Land Degradation and Development* 33(8), 1269–1289.

Muindi, M. M., Muthini, M., Njeru, E. M. and Maingi, J. (2021). Symbiotic efficiency and genetic characterization of rhizobia and non rhizobial endophytes associated with cowpea grown in semi-arid tropics of Kenya. *Heliyon* 7(4), e06867.

Mursinoff, S. and Tack, A. J. (2017). Spatial variation in soil biota mediates plant adaptation to a foliar pathogen. *New Phytologist* 214(2), 644–654.

Musyoka, D. M., Njeru, E. M., Nyamwange, M. M. and Maingi, J. M. (2020). Arbuscular mycorrhizal fungi and Bradyrhizobium co-inoculation enhances nitrogen fixation and growth of green grams (*Vigna radiata* L.) under water stress. *Journal of Plant Nutrition* 43(7), 1036-1047.

Nara, K. (2006). Ectomycorrhizal networks and seedling establishment during early primary succession. *New Phytologist* 169(1), 169-178.

Nasholm, T. and Persson, J. (2001). Plant acquisition of organic nitrogen in boreal forests. *Physiologia Plantarum* 111(4), 419-426.

Njeru, E. M. and Koskey, G. (2021). Using beneficial microorganisms to promote sustainable crop production and resilience of smallholder agroecosystems to changing climate. In: Hebsale Mallappa, V. K. and Shirur, M. (Eds.), *Climate Change and Resilient Food Systems*. Singapore: Springer.

Njeru, E. M., Muthini, M., Muindi, M. M., Ombori, O., Nchore, S. B., Runo, S. and Maingi, J. M. (2020). Exploiting arbuscular mycorrhizal fungi-rhizobia-legume symbiosis to increase smallholder farmers' crop production and resilience under a changing climate. In: Singh, B. R., Safalaoh, A., Amuri, N. A., Eik, L. O., Sitaula, B. K. and Lal, R. (Eds.), *Climate Impacts on Agricultural and Natural Resource Sustainability in Africa*. Cham: Springer International Publishing, pp. 471-488.

Nunes, M. R., Karlen, D. L. and Moorman, T. B. (2020). Tillage intensity effects on soil structure indicators–A US meta-analysis. *Sustainability* 12(5), 2071.

Nyaga, J. W. and Njeru, E. M. (2020). Potential of native Rhizobia to improve cowpea growth and production in semiarid regions of Kenya. *Frontiers in Agronomy* 2, 606293.

O'Callaghan, M. (2016). Microbial inoculation of seed for improved crop performance: Issues and opportunities. *Applied Microbiology and Biotechnology* 100(13), 5729-5746.

Pausas, J. G. and Ribeiro, E. (2017). Fire and plant diversity at the global scale. *Global Ecology and Biogeography* 26(8), 889-897.

Perkins, L. B. and Bennett, J. R. (2018). A field test of commercial soil microbial treatments on native grassland restoration. *Restoration Ecology* 26(5), 851-857.

Pickles, B. J., Wilhelm, R., Asay, A. K., Hahn, A. S., Simard, S. W. and Mohn, W. W. (2017). Transfer of ^{13}C between paired Douglas-fir seedlings reveals plant kinship effects and uptake of exudates by ectomycorrhizas. *New Phytologist* 214(1), 400-411.

Poole, P., Ramachandran, V. and Terpolilli, J. (2018). Rhizobia: From saprophytes to endosymbionts. *Nature Reviews. Microbiology* 16(5), 291-303.

Rashid, M. I., Mujawar, L. H., Shahzad, T., Almeelbi, T., Ismail, I. M. and Oves, M. (2016). Bacteria and fungi can contribute to nutrients bioavailability and aggregate formation in degraded soils. *Microbiological Research* 183, 26-41.

Renard, D. and Tilman, D. (2019). National food production stabilized by crop diversity. *Nature* 571(7764), 257-260.

Roberts, A. J. and Johnson, N. C. (2021). Effects of mob-grazing on soil and range quality vary with plant species and season in a semiarid grassland. *Rangeland Ecology and Management* 79, 139-149.

Roger-Estrade, J., Anger, C., Bertrand, M. and Richard, G. (2010). Tillage and soil ecology: Partners for sustainable agriculture. *Soil and Tillage Research* 111(1), 33-40.

Salomon, M. J., Watts-Williams, S. J., McLaughlin, M. J., Bücking, H., Singh, B. K., Hutter, I., Schneider, C., Martin, F. M., Vosatka, M., Guo, L., Ezawa, T., Saito, M., Declerck, S., Zhu, Y. G., Bowles, T., Abbott, L. K., Smith, F. A., Cavagnaro, T. R. and van der Heijden, M. G.

A. (2022). Establishing a quality management framework for commercial inoculants containing arbuscular mycorrhizal fungi. *iScience* 25(7), 104636.

Schnoor, T. K., Mårtensson, L. M. and Olsson, P. A. (2011). Soil disturbance alters plant community composition and decreases mycorrhizal carbon allocation in a sandy grassland. *Oecologia* 167(3), 809-819.

Selosse, M. A., Richard, F., He, X. and Simard, S. W. (2006). Mycorrhizal networks: Des liaisons dangereuses? *Trends in Ecology and Evolution* 21(11), 621-628.

Semchenko, M., Leff, J. W., Lozano, Y. M., Saar, S., Davison, J., Wilkinson, A., Jackson, B. G., Pritchard, W. J., De Long, J. R., Oakley, S., Mason, K. E., Ostle, N. J., Baggs, E. M., Johnson, D., Fierer, N. and Bardgett, R. D. (2018). Fungal diversity regulates plant-soil feedbacks in temperate grassland. *Science Advances* 4(11), eaau4578.

Sergaki, C., Lagunas, B., Lidbury, I., Gifford, M. L. and Schäfer, P. (2018). Challenges and approaches in microbiome research: From fundamental to applied. *Frontiers in Plant Science* 9, 1205.

Sessions, J., Smith, D., Trippe, K. M., Fried, J. S., Bailey, J. D., Petitmermet, J. H., Hollamon, W., Phillips, C. L. and Campbell, J. D. (2019). Can biochar link forest restoration with commercial agriculture? *Biomass and Bioenergy* 123, 175-185.

Siasou, E., Standing, D., Killham, K. and Johnson, D. (2009). Mycorrhizal fungi increase biocontrol potential of Pseudomonas fluorescens. *Soil Biology and Biochemistry* 41(6), 1341-1343.

Sierra, J. and Nygren, P. (2006). Transfer of N fixed by a legume tree to the associated grass in a tropical silvopastoral system. *Soil Biology and Biochemistry* 38(7), 1893-1903.

Sirami, C., Gross, N., Baillod, A. B., Bertrand, C., Carrié, R., Hass, A., Henckel, L., Miguet, P., Vuillot, C., Alignier, A., Girard, J., Batáry, P., Clough, Y., Violle, C., Giralt, D., Bota, G., Badenhausser, I., Lefebvre, G., Gauffre, B., Vialatte, A., Calatayud, F., Gil-Tena, A., Tischendorf, L., Mitchell, S., Lindsay, K., Georges, R., Hilaire, S., Recasens, J., Solé-Senan, X. O., Robleño, I., Bosch, J., Barrientos, J. A., Ricarte, A., Marcos-Garcia, M. Á., Miñano, J., Mathevet, R., Gibon, A., Baudry, J., Balent, G., Poulin, B., Burel, F., Tscharntke, T., Bretagnolle, V., Siriwardena, G., Ouin, A., Brotons, L., Martin, J. L. and Fahrig, L. (2019). Increasing crop heterogeneity enhances multitrophic diversity across agricultural regions. *Proceedings of the National Academy of Sciences of the United States of America* 116(33), 16442-16447.

Smith, P., Keesstra, S. D., Silva, W. L. and Adhya, T. K. (2021). The role of soils in delivering Nature's Contributions to People. *Philosophical Transactions of the Royal Society B* 376(1834), 20200169.

Smith, R. S., Shiel, R. S., Bardgett, R. D., Millward, D., Corkhill, P., Rolph, G., Hobbs, P. J. and Peacock, S. (2003). Soil microbial community, fertility, vegetation and diversity as targets in the restoration management of a meadow grassland. *Journal of Applied Ecology* 40(1), 51-64.

Smith, R. S., Shiel, R. S., Bardgett, R. D., Millward, D., Corkhill, P., Evans, P., Quirk, H., Hobbs, P. J. and Kometa, S. T. (2008). Long-term change in vegetation and soil microbial communities during the phased restoration of traditional meadow grassland. *Journal of Applied Ecology* 45(2), 670-679.

Soudzilovskaia, N. A., Vaessen, S., Barcelo, M., He, J., Rahimlou, S., Abarenkov, K., Brundrett, M. C., Gomes, S. I. F., Merckx, V. and Tedersoo, L. (2020). FungalRoot: Global online database of plant mycorrhizal associations. *New Phytologist* 227(3), 955-966.

Tisdall, J. M., Smith, S. E. and Rengasamy, P. (1997). Aggregation of soil by fungal hyphae. *Soil Research* 35(1), 55–60.

USDA, N. R. C. S. (2019). *Soil Health*. Washington, DC: Natural Resources Conservation Service, United States Department of Agriculture. Available at: https://www.nrcs.usda.gov/wps/portal/nrcs/main/soils/health/ (Accessed 3 September 2022).

Van der Putten, W. H., Bardgett, R. D., Bever, J. D., Bezemer, T. M., Casper, B. B., Fukami, T., Kardol, P., Klironomos, J. N., Kulmatiski, A., Schweitzer, J. A., Suding, K. N., Van de Voorde, T. F. J. and Wardle, D. A. (2013). Plant-soil feedbacks: The past, the present and future challenges. *Journal of Ecology* 101(2), 265–276.

Van Valen, L. (1973). A new evolutionary law. *Evolutionary Theory* 1, 1–30.

Velasco Ayuso, S., Giraldo Silva, A., Nelson, C., Barger, N. N. and Garcia-Pichel, F. (2017). Microbial nursery production of high-quality biological soil crust biomass for restoration of degraded dryland soils. *Applied and Environmental Microbiology* 83(3), e02179-16.

Verbruggen, E., Veresoglou, S. D., Anderson, I. C., Caruso, T., Hammer, E. C., Kohler, J. and Rillig, M. C. (2013). Arbuscular mycorrhizal fungi – Short-term liability but long-term benefits for soil carbon storage? *New Phytologist* 197(2), 366–368.

von Felten, S., Hector, A., Buchmann, N., Niklaus, P. A., Schmid, B. and Scherer-Lorenzen, M. (2009). Belowground nitrogen partitioning in experimental grassland plant communities of varying species richness. *Ecology* 90(5), 1389–1399.

Wakarera, P. W., Ojola, P. and Njeru, E. M. (2022). Characterization and diversity of native Azotobacter spp. isolated from semi-arid agroecosystems of Eastern Kenya. *Biology Letters* 18(3), 20210612.

Wilkinson, A., Alexander, I. and Johnson, D. (2012). Genotype identity determines productivity and CO_2 efflux across a genotype-species gradient of ectomycorrhizal fungi. *Fungal Ecology* 5(5), 571–580.

Wilkinson, A., Alexander, I. J. and Johnson, D. (2011). Species richness of ectomycorrhizal hyphal necromass increases soil CO_2 efflux under laboratory conditions. *Soil Biology and Biochemistry* 43(6), 1350–1355.

Wilkinson, A., Solan, M., Taylor, A. F., Alexander, I. J. and Johnson, D. (2010). Intraspecific diversity regulates fungal productivity and respiration. *PLoS ONE* 5(9), e12604.

Wubs, E. R. J., van Heusden, T., Melchers, P. D. and Bezemer, T. M. (2019). Soil inoculation steers plant-soil feedback, suppressing ruderal plant species. *Frontiers in Ecology and Evolution* 7, 451.

Yao, Q., Yan, W. and Wei, J. (2018). Establishment of a quality test for arbuscular mycorrhizal inoculum. *International Journal of Agriculture and Biology* 20, 926–932.

Yelenik, S. G. and Levine, J. M. (2011). The role of plant-soil feedbacks in driving native-species recovery. *Ecology* 92(1), 66–74.

Zhang, S., Lehmann, A., Zheng, W., You, Z. and Rillig, M. C. (2019). Arbuscular mycorrhizal fungi increase grain yields: A meta-analysis. *New Phytologist* 222(1), 543-555.

Zhou, G., Zhou, X., He, Y., Shao, J., Hu, Z., Liu, R., Zhou, H. and Hosseinibai, S. (2017). Grazing intensity significantly affects belowground carbon and nitrogen cycling in grassland ecosystems: A meta-analysis. *Global Change Biology* 23(3), 1167–1179.

Chapter 6

The impact and design of field margins in promoting biodiversity in agricultural landscapes

Jane Morrison, Bishop's University, Canada

1 Introduction

The transformation of agricultural landscapes involves the conversion of forests, scrub, and grasslands into fields, and pastures, as well as the expansion and amalgamation of preexisting fields to enhance farming efficiency (Tscharntke et al., 2005). More recent changes have resulted in more homogeneous agricultural landscapes with greater fragmentation of natural habitats and resulting negative effects on biodiversity. Insects and animals displaced by land-use change must find shelter in remaining fragmented patches of semi-natural land, and those that do not perish (because they do not find alternative areas) must travel further to find appropriate unoccupied habitat and food resources (Marshall and Moonen, 2002). While the trend has been to simplify agricultural production systems, heterogeneous agroecosystems with diverse and connected natural areas are potentially more resilient than simple, homogeneous landscapes (e.g. in adapting to disturbances such as more extreme weather events associated with climate change) (Tscharntke et al., 2005). As a result, the importance of uncropped natural and semi-natural areas in agricultural landscapes is better recognized. For example, the European Union Common Agricultural Policy (CAP) for 2023–2027 requires that at least

http://dx.doi.org/10.19103/AS.2024.0134.08

4% of eligible land be devoted to biodiversity and landscape protection on all farms of at least 10 ha (European Commission, 2023).

A cost-effective, minimally invasive, and efficient solution is to set aside field margins, or improve existing field margins (sometimes known as buffer strips) to create a network of semi-natural vegetation across the farming landscape (Lagerlöf et al., 1992). Leaving field margins uncropped can be viewed, in some circumstances, as an advantageous trade-off for farmers, as these areas are typically less fertile and more prone to drought, shading, and lower yields than other parts of the field (Pywell et al., 2005). Nevertheless, the biodiversity supported by field margins is still only a fraction of what would have been present before agriculture.

A field margin has been defined as 'the whole of the crop edge, any margin strip present and the semi-natural habitat associated with the boundary' (Greaves and Marshall, 1987). Field margins have also been defined as linear zones of semi-natural vegetation located at arable field boundaries, usually comprising a boundary structure (Marshall and Moonen, 2002). The boundary may be a fence, line of trees, hedge, wall, terrace, track, watercourse, grass strip, or a combination of these elements. In arable landscapes, there is almost always some form of margin dividing individual fields from one another, as well as separating fields from features such as roads, tracks, and watercourses. In some situations, the boundary is simply another type of habitat (e.g. a wooded area). A variety of plant communities can occur within field margins (Marshall, 2005). Figure 1 shows the common components of an arable field margin, including the boundary structure or barrier, vegetative area or field margin strip, conservation headland (a part of the field planted with a crop but where fertilizer and pesticide application is reduced in order to promote greater floral and faunal diversity and protect the field margin) and the crop edge.

Field margins vary widely in their size, shape, composition, and level of plant diversity. These differences affect the ecosystem services these margins provide (Cirujeda et al., 2019; Pallavicini et al., 2020). Figure 2 provides several different examples of field margins in cereal agroecosystems.

Past studies have utilized simple classification systems for field margins, such as differentiating between woodland and ruderal vegetation (species that colonize uncultivated or unused land) (Marshall and Moonen, 2002). A study by Cirujeda et al. (2019) classified field margins (referred to as 'field boundaries') based on five descriptors:

- Presence of a bank;
- Width;
- Percentage cover of woody and evergreen perennials;
- Presence of a stonewall; and
- Presence of trees.

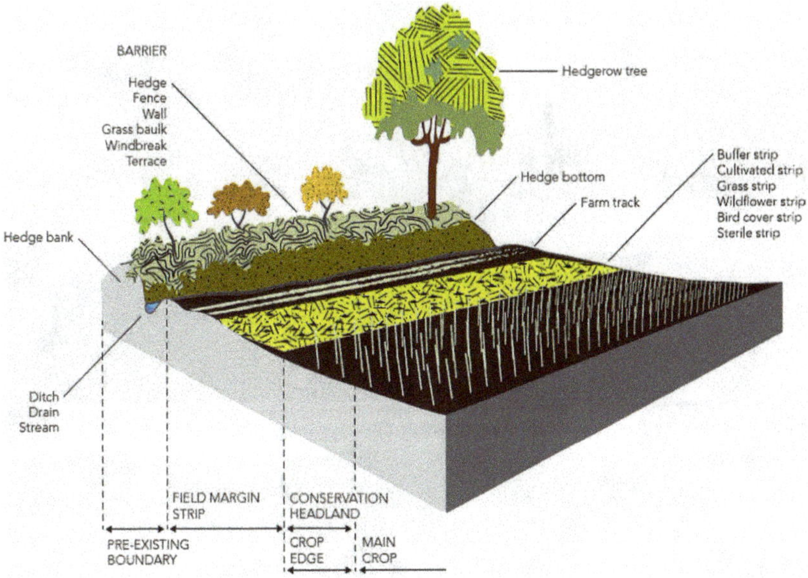

Figure 1 The common components of an arable field margin. Source: Greaves and Marshall (1987) and Marshall and Moonen (2002).

Field margins are categorized based on level of complexity, ranging from structurally simple (flat, narrow margins dominated by annual species) to structurally complex (more than 3 m wide, presence of a bank, dominated by woody and evergreen perennials) (Fig. 3).

Field margins can be permanent or temporary. Field margins comprising a diversity of annual herbaceous species, sometimes referred to as floral strips, can be a lower-investment alternative than permanent field margins comprised of perennials. These types of margins also provide a variety of ecosystem services but have been found to be less effective than perennial-dominated field margins in the delivery of some key ecosystem services (e.g. as a barrier to wind and in providing permanent habitats) (Uyttenbroeck et al., 2016). Floral strips can be rotated throughout the field, potentially as part of a larger crop rotation plan. Strips that are rotated annually may not be as effective in terms of pollinator support (McHugh et al., 2022), but may be a compromise solution when permanent land dedicated to field margins is not available. This chapter reviews the ways field margins benefit biodiversity, their design and management, enhancement with seed mixes, as well as economic aspects.

2 Beneficial impacts

This section focuses on the benefits of field margins in terms of enhancing biodiversity in agroecosystems by supporting plant diversity, increasing the

Figure 2 Examples of field margins in cereal agroecosystems (Catalonia, Spain, 2014–2016). (a) Flat margin between two cereal fields (mean width ≈4 m; ≈5% shrubs and trees). (b) Sloped, rocky margin between two cereal fields (mean width ≈4 m; only herbaceous plants). (c) Sloped margin between fallow land and a cereal field (mean width ≈5 m; ≈10% shrubs and trees). (d) Flat, rocky margin between a low-traffic road and a cereal field (mean width ≈4 m; ≈7% shrubs and trees). (e) Sloped margin between two cereal fields (mean width ≈3 m; ≈30% shrubs and trees). (f) Sloped, rocky margin between two cereal fields (mean width ≈8 m; ≈15% shrubs and trees). (g) Sloped margin between a low-traffic road and a cereal field (mean width ≈5 m; ≈30% shrubs and trees). (h) Flat margin between two cereal fields (mean width ≈3 m; only herbaceous plants). (i) Sloped, rocky margin between two cereal fields (mean width ≈4 m; ≈40% shrubs and trees).

abundance and diversity of beneficial insects, providing wildlife corridors, and contributing to landscape heterogeneity.

Field margins can enhance plant species richness, including rare species, and diversity. They are promoted as a means of increasing biodiversity in an agricultural system, offering many well-documented ecosystem services, including the provision of habitat and resources for beneficial insects (both pollinators and natural enemies of pests), acting as wildlife corridors, as well as other benefits for farming operations, environmental conservation, recreation, and rural development (Altieri, 1995; Balzan et al., 2014; Crowther et al., 2023; Dover, 2019; Hackett and Lawrence, 2014; Hannon and Sisk, 2009; Lagerlöf et al., 1992; Marshall and Moonen, 2002; Marshall, 2005; Ryszkowski, 2001; Sutter et al., 2018). The potential functions and roles of semi-natural field margins are summarized in Table 1.

%WEP <60% / >60%	SLOPE FLAT or BANK	WIDTH <3m >3m	TREE	STONE WALL	FINAL STRUCTURE
<60%	FLAT	<3m	NO	NO	TYPE A 
<60%	BANK	<3m or>3m	NO	NO	TYPE B 
>60%	BANK	<3m or >3m	NO	NO	TYPE C 
>60%	BANK	<3m or >3m	YES	NO	TYPE D 
-	-	-	NO/YES	YES	TYPE E 

Figure 3 Field boundary categories (WEP, woody and evergreen perennials). Source: Cirujeda et al. (2019).

2.1 Plant diversity

Declining plant biodiversity is a widespread phenomenon occurring in different habitats and geographic regions across the world (Eichenberg et al., 2020). Transforming an undisturbed landscape into an agroecosystem displaces naturally occurring plants in that landscape (Dorado and López-Fando, 2006; Mas et al., 2007; Romero et al., 2008). Once a species is removed from a landscape, it is possible that it either takes many years to re-establish or

Table 1 Potential functions and roles of field margins (Altieri, 1995; Balzan et al., 2014; Crowther et al., 2023; Dover, 2019; Hackett and Lawrence, 2014; Hannon and Sisk, 2009; Lagerlöf et al., 1992; Marshall and Moonen, 2002; Marshall, 2005; Ryszkowski, 2001; Sutter et al., 2018)

Function	Role
Agronomy	To define the field edge
	To increase water retention in the field (reduce run-off)
	To promote ecological stability in crops
	To provide shelter for crops, particularly as windbreaks
	To reduce soil erosion by wind or water
	To enhance crop pollination services
	To facilitate biological pest control and reduce crop damage
	To create microclimates
	To provide organic matter and improve soil health
	To optimize crop nutrition (by helping to retain fertilizers in the field)
	To increase crop yields
	To keep animals (or trespassers) in or out
	To provide shelter for animals
	To provide a source of fruits or wood
Environment	To promote biodiversity and farm wildlife conservation
	To act as a refuge or corridor for beneficial plants and animals
	To support the biodiversity of soil microorganisms
	To buffer pesticide drift
	To reduce pesticide use (by facilitating biological pest control)
	To reduce the movement of fertilizer and other pollutants, especially in run-off (i.e. reducing eutrophication)
	To maintain landscape diversity
	To improve soil carbon sequestration
Recreation and rural development	To provide access (i.e. walking or driving paths)
	To promote populations of game species and facilitate hunting activities
	To enhance landscape aesthetics

never becomes re-established at all (Firbank, 1999). Field margins provide an opportunity for threatened wild plant species to persist across the agricultural landscape (Fried et al., 2009). Field margins may often harbor naturally occurring plants that can play a significant role in supporting local biodiversity and ecosystem services (Norris and Kogan, 2005).

Published by Burleigh Dodds Science Publishing Limited, 2024.

2.2 Supporting beneficial insects: pollinator and pest predator species

Intensively managed agriculture has had negative effects on beneficial insects, including pollinators and natural enemies of pests, due to loss of habitat and food resources, increased agrochemical use, and introduction of alien species (Hopwood, 2008; Koh et al., 2016). Field margins potentially provide nesting habitats for beneficial insect species, continuity and diversity of floral resources as a food source, and can act as a potential refuge from pesticides and other agrochemicals (Nicholls and Altieri, 2013; Requier et al., 2015).

We are currently experiencing a 'global pollinator crisis' – a marked reduction in the diversity, density, and distribution of pollinators around the world (Shivanna et al., 2020). Maintaining diverse and abundant communities of effective pollinators is crucial for successful agricultural production and sustaining native plant diversity (Balzan et al., 2014).

In many circumstances, wild pollinators are key for stable crop pollination or are an important supplement to managed honeybees. For example, buzz-pollinated crops, which include species from 65 families and comprise tomatoes and potatoes, cannot be pollinated by honeybees and rely on certain wild bee species with the ability to use vibration to extract pollen from the anthers (De Luca and Vallejo-Marín, 2013). For seed production, specialist bees or bees with preferred diets may be required, such as *Megachile rotundata*, which is used extensively to pollinate alfalfa (Goettel et al., 1991). In addition, because of the instability of honeybee colonies in recent decades as a result of *Varroa* mites, the spread of viruses, and colony collapse disorder, the importance of wild pollinators for stable crop pollination has become increasingly evident (Drummond, 2012; Buchmann, 2022). Field research with a range of crops (e.g. coffee and watermelon) has suggested that abundant and diverse native bee communities can compensate for declining honeybee populations (Klein et al., 2003; Winfree et al., 2007). However, compared to honeybees, solitary bees do not travel nearly as far and have inherent survival challenges in agricultural landscapes, such as reduced spatial flexibility and recolonization ability, and are not able to forage for resources at larger spatial scales (Tscharntke et al., 2005).

The negative effects of agriculture on wild bees and other pollinator populations have been well documented (Koh et al., 2016; Le Féon et al., 2010). Several entomologists and ecologists have hypothesized that an increased separation from critical floral and nesting resources is likely the main cause of wild bee declines and, consequently, reduced pollination services (Nicholls and Altieri, 2013). Monocultures, fields of a single variety of a single species often grown on a large scale (Cooper et al., 2007), have resulted in loss of habitat and reduced floral resources (i.e. lower floral diversity with greater fragmentation)

(Kline and Joshi, 2020). Monocultures of flowering crops usually provide only a few weeks of abundant food for wild pollinators (Nicholls and Altieri, 2013), and in some countries, the expansion of agricultural land devoted to wind-pollinated grain crops has greatly reduced forage options (e.g. the 'corn belt' in the Midwestern USA) (Cane and Tepedino, 2001).

Pollinators require nearby nesting areas and floral resources that bloom before and after the main crop (Nicholls and Altieri, 2013). Numerous studies have shown that field margins help to preserve both pollinator abundance and diversity within agroecosystems, and pollination services in general (Pywell et al., 2005; Carvell et al., 2007; Hannon and Sisk, 2009; Marini et al., 2012; Morandin and Kremen, 2013). Field margins with diverse plant species offer several ecosystem services to support pollinators. Firstly, they have the potential to provide continuous nesting and floral resources. A diversity of flowering plants provides pollinators with a stable supply of pollen, nectar, and floral oils (Hannon and Sisk, 2009). A consistent floral supply is especially critical early and late in the year and when the main crop is not in bloom. For example, most solitary bee species hibernate over winter and need food sources immediately after emerging in the spring. Bumblebee queens also require late-season floral resources to build up their reserves before hibernating (Nicholls and Altieri, 2013). Because different pollinator species have different floral preferences, a high diversity of plant species, preferably flowering all year long depending on the climate, is required to support a high diversity of pollinator species (Nicholls and Altieri, 2013). Field margins indirectly support crop pollination services. Since bees return to fixed nest sites after foraging, the proximity of nesting habitats to agricultural fields is critical for pollinating crops (Ricketts et al., 2006). Indeed, wild bee pollination services have been shown to have a negative relationship with distance from the natural habitat (Kwaiser and Hendrix, 2008). For example, Kremen et al. (2004) found that pollination services from native bees were positively related to the proportion of upland natural habitat in the vicinity of farm sites in California. A model based on the data determined that if farmers were to rely entirely on native bee communities for pollination, their farms would need to be situated in areas containing at least 40% natural habitat within a 2.4 km radius or at least 30% within a 1.2 km radius.

Natural enemies, including parasitoids and predators, can help regulate insect pest populations to levels that minimize plant damage (Coppel and Mertins, 1977). Integrated pest management strategies attempt to increase plant diversity to subsequently increase the abundance of natural enemies (Harvey and Eubanks, 2005). In addition to helping to conserve pollinators, naturally occurring plants in field margins provide hibernation services for natural enemies of pests (Altieri, 1995). By conserving wild plants in field margins, populations of these beneficial insects can be maintained or enhanced by providing shelter, breeding sites, and food, either directly (i.e. with plant matter, pollen, or nectar) or by hosting their preferred insect food sources (Norris and Kogan, 2005). These plants also provide continual shelter during times of the year when the crop is absent or not able to support insects (e.g. during the early stages of growth). For example, a study by Ndakidemi et al. (2022) found that natural enemies (i.e. lady beetles, assassin bugs, hoverflies, and lacewings) benefited from field margin plants in smallholder bean farming systems in sub-Saharan Africa. They observed that certain plant species were preferred by different natural enemy groups since they provided different benefits.

2.3 Wildlife corridors

Field boundaries provide habitat for many species. In addition to plants and beneficial insects, field margins may also support a diverse fauna, including other invertebrates, mammals, and birds (Marshall and Moonen, 2002). Unfortunately, suitable habitats have become more fragmented due to field amalgamation, resulting in large gaps and discontinuous networks (Barr and Gillespie, 2000).

In addressing this problem, another benefit of field margins is that they can act as wildlife corridors, connecting otherwise isolated patches of suitable habitat (Rosenberg et al., 1997). A network of linear field margins can facilitate the movement of fauna and flora throughout the landscape. They can link patches of fragmented habitats to provide sufficient territory for animals to maintain population integrity or facilitate recolonization of a habitat after local extinction (Marshall and Moonen, 2002; Cirujeda and Pardo, 2020). For example, wood mice move out of fields and into woodlands after harvest. A network of field margins would help them to move and disperse safely (Feber and Macdonald, 2013).

Ideally, margins should link to each other and link other habitats across the farm. Increasing these linkages will help to maximize their effectiveness as movement corridors for wildlife around the farm and across the landscape (Feber and Macdonald, 2013). The likelihood of a field margin functioning as a wildlife corridor is dependent on having larger remnant patches in the network. Without the remnant patches, the role of field margins acting as wildlife corridors is greatly diminished (Marshall and Moonen, 2002; Bennett, 2003). It is also important to note that the corridor function provided by field margins may not be equally beneficial for all species. Specific benefits have not been quantified for the majority of species (Marshall and Moonen, 2002), and a corridor for one species may be a barrier for another (Dover and Settele, 2009). Corridors can also have unintended negative effects. Five categories of potential negative ecological effects on native biodiversity have been identified (Haddad et al., 2014):

- Dispersal of species antagonistic to species that are conservation targets (predators or pathogens; e.g. increased rates of seed predation by small mammals);
- Creation of negative edge effects (e.g. a change in corridor habitat leading to an increase in parasitic brown-headed cowbirds *Molothrus ater*);
- Spread and increased abundance of invasive species;
- Spread of disturbances (e.g. fire); and
- Synchronization of population dynamics (because connecting patches increases the likelihood of simultaneous extinction).

However, a literature review and meta-analysis by Haddad et al. (2014) to evaluate the prevalence of negative effects of habitat corridors concluded that any negative effects were relatively small and manageable compared to the large positive effects. For example, they found no evidence that corridors increased non-native species invasion, although they noted that this had not been well studied.

2.4 Ecosystem service multifunctionality

In addition to supporting plant diversity, increasing the abundance and diversity of beneficial insects, and providing wildlife corridors, field margins can provide numerous other ecosystem services within agroecosystems (Table 1). When assessing the impact of field margins in an agroecosystem, it is important to consider the combined contribution of multiple ecosystem services, rather than focusing on a single ecosystem service. The ability of a management strategy to simultaneously provide multiple ecosystem services is referred to as ecosystem service multifunctionality (Manning et al., 2018). This is an important consideration when calculating the return on investment of implementing or restoring field margins.

2.5 Contribution to landscape heterogeneity

It is important to recognize the overarching importance of biodiversity in restoring complexity to landscapes simplified by more intensive, monocultural forms of agricultural production (Tscharntke et al., 2005; Morrison et al., 2017). Field margins should be one component in a wider set of measures to restore heterogeneity. Other measures include permanent set-aside land where natural vegetation can recover (toward the state represented by climax vegetation in uncleared areas), which can then be connected by field margins, as well as practices such as more frequent rotations involving a wider range of crops flowering at differing times, or introducing greater diversity through intercropping and the use of cover crops (Cirujeda and Pardo, 2020).

It is important to be aware that setting aside new field margins may not always be the most appropriate conservation solution. In general, more diverse field margins support more diverse pollinator communities (Brittain et al., 2022). However, less benefit is seen in landscapes that are already heterogeneous (Scheper et al., 2013; Morrison et al., 2017; Wix et al., 2019). In these landscapes, conservation efforts focused on maintaining the quality of existing natural areas may be best (Carvell et al., 2011). More generally, the role of field margins and other practices promoting biodiversity in farming landscapes can be viewed in the context of the wider debate about the relative benefits for biodiversity of a 'land sparing' strategy (increasing efforts to preserve or restore natural landscapes while concentrating farming in a smaller area and focusing on improving productivity) versus a 'land sharing' strategy (where farming practices are modified to accommodate greater biodiversity). At a global level, a significant body of research seems to favor the benefits of land-sparing over land-sharing, though some argue that the two approaches need not be mutually exclusive, with land-sharing practices ensuring greater connectivity between conservation areas separated by farmland, for instance (Grass et al., 2019; Balmford, 2021).

3 Design and management

As noted, field margins come in all shapes and sizes. Some are carefully designed, constructed and maintained by intentionally seeding carefully selected mixes of diverse plant species, adding boundary or other features, and with periodic mowing to control dominant species and maintain floral diversity. Other marginal areas are simply left undisturbed. In general, a one-size-fits-all conservation approach is not an efficient way of spending the limited funds available for preserving biodiversity on farms. The ideal approach should be adapted to the landscape context and local flora, as well as any species groups being targeted (Batáry et al., 2011). Field margins should not compete with the crop for light, moisture, or nutrients, and they should not harbor weeds, pests, or diseases (Marshall and Moonen, 2002).

3.1 Protecting field margins

It is important that field margins be protected from agrochemicals applied to crops in the main field area. Long and Krupke (2016) found that dust or spray drift from pesticides could accumulate in margins, settling on and contaminating flowers and their pollen – potentially poisoning insects (Nicholls and Altieri, 2013). Pesticides making their way to field boundaries can also cause changes in plant communities, with resistant species becoming more dominant (Ryszkowski, 2001), the disappearance of vulnerable species (Schmitz et al., 2014) and overall lower plant species richness (Marshall, 2004).

This can be attributed to the damaging effects of pesticides on the natural plant community, such as flower suppression, reduced seed production, and reduced growth affecting plant fitness and competitive ability (Schmitz et al., 2014). Schippers and Joenje (2002) showed that excessive nutrients such as nitrogen in field boundaries, resulting from spray drift during fertilizer application, had a negative effect on plant species diversity as well as on surrounding water bodies. Fertilizers can favor plants with high nutrient uptake (e.g. tall grasses) and decrease the frequency of small and subordinate species, which are replaced by fast-growing species and species with large leaf canopies (Schmitz et al., 2014). Fertilizers can also have a particularly adverse effect on rare wild plants (Ryszkowski, 2001). Other types of disturbances can be inadvertent, as a result of adjacent cultivation, or intentional (e.g. from tilling or cutting practices) (Ryszkowski, 2001). Margins should be minimally tilled and ideally left *in situ* over multiple years (Fried et al., 2009; McHugh et al., 2022).

Enlarging margins (to at least 2 m in width) is one solution to reduce disturbances, agrochemical drift, and edge effects, as well as setting aside a spray-free zone or conservation headland in the adjacent field (Lagerlöf et al., 1992; Haddad et al., 2014). Moonen and Marshall (2001) found that sown grass strips along the edge of margins protected margin flora and facilitated greater species richness by providing extra distance from farming operations. Gove et al. (2007) found that the impacts of pesticide drift on non-target plants in woodland margins in the United Kingdom were sufficient to affect community structure, but only within the first few meters of the woodland edge. They determined that a 5 m no-spray buffer zone would be very likely to render spray drift and fertilizer overspread impacts negligible and would protect the majority of woodland species from the impacts of agrochemicals applied to adjacent land. They also found that a 2 m buffer would still have considerable benefits but recommended a precautionary approach (5 m buffer) to give valuable sites as much protection as reasonably possible.

3.2 Configuration

In order for beneficial insects and other wildlife to thrive in agroecosystems, they require suitable nesting sites and materials (e.g. vegetation, tree cavities, soil substrates), sufficient and continuous floral resources within their flight range (e.g. pollen, nectar, and floral oils) and, if necessary, suitable sites for overwintering (Hannon and Sisk, 2009; Le Féon et al., 2016). Common substrates found in margins (e.g. bare ground, dry branches, hollow logs, and earth banks) provide ideal nesting sites for various pollinator species (Nicholls and Altieri, 2013).

Boundary structure, such as width and the presence or absence of a bank, has been found to drive plant community composition (Hovd and Skogen,

2005; Schippers and Joenje, 2002). Features such as banks also help to reduce disturbances from farming practices (Cirujeda et al., 2019). Wider margins have been found to host more diverse plant communities than narrower margins (Pallavicini et al., 2020; Schippers and Joenje, 2002). This is not just because wide margins buffer against agrochemical drift (one of the major causes of diversity decline and margin homogenization) but also because species richness tends to increase with increasing area (Schippers and Joenje, 2002). Studies have found that wider field margins better support wild pollinator abundance and diversity (e.g. Rands and Whitney, 2010; Häussler et al., 2016; Morrison et al., 2017).

3.3 Weed control

One common concern of farmers is that margins may encourage pernicious weeds, which can spread into the crop (Marshall and Moonen, 2002). This reflects a wider concern amongst some farmers that field margins are sources of competing species (e.g. competing with the crop for water, nutrients, and light) as well as a potential source of pests (Cirujeda and Pardo, 2020). Such concerns could result, for example, in the application of herbicides to margins, resulting in the perpetuation of species-poor plant communities with negative impacts on species biodiversity, as well as the need for ongoing management (Feber et al., 1996).

Marshall (2004) claimed that the perception of weed species in field margins invading crops is misguided and that the amount of important weed species that originate in margins is limited. However, this is not always the case (Cirujeda et al., 2013). What has become clear is that whether or not weeds thrive in the field margin depends on the level of disturbance and margin width (Cirujeda et al., 2019). High levels of disturbance from machinery or agrochemical drift tend to promote more annual species (often weeds) and impede the growth of perennial species (De Cauwer et al., 2008; Marshall, 2009; Schippers and Joenje, 2002). Cirujeda et al. (2019) found that structurally simple, flat, and narrow field boundaries contained high abundances of many of the weed species found in the center of the adjacent field. They also found that more complex and wider field boundaries with a slope and high percentages of woody and evergreen perennials had a lower probability of hosting the main weeds present throughout adjacent fields.

Marshall (2004) found that, when winter annuals dominate boundary flora, field margins may have a more significant influence on the presence of weed flora in adjacent fields. In these circumstances, specialized margin management strategies should be applied, such as sowing non-invasive perennial species to help provide a barrier against the spread of weeds from the field edge into the crop, or having a grassy strip beside the margin (Marshall and Moonen, 2002).

Removal or mowing of noxious weeds in field margins may be necessary in some cases.

4 Enhancement: sowing with seed mixes

Margins that are simply allowed to regenerate naturally are more cost-effective and involve lower maintenance for farmers. In some situations, maintaining natural field margins may be sufficient to support a robust wildlife community (Williams et al., 2015). However, plant species richness tends to be lower in more intensively managed agricultural areas (Görzen et al., 2021). The impoverished flora persisting in arable lands typically comprises species that can tolerate intensive farming practices, and these species often do not provide suitable forage resources for bees or other pollinators (Pywell et al., 2005). In highly degraded agricultural landscapes, floral density, richness, and specific floral traits can be increased by supplementing naturally occurring vegetation with sown plant species.

Many studies have found that sowing field margins with carefully selected wildflower mixtures increases the abundance of pollinators within intensively farmed agricultural landscapes (Feber et al., 1996; Pywell et al., 2005). For example, Pywell et al. (2005), in the UK, observed that sowing field margins with wildlife seed mixtures had the potential to provide the best foraging habitat for bumblebees. They found that all bumblebee species except one were significantly more abundant on the margins sown with the wildlife seed mixtures (mean of 28.6 ± 6.6 bumblebees per 100 m) compared with conventional cereal field margins (mean of 0.3 ± 0.1). However, they emphasized the importance of including preferred forage species for the target species in the mixture (in this case, *Trifolium pratense*, *Lotus corniculatus*, and *Borago officinalis*). Plant species differ in their potential contribution to ecosystem services, so increasing the diversity of species enhances the ability of the floral margin community to deliver a range of ecosystem services (Tscharntke et al., 2005). Sowing field margins with flowering plants, particularly native species, can help to increase the abundance and diversity of pollinators (Hopwood, 2008; McHugh et al., 2022).

Many seed mixtures marketed for bee or pollinator conservation are available for purchase, some of which contain specific host plants for the larval stages of butterflies, moths and beetles (Tscharntke et al., 2005). However, some studies have found that flowers planted in dense clumps have a stronger attractive potential for pollinators compared to those that are highly intermingled, as is the case with seed mixtures (e.g. Ohashi and Yahara, 1999; Grindeland et al., 2005; Barbir et al., 2014; Morrison et al., 2021). A higher density of species increases the probability that insects detect specific flowers from a distance (e.g. via visual or olfactory cues) while also increasing the ease of travel from one plant to another (Morrison et al., 2021). These results indicate that it would be more beneficial, in terms of attracting pollinators, to sow single species in clusters. However, the relationship between pollinator visitation and floral density is complex. Some studies have found no or even negative effects of high floral density on pollinator visitation rates to flowers (Essenberg, 2012).

According to Carreck and Williams (1997), the practical requirements for a sown floral mixture are that it should establish well, require minimum seedbed preparation, be able to compete with arable weed species, not require agrochemicals (e.g. fertilizers), and be affordable. The established plant system should also be self-sustaining and resilient. When considering which plant species to sow and in what proportion, it is important to consider providing not just overall species diversity, but functional diversity, that is, a high distribution of functional traits within the field margin community (Navas, 2012). Increased biodiversity enhances ecosystem functioning if new species add unique and complementary traits to the overall floral community (Tscharntke et al., 2005). Functional traits include flowering time and duration, floral display size and color, nectar and pollen content, leaf area, and root structure, all of which can be important in supporting insects and other species (Creswell et al., 2019). A community with both functional trait diversity and redundancy (i.e. the presence of multiple species with a similar effect on a given ecosystem function) is more stable and able to reorganize more effectively after a disturbance (de Bello et al., 2021).

In general, a variety of floral colors and shapes will attract a diverse community of beneficial insects (Tscharntke et al., 2005). However, the species composition of a community of beneficial insects is highly affected by fine-scale habitat variations (e.g. the presence or absence of a particular flower species) (Hannon and Sisk, 2009). Flowering plant species possess particular morphological and physiological characteristics that attract certain groups of floral visitors over others (Bosch et al., 1997). The choice of plant species to sow should thus be context specific. Margins can be sown with plant species known to support specific beneficial insects or wildlife species that are important within an agroecosystem, or those in need of conservation (Norris and Kogan, 2005).

The floral rewards offered by particular species are important in determining how flower-visiting insects partition themselves among available flowering plants (Bosch et al., 1997). In order to examine the capacity of certain flower species to attract particular beneficial insects, it is essential to consider which floral traits are important. Hegland and Totland (2005) pointed out that flower-visiting insects do not choose a flower species based on any one floral trait. Their research shows that flower species with similar combinations of traits have a comparable attractiveness to insect visitors, indicating that several floral traits can affect the flower choice of insects. Nectar (for carbohydrates) and pollen (for protein, vitamins, and minerals)

(Sammataro and Yoder, 2011) are usually considered the ultimate attractions driving insect visitors, while color, odor, and shape act more like 'cues' to help insects discriminate between the variety of reward sources available (Real, 2012).

The quantity and accessibility of nectar and pollen vary depending on the flower species. For example, perennials with long corollas tend to have more nectar than annuals (Nicholls and Altieri, 2013). How accessible the nectar is for the insect is based on the architecture of the flower, the concentration of the nectar, the morphology of the insect, and the mechanisms it uses to extract nectar (Wäckers, 2004). Plant height impacts the visibility of flowers to insects, particularly as many pollinating insects prefer to fly and forage at a constant height in order to conserve energy (Dafni and Potts, 2004).

Studies have shown that margins and other restored semi-natural areas with perennial plants may help to improve pollinator diversity and support less-common insect species (e.g. Morandin and Kremin, 2013; Erickson et al., 2021; McHugh et al., 2022). Cirujeda et al. (2019) found that field margins with high percentages of woody and evergreen perennials also demonstrated a lower probability of hosting crop weeds that might outcompete floral species.

Another important aspect to consider when examining the relationships between flower species and their visitors is the overlap between the flowering period of the species and the activity periods of insects (Bosch et al., 1997). In general, it is beneficial to have continuous flowering throughout the margin, choosing species with different and overlapping flowering periods without gaps. This can be further refined if a specific insect species is being targeted (i.e. sowing flower species that bloom when the insect species is active).

5 The economics of field margins

Field margins can deliver a number of financial benefits to farmers (see Table 1, Agronomy function). However, field margins also occupy space that is often limited on farms, and they require investment of capital and time for their implementation and maintenance. Farmers may be hesitant to invest in margins because of the lack of clarity concerning the economic benefits (Morandin et al., 2016) or concerns with regard to weeds or pests. The limited available research suggests that, when viewed in terms of lost revenue (from land not used for crop production) and expenditure to create and maintain them, field margins resulted in overall financial losses for most crop types (Segre et al., 2019, 2022). This research also suggests an inverse relationship between features promoting biodiversity (e.g. increasing field margin width) and the level of potential losses. This means government policies that encourage and support farmers to actively establish and manage field margins may be required. Farmers can be rewarded for implementing ecological projects or compensated for losses in income that arise as a result of sacrificing fertile land for field margins and other semi-natural areas on farms.

6 Conclusion and future trends in research

Increasing biodiversity is key for improving both productivity and resilience within agricultural systems. Field margins support biodiversity on farms, offering multiple ecosystem services simultaneously. This includes the provision of habitat and resources for beneficial insects, the creation of wildlife corridors, as well as other agronomic benefits for farmers, and contributions to landscape heterogeneity. In order to compensate for any financial losses to farmers, government policies that encourage and support farmers to actively establish and manage field margins may be required.

Future research should aim to determine the most effective and practical approaches to utilizing field margins as a means of maintaining or increasing biodiversity in agriculture. More studies are needed that focus on the characteristics of the margin and the traits of the vegetation communities that are important, based on conservation goals and the objectives of the farmer (e.g. how field margins can contribute toward soil carbon sequestration). Protocols are required to provide practical guidance to farmers on whether a new or regenerated field margin would be a wise investment for their specific context and goals, and how to achieve this (e.g. recommended sown plant species).

Region-specific studies are needed in order to determine recommended sown plant species and to optimize the capacity of plant species to thrive in given combinations of soil, climate, and altitude. The information available about many native plant species (i.e. traits, attractiveness to pollinators, and invasiveness) is limited and often out of date. In order to optimize sown plant recommendations, new open-access data are needed about a wide array of species.

Research will continue to measure the contribution and practicality of field margins in different applied scenarios. However, in order to examine the full impact of agricultural field margins (combined benefits as well as trade-offs), future research should consider amalgamated contributions of multiple ecosystem services together, rather than focusing on a single ecosystem service (Van Vooren et al., 2018).

Finally, integrating biodiversity into a productive farm is extremely challenging, partly because of increased system complexity (Petit et al., 2011). More research is needed in order to develop reliable, context-based protocols and management plans which ensure efficient farm operations and high yields while supporting optimal ecosystem health. Research is also needed to determine the best ways of imparting knowledge to farmers and encouraging their acceptance of biodiversity enhancement practices through the effective use of field margins.

7 Where to look for further information

Related scientific articles can be found in a range of journals, including *Agriculture Ecosystems and Environment*, *Biological Conservation*, *Ecological Applications*, *Ecology Letters*, and *Journal of Applied Ecology*. Some recommended recent books are *Reconciling Agricultural Production with Biodiversity Conservation* (Bàrberi and Moonen, 2020) and *The Ecology of Hedgerows and Field Margins* (Dover, 2019). Research centers studying field margins are located across the world, such as in the United Kingdom (e.g. Centre for Ecology and Hydrology, Staffordshire University, University of Bristol), Spain (e.g. Agri-Food Research and Technology Centre of Aragon, Institute for Sustainable Agriculture), Italy (e.g. Scuola Superiore Sant'Anna), the United States (e.g. University of California, Berkeley), and Canada (e.g. University of British Columbia).

8 References

Altieri, M. A. (1995), *Agroecology: The Science of Sustainable Agriculture*, Westview Press, Boulder, CO.

Balmford, A. (2021), 'Concentrating vs. spreading our footprint: how to meet humanity's needs at least cost to nature', *J. Zool.* 315(2), 79–109.

Balzan, M. V., Bocci, G. and Moonen, A. C. (2014), 'Augmenting flower trait diversity in wildflower strips to optimise the conservation of arthropod functional groups for multiple agroecosystem services', *J. Insect Conserv.* 18(4), 713–728.

Bàrberi, P. and Moonen, A. C. (Eds.) (2020), *Reconciling Agricultural Production with Biodiversity Conservation*, Burleigh Dodds Science Publishing Limited, Cambridge, UK.

Barbir, J., Badenes-Pérez, F. R., Fernández-Quintanilla, C. and Dorado, J. (2014), 'The attractiveness of flowering herbaceous plants to bees (Hymenoptera: Apoidea) and hoverflies (Diptera: Syrphidae) in agro-ecosystems of Central Spain', *Agric. For. Entomol.* 17(1), 20–28.

Barr, C. J. and Gillespie, M. K. (2000), 'Estimating hedgerow length and pattern characteristics in Great Britain using Countryside Survey data', *J. Environ. Manag.* 60(1), 23–32.

Batáry, P., Báldi, A., Kleijn, D. and Tscharntke, T. (2011), 'Landscape-moderated biodiversity effects of agri-environmental management: a meta-analysis', *Proc. Biol. Sci.* 278(1713), 1894–1902.

Bennett, A. F. (2003), *Linkages in the Landscape: The Role of Corridors and Connectivity in Wildlife Conservation*, International Union for Conservation of Nature and Natural Resources, Gland, Switzerland and Cambridge, UK.

Bosch, J., Retana, J. and Cerdá, X. (1997), 'Flowering phenology, floral traits and pollinator composition in a herbaceous Mediterranean plant community', *Oecologia* 109(4), 583–591.

Published by Burleigh Dodds Science Publishing Limited, 2024.

Brittain, C., Benke, S., Pecze, R., Potts, S. G., Peris-Felipo, F. J. and Vasileiadis, V. P. (2022), 'Flower margins: attractiveness over time for different pollinator groups', *Land* 11(11), 1933.

Buchmann, S. (2022), 'The role of habitat conservation and restoration in protecting pollinators in agricultural landscapes', In Kevan, P. and Chan, S. W. (Eds.), *Promoting Pollination and Pollinators in Farming*, Burleigh Dodds Science Publishing, Cambridge, UK, 259–282.

Cane, J. H. and Tepedino, V. J. (2001), 'Causes and extent of declines among native North American invertebrate pollinators: detection, evidence, and consequences', *Conserv. Ecol.* 5(1), 1.

Carreck, N. L. and Williams, I. H. (1997), 'Observations on two commercial flower mixtures as food sources for beneficial insects in the UK', *J. Agric. Sci.* 128(4), 397–403.

Carvell, C., Meek, W. R., Pywell, R. F., Goulson, D. and Nowakowski, M. (2007), 'Comparing the efficacy of agri-environment schemes to enhance bumble bee abundance and diversity on arable field margins', *J. Appl. Ecol.* 44(1), 29–40.

Carvell, C., Osborne, J. L., Bourke, A. F. G., Freeman, S. N., Pywell, R. F. and Heard, M. S. (2011), 'Bumble bee species' responses to a targeted conservation measure depend on landscape context and habitat quality', *Ecol. Appl.* 21(5), 1760–1771.

Cirujeda, A., Marí, A., Murillo, S., Aibar, J. and Zaragoza, C. (2013), 'The flora in the intern field margins is not a threat for the adjacent fields from the weed point of view in the cereal cropping region of Aragón' Congress of the Spanish Weed Science Society, Spain (vol. XIV).

Cirujeda, A. and Pardo, G. (2020), 'The role of field margins in biodiversity conservation in agroecosystems', In Bàrberi, P. and Moonen, A. C. (Eds.), *Reconciling Agricultural Production with Biodiversity Conservation*, Burleigh Dodds Science Publishing, Cambridge, UK, 149–176.

Cirujeda, A., Pardo, G., Marí, A. I., Aibar, J., Pallavicini, Y., González-Andújar, J. L., Recasens, J. and Solé-Senan, X. O. (2019), 'The structural classification of field boundaries in Mediterranean arable cropping systems allows the prediction of weed abundances in the boundary and in the adjacent crop', *Weed Res.* 59(4), 300–311.

Cooper, J., Leifert, C. and Niggli, U. (Eds.) (2007), *Handbook of Organic Food Safety and Quality*, Woodhead Publishing Limited.

Coppel, H. C. and Mertins, J. W. (1977), *Biological Insect Pest Suppression*, Springer, Berlin/Heidelberg, Germany.

Creswell, C. J., Cunningham, H. M., Wilcox, A. and Randall, N. P. (2019), 'A trait-based approach to plant species selection to increase functionality of farmland vegetative strips', *Ecol. Evol.* 9(8), 4532–4543.

Crowther, L. I., Wilson, K. and Wilby, A. (2023), 'The impact of field margins on biological pest-control: a meta-analysis', *BioControl* 68(4), 387–396.

Dafni, A. and Potts, S. G. (2004), 'The role of flower inclination, depth, and height in the preferences of a pollinating beetle (Coleoptera: Glaphyridae)', *J. Insect Behav.* 17(6), 823–834.

De Bello, F., Lavorel, S., Hallett, L. M., Valencia, E., Garnier, E., Roscher, C., Conti, L., Galland, T., Goberna, M., Májeková, M. and Montesinos-Navarro, A. (2021), 'Functional trait effects on ecosystem stability: assembling the jigsaw puzzle', *Trends Ecol. Evol.* 36(9), 822–836.

De Cauwer, B., Reheul, D., Nijs, I. and Milbau, A. (2008), 'Management of newly established field margins on nutrient-rich soil to reduce weed spread and seed rain into adjacent crops', *Weed Res.* 48(2), 102–112.

De Luca, P. A. and Vallejo-Marín, M. (2013), 'What's the "buzz" about? The ecology and evolutionary significance of buzz-pollination', *Curr. Opin. Plant Biol.* 16(4), 429–435.

Dorado, J. and López-Fando, C. (2006), 'The effect of tillage system and use of a paraplow on weed flora in a semiarid soil from Central Spain', *Weed Res.* 46(5), 424–431.

Dover, J. and Settele, J. (2009), 'The influences of landscape structure on butterfly distribution and movement: a review', *J. Insect Conserv.* 13(1), 3–27.

Dover, J. W. (Eds.) (2019), *The Ecology of Hedgerows and Field Margins*, Routledge, London.

Drummond, F. (2012), 'Commercial bumble bee pollination of lowbush blueberry', *Int. J. Fruit Sci.* 12(1–3), 54–64.

Eichenberg, D., Bowler, D. E., Bonn, A., Bruelheide, H., Grescho, V., Harter, D., Jandt, U., May, R., Winter, M. and Jansen, F. (2020), 'Widespread decline in Central European plant diversity across six decades', *Glob. Change Biol.* 27(5), 1097–1110.

Erickson, E., Patch, H. M. and Grozinger, C. M. (2021), 'Herbaceous perennial ornamental plants can support complex pollinator communities', *Sci. Rep.* 11(1), 17352.

Essenberg, C. J. (2012), 'Explaining variation in the effect of floral density on pollinator visitation', *Am. Nat.* 180(2), 153–166.

European Commission. (2023), Common agricultural policy. Available at: (https://agriculture.ec.europa.eu/common-agricultural-policy_en) (accessed September 1, 2023).

Feber, R. E. and Macdonald, D. W. (2013), *Wildlife & Farming: Conservation on Lowland Farms*, Wildlife Communications Conservation Research Unit, University of Oxford, Oxford.

Feber, R. E., Smith, H. and MacDonald, D. W. (1996), 'The effects on butterfly abundance of the management of uncropped edges of arable fields', *J. Appl. Ecol.* 33(5), 1191.

Firbank, L. G. (1999), 'The diversity of arable plants – past, present and some futures', In *Brighton Crop Protection Conference*, Weeds, England, pp. 251–260.

Fried, G., Petit, S., Dessaint, F. and Reboud, X. (2009), 'Arable weed decline in Northern France: crop edges as refugia for weed conservation?', *Biol. Conserv.* 142(1), 238–243.

Goettel, M. S., Richards, K. W. and Schaalje, G. B. (1991), 'Bioassay of selected fungicides for control of chalkbrood in alfalfa leafcutter bees, *Megachile rotundata*', *Apidologie* 22(5), 509–522.

Görzen, E., Diekötter, T., Meyerink, M., Kretzschmar, H. and Donath, T. W. (2021), 'The potential to save agrestal plant species in an intensively managed agricultural landscape through organic farming–A case study from Northern Germany', *Land* 10(2), 219.

Gove, B., Power, S. A., Buckley, G. P. and Ghazoul, J. (2007), 'Effects of herbicide spray drift and fertilizer overspread on selected species of woodland ground flora: comparison between short-term and long-term impact assessments and field surveys', *J. Appl. Ecol.* 44(2), 374–384.

Grass, I., Loos, J., Baensch, S., Batáry, P., Librán-Embid, F., Ficiciyan, A. and Tscharntke, T. (2019), 'Land-sharing/sparing- connectivity landscapes for ecosystem services and biodiversity conservation', *People Nat.* 1(2), 262–272.

Published by Burleigh Dodds Science Publishing Limited, 2024.

Greaves, M. P. and Marshall, E. J. P. (1987), 'Field margins: definitions and statistics', *Br. Crop Prot. Counc.* 35, 3–10.

Grindeland, J. M., Sletvold, N. and Ims, R. A. (2005), 'Effects of floral display size and plant density on pollinator visitation rate in a natural population of *Digitalis purpurea*', *Funct. Ecol.* 19(3), 383–390.

Hackett, M. and Lawrence, A. (2014), *Multifunctional Role of Field Margins in Arable Farming*, Report #CEA 1118, European Crop Protection Association. Available at: (https://croplifeeurope.eu/multifunctional-role-of-field-margins-in-arable-farming/) (accessed February 9, 2024).

Haddad, N. M., Brudvig, L. A., Damschen, E. I., Evans, D. M., Johnson, B. L., Levey, D. J. and Weldon, A. J. (2014), 'Potential negative ecological effects of corridors', *Conserv. Biol.* 28(5), 1178–1187.

Hannon, L. E. and Sisk, T. D. (2009), 'Hedgerows in an agri-natural landscape: potential habitat value for native bees', *Biol. Conserv.* 142(10), 2140–2154.

Harvey, C. T. and Eubanks, M. D. (2005), 'Intraguild predation of parasitoids by *Solenopsis invicta*: a non-disruptive interaction', *Entomol. Exp. Appl.* 114(2), 127–135.

Häussler, J., Sahlin, U., Baey, C., Smith, H. G. and Clough, Y. (2016), 'Pollinator population size and pollination ecosystem service responses to enhancing floral and nesting resources', *Ecol. Evol.* 7(6), 1898–1908.

Hegland, S. J. and Totland, Ø. (2005), 'Relationships between species' floral traits and pollinator visitation in a temperate grassland', *Oecologia* 145(4), 586–594.

Hopwood, J. L. (2008), 'The contribution of roadside grassland restorations to native bee conservation', *Biol. Conserv.* 141(10), 2632–2640.

Hovd, H. and Skogen, A. (2005), 'Plant species in arable field margins and road verges of central Norway', *Agric. Ecosyst. Environ.* 110(3–4), 257–265.

Klein, A. M., Steffan-Dewenter, I. and Tscharntke, T. (2003), 'Fruit set of highland coffee increases with the diversity of pollinating bees', *Proc. Biol. Sci.* 270(1518), 955–961.

Kline, O. and Joshi, N. K. (2020), 'Mitigating the effects of habitat loss on solitary bees in agricultural ecosystems', *Agriculture* 10(4), 115.

Koh, I., Lonsdorf, E. V., Williams, N. M., Brittain, C., Isaacs, R., Gibbs, J. and Ricketts, T. H. (2016), 'Modeling the status, trends, and impacts of wild bee abundance in the United States', *Proc. Natl. Acad. Sci. U. S. A.* 113(1), 140–145.

Kremen, C., Williams, N. M., Bugg, R. L., Fay, J. P., and Thorp, R. W. (2004), 'The area requirements of an ecosystem service: crop pollination by native bee communities in California', *Ecol. Lett.* 7(11), 1109–1119.

Kwaiser, K. S. and Hendrix, S. D. (2008), 'Diversity and abundance of bees (Hymenoptera: Apiformes) in native and ruderal grasslands of agriculturally dominated landscapes', *Agric. Ecosyst. Environ.* 124(3–4), 200–204.

Lagerlöf, J., Stark, J. and Svensson, B. (1992), 'Margins of agricultural fields as habitats for pollinating insects', *Agric. Ecosyst. Environ.* 40(1–4), 117–124.

Le Féon, V., Poggio, S. L., Torretta, J. P., Bertrand, C., Molina, G. A. R., Burel, F., Baudry, J. and Ghersa, C. M. (2016), 'Diversity and life-history traits of wild bees (Insecta: Hymenoptera) in intensive agricultural landscapes in the Rolling Pampa, Argentina', *J. Nat. Hist.* 50(19–20), 1175–1196.

Le Féon, V., Schermann-Legionnet, A., Delettre, Y., Aviron, S., Billeter, R., Bugter, R., Hendrickx, F. and Burel, F. (2010), 'Intensification of agriculture, landscape composition and wild bee communities: a large scale study in four European countries', *Agric. Ecosyst. Environ.* 137(1–2), 143–150.

Long, E. Y. and Krupke, C. H. (2016), 'Non-cultivated plants present a season-long route of pesticide exposure for honey bees', *Nat. Commun.* 7, 11629.

Manning, P., Van Der Plas, F., Soliveres, S., Allan, E., Maestre, F. T., Mace, G., Whittingham, M. J. and Fischer, M. (2018), 'Redefining ecosystem multifunctionality', *Nat. Ecol. Evol.* 2(3), 427–436.

Marini, L., Quaranta, M., Fontana, P., Biesmeijer, J. C. and Bommarco, R. (2012), 'Landscape context and elevation affect pollinator communities in intensive apple orchards', *Basic Appl. Ecol.* 13(8), 681–689.

Marshall, E. J. P. (2004), 'Agricultural landscapes: field margin habitats and their interaction with crop production', *J. Crop Improv.* 12(1–2), 365–404.

Marshall, E. J. P. (2005), 'Field margins in northern Europe: Integrating agricultural, environmental and biodiversity functions', In Thomas, A. G. (Ed.), *Field Boundary Habitats: Implications for Weed, Insect and Disease Management*, Canadian Weed Science Society, Canada, pp. 39–67.

Marshall, E. J. P. (2009), 'The impact of landscape structure and sown grass margin strips on weed assemblages in arable crops and their boundaries', *Weed Res.* 49(1), 107–115.

Marshall, E. J. P. and Moonen, A. C. (2002), 'Field margins in northern Europe: their functions and interactions with agriculture', *Agric. Ecosyst. Environ.* 89(1–2), 5–21.

Mas, M. T., Poggio, S. L. and Verdu, A. M. C. (2007), 'Weed community structure of mandarin orchards under conventional and integrated management in northern Spain', *Agric. Ecosyst. Environ.* 119(3–4), 305–310.

McHugh, N. M., Bown, B., McVeigh, A., Powell, R., Swan, E., Szczur, J., Wilson, P. and Holland, J. (2022), 'The value of two agri-environment scheme habitats for pollinators: annually cultivated margins for arable plants and floristically enhanced grass margins', *Agric. Ecosyst. Environ.* 326, 107773.

Moonen, A. C. and Marshall, E. J. P. (2001), 'The influence of sown margin strips, management and boundary structure on herbaceous field margin vegetation in two neighbouring farms in southern England', *Agric. Ecosyst. Environ.* 86(2), 187–202.

Morandin, L. A. and Kremen, C. (2013), 'Hedgerow restoration promotes pollinator populations and exports native bees to adjacent fields', *Ecol. Appl.* 23(4), 829–839.

Morandin, L. A., Long, R. F. and Kremen, C. (2016), 'Pest control and pollination cost–benefit analysis of hedgerow restoration in a simplified agricultural landscape', *J. Econ. Entomol.* 109(3), 1020–1027.

Morrison, J., Izquierdo, J., Plaza, E. H. and González-Andújar, J. L. (2021), 'The role of field margins in supporting wild bees in Mediterranean cereal agroecosystems: which biotic and abiotic factors are important?', *Agric. Ecosyst. Environ.* 247, 216–224.

Navas, M.-L. (2012), 'Trait-based approaches to unravelling the assembly of weed communities and their impact on agro-ecosystem functioning', *Weed Res.* 52(6), 479–488.

Ndakidemi, B. J., Mbega, E. R., Ndakidemi, P. A., Belmain, S. R., Arnold, S. E. J., Woolley, V. C. and Stevenson, P. C. (2022), 'Field margin plants support natural enemies in Sub-Saharan Africa smallholder common bean farming systems', *Plants (Basel)* 11(7), 898.

Nicholls, C. I. and Altieri, M. A. (2013), 'Plant biodiversity enhances bees and other insect pollinators in agroecosystems. A review', *Agron. Sustain. Dev.* 33(2), 257–274.

Norris, R. F. and Kogan, M. (2005), 'Ecology of interactions between weeds and arthropods', *Annu. Rev. Entomol.* 50, 479–503.

Ohashi, K. and Yahara, T. (1999), 'How long to stay on, and how often to visit a flowering plant? A model for foraging strategy when floral displays vary in size', *Oikos* 86(2), 386–392.

Pallavicini, Y., Bastida, F., Hernández-Plaza, E., Petit, S., Izquierdo, J. and Gonzalez-Andujar, J. L. (2020), 'Local factors rather than the landscape context explain species richness and functional trait diversity and responses of plant assemblages of Mediterranean cereal field margins', *Plants (Basel)* 9(6), 778.

Petit, S., Boursault, A., Le Guilloux, M., Munier-Jolain, N. and Reboud, X. (2011), 'Weeds in agricultural landscapes. A review', *Agron. Sustain. Dev.* 31(2), 309–317.

Pywell, R. F., Warman, E. A., Carvell, C., Sparks, T. H., Dicks, L. V., Bennett, D., Wright, A., Critchley, C. N. R. and Sherwood, A. (2005), 'Providing foraging resources for bumblebees in intensively farmed landscapes', *Biol. Conserv.* 121(4), 479–494.

Rands, S. A. and Whitney, H. M. (2010), 'Effects of pollinator density-dependent preferences on field margin visitations in the midst of agricultural monocultures: a modelling approach', *Ecol. Modell.* 221(9), 1310–1316.

Real, L. (2012), *Pollination Biology*, Elsevier, London.

Requier, F., Odoux, J. F., Tamic, T., Moreau, N., Henry, M., Decourtye, A. and Bretagnolle, V. (2015), 'Honey bee diet in intensive farmland habitats reveals an unexpectedly high flower richness and a major role of weeds', *Ecol. Appl.* 25(4), 881–890.

Ricketts, T. H., Williams, N. M. and Mayfield, M. M. (2006), 'Connectivity and ecosystem services: crop pollination in agricultural landscapes', In Crooks, K. R. and Sanjayan, M. (Eds.), *Connectivity Conservation*, Cambridge University Press, Cambridge, UK, pp. 255–289.

Romero, A., Chamorro, L. and Sans, F. X. (2008), 'Weed diversity in crop edges and inner fields of organic and conventional dryland winter cereal crops in NE Spain', *Agric. Ecosyst. Environ.* 124(1–2), 97–104.

Rosenberg, D. K., Noon, B. R. and Meslow, E. C. (1997), 'Biological corridors: form, function, and efficacy', *BioScience* 47(10), 677–687.

Ryszkowski, L. (2001), *Landscape Ecology in Agroecosystems Management*, CRC Press, Boca Raton, FL.

Sammataro, D. and Yoder, J. (2011), *Honey Bee Colony Health: Challenges and Sustainable Solutions*, CRC Press, Boca Raton, FL.

Scheper, J., Holzschuh, A., Kuussaari, M., Potts, S. G., Rundlöf, M., Smith, H. G. and Kleijn, D. (2013), 'Environmental factors driving the effectiveness of European agri-environmental measures in mitigating pollinator loss–A meta-analysis', *Ecol. Lett.* 16(7), 912–920.

Schippers, P. and Joenje, W. (2002), 'Modelling the effect of fertiliser, mowing, disturbance and width on the biodiversity of plant communities of field boundaries', *Agric. Ecosyst. Environ.* 93(1–3), 351–365.

Schmitz, J., Hahn, M. and Brühl, C. A. (2014), 'Agrochemicals in field margins – an experimental field study to assess the impacts of pesticides and fertilizers on a natural plant community', *Agric. Ecosyst. Environ.* 193, 60–69.

Segre, H., Carmel, Y., Segoli, M., Tchetchik, A., Renan, I., Perevolotsky, A., Rotem, D. and Shwartz, A. (2019), 'Cost-effectiveness of uncultivated field margins and semi-natural patches in Mediterranean areas', *Biol. Conserv.* 240, 108262.

Segre, H., Carmel, Y. and Shwartz, A. (2022), 'Economic and not ecological variables shape the sparing-sharing trade-off in a mixed cropping landscape', *J. Appl. Ecol.* 59(3), 779–790.

Shivanna, K. R., Tandon, R. and Koul, M. (2020), 'Global pollinator crisis' and its impact on crop productivity and sustenance of plant diversity', In Tandon, R., Shivanna, K. and Koul, M. (Eds.), *Reproductive Ecology of Flowering Plants: Patterns and Processes*, Springer, Singapore, pp. 395–413.

Sutter, L., Albrecht, M. and Jeanneret, P. (2018), 'Landscape greening and local creation of wildlife strips and hedgerows promote multiple ecosystem services', *J. Appl. Ecol.* 55(2), 612–620.

Tscharntke, T., Klein, A. M., Kruess, A., Steffan-Dewenter, I. and Thies, C. (2005), 'Landscape perspectives on agricultural intensification and biodiversity - ecosystem service management', *Ecol. Lett.* 8(8), 857–874.

Uyttenbroeck, R., Hatt, S., Paul, A., Boeraeve, F., Piqueray, J., Francis, F., Danthine, S., Frederich, M., Dufrêne, M., Bodson, B. and Monty, A. (2016), 'Pros and cons of flowers strips for farmers. A review', *Biotechnol. Agron. Soc.* 20(s1), 225–235.

Van Vooren, L., Reubens, B., Ampoorter, E., Broekx, S., Pardon, P., Van Waes, C. and Verheyen, K. (2018), 'Monitoring the impact of hedgerows and grass strips on the performance of multiple ecosystem service indicators', *Environ. Manag.* 62(2), 241–259.

Wäckers, F. L. (2004), 'Assessing the suitability of flowering herbs as parasitoid food sources: flower attractiveness and nectar accessibility', *Biol. Control* 29(3), 307–314.

Williams, N. M., Ward, K. L., Pope, N., Isaacs, R., Wilson, J., May, E. A., Ellis, J., Daniels, J., Pence, A., Ullmann, K. and Peters, J. (2015), 'Native wildflower plantings support wild bee abundance and diversity in agricultural landscapes across the United States', *Ecol. Appl.* 25(8), 2119–2131.

Winfree, R., Williams, N. M., Dushoff, J. and Kremen, C. (2007), 'Native bees provide insurance against ongoing honey bee losses', *Ecol. Lett.* 10(11), 1105–1113.

Wix, N., Reich, M. and Schaarschmidt, F. (2019), 'Butterfly richness and abundance in flower strips and field margins: the role of local habitat quality and landscape context', *Heliyon* 5(5), e01636.

Published by Burleigh Dodds Science Publishing Limited, 2024.

Chapter 7

The impact and management of hedgerows in promoting biodiversity in agricultural landscapes

Ian Montgomery and Neil Reid, Queen's University of Belfast, UK

1 Introduction

Hedgerows in agricultural and horticultural systems occur worldwide with great diversity in their form and function (Montgomery et al. 2020). We focus on the role of hedgerows in promoting biodiversity and delivery of ecosystem services (Millennium Ecosystem Assessment 2005) in temperate agriculture. Hedgerows around the edges of fields, comprised of woody plant species, are common particularly in UK, Ireland and continental Europe. They are prominent features around the margins of cereal and non-cereal arable crops, pastoral fields, orchards, vineyards and other horticultural systems, and bioenergy crops. Hedgerows may be associated spatially with drains, banks, stone walls, conservation headlands, roadside verges, and woodland and grassland ecotones (Dover 2019). Hedges occupy 10% of the total land area and 43% of

http://dx.doi.org/10.19103/AS.2024.0134.09

wildlife areas in northwest Europe (Larkin et al. 2019), and occur in temperate parts of North and South America.

European hedgerows are typically planted, continuous structures around fields and along rural roads. Field size plays an important role in determining overall hedge length (km) and density (km/ km^2). Hedge structure and opacity (hedgerow foliage and wood denseness) is determined by frequency and form of cutting and varies between low, narrow, vertical-sided hedges (Fig. 1a and b) and tall hedges with a broad base and angled sides (Fig. 2a and b). Hedges in the British Isles, for example, are formed by woody shrubs and smaller trees such as hawthorn *Crataegus monogyna* and blackthorn *Prunus spinosa* but often contain sapling trees such as ash *Fraxinus excelsior*, elm *Ulmus* spp, alder *Alnus glutinosa*, horse chestnut *Aesculus hippocastanum* and oak *Quercus* spp. Smaller woody species including hazel *Corylus avellana*, birch *Betula* spp, elder *Sambucus nigra*, holly *Ilex* spp, ivy *Hedera* spp, gorse *Ulex* spp, willow *Salix* spp, bramble *Rubus* spp, wild roses *Rosa* spp, spindle *Euonymus europaeus* and guelder rose *Viburnum opulus*, may also be present. However, there is much local and regional variation in the composition and management of hedgerows. In some regions, non-native species may also be common. For example, the deciduous shrub fuchsia *Fuchsia magellanica*, a native of South America, has widely naturalised in hedgerows in coastal areas of Ireland. Hedges frequently contain emergent trees which project above the main part of the hedgerow (Figs 1a and b, 2b and d).

The role of hedgerows and their management is critical in promoting biodiversity at the landscape, as well as the habitat level in agricultural systems. Thus, we consider biodiversity at the patch (within a habitat or alpha diversity), farm (between habitats or beta diversity) and landscape (overall or gamma diversity) levels. The contribution of hedgerows to farmland biodiversity should be considered in the context of other semi-natural habitats such as small woods, meadows and ponds. Hedgerows are particularly important for biodiversity in conventional agriculture characterised by monocultures, use of pesticides and fertilisers, and loss of other semi-natural habitats (Boutin et al. 2011, Collier 2021, Raven and Wagner 2021).

2 Origin and management history of hedgerows

Some very old hedgerows in Britain and elsewhere are remnants of ancient woodland surrounding arable, mixed or grassland-based agriculture. Other hedges were planted as field margins enclosing fields. The main period of enclosure of farmed strips in England occurred from the early seventeenth century onwards (Pollard et al. 1974) with the gradual replacement of subsistence farming by a more market-based rural economy. Traditional management of hedgerows involves cutting and hedge laying by hand, a process where most stems are partially severed near ground level, bent at an angle and pinned in place. These stems resprout to form a thick, livestock-proof hedge. This labour-intensive process has been replaced by mechanised cutting using tractor-mounted flails and saws. Mechanisation of field operations requires larger fields leading to hedge removal and field amalgamation. Hedgerows may also be lost through neglect, phosphate enrichment, herbicide spillover, soil compaction and overgrazing (Fig. 1a; Wright 2016). Reduction in hedgerow length and quality leads to diminished ecosystem services delivery and undermines overall biodiversity in agroecosystems. Deterioration of hedge quality and floral diversity is linked directly to management (Staley et al. 2013). England lost 5,600 km of hedgerows per year in the 1960s (Pollard et al. 1974). Belgium lost 30% of its hedges in the last 40 years (Van den Berge et al. 2019). Losses of ground flora due to overspill by nitrogen-rich fertiliser and manure from adjacent fields (French and Cummins, 2001) and removal of structures such as 'green lanes', comprising parallel hedgerows separated by a narrow track (Fig. 2c; Walker et al. 2006), compound the effects of hedgerow loss.

Tilzey (2022) argued that prevailing hedgerow management in the UK undermines the delivery of ecosystem services supporting national food production and urged the restoration of farmland heterogeneity, increased habitat diversity and wildlife abundance. Recent changes in agri-environment schemes favour hedgerow restoration to ameliorate the negative impact of conventional agriculture. The UK aims to create or restore 30 000 miles (~48 000 km) by 2037 and 45 000 miles (~72 000 km) by 2050 under the Environmental

Figure 1 (A). Low hawthorn-based hedgerow in late summer. Grazed field (left) allows stock access to the base of the hedge. Silage cut to the base of the hedge in the field to the right. (B). Roadside verge and mixed species hedge on steep bank cut in late summer. (C). Hawthorn hedgerow on a bank with gaps resulting from poor management of grazing and slurry application.

Published by Burleigh Dodds Science Publishing Limited, 2024.

Improvement Plan 2023 (DEFRA 2023). New hedges may restore some functions of older hedges quickly, but time and sympathetic management are required to restore biodiversity and related ecosystem services as fully as possible (Litza and Diekmann 2019).

3 Ecological processes and biodiversity in hedgerows

Biodiversity in hedgerows is often compared to and connected with other landscape features. Hedgerow structure and microclimate differ from surrounding fields and other semi-natural habitats potentially influencing the geographical ranges of species. Hedgerows, e.g. have thermal advantages for reptiles (Meek and Luiselli 2021) and hedge removal can lead to a decline in reptile abundance (Guiller et al. 2022). Vanneste et al. (2020b) recorded a more moderate range of temperatures in hedges than in their immediate surroundings, along transects spanning 1600 km from France to Norway. Temperatures in hedgerows did not reach the extreme high or low temperatures in neighbouring habitats, potentially favouring the northward expansion of southern species and southward expansion of northern species. Hedgerows, however, were less efficient as an insulator than woodland, especially on days with high ambient temperatures. Two-metre-high hedges around vineyards in South Africa reduced windspeed by 39% up to 18 m downwind from the hedge, lowering evapotranspiration of vines by over 18% and conserving water (Veste et al. 2020). Such effects may assist native plants growing in or downwind of hedgerows, especially those vulnerable to drought, in avoiding local extinction (Um et al. 2020).

The effects of hedgerows on ecological processes are often variable when considered on a whole ecosystem basis. This is not unusual. Genes and Dirzo (2022) conducted a meta-analysis of the effects of restoration and rewilding of terrestrial ecosystems on herbivory, pollination, seed predation and dispersal. They concluded that terrestrial habitat restoration and trophic rewilding projects in general improved only seed dispersal and pollination. The interaction of habitat (management) and landscape factors on bird and pollinator abundance and species richness also emerged in analysis of cider orchards (Martinez-Sastre et al. 2020). Albrecht et al. (2020) reviewed the efficacy of wildflower strips (17 studies) and hedgerows (18 studies) in enhancing pollination and natural enemy services; wildflower strips but not hedgerows enhanced pest control whilst the effects on pollination and crop yield were inconsistent.

Most research on hedgerow ecosystems has focussed on particular taxa and aboveground components due to their abundance, diversity, body size, taxon conspicuousness, the interests of researchers, and potential ecological and economic importance. Herbaceous and woody plants, Coleoptera,

Aculeata (stinging and social bees and wasps), Lepidoptera, Diptera, birds and mammals, feature more frequently than soil bacteria, fungi and microarthropods, epiphytes, spiders, parasitic Hymenoptera and most other insect orders. This bias is being addressed with increased interest in soil restoration (Spaans et al. 2019) and carbon storage (Drexler et al. 2021), food webs (Montgomery et al. 2023) and the ecological role of pathogens and parasites (Morandin et al. 2014). Organisms often have a particular role in terrestrial systems. For example, many Hymenoptera are pollinators, and spiders, harvestmen and most ground beetles (Carabidae), are predators. Here we combine ecological and taxonomic approaches to review hedgerow biodiversity focussing mainly on plants, soils, insects, birds and mammals.

3.1 Hedge structure and woody species

Trees and shrubs dictate overall hedge height, width, thickness and opacity. Ancient hedges have more woody species than recent hedges: in southern Britain, species richness per 30 m increased by one species for every 100 years in existence (Pollard et al. 1974). The frequency of cutting, hedge length, continuity and connectedness, are also important factors in determining overall plant and animal diversity, particularly at the landscape level. Van den Berge et al. (2019) showed that the loss of 30% of hedgerow length in northern Belgium between 1974 and 2015 was associated with an increase in fragmented, less opaque hedges amongst those that were retained. This

Figure 2 (A). Tall, wide, opaque mixed-species hedge in late summer. (B). Mature hedge with standard trees and fence protecting base from stock and machinery. (C). Hedges on banks facing each other across access track forming a 'green lane'. (D). Hawthorn hedgerow establishment with additional mixed native species and protective fences.

Published by Burleigh Dodds Science Publishing Limited, 2024.

in turn led to greater hedgerow plant species richness due to increased light penetration. Forests in contrast contained less light with more nutrient rich soils, an increase in community homogeneity and reduced plant species richness. Therefore, it is important to conserve hedgerow remnants which may contain woodland plant species and genotypes no longer present elsewhere in the vicinity, and also provide dispersal routes between otherwise isolated woodland fragments.

The effects of landscape structure on the genetic components of biodiversity are seldom considered but increased length and connectivity of linear features may help sustain genetic as well as ecological diversity. The sustainability of genetic diversity is complicated further by the ability of many woody trees to reproduce asexually. Blackthorn, e.g. has spatially structured, multi-ramet clones (Brown et al. 2022). There can also be problems identifying woody species, hybrids and plants derived from horticultural stock especially where hedgerows have been planted. For example, two readily identifiable species of hawthorn occur in Britain and Ireland, but taxonomic expertise is required to identify hybrids between closely related species and cultivars (Fichtner and Wissemann 2021).

3.2 Ground flora

Hedgerows have a ground flora comprised of bryophytes, ferns, and herbs. These are important components of biodiversity providing resources for pollinators as well as sheltering natural enemies of crop pests. Ground flora composition and diversity is affected by hedgerow and wider landscape characteristics and management as well as plant species traits. Bocage (landscapes of mixed woodland, hedges and small pasture fields) have a relatively high diversity but not abundance of ground flora (Boinot et al. 2023). Boinot and Alignier (2023) showed that plant taxonomic and functional diversity are influenced by proximate hedgerow conditions and wider landscape parameters. For example, plant communities in hedgerows in areas with low spatial density of hedgerows and greater use of herbicide, were more homogeneous and dominated by common weed species. Reduced use of herbicides and the presence of pesticide-free, conservation headlands adjacent to hedges can increase floral diversity beneath hedgerows (Aude et al. 2003). Hedgerow flora was most diverse where hedges were more opaque and not exposed to herbicides. On a gradient from southern Sweden to northern France, forest plant species that are thermophilic, tolerant of disturbance and effective dispersers, are more numerous in wider hedgerows that provide more moderate temperatures than narrower hedgerows bordered by more intensive land use that conserve heat poorly (Litza et al. 2022).

Stanbury et al. (2020) manipulated hedgerow cutting regimes over 6 years. Different cutting regimes can change species richness and community composition of the ground flora quickly (see also Fig. 3a and 3b). Compared to cutting back to a standard hedge height and width (flail set to prevent growth), hedges cut less severely in early autumn to allow 10-cm growth in height and width at each cut lost one herbaceous plant species, on average, over time probably due to increased shading. The ground flora also shifted towards plants characteristic of less fertile soils probably due to the reduction in woody plant biomass returned to the soil. The ground flora of hedges cut late in winter rather than autumn, shifted toward communities typical of more shaded conditions.

The habitat mosaic supported by the wider landscape also affects hedgerow ground flora. Removal of forest leaving a network of hedgerows supporting forest species may lead to 'extinction debt' (i.e. a continued loss of species after the event). However, this was not evident in northern Germany where the best predictor of forest plant species richness in hedges was hedgerow width rather than time since forest clearance (Litza and Diekmann 2020). The best explanatory variable for ancient woodland indicator plant species, however, was proximity to the nearest forest suggesting that extinction debt in hedgerows may be important for rarer species. Nelson et al. (2021) showed that ancient woodland indicator species colonise new woodlands more slowly than less specialised woodland ground flora. Plant dispersal and species richness are enhanced by interconnected hedgerows (Thiele et al. 2018).

Litza and Diekmann (2019) found that the composition of plant communities in recent and ancient hedgerows in Germany becomes more similar over time. Recent hedges function as habitats for forest ground cover but have fewer forest species, are dominated by plants with epizoochorous seed dispersal, and have soils with a higher pH than those of ancient hedges. Ancient hedges should be protected and recent hedges should be managed less intensively to become valuable refuges for biodiversity. Species diversity, but not functional diversity, of ground flora was improved in hedges on organic farms adjacent to conventional farms (Boinot and Alignier 2022). Organic farming may promote conservation of hedgerow plant communities better than conventional farming at the farm level but may have little impact across intensively farmed landscapes as a whole (Boinot and Alignier 2022).

Seed banks and ground vegetation in old hedges are dominated by stress-tolerant woodland species and seed banks are more species rich in uncultivated areas such as hedgerows, than cultivated areas (Hegarty and Cooper 1994, Evans et al. 2011). Seed banks are a vital resource exploited by some 330 animal species in the UK. Vanneste et al. (2021) showed that some forest herbs respond phenotypically to experimentally increased temperatures in hedgerows with increased biomass, height and production of seeds, and thus

Published by Burleigh Dodds Science Publishing Limited, 2024.

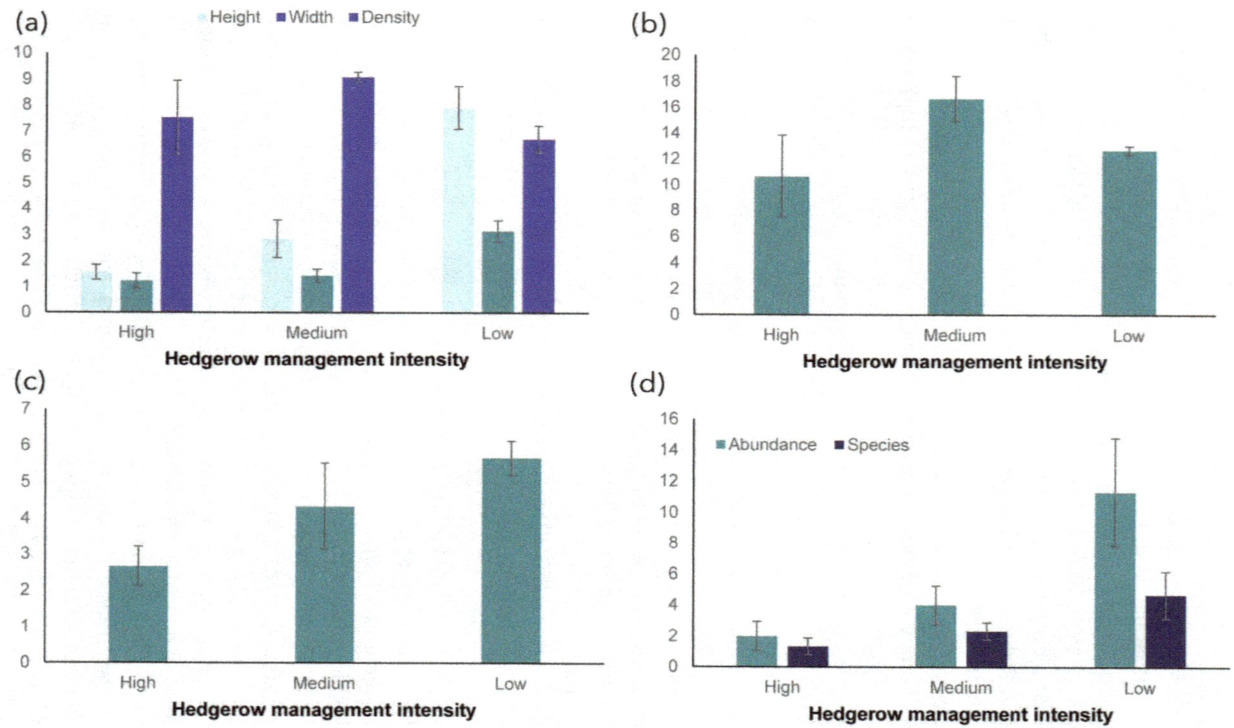

Figure 3 Effect of management on nine 30-m lengths of species-rich hedgerows subject to cutting in the previous winter (high), cutting two winters previously (medium) and uncut for at least 5 years (low). All data collected in September. (A). Mean (±SE) height (m), width (m) and foliage density (opacity of hedge at waist height on scale 1 to 10 from least to most). (B). Mean (±SE) number of plant species. (C). Mean (±SE) number of individual wood mice, *Apodemus sylvaticus*, caught over two nights. (D). Mean (±SE) bird abundance and number of species recorded at observation points. Unpublished data, WI Montgomery.

overall abundance, which may help them benefit from hedgerows as corridors. Little is known about seed predation in hedgerows. Seeds in hedges in southern Chile are eaten mostly by rodents, then invertebrates and birds (Gantz and Valdivia 2020). Dispersal of seeds and hence, dispersal of plant populations in relation to hedgerows, are poorly understood and may be affected by climate change (Fernandez-Fernandez et al. 2022). The latter authors found the effect of experimental warming on population growth rates in wood avens *Geum urbanum* was negative in forests but not hedgerows, which may reflect lower experimental increase in temperatures achieved in the latter.

Hedgerows with contiguous roadside verges provide an ecotone where two contrasting communities meet and thus support a wider variety of plant species (Figs 1b, 2b). Across temperate Europe, 55% and 52% of plant species, respectively, are shared between roadside hedges and forests, and verges and grassland (Vanneste et al. 2020b). Forest plants in hedgerows are enhanced by the presence of trees, increased hedgerow height and proximity to forests whilst grassland plants in verges are associated with physicochemical soil conditions. Local habitat features (canopy structure, subcanopy microclimate, soil properties, mowing regime) also enhance colonisation success of specialist plant species and, thus, overall biodiversity (Vanneste et al. 2020b).

3.3 Belowground biodiversity, decomposition and nutrient cycling

Hedgerows have positive effects on the biodiversity and ecosystem services provided by soil organisms (Monokrousos et al. 2006, Holden et al. 2019). Hedgerows increase the diversity and functional composition of arbuscular mycorrhizal fungi although not their phylogenetic diversity in Mediterranean agroecosystems (González-Fradejas et al. 2022). The effects of hedgerows on soil extend a short distance from the hedgerow itself. For example, soil fungal diversity peaks 8 m from windbreaks in Quebec, Canada whilst bacterial diversity and earthworm abundance peak 50 m into pastoral fields suggesting that hedges benefit fungi, but earthworms and bacteria are associated more with agricultural activity (Rivest et al. 2020). Earthworm abundance, however, is four times higher in hedgerows, other margins with undisturbed soils, and 2-year-old pasture, than in arable fields (Prendergast et al. 2021), suggesting soil beneath hedgerows in arable areas may be useful in the restoration of earthworm abundance in fields following soil disturbance. Hedgerows, therefore, add considerably to the biodiversity of soil communities in arable landscapes where soils are continually degraded by tillage in intensive agriculture. Hoeffner et al. (2021) showed that hedgerows in the grasslands of Brittany are associated with high earthworm species richness, especially of the epigeic (surface dwelling) and anecic (vertical burrows) species characteristic of soil rich in organic matter. Hedgerow management and presence of ash trees increases diversity of soil mite communities below and adjacent to hedgerows in intensively managed pastoral land (Spaans et al. 2019). Thus, hedgerows create soil conditions that promote biodiverse soil biota, presumably due to leaf litter inputs and are potentially a source of organisms in agricultural systems where these are depleted. Further research is required to establish how best to exploit hedgerow soils during soil restoration programs.

3.4 Invertebrate predators and parasites

Hedgerows are important habitats for beetles, spiders, eusocial and parasitic wasps, and flies which are the main natural enemies of agricultural pests. Global decline in the abundance of insects and other arthropods in farmland is attributed to agricultural intensification, habitat loss and climate change (Wagner 2020). Hedgerow insects and other invertebrates are more numerous on organic than conventional farms, often declining inwards from field margins and increasing where there is less frequent hedge cutting (Fig. 4).

Natural enemies of agricultural pests frequently show strong affiliation with hedgerows. Beetles are more numerous in hedgerows than in adjacent

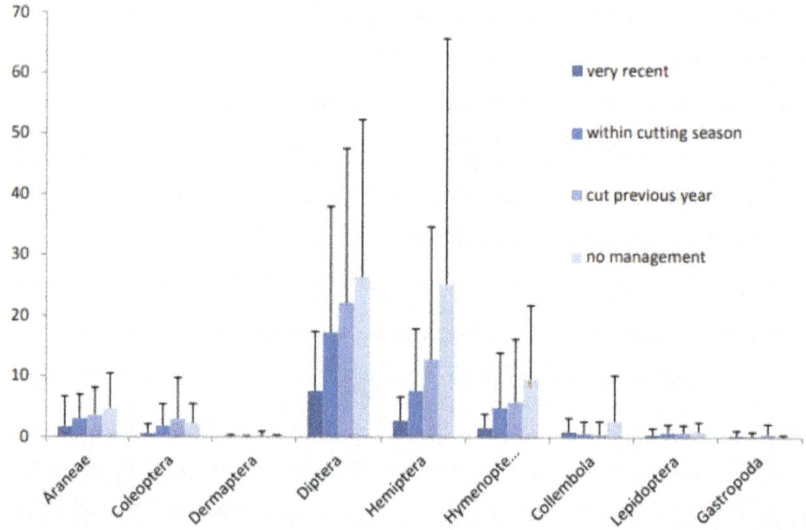

Figure 4 Mean abundance (±SD) of different invertebrate Orders or subclasses found in sweep net samples in mid-summer, for different management levels. Very recent represents hedgerows cut during the closed cutting period (March–August). Cutting season represents hedgerows cut during the previous open cutting season (September–February). Cut previous year represents hedgerows that were cut sometime during the previous year. 'No management' represents hedgerows that showed no obvious signs of management in the last 2 years (Findlay 2017).

Published by Burleigh Dodds Science Publishing Limited, 2024.

grassy field margins (Brooks et al. 2012). Carabid species richness is greater and species composition differs in hedgerows compared to fields under low input, conventional management in central Italy (Rossi et al. 2019). Older hedges have greater carabid functional diversity (Theves and Zebitz 2012). Large forest beetle species tend to dominate hedgerows whilst more open habitats are favoured by smaller species (De la Pena et al. 2003). Hedgerows support more staphylinid (rove) beetles and spiders than non-hedgerow field margins in arable farmland (Pywell et al. 2005). Harvestman (Opiliones) species richness is high in hedgerows in agricultural land of the western Carpathians, and varies positively with hedgerow length, area and width, tree diversity and species richness (Stasiov et al. 2020). Aculeate hymenopterans, mainly bees and eusocial wasps (Vespinae), are more abundant and species richness is higher in Malaise traps adjacent to hedgerows compared to those beside open fields (Volpato et al. 2020). The species richness of these taxa and the abundance of red wasp *Vespula rufa* and common wasp *V. vulgaris* are lower on intensively farmed land compared to less intensive farms.

Larval flies are significant predators of other insects and invertebrates. Ahmed et al. (2021) showed that syrphids (hoverflies) and sciomyzids (snail or marsh flies) are more abundant in more opaque hedgerows and open hedges with adjacent watercourses, respectively, than where these features are absent. A potential biocontrol agent for terrestrial slugs in Ireland, *Tetanocera elata* (a sciomyzid), e.g. is more abundant close to hedges (Bistline et al. 2020). Syrphids encroach from hedgerows into fields of oilseed rape and wheat, particularly where hedgerows are in contact with forest (Haenke et al. 2014). They are a major natural enemy of aphids; they are more diverse in organic than conventional fields and are more numerous where hedgerow length and structural diversity are greater (Puech et al 2015).

Experimental investigations of the impact of natural enemies in field boundaries on pests in crops are scarce. The efficacy of natural enemies in heterogeneous agricultural landscapes may be tested by excluding them or using 'sentinel' prey presented in different habitat types such as hedgerows and marginal grassland. However, care must be taken interpreting results as a selection of prey by predators varies with prey type and habitat, suggesting multiple sentinel species may be required (McHugh et al. 2020). Lajos et al. (2020) showed that the exclusion of natural enemies increased leaf loss due to cereal leaf beetles *Oulema melanopus* in wheat fields in lowland Hungary. The effect was strongest within 150–200 m of hedgerows and shelterbelts (similar linear features to hedgerows but taller and wider). Crowther et al. (2023) demonstrated that sown and naturally regenerated wildflower strips on the edge of fields are also linked to greater abundance and diversity of natural enemies, fewer invertebrate pests, and less damage to adjacent crops than grass field margins. Naturally regenerated floral field margins support fewer pests than sown wildflower strips.

Models of predator–prey populations suggest that hedge presence alone is insufficient to maintain biological control of pests. Spatial heterogeneity and species traits and their interaction are critical (Zamberletti et al. 2021). For example, modelling suggests that presence of hedgerows provides biological control of pests, but this effect increases with increasing dispersal ability of predators. Analysis of a 9-year dataset for 30 major arable pests and pathogens also showed a relationship with concurrent host plant extent but no consistent effect of semi-natural components of the landscape including hedges, without considering traits of each pathogen and pest species (Delaune et al. 2021).

Abundance of natural enemies typically declines away from hedgerows towards central areas of apple orchards. Kranz et al. (2019) examined the efficacy of using hazel and raspberry *Rubus* spp scrub to increase numbers of natural enemies mid-orchard. Shrub crops (raspberries or hazel or both) were an important predictor of arthropod predators of apple pest species. Hedgerows affect size and capture rates of apple insect pests by the generalist spider *Phdodromus cespitum* in apple orchards. Larger spiders disperse further into the orchard from hedgerows and take more and larger prey. Smaller spiders remain close to the hedgerow and take fewer and smaller prey. Thus, biocontrol in this case is more effective in the middle of the orchard (Michalko and Dvoryankina 2019). Californian tomato fields are penetrated by parasitoids of aphids up to 100 m whilst predatory beetles reached the middle of fields 200 m from a hedgerow (Morandin et al. 2014). These authors suggested that greater biocontrol could be achieved where there are multiple hedgerows around fields. Hedgerows in northern Italy are rich in predatory mites that predate spider mites (Tetranychidae spp) on vines. Replacement of living supports for vines, such as maple *Acer* spp and elm by cement and steel pillars, reduces the inward dispersal of predatory mites (Gavinelli et al. 2020).

The abundance of natural enemies of pests of strawberries *Fragaria x ananassa* increases with the proportion of fields having woody hedgerows (Lu et al. 2022). Increased abundance of natural enemies associated with hedgerows also has an impact on pest numbers at the landscape scale. Pesticide use and indiscriminate tractor-mounted vacuuming of arthropods, increase costs and deplete numbers of pest predators and parasites (Lu et al. 2022).

Biocontrol of agricultural pests by arthropods in hedgerows has been demonstrated in a variety of temperate horticultural systems including orchards and glasshouse systems (Rodriguez et al. 2023). Specific traits as well as abundance and community composition of natural enemies occupying hedgerows determine their capacity for biocontrol of agricultural pests. Hedgerows with aromatic plants surrounding citrus species, e.g. attract parasitic

Published by Burleigh Dodds Science Publishing Limited, 2024.

wasps, which supplement wasp populations living among the citrus, potentially limiting numbers of pests (Stathakis et al. 2023). Control of the woolly apple aphid *Eriosoma lanigerum* by the parasitoid wasp *Aphelinus mali* in orchards was enhanced by the presence of firethorn *Pyracantha* spp hedgerows surrounding orchards, with levels of aphid infection by wasps increasing as the season progressed (Penalver-Cruz et al. 2020).

Different field margins facilitate biological control of different pests. For example, abundance of parasitic insects and levels of parasitism increase with proportion of margins of cereal fields featuring a woody hedge (Salat-Molto et al. 2023). The candy-striped spider *Enoplognatha ovata*, which lays its eggs on herbaceous plants in hedgerows, is a potential biocontrol agent of pests affecting arable crops (Cuff et al. 2022). Grass strips adjacent to winter wheat fields (planted in the previous autumn) support more ground beetles and spiders than hedges whilst crop pests are more abundant in the latter (Tougeron et al. 2022). Comparisons between three regions in northwest Europe indicated that biological control in winter wheat fields may increase where winters are milder and decrease with increasing interconnections between hedgerows, crops and woodland. Tougeron et al. (2022) concluded that a combination of grassy strips, hedges and woodlots may deliver control of pests by natural enemies more consistently than monocultural agroecosystems because of taxon-specific responses to environmental variation at multiple scales. A landscape maintaining multiple populations of predators and prey must be able to support multiple predators in pursuit of prey and multiple prey that can avoid predators (Sih et al. 2005).

Sunflowers *Helianthus annuus* are grown for their seeds but are attacked by insect pests such as the sunflower moth *Homoeosoma electellum.* Avian predation on moths reduces moth abundance but also damages sunflowers. Kross et al. (2020) showed that damage by the moth is reduced where fields are bounded by hedgerows whilst bird damage is less than that by insects and unaffected by the presence or absence of hedgerows. Ecosystem services provided by hedgerows thus outweigh the disservices in economic costs and benefits. Badenhausser et al. (2020) showed that weed seed predation in sunflower fields depends on hedge length whilst predation of aphids is associated with more spiders and is also influenced positively by hedge length.

Hedgerows, thus, play a major role in numerous interactions involving multiple species of pests, predators and parasites along field margins as well as in field interiors. The efficacy of natural enemies in controlling pest populations below a threshold where economic damage occurs is affected by hedgerow length, management and spatial density as well as crop system and other field margin elements. Further studies are required elucidating the role of hedgerows in the biocontrol of agricultural and horticultural pests by natural enemies without using pesticides. It is also vital to investigate the economic sustainability of such systems.

3.5 Pollinators

Pollinators, mainly Hymenoptera, Lepidoptera and Diptera, support 35% of global crop production involving >800 plant species (Nicholls and Altieri 2013). Intensive agriculture has a strong negative effect on pollinators due to habitat loss and use of agrochemicals (Happe et al. 2018). Older hedges have more pollinator species, a greater abundance of pollinators and support rare and specialist species, an effect that levels off 5–10 years after cutting in recently planted and longer established hedgerows (Kremen and McGonigle 2015, Ponisio et al. 2016, Kremen et al. 2018). Hedgerows whether planted or remnants of wooded vegetation are a main source of pollinating insects (Phillips et al. 2019). Dover and Sparks (2000) recorded 39 butterfly species, 64% of the British list, in hedgerows. However, hedgerows support fewer pollinators than wildflower strips or conservation headlands. For example, Clausen et al. (2022) found wild bee species richness and abundance to be greater on flowers in grass verges than hedges in British Columbia despite hedges providing additional resources such as woody material, nest sites and hibernacula. A major meta-analysis demonstrated that local floral abundance and diversity has a positive effect on bee communities and are integral to the management of pollinators in general, and design of flower rich strips and hedgerows in particular (Hyjazie and Sargent 2022). Von Konigslow et al. (2022) found bee and hoverfly abundance in apple orchards to be greatest in perennial flower strips followed by hedges with a sown herb layer and then established hedgerows, in comparison to no interventions. Composition of bee communities and, to a lesser extent, hoverflies differed between control sites (no habitat intervention) and floral strips, mature hedges and hedgerows with additional floral resources. Some bee species are dependent on nectar and pollen of willow trees, which are common in hedges. Von Königslöw et al. (2022) also suggested that the combination of floral strip and flower-rich hedgerows would promote populations of pollinating insects by extending the pollination season and connecting habitats suitable for pollinators.

Hedgerows are an important source of nectar and pollen as well as providing shelter for pollinators. Using both hedges and wildflower strips helps stabilise pollination services from 1 year to the next where crops are rotated (Image et al. 2023a). The latter authors concluded that hedgerow and woodland edge management adjacent to oilseed rape, field beans and non-cropped areas, is the best way to enhance pollination services (Image et al. 2023b). Hedgerow creation increases the numbers of bumblebees with more consistent pollination of oilseed rape and field beans than using fruit trees (such as apple and cherry) or willow species in silvoarable alley systems. Gardiner et al. (2021) showed that combinations of field boundary features such as hedges, grassy margins or floral strips, help stabilise crop visitation by ground-nesting bumblebees and thus pollination in oilseed rape and field beans in rotational systems in England.

Wildflower strips and woody hedges must be managed appropriately to benefit pollinators. Kaur et al. (2019) found that verges and hedges along roadways in southwest England have more flowers and greater flower species richness and pollinator abundance than fields and field hedges. Cutting of roadside verges in summer prolongs the period with fewer flowers and pollinators (Phillips et al. 2019). Roadside verges restored using native plants are relatively chemical free and being unploughed, support more ground-nesting bees than verges dominated by weed species (Hopwood 2008).

Sources of nectar in farmland vary throughout the season (Image et al. 2023a). In Belgium, flowering woody species in hedgerows and forest edges constitute the main nectar source in April; intensive grassland and mass-flowering crops are the main sources in May, and extensive grassland and organic crops contribute to nectar resources in June (Langlois et al. 2020). Timberlake et al. (2019) and Gervais et al. (2021) advised combining hedgerows and wildflower strips to increase species richness and abundance of wild bumblebees, thus saving the cost of pollination by bought honeybees *Apis mellifera*. Apples from orchards with hedgerows, however, had fewer seeds and were lighter than apples from conventional orchards. Body size of pollinators may also be important. MacInnis et al. (2020) found that abundance of small bees was reduced by 0% within 50 m into strawberry fields compared to the edge, but large bees did not decline significantly in abundance up to 150 m, suggesting that smaller bees penetrated less into the centre of strawberry fields than larger bumblebees and honeybees.

Enhanced abundance and diversity of pollinators in hedges and flower strips does not necessarily result in an increase in numbers in field interiors or enhanced crop pollination (Zamorano et al. 2020). For example, hedgerows have little direct impact on seed-set of sunflowers (Sardinas et al. 2016). The efficacy of conservation measures promoting pollinators may also depend on the wider landscape. Hedgerows, arable fields and woodland promote non-*Apis* pollinating insects in areas dominated by pasture such as Wales (Alison et al. 2022). Increase in the abundance of non-*Apis* pollinating insects gradually diminished with greater percentage flower cover within 5 m of transects (Groom et al. 2010).

Nectar provides pollinators with energy and pollen provides them with proteins, carbohydrates, lipids, fatty acids, vitamins and minerals (El Ghouizi et al. 2023). The nutritive requirements of bees and other pollinators, however, may not be met even where floral resources have been maintained in hedgerows and other habitats. Filipiak (2019) demonstrated that providing habitats and plant species as a source of nectar and pollen for bees is insufficient as they require at least nine chemical elements in their diet for development, and the relative amount of these elements varies between plant species. Female bees have higher demands than males for P, Cu and Zn but Na, Mn, Mg, K, Fe and Ca may be limiting for both sexes. Most conservation measures promoting pollination by bees do not consider non-nectar resources. Up to 30% of social and solitary bees and wasps worldwide require resin, leaf pieces, trichome exudate and wood, for nest building, defence and health. These are largely provided by woody vegetation (Requier and Leonhardt 2020) and may explain the complementary benefits of hedgerows and floral strips for many pollinators.

3.6 Birds

The negative impacts of conventional farming on farmland birds are well-established (Rigal et al. 2023). Decline in farmland birds in southwest France over 30 years, e.g., is linked to changes in climate and agricultural land use (Gauziere et al. 2020). Since hedgerows provide food and cover for birds, restoration of semi-natural habitats such as hedgerows in farmland could redress the marked population declines in both specialist and generalist farmland bird species (Rigal et al. 2023). Agrochemicals have direct negative effects on birds. Comparison of individual vitality in six passerine species breeding in hedges revealed more flee attempts, aggression, pecks 'to the hand' and distress calls in captive birds caught in hedges on organic farms than conventional farms (Moreau et al. 2022). It is not yet clear how individual behaviours translate to the higher levels of the population or the community.

Avian species richness and abundance generally are greater in hedgerows than in other field margins (Heath et al. 2017) with an estimated 5 million breeding pairs of birds in hedges in the UK (Newton 2017). Insectivorous birds benefit more than granivorous species from woody habitats along field margins including wider, more opaque hedgerows (Otieno et al. 2019). Hedgerows support farmland and woodland bird species. Abundance and species richness of farmland birds in southern Sweden were greater than corresponding measures for woodland birds in hedges especially where hedges were isolated from forest, suggesting the composition of woodland and hedgerow avian communities differ (Batáry et al. 2012). Farmland bird species, Linnet *Carduelis cannabina*, Greenfinch *C. chloris*, Yellowhammer *Emberiza citrinella* and Common Whitethroat *Sylvia communis*, were numerous in hedges but shunned forest edges. Blackbird *Turdus merula* was the most common woodland bird species frequenting hedgerows. The surrounding habitat may also be important. In the Czech Republic and Austria, open scrubland in farmland was favoured by species of open habitat such as Corn bunting *Emberiza calandra*,

Skylark *Alauda arvensis* and Quail *Coturnix coturnix*, whereas hedges were preferred by woodland birds such as Chaffinch *Fringilla coelebs*, Nightingale *Luscinia megarhynchos* and Blackcap *Sylvia atricapilla* (Salek et al. 2022). This suggests that hedges may enhance the role of scrubland in promoting overall avian species richness and abundance.

Farmland birds which nest in hedges at ground level and above ground level are vulnerable to egg and nestling predation. Dunn et al. (2016) found that nestling mortality due to avian predators such as corvids was lower in managed hedges which were more difficult to penetrate than unmanaged hedges and, hence, prevented predator access to the nests. Iberian populations of the Eurasian bullfinch *Pyrrhula pyrrhula* favour nest sites in thorny plants surrounded by meadows and hedgerows which deter predators (Hernandez and Zaldivar 2021).

Many of the birds associated with hedgerows feed on insect pests of agricultural and horticultural crops. The potential impact of birds in reducing insect pests in maize fields was demonstrated experimentally using 3 x 3 m cages to exclude birds (Otieno et al. 2019). Herbivorous arthropod abundance was greater on conventional than on organic farms. When birds were excluded, the abundance of arthropods was up to three times greater than where birds were present. This cage effect was greater on organic farms than conventional farms. The willingness of birds to forage in adjacent fields away from the safety of hedgerows, which provide cover and escape routes but also act as corridors for nocturnal predators, is affected by individual behavioural and health traits. For example, bolder, healthier Grey partridge *Perdix perdix* strayed further from hedgerows at night than shy, less healthy conspecifics, presumably to avoid predators (Harmange et al. 2021).

Avian diversity and abundance are generally positively related to hedge size (MacDonald and Johnson 1995; Hinsley and Bellamy 2000). Most hedgerow bird species favour taller, wider hedges with trees, with a small minority favouring short, narrow, treeless hedges (Green et al. 1994). Hedgerow removal reduces nesting and foraging opportunities (Peach et al. 2004). Avian species richness increases in taller hedges even at the rural/urban interface (Pithon et al. 2021). Populations of some bird species are related directly to a single hedgerow variable. For example, Turtle dove *Streptopelia turtur* numbers increase with hedgerow length (Stanbury et al. 2021, Sauser et al. 2022). Populations of most birds, however, depend on multiple hedgerows and other landscape factors: 71% of the variation in bird species richness in hedgerows in Germany was explained by hedgerow length, width and height (Sybertz et al. 2019). Batáry et al. (2012), however, found no effect of hedge height or width on farmland and woodland bird species richness and abundance in Lower Saxony, where connected and isolated hedges were large and similar in size.

The relationship between farmland bird communities and populations and hedgerow characteristics varies with location. Breeding bird abundance increased with hedgerow density and volume in the UK and elsewhere, but shrub, bramble and nettle cover, crown and base width, and number of hedge intersections within 150 m, enhanced breeding bird abundance in East Fryslan, in The Netherlands (Oosterveld et al. 2022). Studies at large spatial scales employing remote sensing further elucidate the interplay among bird diversity, abundance and habitat. For example, Broughton et al. (2021b) used LIDAR to measure the height, length and width of hedgerows, and assessed how these landscape features influenced breeding farmland birds in timed tetrad surveys conducted on a regional scale by experienced observers. They showed that the abundance of songbirds is positively related to combinations of land cover and height of linear woody features (hedgerows 1.5–2.9 m tall and treelines >6 m in height). Maintaining diverse heights in a hedgerow network, with pockets of arable land in pastoral-dominated landscapes and *vice versa*, enhanced the richness and abundance of songbird species. Wildflower strips and drainage ditches associated with hedgerows, also improved habitat quality if they provided food and cover throughout the year (Hinsley and Bellamy 2000; Rieger et al. 2022).

Whilst generally benefitting biodiversity of songbirds, hedges may have negative effects on avian conservation and agricultural production. Rodent increase in less managed hedgerows (Fig. 3c) and are likely to predate eggs and nestlings during the breeding season (Gantz and Valdivia 2021). Granivorous birds can also be economically important pests of cereal crops. Otieno and Pryke (2022) showed that reducing the height of hedges and size of woodlots as well as planting isolated trees limits size of aggregations of granivorous birds and, hence, damage to grain crops but at the expense of birds more generally.

Apple orchards are often surrounded by hedgerows of a similar height. Bird communities are adversely affected by homogeneous landscapes of intensive fruit orchards but enhanced in terms of richness and abundance where there are hedgerows, isolated trees and woods (Rime et al. 2020). Birds favour orchard features including unharvested fruit in winter, low vegetation in spring and the presence of old trees. Breeding Great tits *Parus major* in conventional apple orchards where insecticides were used to control pest insects, depended on insects from adjacent hedgerows to breed successfully (Bouvier et al. 2022a). Single-row netting protects orchards against bird attacks whilst sustaining bird abundance and species richness where orchards have adjacent hedgerows (Bouvier et al. 2022b). Apple orchards support larger numbers of overwintering birds where more fruit is unharvested, and fruiting ivy plants (but not other berries) are abundant in surrounding hedgerows (Bouvier et al. 2020).

Birds may help to control numbers of agricultural pests but may also be responsible for damaging crops potentially creating a conflict between farming and conservation interests. There may be a trade-off between harm done directly to crops and biocontrol of pests by birds. Gonthier et al. (2019) demonstrated experimentally that insectivorous birds suppressed insect damage to strawberries in fields surrounded by hedgerows, by 3.8%, but 3.2% of all damage to the crop was due to birds. Features such as hedgerows increased numbers of beneficial and non-beneficial birds with regards to strawberries, but the quantity and quality of strawberries were improved where there were interconnected hedges nearby compared to isolated hedges and grassy banks (Castle et al. 2019).

3.7 Bats

Bats are ecologically important but less studied than birds with regards to hedgerows. Field boundaries are important for bats, not only for foraging for insects but also commuting to other feeding sites (Coulthard et al. 2016). Krings et al. (2022) found bat abundance but not feeding activity along hedgerows was significantly greater than in cereal fields, perennial flower fields and mixed flower fields. Foxley et al. (2023) showed greater bat activity over hedges with trees than without trees. Similarly, Bellamy et al. (2020) found that the activity of lesser horseshoe bats *Rhinolophus hipposideros* increased with the area of deciduous woodland, hedgerows and access to rural buildings. Herbaceous linear features of farmland are less important than hedgerows for bat species which forage on open edges around fields and narrow-space foraging species (Blary et al. 2021).

The level of association between bats and hedgerows varies with hedgerow characteristics and surrounding land use, and between bat species (Lacoeuilhe et al. 2018). Hedgerows are an important source of animal food for bats but diet varies between bat species. Thus, the barbastelle bat *Barbastella barbestellus* requires the food plants of the moth larvae it feeds on in its foraging habitats (e.g. woods, riparian woodland and hedgerows), within 6.5 km of its maternity roost (Carr et al. 2020). Woody linear features and meadows are important in maintaining a diversity of moths. Forest moth abundance and species richness are associated with hedgerow length and forest area within a 250-m radius whereas grassland moth abundance is only weakly associated with hedgerow length (Kuhne et al. 2022). Planting wildflowers increased abundance of insects, especially moths, up to fourfold in comparison to surrounding crops, and increased the numbers of bats (Peter et al. 2021). The positive effect of planting wildflowers was evident at the start of summer when the demand for food to support mothers nursing juvenile bats is greatest, but not later in the summer.

Diversity and abundance of food resources of bats is related to hedgerow management and its timing. Russ and Montgomery (2002) found that bats avoided hedges that are cut at regular, short intervals. Froidevaux et al. (2019) demonstrated that delaying hedge trimming by up to 10 years increased hedge height and insect prey diversity and abundance. In turn, these effects benefited horseshoe bats (*Rhinolophus* spp) and long-eared bats (*Plecotus* spp). Extending the trimming interval from 3 to 10 years also increases many other conservation benefits. In an investigation of the benefits of ponds for bats in farmland, activity of all bat species increased along linear features including hedgerows near ponds (Ancillotto et al. 2019).

3.8 Other mammals

Hedgerows are important habitats for many non-volant mammal species for shelter, food or as a thoroughfare in agricultural landscapes. Those maintaining populations in hedgerows tend to be generalists (Schlinkert et al. 2016). Wide, continuous, well-connected hedges of native plant species are favoured by arboreal and fossorial mammals (Dondina et al. 2016) but species differences abound, with some favouring hedges adjacent to intensively managed fields and others less intensive agriculture (Michel et al. 2006). In an innovative study in Quebec, Canada, use of hedgerows by seven large and medium-sized mammal species including, deer, carnivores, squirrels and a lagomorph, indicated that mammals in areas of intensive agriculture, used hedgerows as corridors and habitat, with activity increasing in long hedges free from human disturbance (Pelletier-Guittier et al. 2020). Hedgehogs *Erinaceus europaeus* in England, hibernate close to hedgerows, roads and woodlands but avoid pastoral fields (Bearman-Brown et al. 2020). Phillips et al. (2022) concluded that large, species rich, well-connected hedgerows could help restore populations of hazel dormice *Muscardinus avellanarius* in Great Britain.

Small mammals use hedgerows and other woody linear features to disperse (Montgomery and Dowie 1993; Sullivan and Sullivan 2009). Their abundance is related to immediate habitat conditions, as well as to landscape factors such as distance from woodland and land use (Montgomery et al. 2012). In a comparative study (Dondina et al. 2016), badgers *Meles meles* dispersed using wider hedgerows preferably without grassy ground cover. In contrast, hedgerows are more important as habitats for small mammals which require gap-free hedgerows and dense, low-level cover. Hazel dormice disperse 150 to 300 m and, thus, are more sedentary than badgers with dispersal distances of up to 8 km (Dondina et al. 2016). Radiotelemetry revealed that wood mice *Apodemus sylvaticus* living in hedgerows venture into arable crops (Tew et al. 2000) but not pasture (O'Neill 1998).

Landscapes containing well-connected hedgerows can mitigate the negative genetic effects of fragmentation due to other sources of land disturbance (Biello et al. 2023). Hedges, however, may have unexpected negative ecological and genetic impacts on biodiversity. For example, hedges may facilitate dispersal of invasive alien mammals leading to reduction in abundance, species richness and biomass of native species (Montgomery et al. 2012, 2015, 2023, Caravaggi et al. 2015). Transmission of tick-borne diseases may also be facilitated by hedgerows as nymphs infect small mammals that may be more abundant along field boundaries with hedgerows than in grassy field margins bordering fields (Medlock et al. 2020).

4 Landscape factors influencing hedgerow biodiversity

Higher biodiversity in farmland measured as species richness, abundance and evenness, is associated with more complex agricultural landscapes but effect size is nearly twice as large where both linear (e.g. hedgerows) and areal features (e.g. woodlots) are combined (Pearson, $r = 0.15$ linear only vs $r = 0.3$ linear plus areal; Estrada-Carmona et al. 2022). Landscape variability affects the abundance of pollinators and natural enemies of agricultural pests and thus biodiversity influences sustainable agricultural production and food security (Seppelt et al. 2020, Ricciardi et al. 2021). Regional variations in historical hedgerows, rural roads and settlements are highly valued especially where these distinctive features are disappearing with changes in agriculture (Fig. 1a and c). Even small habitat patches (<5% of total farm area) given over to hedgerows and 'waste ground' may contribute disproportionately to biodiversity (Evans et al. 2013). Conservation and restoration of these historical landscapes is desirable as part of overall landscape management promoting sustainable agricultural production and biodiversity.

Productive land area is limited and subject to diverse, expanding demands (Fróna et al. 2019). Whilst land sparing and sharing may be appropriate in some places, global food security for a larger human population will require almost all potentially productive land to support increased yields as well as biodiversity (see Bateman and Balmford 2023). Thus, productive land should provide connected patches of semi-natural habitat, such as hedgerows, to maintain immediate, critical ecosystem services such as pollination and pest control without overdependence on agrochemicals (Bateman and Balmford 2023). Balmford (2021) concluded that farming with higher yields combined with land-sparing results in larger wildlife populations than land sharing, whilst Kremen (2020) advocated restoration and creation of farming landscape mosaics to satisfy food production on the one hand and provide the complexity and heterogeneity of semi-natural habitats to support biodiversity on the other. Hedgerows will be a critical part of this process enhancing biodiversity and, thus, promoting the ecological services that support food production (Montgomery et al. 2020).

Plant communities in woods in farmland reflect species richness, age and width of connecting hedgerows (Roy and de Blois 2008). Total area of habitat and landscape heterogeneity affect the functional diversity of plants in hedgerows, selecting for plant dispersal, seasonality and competitiveness (Lecoq et al. 2022). Fijen et al. (2022) showed that simplification of dairy-based agroecosystems by reduction of ecologically functional features such as hedgerows to <5% of total area, has a negative effect on crop pollination services; pollen delivery of pak choi *Brassica rapa chinensis* grown for seed, was 80% higher next to farm gardens compared to fence lines. Pollinating species can be reinstated by newly planted hedges of native or exotic species. Mony et al. (2022) monitored seed deposition in seed traps and experimentally quantified the removal of seeds by seed predators (beetles, birds and mammals) in hedgerows in western France. The abundance of viable seeds and predation of smaller seeds was greater in connected than unconnected hedgerows, helping to maintain plant species diversity in woodland patches directly by predators scattering and hiding seeds, and indirectly, by reducing the abundance of common plant species which otherwise outcompete less common species for space.

Vanneste et al. (2020b) compared growth and reproductive traits in four herbaceous plant species in hedgerows and forests, to better understand hedgerows at the landscape level. All four species had higher biomass in hedges than woodland and three species had higher reproductive performance in hedges than woodland. Wood anemone *Anemone nemorosa* and other ancient woodland indicator plant species, conversely, had lower reproductive performance in hedgerows (Vanneste et al. 2020a, Nelson et al. 2021). Biodiversity at the landscape level, therefore, is best served by a mosaic of woody habitats, each supporting different communities of herbaceous species.

Hoffmann et al. (2021) described a non-linear relationship between ground-dwelling arthropods and the extent of habitats on the edge of fields such as hedgerows. Landscape diversity and density of field borders, for example, best explain ground beetle richness at the landscape scale. The abundance of ground beetle species in bocage is related positively to green lanes, density of trees and herbaceous layers and proximity to large areas of woodland at a landscape level (Burel 1989). Large ground beetles travel further in hedges with less cover suggesting that hedgerow structure and condition affect landscape connectivity for these beetles (Charrier et al. 1997). Species richness of common carabid species in areas of cereal production is enhanced by increased areas of permanent grassland (Massaloux et al. 2020). Spider communities show strong habitat affinities such that spillover into temporary crops occurs only where these different habitats are adjacent. The spider communities associated with woodland and hedgerows are more likely to increase biocontrol where crops are perennial, and crops are connected to

Published by Burleigh Dodds Science Publishing Limited, 2024.

multiple semi-natural habitat patches, including hedgerows (Nardi et al. 2019). Greater cover by hedgerows in intensive farmland in Italy was associated with a 12–18% increase in biocontrol of cereal aphids by parasitic wasps. Potential pollination rate (pollinator visitation rate and number of seeds set in potted radish plants exposed to insects adjacent to three types of field margin) increased by up to 70% compared to plants excluded from insects, independent of field margin condition (Dainese et al. 2017).

There are relatively few studies of how networks of semi-natural features benefit overall biodiversity. Ponisio et al. (2016, 2019) in a 10-year study of restored hedgerows and unrestored field margins in an intensively farmed landscape, showed that restored hedgerows interact with remnant woodland patches to improve floral diversity, bee diversity and body size, and increase site occupancy and persistence of bee species. This study revealed the biodiversity benefits of restoring hedges in a matrix of less hospitable habitats. The importance of hedgerows connecting to other features promoting overall landscape connectivity is illustrated further by bats avoiding major roads which disrupt landscape continuity over an estimated 35% of the land area of Europe (Claireaux et al. 2019). Deaths of bats due to road traffic may be substantial; Medinas et al. (2013) recovered 154 bats of 11 species along a 51 km daily road transect over 8 months in Portugal. Concerns for bat conservation and welfare thus prompted the introduction of underpasses on main roads to separate bats and road traffic. The use of underpasses by myotid and barbastelle bats increased with forest area. However, there were negative relationships between distance to the nearest forest and underpass use in horseshoe bats, and underpass use in myotid bats and nearby hedgerow length (Laforge et al. 2019). This suggests that landscape features such as hedgerows have subtle effects on the interaction between bats and roads.

5 Regionality: a case study of hedgerows in Northern Ireland

Wooded habitats particularly those featuring native biodiversity are scarce in the UK, especially in Northern Ireland. The importance of hedgerows to biodiversity at the regional scale is greatest where there is less alternative tree cover available (Fig. 1a). Hedgerows are vital to the maintenance of biodiversity in Ireland as a whole, as it lost most of its primary tree cover between the Neolithic (4000–2500 BC) and Medieval (400–1550 AD) periods, by which time only small remnants of forest remained. These remnants were cleared in Tudor times (1485–1603 AD) creating a largely treeless landscape subject to subsistence cultivation. Hedgerows in Ireland were established in the seveenteenth century AD for agricultural improvement involving field enclosure and hedge planting. New woodland was also planted at this time as a source of timber on estates granted to English and Scottish incomers (1550–1620 AD). Further woodland developed as deciduous scrub from the early twentieth century AD onwards. Farming was mixed until the 1960s after which there was greater specialisation in intensive dairy and beef cattle farming in the lowlands and sheep production in the uplands, with small, localised areas of cereal, fruit and vegetable crops. By the end of the twentieth century AD, lowland Ireland was predominantly small fields of grass surrounded by hedgerows with small, isolated areas of woodland (State of Europe's Forests 2020).

Irish hedges often involve banks, low walls and drains, and comprise mainly hawthorn, blackthorn, gorse, bramble, ash and hazel. The length of hedgerows in Northern Ireland is approximately 143 000 km with an average density of 9 km/km^2, greater than any other region of the UK or Ireland. Average hedge width is 1.6 m and height 1.5 m (Finlay 2017). Passive rewilding of abandoned farmland in England led to 132 and 390 tree stems per hectare after 24 and 59 years, respectively (Broughton et al. 2021a). Using these data for the density of trees on abandoned farmland, and an estimate of 5.3 million, mature, mostly deciduous trees in hedgerows in Northern Ireland (Spaans et al. 2018), hedgerows support an equivalent of more than 40 000 ha and 13 500 ha, or 1–3% total land area, of younger and older trees in Northern Ireland, respectively. Total deciduous woodland area in Northern Ireland in 2022 was estimated as 54 000 ha (Forest Research 2023). Thus, hedgerows in Northern Ireland support between 25% and 74% additional deciduous woody habitat and related biodiversity depending on age structure, as well as connecting isolated stands of woodland.

Management involves cutting by tractor-mounted flail or saw. Length and condition of hedgerow have declined as levels of mechanisation have increased. Enlargement of fields, use of herbicides and inorganic fertilisers, and grazing and slurry (liquid animal faeces) spread along the base of hedges (Fig. 1c), have led to a decline in flowering herbs, pollinating insects, natural enemies and farmland birds (Figs 3 and 4). Invertebrates (e.g. moth larvae and flies) respond positively to less frequent cutting (Facey et al. 2014, Froidevaux et al. 2019). Illegal cutting during the breeding season for birds, March to August inclusive (Fig. 1b), removes flowers of woody species critical to bees as well as cover for nesting birds (Porter 2017). Autumnal hedge cutting is legal but reduces berry abundance and sugar content (Fig. 5; Porter 2017). Hedges with a complex structure and less frequent cutting produce more berries that are critical for overwintering birds (Chamberlain et al. 2001). Dover (2019) suggested cutting every third year to maximise the fruit of hawthorn and bramble.

The typical Northern Irish hedgerow surrounding pasture and cut annually supports an impoverished biota. The beneficial effects of good hedgerow management on natural enemies and pollinators are evident even where

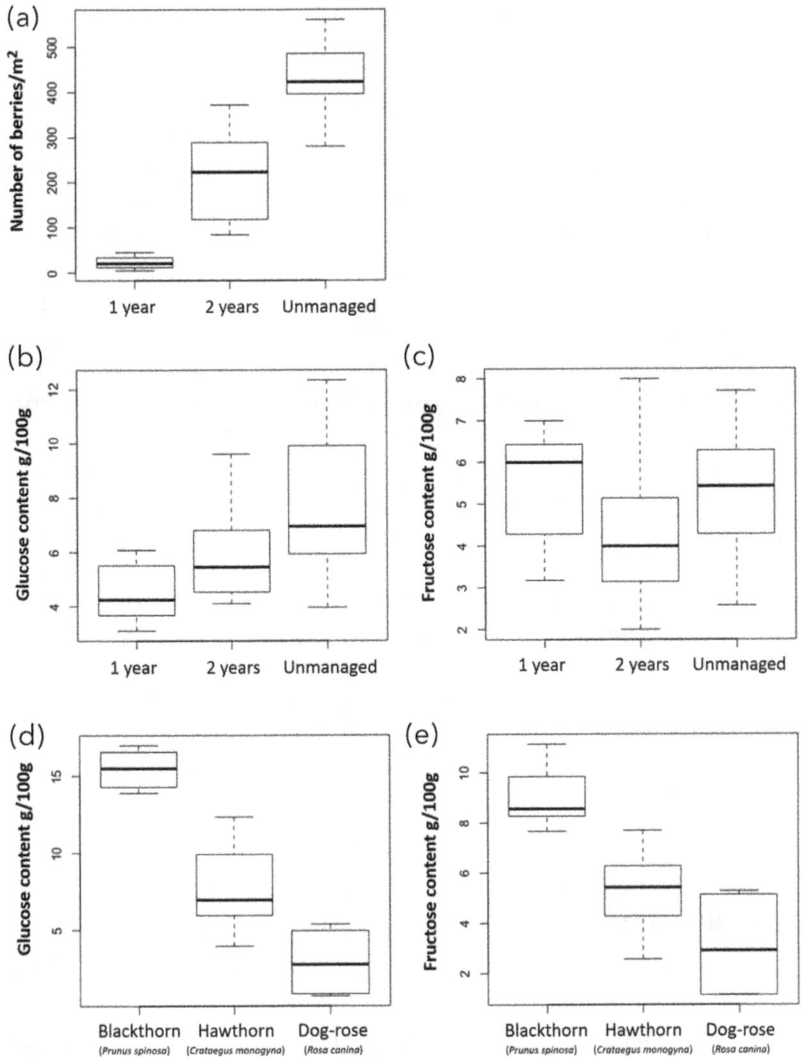

Figure 5 Median values (50th percentile bold line), upper and lower quartile boxes (25th–75th percentiles) and 95% Confidence Intervals whiskers (5th–95th percentiles) showing hawthorn berry (a) abundance, (b) glucose and (c) fructose content (g/100 g) for hedgerows cut every year, every two years or left unmanaged. (d) Glucose and (e) fructose content in berries from blackthorn, hawthorn and dog-rose (Porter 2017).

hedges are at high density (arbitrarily, low, medium and high hedgerow density set at <4, 4 –7 and >7 km/km^2, respectively). Extending intervals between cutting hedges, restoration to fill gaps, and new hedges could have a rapid effect on the farmland biodiversity of Northern Ireland without significant extra financial costs.

6 Wider benefits of increased biodiversity in hedgerows

Hedgerow management safeguarding one ecosystem service usually benefits other ecosystem services (Wratten et al. 2012, Montgomery et al. 2020). Restoration and better management of hedgerows in temperate agroecosystems is a nature-based solution helping to mitigate anthropogenic climate change (Drexler et al. 2021), biodiversity loss (Graham et al. 2018) and water pollution (Wallace et al. 2021) without compromising agricultural production. Prevention of soil erosion (Rajbanshi et al. 2023), shelter of crops and grazing animals (Couthouis et al. 2023), livestock biosecurity (Delgadillo et al. 2021), provision of food, fuel and wood (Montgomery et al. 2020), carbon sequestration (Drexler et al. 2021), flood control (Thomas et al. 2012) and benefits in health and well-being (Dover 2019), in combination with those related directly to biodiversity (e.g. crop pollination and natural pest control), provide a strong basis for hedgerows to be part of sustainable agriculture and a foundation for rural communities and their economies. For example, hedgerows could play a significant role in carbon sequestration especially where they are a dominant part of the landscape (Black et al. 2023). Drexler et al. (2021) calculated that hedgerows contain an average (±1 SD) of 104 ± 42 Mg C ha^{-1} more than croplands and 81 ± 40 Mg C ha^{-1} more than grassland. They also concluded that carbon stocks in hedgerow soil and biomass above and below ground, are comparable to forests, and that creation of hedges on cropland could capture carbon at an annual rate of between 2.1 Mg C $ha^{-1}.yr^{-1}$ and 5.2 Mg C $ha^{-1}.yr^{-1}$ for a period of 50 and 20 years, respectively.

7 Key hedgerow management techniques to enhance biodiversity

Conservation of ancient hedgerows and their continuity with woodlands are a priority as they contain forest plant species and act as corridors for dispersal of plants and animals (Lenoir et al. 2021). New hedgerows should connect with woodland promoting dispersal of plants and animals across farmland. However, hedge planting alone may not increase species richness (Le Viol et al. 2008) and it takes time to restore landscape connectivity. Recommendations for management of established hedgerows focus on hedgerow length, connectedness and frequency of cutting, but there are additional management prescriptions for smaller schemes and particular taxa (Kremen and McGonigle 2015). Schemes aimed at restoring hedgerow biodiversity should consider how taxa use hedgerows. Honeybees, for example, forage in hedges but also use them in orientation and navigation (Menzel et al. 2019). Hedges act as a corridor and a semi-permeable barrier to honeybees and hence, well-connected and

Published by Burleigh Dodds Science Publishing Limited, 2024.

continuous hedgerows channel honeybee movements (Klaus et al. 2015). Ground flora species composition in hedgerows can be manipulated by excluding fertilisers, increasing or reducing shade, use of selective herbicides and sowing seeds of perennial species (Maudsley et al. 1998, Minarro and Prida 2013). Introducing local soil to a new hedge may increase species richness of a wide range of taxa including fungi and arthropods (Lecq et al. 2018, Holden et al. 2019).

Rotational cutting of hedges at intervals from 3 to 10 years creates a mosaic of habitat patches at the landscape scale and trimming sides alternatively, further enhances biodiversity. Cutting affects not only hedge structure but also the availability and quality of food for wildlife. Extending periodicity of cutting hawthorn hedges, e.g. increases abundance and sugar content of berries (Fig. 5a and b). Mechanical cutting at longer intervals can be just as effective as hedge laying in intensive systems (Staley et al. 2015). Native, multispecies plant communities are more biodiverse and have greater productivity than non-native monocultures (Tilman et al. 2014). Recently planted hedges based on several native species rapidly become a haven for biodiversity (Boutin et al. 2002; Le Viol et al. 2008). Further, hedgerows have more benefit for biodiversity when managed in conjunction with other semi-natural landscape features such as wildflower margins, roadside verges, woodlands and ponds. Farm productivity is reduced close to field margins due to shading by hedgerows and soil compaction by farm machinery. Shading effects may be managed by reducing the height of hedges running east-west (Findlay 2017), whilst compaction adjacent to hedges can be managed by adapting machinery, reducing traffic frequency, and deep ripping or introducing wildflower strips, bird cover crops, beetle banks or fallow field margins.

Taller, wider, denser hedgerows in temperate regions mitigate the effects of increased temperature, destructive levels of rainfall and windspeed on crop production and grazing livestock associated with anthropogenic climate change. In the absence of hedgerow management, however, abandoned farmland undergoes slow 'passive rewilding' where hedgerows tend to be subsumed into scrub or degraded with increasing gaps formation (Broughton et al. 2022). Abandoning hedgerows as opposed to managing them to maximise ecosystem services undermines agricultural sustainability. Gaps in hedges where there is no scrub encroachment generally reduce animal and plant diversity (McDonald and Johnson 1995). By comparison, windbreaks and shelterbelts are less well-researched than hedges and often overlooked in policy and the management of agroecosystems (Wenniger et al. 2021).

Corridors of undeveloped land set aside for recreational or environmental purposes, often called greenways, contribute to landscape restoration by adapting redundant infrastructure such as abandoned railways and frequently take the form of pathways between parallel adjacent hedgerows. Carlier and Moran (2019) sampled vegetation in hedgerows along 70 km of greenway in Ireland, identifying four groups of optimum plant species composition and structure which vary locally with surrounding land use. They concluded that appropriately managed greenways and green lanes where present could play a major role in restoring hedgerow ecosystem services locally and regionally, and as part of a larger European network.

Spatial continuities between different crops and semi-natural habitats including hedgerows, annual crops and permanent and semi-permanent structures, may affect biodiversity at a scale greater than the farm (McKenzie et al. 2013). Modelling ground beetle biodiversity at field, farm and landscape levels with different densities of hedgerows, Martel et al. (2019), for example, suggested maize supported more species in landscapes dominated by pig farming due to spatial links between winter and spring crops, whereas woody habitats (including hedgerows) supported more species in mixed (pig and dairy farms) landscapes with greater crop diversity. Species-specific factors affect the network of semi-natural sites utilised by different species across an otherwise intensively farmed landscape. For example, badgers in northern Italy used woodland, poplar plantations and hedgerows as habitat patches and short rotation coppices as connecting corridors, whilst roe deer *Capreolus capreolus* were limited mostly to the main forest areas with hedgerows and plantations as habitat patches, and coppices and hedgerows as corridors (Dondina et al. 2019). In contrast to Britain where badgers generally have their main setts (underground dens) in woodland, in the absence of woodland in Ireland, setts are often located in hedgerows (O'Brien et al. 2016) where they are accessible to cattle. This has serious implications for bovine tuberculosis (bTB) transmission between badgers and cattle (Byrne et al. 2024). Double-fenced, intact, hedgerows at a high spatial density, however, are associated with lower levels of bTB in cattle presumably due to a reduction in 'nose-to-nose' contact (Matthews et al. 2006). Planning issues with regards to renewable energy are also important in developing hedgerow systems. Wind turbines close to hedgerows, e.g. repel bats whilst turbines located further away from hedgerows in open fields, attract bats, increasing their vulnerability to collision (Leroux et al. 2022). Local factors, therefore, indicate the need for a regional approach to hedgerow restoration and management (Barr and Petit 2001).

8 Barriers to conservation management of hedgerows

Reversing intensive management of hedgerows promotes sustainable production and increases biodiversity on multiple scales in arable and pastoral sectors of agriculture and horticulture. Warnings of diminished food production, the cost of change, economic losses and unemployment, however, may be used to block or delay change. Inertia and overt opposition from conventional farming interests in favour of continued use of agrochemicals (Kremen 2020), encourage farmers to continue with conventional management. Farmers are reluctant to change their behaviour regarding field boundary management even where there is demonstrable agronomic value from enhanced pollination or biological pest control (Morandin et al. 2016). However, advice, on-farm training and financial incentives for delivery of ecosystem services can change attitudes (Rey Benayas et al. 2020, Lopez-Felices et al. 2022).

Farmers frequently express concern that there are crop losses close to semi-natural habitats such as hedgerows. Raatz et al. (2021) examined the basis of this effect in wheat fields. They found that hedgerows might lead to more weeds between 1 and 5 m distance from the field boundary, negatively affecting yields. However, there was no effect at field and landscape levels suggesting that any effect is weak and reduced by existing farming practices or compensated for by increased yields due to reduced wind speeds (Weninger et al. 2021). Modelling the effect of forest edge and hedgerows on wheat production in Ethiopia suggests there is an asymptotic relationship between increasing hedgerow length and yield (Yang et al. 2020).

Precigout and Robert (2022) demonstrated a scientific consensus that hedgerows are unrivalled habitats for the maintenance of farmland biodiversity whilst providing pest and pollination services in adjacent crops. However, these services are subject to particular conditions such as species traits, landscape components and spatial scale which require further research (Precigout and Robert 2022). Nevertheless, many farmers in many countries remain skeptical finding the scientific complexity involved daunting. Demonstrations of improved hedgerow management, drawing on experience of local farmers, is more effective than convoluted ecology and advice from other sources (Marinus et al. 2021) in persuading farmers to abandon intensive practices that are no longer appropriate. This approach might, e.g. reverse a prevailing attitude that freshly cut, neat hedges are desirable in mid-summer even though this is illegal, undermines biodiversity generally, disrupts bird breeding and reduces winter food supply.

Resistance to offsetting impacts on biodiversity due to specific developments and changes in land use (Millon et al. 2021) could also be reduced by offering multiple options with similar biodiversity benefits and, thus, choice of management prescriptions. Hedgerows always provide most biodiversity benefit for birds and bats and many invertebrates compared to other forms of field margin. Conflicting conservation interests, however, may arise where there are conflicting needs amongst species. For example, grassland and woodland specialists may experience opposite effects where land use changes from forest to grassland. De Zwaan et al. (2022) showed that populations of threatened grassland and woodland specialist bird species benefit from creating a mosaic

Published by Burleigh Dodds Science Publishing Limited, 2024.

landscape featuring hedgerows as well as retaining woodland and grassland, increasing the number of management options and offering a more flexible approach to conservation.

Costs of purchasing equipment to implement less frequent hedgerow cutting regimes are often a cause for concern amongst managers but could be offset by harvesting wood for fuel and the lower labour costs involved in hedge maintenance. Smith et al. (2023) estimated that the sale of woodchip could be profitable and sustainable using tree shears. Aerial imaging is being explored as a means of estimating and monitoring carbon sequestration by hedgerows (Black et al. 2023). Long et al. (2017) showed that hedge restoration enabled better pest control and crop pollination and paid for itself. Similarly, Morandin et al. (2016) calculated that reduced costs and improved pollination tilts the economics in favour of hedge restoration. In conclusion, the present review and Precigout and Roberts (2022) provide further support for the view that hedgerows, alone and in conjunction with other semi-natural habitats, provide diverse ecosystem services, critically pollination and pest regulation, that underpin economic sustainability of diverse farming enterprises.

9 Policy affecting hedgerows

The history of agri-environmental schemes (AESs) and environmentally friendly agriculture in Europe and North America is exemplified by the Common Agricultural Policy (CAP) in the EU. The original objective was food security with financial incentives linked to farm size and agricultural condition. Additional components of CAP promote farm management intended to maintain or enhance farmland biodiversity. CAP, however, is widely regarded as having failed to reverse declines, for example, in bird populations (Rieger et al. 2022, Rigal et al. 2023). AESs may also have unintended consequences. For example, in Northern Ireland, management of Environmentally Sensitive Areas (ESAs) promoted less intensive farming and was expected to benefit Irish hares *Lepus timidus hibernicus*, a species of conservation concern (EHS 2000, 2005). Hare numbers did not increase but rabbit *Oryctolagus cuniculus* and fox *Vulpes vulpes*, widely regarded as agricultural pests, were two to three times more abundant in ESAs than conventionally managed areas due to increased hedgerow cover (Reid et al. 2007).

Not all species benefit from AES prescriptions. Thirteen years of territorial mapping of hedgerow birds, cropping and hedge management on an AES farm in southeast England suggested some species were associated positively with roadside hedgerows, hedgerow trees and unsown fields whilst others favoured areas of oilseed rape (Tresise et al. 2021). In Wales, the AES (known as Tis Gofal) was associated with increased growth rates in populations of woodland and hedgerow birds but effects on priority bird species were inconsistent (Dadam and Sriwardena 2019). Thus, management prescriptions must be specific, evidence-based and address unintended consequences and geographical variation, for the best biodiversity outcomes.

In the EU, CAP reform aims to improve agricultural productivity whilst tackling climate change, sustainable management of natural resources, and protecting landscapes and areas (Cuadros-Casanova et al. 2023). CAP prescriptions have been partially effective in addressing loss of biodiversity across the EU but are being revised in line with more emphasis on environmental concerns and recent scientific research (Tarjuelo et al. 2021, Pe'er et al. 2022). For example, hedgerows contain brambles *Rubus* spp which are an important flowering resource for insects in summer. Crowther and Gilbert (2020) found significantly more bramble and bees on farms under an earlier AES than on those under conventional management. AES management is usually focussed on target species but may also benefit non-target species. Pollinating insects, e.g. may respond positively to flower-rich habitats but so do soprano pipistrelle *Pipistrellus pygmaeus* and barbastelle bats which prey on such insects (McHugh et al. 2019).

The devolved jurisdictions in the UK are introducing their own AES to replace CAP. These vary in terminology and specifications but generally require farms to choose one of three levels: Conventional, and Mid and Higher Tier Stewardship (UK Government 2024a). Conventional level reflects current practices with regard to monocultural agriculture and use of chemicals, and the Stewardship levels require hedgerow mapping, monitoring, creation and restoration. Farmers enter a contract to manage areas of higher conservation value features on their farms sensitively for payments based on an area (e.g. flower rich meadows) or length (e.g. hedgerows) for 5 years. English Nature provide a detailed set of actions for hedgerow management at the Mid and Higher Tiers (UK Government 2024b).

Organic farming and hedgerow density (km/km^2) have a positive effect on biodiversity conservation and pest control on AES farms in comparison to non-AES farms in northwest France (Couthouis et al. 2023). However, benefits of hedgerows and semi-natural habitats in terms of ecological, agronomic and socio-economic performance were not apparent at the landscape scale, suggesting that local management of fields and adjacent margins represent appropriate targets for implementation of AESs. Oosterveld et al. (2022) demonstrated that AESs had larger populations of shrubland birds in Dutch hedgerows mostly due to hedgerow characteristics, notably bramble and nettle cover, and width of the hedgerow at ground level. Swiss farmers receive direct payments provided 7% of their land is managed as a Biodiversity Promotion Area (BPA) including hedgerows (Ravetto Enri et al. 2020). Jeanneret et al. (2022) found that diversity of spiders was greater in areas of trees, especially hedgerows, in BPAs than in non-BPAs.

Environmental heterogeneity and variation between AESs and their benefits, vary geographically suggesting that planning should be at a national or regional level. This variability is also evident in the trade-off between food security and environmental objectives. Jeanneret et al. (2021) examined a hypothetical proposal to convert hedgerows and other semi-natural habitats into agricultural land to help meet global food requirements. Analysis of data from 169 farms in 10 European regions suggested that this would incur substantial losses of biodiversity because 49% of taxa were restricted to semi-natural habitats. Species loss due to hedgerow conversion would be greater in arable systems and least in grassland-based agroecosystems which may have inherently low levels of diversity. Food production benefitted by only 10% suggesting that agricultural policy decisions must consider potential production and environmental benefits. Staley et al. (2023) suggested that extending mean hedgerow length in the UK from 4.2 to 10 km/km^2 would enhance wildlife support, carbon sequestration and connectivity with other semi-natural habitats. It is likely that increased hedgerow length could have both direct and indirect positive and negative effects on food production in arable, mixed and pastoral agriculture. However, this has not yet been assessed and there is also a lack of data on the current condition of hedges, where best to plant, potential for climate change mitigation and vulnerability to invasive species.

Perverse outcomes of AES affecting biodiversity suggest that some programs may have been poorly conceived, over-optimistic or implemented badly. Fragmentation of land holdings may prohibit development of agriculture. Consolidation of small farms into larger ones in Slovenia, on the other hand, is associated with intensification, and increased use of agrochemicals, undermining farmland bird diversity and abundance (Denac and Kmecl 2021). Central Spain lost nearly half of its hedgerows in the last 70 years leading to hedgerow soils drying out and a loss of functionality exacerbating the effect of a succession of droughts due to anthropogenic climate change. Alvarez et al. (2021a) concluded that these losses occurred despite European legislation on biodiversity, cultural heritage and rural development, which failed to prevent the loss of hedgerows at regional scales.

The EU CAP and other AESs are under reform with greater emphasis on environmental outcomes. This requires replacing conventional views that conservation conflicts with farm production and establishing a joint approach to increasing the sustainability of ecosystems that support food production. Farming systems, landscape and habitats such as hedgerows are heterogeneous often making it difficult to apply overarching, generalised prescriptions. However, the principle of protecting the ecosystem services provided by biodiversity and environmental sustainability, to support enduring food production, should be universal in agriculture. Performance-based or results-based payments have been proposed as an alternative to area-based or action-based AESs (Ruas et al. 2021).

10 Future directions in monitoring hedgerow condition

AESs require regular monitoring using standardised methods to enable comparison within and between regions. Due to the complexity and variability of landscapes, habitats and mobility of species, there is no single monitoring system. Staley et al. (2021) presented a plan for monitoring the impact of AES prescriptions across multiple scales, involving site visits and scoring of taxa including mobile taxa, against predicted gradients. Time and labour costs as well as the expertise required, however, are likely to limit the use of such an approach. A simplified approach is to use bioindicator species that farmers and others can easily identify. In contrast to grassland, putative indicator species in hedgerows are not strongly related to hedgerow or other environmental variables. Performance-based or results-based AESs require further research and consultation with farmers (Ruas et al. 2021).

Hedgerows are difficult targets for remote sensing because of their variability and 3D structure. Thompson et al. (2023) used satellite and aerial imagery to measure percentage of land and farms comprising habitats including hedgerows and windbreaks in California. Smigaj and Gaulton (2021) showed the potential of drones to capture images of hedgerows in red, green and blue (RBG), and multispectral nadir and oblique imagery for estimating height and width. Ahlswede et al. (2021) used machine learning and neural network methods to map different hedgerow types in southeast Germany. Similarly, Alvarez et al. (2021b) used aerial and satellite imagery to categorise moist, semi-moist and dry hedgerows in Mediterranean countries. Integrating technologies with coding and cloud-based supercomputing platforms such as Google Earth Engine and Microsoft Planetary Computer allow for real-time monitoring of hedgerows, their cover and management. Full LIDAR coverage of hedgerows in England was announced in January 2024 (Centre for Ecology and Hydrology 2024). This three-dimensional mapping has the potential to monitor implementation of environmental prescriptions as well as sampling biodiversity benefits and hedge management and structure from the field through farm, landscape and regional levels.

11 Where to look for further information

Pollard et al. (1974) is an excellent book on hedgerow ecology and management while Dover (2019) is a more recent general treatment of the ecology of hedges. Dicks et al. (2014) presented management prescriptions addressing conservation of biodiversity and environmental degradation in diverse farmland systems. Montgomery et al. (2020) and Précigout and Robert (2022) reviewed hedgerow ecosystem services, sustainable hedgerow management and the global role of hedgerows in meeting food requirements and mitigating the environmental degradation due to climate change and loss of biodiversity.

12 References

Ahlswede, S., Asam, S. and Roeder, A., 2021. Hedgerow object detection in very high-resolution satellite images using convolutional neural networks. *Journal of Applied Remote Sensing*, *15*(1), pp.018501–018501.

Ahmed, K.S., Volpato, A., Day, M.F., Mulkeen, C.J., O'Hanlon, A., Carey, J., Williams, C., Ruas, S., Moran, J., Rotchés-Ribalta, R. and ÓhUallacháin, D., Stout, J.C., Hodge, S., White, B. and Gormally, M.J., 2021. Linear habitats across a range of farming intensities contribute differently to dipteran abundance and diversity. *Insect Conservation and Diversity*, *14*(3), pp.335–347.

Albrecht, M., Kleijn, D., Williams, N.M., Tschumi, M., Blaauw, B.R., Bommarco, R., Campbell, A.J., Dainese, M., Drummond, F.A., Entling, M.H. and Ganser, D., 2020. The effectiveness of flower strips and hedgerows on pest control, pollination services and crop yield: a quantitative synthesis. *Ecology Letters*, *23*(10), pp.1488–1498.

Alison, J., Botham, M., Maskell, L.C., Garbutt, A., Seaton, F.M., Skates, J., Smart, S.M., Thomas, A.R., Tordoff, G., Williams, B.L. and Wood, C.M., 2022. Woodland, cropland and hedgerows promote pollinator abundance in intensive grassland landscapes, with saturating benefits of flower cover. *Journal of Applied Ecology*, *59*(1), pp.342–354.

Alvarez, F.A., Gomez-Mediavilla, G., López-Estébanez, N. and Holgado, P.M., 2021a. Classification of Mediterranean hedgerows: a methodological approximation. *MethodsX*, *8*, p.101355.

Álvarez, F.A., Mediavilla, G.G. and Estébanez, N.L., 2021b. Environmental, demographic and policy drivers of change in mediterranean hedgerow landscape (Central Spain). *Land Use Policy*, *103*, p.105342.

Aude, E., Tybirk, K. and Pedersen, M.B., 2003. Vegetation diversity of conventional and organic hedgerows in Denmark. *Agriculture, Ecosystems & Environment*, *99*(1–3), pp.135–147.

Ancillotto, L., Bosso, L., Salinas-Ramos, V.B. and Russo, D., 2019. The importance of ponds for the conservation of bats in urban landscapes. *Landscape and Urban Planning*, *190*, p.103607.

Badenhausser, I., Gross, N., Mornet, V., Roncoroni, M., Saintilan, A. and Rusch, A., 2020. Increasing amount and quality of green infrastructures at different scales promotes biological control in agricultural landscapes. *Agriculture, Ecosystems & Environment*, *290*, p.106735.

Balmford, A., 2021. Concentrating vs. spreading our footprint: how to meet humanity's needs at least cost to nature. *Journal of Zoology*, *315*(2), pp.79–109.

Barr, C. and Petit, S., 2001. *Hedgerows of the world: their ecological functions in different landscapes*. Proceedings of the 10th Annual Conference of the International

Association for Landscape Ecology, held at Birmingham University, 5th-8th September 2001. International Association for Landscape Ecology (IALE (UK)).

Batáry, P., Kovács-Hostyánszki, A., Fischer, C., Tscharntke, T. and Holzschuh, A., 2012. Contrasting effect of isolation of hedges from forests on farmland vs. woodland birds. *Community Ecology*, *13*(2), pp.155–161.

Bateman, I. and Balmford, A., 2023. Current conservation policies risk accelerating biodiversity loss. *Nature*, *618*(7966), pp.671–674.

Bearman-Brown, L.E., Baker, P.J., Scott, D., Uzal, A., Evans, L. and Yarnell, R.W., 2020. Over-winter survival and nest site selection of the west-European hedgehog (*Erinaceus europaeus*) in arable dominated landscapes. *Animals*, *10*(9), p.1449.

Bellamy, C., Boughey, K., Hawkins, C., Reveley, S., Spake, R., Williams, C. and Altringham, J., 2020. A sequential multi-level framework to improve habitat suitability modelling. *Landscape Ecology*, *35*(4), pp.1001–1020.

Biello, R., Brunelli, A., Sozio, G., Havenstein, K., Mortelliti, A., Ketmaier, V., Vilaça, S.T. and Bertorelle, G., 2023. The genetic structure and connectivity in two sympatric rodent species with different life histories are similarly affected by land use disturbances. *Conservation Genetics*, *24*(1), pp.59–72.

Bistline-East, A., Burke, D., Williams, C.D. and Gormally, M.J., 2020. Habitat requirements of *Tetanocera elata* (Diptera: Sciomyzidae): Case study of a dry meadow in western Ireland. *Agricultural and Forest Entomology*, *22*(3), pp.250–262.

Black, K., Lanigan, G., Ward, M., Kavanagh, I. and Sullivan, L.O., 2023. Biomass carbon stocks and stock changes in managed hedgerows. *Science of The Total Environment*, *871*, p.162073.

Blary, C., Kerbiriou, C., Le Viol, I. and Barré, K., 2021. Assessing the importance of field margins for bat species and communities in intensive agricultural landscapes. *Agriculture, Ecosystems & Environment*, *319*, p.107494.

Boinot, S. and Alignier, A., 2022. On the restoration of hedgerow ground vegetation: local and landscape drivers of plant diversity and weed colonization. *Journal of Environmental Management*, *307*, p.114530.

Boinot, S. and Alignier, A., 2023. Discrepancies between the drivers of alpha and beta plant diversity in arable field margins. *Proceedings of the Royal Society B*, *290*(1992), p.20222179.

Boinot, S., Mony, C., Fried, G., Ernoult, A., Aviron, S., Ricono, C., Couthouis, E. and Alignier, A., 2023. Weed communities are more diverse, but not more abundant, in dense and complex bocage landscapes. *Journal of Applied Ecology*, *60*(1), pp.4–16.

Boutin, C., Baril, A., McCabe, S.K., Martin, P.A. and Guy, M., 2011. The value of woody hedgerows for moth diversity on organic and conventional farms. *Environmental Entomology*, *40*(3), pp.560–569.

Boutin, C., Jobin, B., Bélanger, L. and Choinière, L., 2002. Plant diversity in three types of hedgerows adjacent to cropfields. *Biodiversity & Conservation*, *11*, pp.1–25.

Bouvier, J.C., Boivin, T. and Lavigne, C., 2020. Conservation value of pome fruit orchards for overwintering birds in southeastern France. *Biodiversity and Conservation*, *29*, pp.3169–3189.

Bouvier, J.C., Delattre, T., Boivin, T., Musseau, R., Thomas, C. and Lavigne, C., 2022a. Great tits nesting in apple orchards preferentially forage in organic but not conventional orchards and in hedgerows. *Agriculture, Ecosystems & Environment*, *337*, p.108074.

Bouvier, J.C., Boivin, T. and Lavigne, C., 2022b. Single-row exclusion nets: an alternative pest control method with no detectable impact on breeding bird assemblages in orchards bordered by hedgerows. *Agronomy for Sustainable Development*, *42*(2), p.23.

Brooks, D.R., Bater, J.E., Clark, S.J., Monteith, D.T., Andrews, C., Corbett, S.J., Beaumont, D.A. and Chapman, J.W., 2012. Large carabid beetle declines in a United Kingdom monitoring network increases evidence for a widespread loss in insect biodiversity. *Journal of Applied Ecology*, *49*(5), pp.1009–1019.

Broughton, R.K., Bullock, J.M., George, C., Hill, R.A., Hinsley, S.A., Maziarz, M., Melin, M., Mountford, J.O., Sparks, T.H. and Pywell, R.F., 2021a. Long-term woodland restoration on lowland farmland through passive rewilding. *PLoS One*, *16*(6), p.e0252466.

Broughton, R.K., Bullock, J.M., George, C., Gerard, F., Maziarz, M., Payne, W.E., Scholefield, P.A., Wade, D. and Pywell, R.F., 2022. Slow development of woodland vegetation and bird communities during 33 years of passive rewilding in open farmland. *Plos One*, *17*(11), p.e0277545.

Broughton, R.K., Chetcuti, J., Burgess, M.D., Gerard, F.F. and Pywell, R.F., 2021b. A regional-scale study of associations between farmland birds and linear woody networks of hedgerows and trees. *Agriculture, Ecosystems & Environment*, *310*, p.107300.

Brown, J.A., Montgomery, W.I. and Provan, J., 2022. Strong spatial structuring of clonal genetic diversity within blackthorn (*Prunus spinosa*) hedgerows and woodlands. *Tree Genetics & Genomes*, *18*, pp.1–9.

Burel, F., 1989. Landscape structure effects on carabid beetles' spatial patterns in western France. *Landscape Ecology*, *2*(4), pp.215–226.

Byrne, A.W., Allen, A., Ciuti, S., Gormley, E., J. Kelly, D.J., Marks, N.J., Marples, N.M., Menzies, F., Montgomery, I., Newman, C., O'Hagan, M., Reid, N., Scantlebury, D.M., Stuart, P. and Tsai, M., 2024. Badger ecology, bovine tuberculosis, and population management: lessons from the Island of Ireland. *Transboundary and Emerging Diseases, 2024*, Article ID 8875146, 18 pages. https://doi.org/10.1155/2024/8875146

Caravaggi, A., Montgomery, W.I. and Reid, N., 2015. Range expansion and comparative habitat use of insular, congeneric lagomorphs: invasive European hares *Lepus europaeus* and endemic Irish hares *Lepus timidus hibernicus*. *Biological Invasions*, *17*, pp.687–698.

Carr, A., Weatherall, A., Fialas, P., Zeale, M.R., Clare, E.L. and Jones, G., 2020. Moths consumed by the Barbastelle *Barbastella barbastellus* require larval host plants that occur within the bat's foraging habitats. *Acta Chiropterologica*, *22*(2), pp.257–269.

Carlier, J. and Moran, J., 2019. Hedgerow typology and condition analysis to inform greenway design in rural landscapes. *Journal of Environmental Management*, *247*, pp.790–803.

Castle, D., Grass, I. and Westphal, C., 2019. Fruit quantity and quality of strawberries benefit from enhanced pollinator abundance at hedgerows in agricultural landscapes. *Agriculture, Ecosystems & Environment*, *275*, pp.14–22.

Centre for Ecology and Hydrology, 2024. High tech aerial mapping reveals England's hedgerow landscape. http://www.ceh.ac.uk/press/ Accessed February 1 2024.

Chamberlain, D.E., Vickery, J.A., Marshall, E.J.P. and Tucker, G.M., 2001. The effect of hedgerow characteristics on the winter hedgerow bird community. In *Hedgerows of the world: their ecological functions in different landscapes. Proceedings of the*

10th Annual Conference of the International Association for Landscape Ecology, held at Birmingham University, 5th-8th September 2001 (pp. 197–206). International Association for Landscape Ecology (UK).

Charrier, S., Petit, S. and Burel, F., 1997. Movements of *Abax parallelepipedus* (Coleoptera, Carabidae) in woody habitats of a hedgerow network landscape: a radio-tracing study. *Agriculture, Ecosystems & Environment*, *61*(2–3), pp.133–144.

Claireau, F., Bas, Y., Pauwels, J., Barré, K., Machon, N., Allegrini, B., Puechmaille, S.J. and Kerbiriou, C., 2019. Major roads have important negative effects on insectivorous bat activity. *Biological Conservation*, *235*, pp.53-62.

Clausen, M.A., Elle, E. and Smukler, S.M., 2022. Evaluating hedgerows for wild bee conservation in intensively managed agricultural landscapes. *Agriculture, Ecosystems & Environment*, *326*, p.107814.

Collier, M.J., 2021. Are field boundary hedgerows the earliest example of a nature-based solution?. *Environmental Science & Policy*, *120*, pp.73–80.

Coulthard, E., McCollin, D. and Littlemore, J., 2016. The use of hedgerows as flight paths by moths in intensive farmland landscapes. *Journal of Insect Conservation*, *20*, pp.345–350.

Couthouis, E., Aviron, S., Pétillon, J. and Alignier, A., 2023. Ecological performance underlying ecosystem multifunctionality is promoted by organic farming and hedgerows at the local scale but not at the landscape scale. *Journal of Applied Ecology*, *60*(1), pp.17–28.

Crowther, L.I. and Gilbert, F., 2020. The effect of agri-environment schemes on bees on Shropshire farms. *Journal for Nature Conservation*, *58*, p.125895.

Crowther, L.I., Wilson, K. and Wilby, A., 2023. The impact of field margins on biological pest control: a meta-analysis. *BioControl*, pp.1–10.

Cuadros-Casanova, I., Cristiano, A., Biancolini, D., Cimatti, M., Sessa, A.A., Mendez Angarita, V.Y., Dragonetti, C., Pacifici, M., Rondinini, C. and Di Marco, M., 2023. Opportunities and challenges for Common Agricultural Policy reform to support the European Green Deal. *Conservation Biology*, *37*, p.e14052.

Cuff, J.P., Evans, S.A., Porteous, I.A., Quiñonez, J. and Evans, D.M., 2022. Candy-striped spider leaf and habitat preferences for egg deposition. *Agricultural and Forest Entomology*, *24*(3), pp.422–431.

Dadam, D. and Siriwardena, G.M., 2019. Agri-environment effects on birds in Wales: Tir Gofal benefited woodland and hedgerow species. *Agriculture, Ecosystems & Environment*, *284*, p.106587.

Dainese, M., Montecchiari, S., Sitzia, T., Sigura, M. and Marini, L., 2017. High cover of hedgerows in the landscape supports multiple ecosystem services in Mediterranean cereal fields. *Journal of Applied Ecology*, *54*(2), pp.380–388.

DEFRA. (2023) Ambitious roadmap for a cleaner, greener country. Department for Environment, Food & Rural Affairs, Natural England, Forestry Commission, The Rt Hon Rishi Sunak MP, and The Rt Hon Thérèse Coffey MP. https://www.gov.uk/government/news/ambitious-roadmap-for-a-cleaner-greener-country

Delaune, T., Ouattara, M.S., Ballot, R., Sausse, C., Felix, I., Maupas, F., Chen, M., Morison, M., Makowski, D. and Barbu, C., 2021. Landscape drivers of pests and pathogens abundance in arable crops. *Ecography*, *44*(10), pp.1429–1442.

de la Peña, N.M., Butet, A., Delettre, Y., Morant, P. and Burel, F., 2003. Landscape context and carabid beetles (Coleoptera: Carabidae) communities of hedgerows in western France. *Agriculture, Ecosystems & Environment*, *94*(1), pp.59–72.

Delgadillo, E., Glidden, C., Pollak, M., Rysenga, H., Jolles, A. and Beechler, B., 2021. The benefit of hedgerow access on the health and growth rate of pasture raised broiler chickens. *Frontiers in Animal Science, 2,* p.649924.

Denac, K. and Kmecl, P., 2021. Land consolidation negatively affects farmland bird diversity and conservation value. *Journal for Nature Conservation, 59,* p.125934.

de Zwaan, D.R., Alavi, N., Mitchell, G.W., Lapen, D.R., Duffe, J. and Wilson, S., 2022. Balancing conservation priorities for grassland and forest specialist bird communities in agriculturally dominated landscapes. *Biological Conservation, 265,* p.109402.

Dicks, L.V., Ashpole, J.E., Dänhardt, J., James, K., Jönsson, A.M., Randall, N., Showler, D.A., Smith, R.K., Turpie, S., Williams, D.R. and Sutherland, W.J., 2014. *Farmland Conservation: Evidence for the effects of interventions in Northern and Western Europe* (Vol. 3). Pelagic Publishing Ltd.

Dondina, O., Kataoka, L., Orioli, V. and Bani, L., 2016. How to manage hedgerows as effective ecological corridors for mammals: a two-species approach. *Agriculture, Ecosystems & Environment, 231,* pp.283–290.

Dondina, O., Orioli, V., Chiatante, G., Meriggi, A. and Bani, L., 2019. Species specialization limits movement ability and shapes ecological networks: the case study of 2 forest mammals. *Current Zoology, 65*(3), pp.237–249.

Dover, J.W., 2019. Introduction to hedgerows and field margins. In J. W. Dover (Ed.), *The ecology of hedgerows and field margins* (pp. 1-34). Routledge, Abingdon, UK.

Dover, J.W. and Sparks, T., 2000. A review of the ecology of butterflies in British hedgerows. *Journal of Environmental Management, 60*(1), pp.51–63.

Drexler, S., Gensior, A. and Don, A., 2021. Carbon sequestration in hedgerow biomass and soil in the temperate climate zone. *Regional Environmental Change, 21*(3), p.74.

Dunn, J.C., Gruar, D., Stoate, C., Szczur, J. and Peach, W.J., 2016. Can hedgerow management mitigate the impacts of predation on songbird nest survival? *Journal of Environmental Management, 184,* pp.535–544.

EHS. (2000) *Biodiversity in Northern Ireland: Species Action Plans – Irish Hare, Chough and Curlew,* Environment and Heritage Service NI, Department of Environment, Belfast, UK.

EHS. (2005) *All-Ireland Species Action Plan: Irish Hare (*Lepus timidus hibernicus). Environment and Heritage Service NI, Department of Environment, Belfast, UK.

El Ghouizi, A., Bakour, M., Laaroussi, H., Ousaaid, D., El Menyiy, N., Hano, C. and Lyoussi, B., 2023. Bee Pollen as Functional Food: Insights into Its Composition and Therapeutic Properties. *Antioxidants, 12*(3), p.557.

Estrada-Carmona, N., Sánchez, A.C., Remans, R. and Jones, S.K., 2022. Complex agricultural landscapes host more biodiversity than simple ones: a global meta-analysis. *Proceedings of the National Academy of Sciences, 119*(38), p.e2203385119.

Evans, D.M., Pocock, M.J., Brooks, J. and Memmott, J., 2011. Seeds in farmland food-webs: resource importance, distribution and the impacts of farm management. *Biological Conservation, 144*(12), pp.2941–2950.

Evans, D.M., Pocock, M.J. and Memmott, J., 2013. The robustness of a network of ecological networks to habitat loss. *Ecology Letters, 16*(7), pp.844–852.

Facey, S.L., Botham, M.S., Heard, M.S., Pywell, R.F. and Staley, J.T., 2014. Moth communities and agri-environment schemes: Examining the effects of hedgerow cutting regime on diversity, abundance, and parasitism. *Insect Conservation and Diversity, 7*(6), pp.543–552.

Fernández-Fernández, P., Sanczuk, P., Vanneste, T., Brunet, J., Ehrlén, J., Hedwall, P.O., Hylander, K., Van Den Berge, S., Verheyen, K. and De Frenne, P., 2022. Different effects of warming treatments in forests versus hedgerows on the understorey plant *Geum urbanum*. *Plant Biology*, *24*(5), pp.734–744.

Fichtner, A. and Wissemann, V., 2021. Biological flora of the British Isles: *Crataegus monogyna*. *Journal of Ecology*, *109*(1), pp.541–571.

Fijen, T.P., Read, S.F., Walker, M.K., Gee, M., Nelson, W.R. and Howlett, B.G., 2022. Different landscape features within a simplified agroecosystem support diverse pollinators and their service to crop plants. *Landscape Ecology*, *37*(7), pp.1787–1799.

Filipiak, M., 2019. Key pollen host plants provide balanced diets for wild bee larvae: a lesson for planting flower strips and hedgerows. *Journal of Applied Ecology*, *56*(6), pp.1410–1418.

Finlay C., 2017. *Scale, production and biodiversity on farmland: Small fields and small hedges or large fields and large hedgerows*. Unpublished PhD thesis, Queen's University of Belfast.

Forest Research 2023. https://www.forestresearch.gov.uk/publications/forestry-facts-figures-2023/.

Foxley, T., Lintott, P. and Stone, E., 2023. What drives bat activity at field boundaries? *Journal of Environmental Management*, *329*, p.117029.

French, D.D. and Cummins, R.P., 2001. Classification, composition, richness and diversity of British hedgerows. *Applied Vegetation Science*, *4*(2), pp.213–228.

Froidevaux, J.S., Boughey, K.L., Hawkins, C.L., Broyles, M. and Jones, G., 2019. Managing hedgerows for nocturnal wildlife: do bats and their insect prey benefit from targeted agri-environment schemes? *Journal of Applied Ecology*, *56*(7), pp.1610–1623.

Fróna, D., Szenderák, J. and Harangi-Rákos, M., 2019. The challenge of feeding the world. *Sustainability*, *11*(20), p.5816.

Gantz, A.G. and Valdivia, C.E., 2020. Post-dispersal seed predation in the hedgerows of agroecosystems in southern Chile. *Gayana*, *84*(2), pp.112–117.

Gantz, A.G. and Valdivia, C.E., 2021. A high rate of rodent-mediated predation on artificial nests in hedgerows of an agricultural landscape in southern Chile. *Bird Study*, *68*(2), pp.284–288.

Gardner, E., Breeze, T.D., Clough, Y., Smith, H.G., Baldock, K.C., Campbell, A., Garratt, M.P., Gillespie, M.A., Kunin, W.E., McKerchar, M. and Potts, S.G., 2021. Field boundary features can stabilise bee populations and the pollination of mass-flowering crops in rotational systems. *Journal of Applied Ecology*, *58*(10), pp.2287–2304.

Gavinelli, F., Fusaro, S., Ivan, D., Ragusa, S. and Paoletti, M.G., 2020. Unpredicted ecological and ecosystem services of biodiversity. Spontaneous vegetation, hedgerows, and maple trees as useful landscape components to increase predatory mite population in agroecosystems. *Applied Soil Ecology*, *154*, p.103584.

Gaüzère, P., Barbaro, L., Calatayud, F., Princé, K., Devictor, V., Raison, L., Sirami, C. and Balent, G., 2020. Long-term effects of combined land-use and climate changes on local bird communities in mosaic agricultural landscapes. *Agriculture, Ecosystems & Environment*, *289*, p.106722.

Genes, L. and Dirzo, R., 2022. Restoration of plant-animal interactions in terrestrial ecosystems. *Biological Conservation*, *265*, p.109393.

Gervais, A., Bélisle, M., Mazerolle, M.J. and Fournier, V., 2021. Landscape enhancements in apple orchards: higher bumble bee queen species richness, but no effect on apple quality. *Insects*, *12*(5), p.421.

Gonthier, D.J., Sciligo, A.R., Karp, D.S., Lu, A., Garcia, K., Juarez, G., Chiba, T., Gennet, S. and Kremen, C., 2019. Bird services and disservices to strawberry farming in Californian agricultural landscapes. *Journal of Applied Ecology*, *56*(8), pp.1948–1959.

González Fradejas, G., García de León, D., Vasar, M., Koorem, K., Zobel, M., Öpik, M., Moora, M. and Rey Benayas, J.M., 2022. Hedgerows increase the diversity and modify the composition of arbuscular mycorrhizal fungi in Mediterranean agricultural landscapes. *Mycorrhiza*, *32*(5–6), pp.397–407.

Gov.UK. 2024a. Countryside-stewardship-get-funding-to-protect-and-improve-the-land-you-manage https://www.gov.uk/guidance/countryside-stewardship-get-funding-to-protect-and-improve-the-land-you-manage. Accessed February 1 2024.

Gov.UK. 2024b Management of hedgerows. https://www.gov.uk/countryside-stewardship-grants/management-of-hedgerows-be3. Accessed February 1 2024.

Graham, L., Gaulton, R., Gerard, F. and Staley, J.T., 2018. The influence of hedgerow structural condition on wildlife habitat provision in farmed landscapes. *Biological Conservation*, *220*, pp.122–131.

Green, R.E., Osborne, P.E. and Sears, E.J., 1994. The distribution of passerine birds in hedgerows during the breeding season in relation to characteristics of the hedgerow and adjacent farmland. *Journal of Applied Ecology*, pp.677–692.

Groom, Q., Northumberland, S., Walker, K., Yorkshire, N.W. and Mcintosh, J., 2010. BSBI recording the British and Irish flora 2010–2020 Annex 1: guidance on sampling approaches. *BSBI Recording the British and Irish Flora*, *2020*, pp.1–9.

Guiller, G., Legentilhomme, J., Boissinot, A., Blouin-Demers, G., Barbraud, C. and Lourdais, O., 2022. Response of farmland reptiles to agricultural intensification: Collapse of the common adder *Vipera berus* and the western green lizard *Lacerta bilineata* in a hedgerow landscape. *Animal Conservation*, *25*(6), pp.849–864.

Happe, A.K., Riesch, F., Rösch, V., Gallé, R., Tscharntke, T. and Batáry, P., 2018. Small-scale agricultural landscapes and organic management support wild bee communities of cereal field boundaries. *Agriculture, Ecosystems & Environment*, *254*, pp.92–98.

Haenke, S., Kovács-Hostyánszki, A., Fründ, J., Batáry, P., Jauker, B., Tscharntke, T. and Holzschuh, A., 2014. Landscape configuration of crops and hedgerows drives local syrphid fly abundance. *Journal of Applied Ecology*, *51*(2), pp.505–513.

Harmange, C., Bretagnolle, V., Pays, O., Sarasa, M. and Moreau, J., 2021. Linking personality and health to use of space in the gray partridge: perspectives for management. *European Journal of Wildlife Research*, *67*(5), p.81.

Heath, S.K., Soykan, C.U., Velas, K.L., Kelsey, R. and Kross, S.M., 2017. A bustle in the hedgerow: Woody field margins boost on farm avian diversity and abundance in an intensive agricultural landscape. *Biological Conservation*, *212*, pp.153–161.

Hegarty, C.A. and Cooper, A., 1994 Regional variation of hedge structure and composition in Northern Ireland in relation to management and land use. *Biology and Environment: Proceedings of the Royal Irish Academy* (vol. 94B(3), pp.223–236). Royal Irish Academy.

Hernández, Á. and Zaldívar, P., 2021. Nest-site selection and nest design of Iberian bullfinches *Pyrrhula pyrrhula iberiae* in northwestern Spain. *Avian Biology Research*, *14*(4), pp.124–142.

Hinsley, S.A. and Bellamy, P.E., 2000. The influence of hedge structure, management and landscape context on the value of hedgerows to birds: a review. *Journal of Environmental Management*, *60*(1), pp.33–49.

Hoffmann, H., Peter, F., Herrmann, J.D., Donath, T.W. and Diekoetter, T., 2021. Benefits of wildflower areas as overwintering habitats for ground-dwelling arthropods depend on landscape structural complexity. *Agriculture, Ecosystems & Environment*, *314*, p.107421.

Hoeffner, K., Santonja, M., Monard, C., Barbe, L., Le Moing, M. and Cluzeau, D., 2021. Soil properties, grassland management, and landscape diversity drive the assembly of earthworm communities in temperate grasslands. *Pedosphere*, *31*(3), pp.375–383.

Holden, J., Grayson, R.P., Berdeni, D., Bird, S., Chapman, P.J., Edmondson, J.L., Firbank, L.G., Helgason, T., Hodson, M.E., Hunt, S.F.P. and Jones, D.T., 2019. The role of hedgerows in soil functioning within agricultural landscapes. *Agriculture, Ecosystems & Environment*, *273*, pp.1–12.

Hopwood, J.L., 2008. The contribution of roadside grassland restorations to native bee conservation. *Biological Conservation*, *141*(10), pp.2632–2640.

Hyjazie, B.F. and Sargent, R.D., 2022. Floral resources predict the local bee community: implications for conservation. *Biological Conservation*, *273*, p.109679.

Image, M., Gardner, E. and Breeze, T.D., 2023a. Co-benefits from tree planting in a typical English agricultural landscape: Comparing the relative effectiveness of hedgerows, agroforestry and woodland creation for improving crop pollination services. *Land Use Policy*, *125*, p.106497.

Image, M., Gardner, E., Clough, Y., Kunin, W.E., Potts, S.G., Smith, H.G., Stone, G.N., Westbury, D.B. and Breeze, T.D., 2023b. Which interventions contribute most to the net effect of England's agri-environment schemes on pollination services? *Landscape Ecology*, *38*(1), pp.271–291.

Jeanneret, P., Lüscher, G., Schneider, M.K., Pointereau, P., Arndorfer, M., Bailey, D., Balázs, K., Báldi, A., Choisis, J.P., Dennis, P. and Diaz, M., 2021. An increase in food production in Europe could dramatically affect farmland biodiversity. *Communications Earth & Environment*, *2*(1), p.183.

Jeanneret, P., Pozzi, S. and Martinez Nuñez, C., 2022. Spiders indicate delivery of an agri-environment scheme at multiple diversity levels. *Frontiers in Ecology and Evolution*, *10*, p.866947.

Kaur, H., Torma, A., Gallé-Szpisjak, N., Šeat, J., Lőrinczi, G., Módra, G. and Gallé, R., 2019. Road verges are important secondary habitats for grassland arthropods. *Journal of Insect Conservation*, *23*, pp.899–907.

Klaus, F., Bass, J., Marholt, L., Müller, B., Klatt, B. and Kormann, U., 2015. Hedgerows have a barrier effect and channel pollinator movement in the agricultural landscape. *Journal of Landscape Ecology*, *8*(1), pp.22–31.

Kranz, A.J., Wolz, K.J. and Miller, J.R., 2019. Effects of shrub crop interplanting on apple pest ecology in a temperate agroforestry system. *Agroforestry Systems*, *93*, pp.1179–1189.

Kremen, C., 2020. Ecological intensification and diversification approaches to maintain biodiversity, ecosystem services and food production in a changing world. *Emerging Topics in Life Sciences*, *4*(2), pp.229–240.

Kremen, C. and M'Gonigle, L.K., 2015. Small-scale restoration in intensive agricultural landscapes supports more specialized and less mobile pollinator species. *Journal of Applied Ecology*, *52*(3), pp.602-610.

Kremen, C., M'Gonigle, L.K. and Ponisio, L.C., 2018. Pollinator community assembly tracks changes in floral resources as restored hedgerows mature in agricultural landscapes. *Frontiers in Ecology and Evolution*, *6*, p.170.

Krings, C.H., Darras, K., Hass, A., Batáry, P. and Fabian, Y., 2022. Not only hedgerows, but also flower fields can enhance bat activity in intensively used agricultural landscapes. *Basic and Applied Ecology*, *63*, pp.23–35.

Kross, S.M., Martinico, B.L., Bourbour, R.P., Townsend, J.M., McColl, C. and Kelsey, T.R., 2020. Effects of field and landscape scale habitat on insect and bird damage to sunflowers. *Frontiers in Sustainable Food Systems*, *4*, p.40.

Kühne, I., Arlettaz, R. and Humbert, J.Y., 2022. Landscape woody features, local management and vegetation composition shape moth communities in extensively managed grasslands. *Insect Conservation and Diversity*, *15*(6), pp.739–751.

Lacoeuilhe, A., Machon, N., Julien, J.F. and Kerbiriou, C., 2018. The relative effects of local and landscape characteristics of hedgerows on bats. *Diversity*, *10*(3), p.72.

Laforge, A., Archaux, F., Bas, Y., Gouix, N., Calatayud, F., Latge, T. and Barbaro, L., 2019. Landscape context matters for attractiveness and effective use of road underpasses by bats. *Biological Conservation*, *237*, pp.409–422.

Langlois, A., Jacquemart, A.L. and Piqueray, J., 2020. Contribution of extensive farming practices to the supply of floral resources for pollinators. *Insects*, *11*(11), p.818.

Lajos, K., Csaszar, O., Sárospataki, M., Samu, F. and Toth, F., 2020. Linear woody landscape elements may help to mitigate leaf surface loss caused by the cereal leaf beetle. *Landscape Ecology*, *35*, pp.2225–2238.

Larkin, J., Sheridan, H., Finn, J.A. and Denniston, H., 2019. Semi-natural habitats and ecological focus areas on cereal, beef and dairy farms in Ireland. *Land Use Policy*, *88*, p.104096.

Lecoq, L., Mony, C., Saiz, H., Marsot, M. and Ernoult, A., 2022. Investigating the effect of habitat amount and landscape heterogeneity on the gamma functional diversity of grassland and hedgerow plants. *Journal of Ecology*, *110*(8), pp.1871–1882.

Lecq, S., Loisel, A., Mullin, S.J. and Bonnet, X., 2018. Manipulating hedgerow quality: embankment size influences animal biodiversity in a peri-urban context. *Urban Forestry & Urban Greening*, *35*, pp.1–7.

Lenoir, J., Decocq, G., Spicher, F., Gallet-Moron, E., Buridant, J. and Closset-Kopp, D., 2021. Historical continuity and spatial connectivity ensure hedgerows are effective corridors for forest plants: Evidence from the species–time–area relationship. *Journal of Vegetation Science*, *32*(1), p.e12845.

Leroux, C., Kerbiriou, C., Le Viol, I., Valet, N. and Barré, K., 2022. Distance to hedgerows drives local repulsion and attraction of wind turbines on bats: Implications for spatial siting. *Journal of Applied Ecology*, *59*(8), pp.2142–2153.

Le Viol, I., Julliard, R., Kerbiriou, C., De Redon, L., Carnino, N., Machon, N. and Porcher, E., 2008. Plant and spider communities benefit differently from the presence of planted hedgerows in highway verges. *Biological Conservation*, *141*(6), pp.1581–1590.

Litza, K., Alignier, A., Closset-Kopp, D., Ernoult, A., Mony, C., Osthaus, M., Staley, J., Van Den Berge, S., Vanneste, T. and Diekmann, M., 2022. Hedgerows as a habitat for forest plant species in the agricultural landscape of Europe. *Agriculture, Ecosystems & Environment*, *326*, p.107809.

Litza, K. and Diekmann, M., 2019. Hedgerow age affects the species richness of herbaceous forest plants. *Journal of Vegetation Science*, *30*(3), pp.553–563.

Published by Burleigh Dodds Science Publishing Limited, 2024.

Litza, K. and Diekmann, M., 2020. The effect of hedgerow density on habitat quality distorts species-area relationships and the analysis of extinction debts in hedgerows. *Landscape Ecology*, *35*, pp.1187–1198.

López-Felices, B., Aznar-Sánchez, J.A., Velasco-Muñoz, J.F. and Mesa-Vázquez, E., 2022. Installation of hedgerows around greenhouses to encourage biological pest control: Farmers' perspectives from Southeast Spain. *Journal of Environmental Management*, *323*, p.116210.

Long, R., Garbach, K. and Morandin, L., 2017. Hedgerow benefits align with food production and sustainability goals. *California Agriculture*, *71*(3), pp.117–119.

Lu, A., Gonthier, D.J., Sciligo, A.R., Garcia, K., Chiba, T., Juarez, G. and Kremen, C., 2022. Changes in arthropod communities mediate the effects of landscape composition and farm management on pest control ecosystem services in organically managed strawberry crops. *Journal of Applied Ecology*, *59*(2), pp.585–597.

MacDonald, D.W. and Johnson, P.J., 1995. The relationship between bird distribution and the botanical and structural characteristics of hedges. *Journal of Applied Ecology*, pp.492–505.

MacInnis, G., Buddle, C.M. and Forrest, J.R., 2020. Small wild bee abundance declines with distance into strawberry crops regardless of field margin habitat. *Basic and Applied Ecology*, *44*, pp.14–23.

Marinus, W., Descheemaeker, K.K., van de Ven, G.W., Waswa, W., Mukalama, J., Vanlauwe, B. and Giller, K.E., 2021. "That is my farm" – An integrated co-learning approach for whole-farm sustainable intensification in smallholder farming. *Agricultural Systems*, *188*, p.103041.

Martel, G., Aviron, S., Joannon, A., Lalechère, E., Roche, B. and Boussard, H., 2019. Impact of farming systems on agricultural landscapes and biodiversity: From plot to farm and landscape scales. *European Journal of Agronomy*, *107*, pp.53–62.

Martínez-Sastre, R., Miñarro, M. and García, D., 2020. Animal biodiversity in cider apple orchards: simultaneous environment
al drivers and effects on insectivory and pollination. *Agriculture, Ecosystems & Environment*, *295*, p.106918.

Massaloux, D., Sarrazin, B., Roume, A., Tolon, V. and Wezel, A., 2020. Landscape diversity and field border density enhance carabid diversity in adjacent grasslands and cereal fields. *Landscape Ecology*, *35*, pp.1857–1873.

Mathews, F., Lovett, L., Rushton, S. and Macdonald, D.W., 2006. Bovine tuberculosis in cattle: reduced risk on wildlife-friendly farms. *Biology Letters*, *2*(2), pp.271–274.

Maudsley, M.J., West, T.M., Rowcliffe, H.P. and Marshall, E.J.P., 1998. Approaches to the restoration of degraded field boundary habitats in agricultural landscapes. In J.W. Dover and R.G.H. Bunce (Eds.), *Approaches to the restoration of degraded field boundary habitats in agricultural landscapes*(pp. 387–392), Key Concepts in Landscape Ecology, International Association for Landscape Ecology, UK, Preston.

McHugh, N.M., Bown, B.L., Hemsley, J.A. and Holland, J.M., 2019. Relationships between agri-environment scheme habitat characteristics and insectivorous bats on arable farmland. *Basic and Applied Ecology*, *40*, pp.55–66.

McHugh, N.M., Moreby, S., Lof, M.E., Van der Werf, W. and Holland, J.M., 2020. The contribution of semi-natural habitats to biological control is dependent on sentinel prey type. *Journal of Applied Ecology*, *57*(5), pp.914–925.

McKenzie, A.J., Emery, S.B., Franks, J.R. and Whittingham, M.J., 2013. Landscape-scale conservation: collaborative agri-environment schemes could benefit both biodiversity and ecosystem services, but will farmers be willing to participate? *Journal of Applied Ecology*, *50*(5), pp.1274–1280.

Medinas, D., Marques, J.T. and Mira, A., 2013. Assessing road effects on bats: the role of landscape, road features, and bat activity on road-kills. *Ecological Research*, *28*, pp.227–237.

Medlock, J.M., Vaux, A.G.C., Hansford, K.M., Pietzsch, M.E. and Gillingham, E.L., 2020. Ticks in the ecotone: the impact of agri-environment field margins on the presence and intensity of *Ixodes ricinus* ticks (Acari: Ixodidae) in farmland in southern England. *Medical and Veterinary Entomology*, *34*(2), pp.175–183.

Meek, R. and Luiselli, L., 2021. Living in Patchy Habitats: Substrate Selection for Basking by Sympatric Lizards in Contrasted Anthropogenic Habitats in Western France. www.researchsquare.com/article/rs-300197/v1.

Menzel, R., Tison, L., Fischer-Nakai, J., Cheeseman, J., Balbuena, M.S., Chen, X., Landgraf, T., Petrasch, J., Polster, J. and Greggers, U., 2019. Guidance of navigating honeybees by learned elongated ground structures. *Frontiers in Behavioral Neuroscience*, *12*, p.322.

Michalko, R. and Dvoryankina, V., 2019. Intraspecific phenotypic variation in functional traits of a generalist predator in an agricultural landscape. *Agriculture, Ecosystems & Environment*, *278*, pp.35–42.

Michel, N., Burel, F. and Butet, A., 2006. How does landscape use influence small mammal diversity, abundance and biomass in hedgerow networks of farming landscapes? *Acta Oecologica*, *30*(1), pp.11–20.

Millennium Ecosystem Assessment, 2005. *Ecosystems and human well-being* (Vol. 5, p. 563). Island Press, Washington, DC.

Millon, L., Barré, K., Julliard, R., Compère, P. and Kerbiriou, C., 2021. Calculation of biodiversity level between different land-uses to improve conservation outcomes of biodiversity offsetting. *Land Use Policy*, *101*, p.105161.

Miñarro, M. and Prida, E., 2013. Hedgerows surrounding organic apple orchards in north-west S pain: potential to conserve beneficial insects. *Agricultural and Forest Entomology*, *15*(4), pp.382–390.

Monokrousos, N., Papatheodorou, E.M., Diamantopoulos, J.D. and Stamou, G.P., 2006. Soil quality variables in organically and conventionally cultivated field sites. *Soil Biology and Biochemistry*, *38*(6), pp.1282–1289.

Montgomery, W.I., Caruso, T. and Reid, N., 2020. Hedgerows as ecosystems: service delivery, management, and restoration. *Annual Review of Ecology, Evolution, and Systematics*, *51*, pp.81–102.

Montgomery, W.I. and Dowie, M., 1993. The distribution and population regulation of the wood mouse *Apodemus sylvaticus* on field boundaries of pastoral farmland. *Journal of Applied Ecology*, *23*, pp.783–791.

Montgomery, W.I., Lundy, M.G. and Reid, N., 2012. 'Invasional meltdown': evidence for unexpected consequences and cumulative impacts of multispecies invasions. *Biological Invasions*, *14*, pp.1111–1125.

Montgomery, W.I., Montgomery, S.S., O'Neill, J. and Reid, N., 2023. Invasion of farmland hedgerows by non-native small mammals is associated with lower soil surface invertebrate diversity, abundance, body size and biomass. *Biological Invasions*, *23*(3), pp. 671–684.

Montgomery, W.I., Montgomery, S.S. and Reid, N., 2015. Invasive alien species disrupt spatial and temporal ecology and threaten extinction in an insular, small mammal community. *Biological Invasions*, *17*(1), pp.179–189.

Mony, C., Landais, Q., Georges, R., Butet, A., Burel, F., Jambon, O., Gouesbet, V. and Ernoult, A., 2022. Effects of connectivity on seed dispersal patterns in hedgerows. *Journal of Vegetation Science, 33*(1), p.e13113.

Morandin, L.A., Long, R.F. and Kremen, C., 2014. Hedgerows enhance beneficial insects on adjacent tomato fields in an intensive agricultural landscape. *Agriculture, Ecosystems & Environment, 189*, pp.164–170.

Morandin, L.A., Long, R.F. and Kremen, C., 2016. Pest control and pollination cost–benefit analysis of hedgerow restoration in a simplified agricultural landscape. *Journal of Economic Entomology, 109*(3), pp.1020–1027.

Moreau, J., Monceau, K., Gonnet, G., Pfister, M. and Bretagnolle, V., 2022. Organic farming positively affects the vitality of passerine birds in agricultural landscapes. *Agriculture, Ecosystems & Environment, 336*, p.108034.

Nardi, D., Lami, F., Pantini, P. and Marini, L., 2019. Using species-habitat networks to inform agricultural landscape management for spiders. *Biological Conservation, 239*, p.108275.

Nelson, K., Nelson, R. and Montgomery, W.I., 2021. Colonisation of farmland deciduous plantations by woodland ground flora. *Arboricultural Journal, 43*(2), pp.115–133.

Newton, I., 2017. *Farming and birds (Collins New Naturalist Library, Book 135)* (Vol. 135). HarperCollins UK.

Nicholls, C.I. and Altieri, M.A., 2013. Plant biodiversity enhances bees and other insect pollinators in agroecosystems. A review. *Agronomy for Sustainable Development, 33*, pp.257–274.

O'Brien, J., Elliott, S. and Hayden, T.J., 2016. Use of hedgerows as a key element of badger (Meles meles) behaviour in Ireland. *Mammalian Biology, 81*, pp.104–110.

O'Neill K. 1998 *Population dynamics of Wood mice Apodemus sylvaticus (L) in pastoral farmland*. Unpub. PhD thesis, Queen's Univ Belfast, UK.

Oosterveld, E.B., Klop, E. and van der Zee, E., 2022. Key habitat factors of breeding birds in agricultural hedgerow landscapes in East-Fryslân, The Netherlands, in European perspective–ecological evaluation and relation to agri-environmental schemes. *Ardea, 110*(2), pp.111–124.

Otieno, N.E., Jacobs, S.M. and Pryke, J.S., 2019. Maize-field complexity and farming system influence insectivorous birds' contribution to arthropod herbivore regulation. *Biotropica, 51*(6), pp.851–861.

Otieno, N.E. and Pryke, J.S., 2022. Tree and hedgerow configurations on maize farms are key drivers of granivorous passerine bird assemblage patterns. *Agriculture, Ecosystems & Environment, 338*, p.108104.

Peach, W.J., Denny, M., Cotton, P.A., Hill, I.F., Gruar, D., Barritt, D., Impey, A. and Mallord, J., 2004. Habitat selection by song thrushes in stable and declining farmland populations. *Journal of Applied Ecology, 41*(2), pp.275–293.

Pe'er, G., Finn, J.A., Díaz, M., Birkenstock, M., Lakner, S., Röder, N., Kazakova, Y., Šumrada, T., Bezák, P., Concepción, E.D. and Dänhardt, J., 2022. How can the European Common Agricultural Policy help halt biodiversity loss? Recommendations by over 300 experts. *Conservation Letters, 15*(6), p.e12901.

Pelletier-Guittier, C., Théau, J. and Dupras, J., 2020. Use of hedgerows by mammals in an intensive agricultural landscape. *Agriculture, Ecosystems & Environment, 302*, p.107079.

Peñalver-Cruz, A., Alvarez, D. and Lavandero, B., 2020. Do hedgerows influence the natural biological control of woolly apple aphids in orchards? *Journal of Pest Science*, *93*(1), pp.219–234.

Peter, F., Brucia, L., Carstens, F., Göttsche, M. and Diekötter, T., 2021. Sown wildflower fields create temporary foraging habitats for bats in agricultural landscapes. *Biological Conservation*, *264*, p.109364.

Phillips, B.B., Crowley, S.L., Bell, O. and McDonald, R.A., 2022. Harnessing practitioner knowledge to inform the conservation of a protected species, the hazel dormouse *Muscardinus avellanarius*. *Ecological Solutions and Evidence*, *3*(4), p.e12198.

Phillips, B.B., Gaston, K.J., Bullock, J.M. and Osborne, J.L., 2019. Road verges support pollinators in agricultural landscapes, but are diminished by heavy traffic and summer cutting. *Journal of Applied Ecology*, *56*(10), pp.2316–2327.

Pithon, J.A., Duflot, R., Beaujouan, V., Jagaille, M., Pain, G. and Daniel, H., 2021. Grasslands provide diverse opportunities for bird species along an urban-rural gradient. *Urban Ecosystems*, *24*(6), pp.1281–1294.

Pollard, E., Hooper, M.D. and Moore, N.W., 1974. *Hedges. The New Naturalist. A Survey of British Natural History*. William Collins Sons & Limited, Glasgow, UK.

Ponisio, L.C., de Valpine, P., M'Gonigle, L.K. and Kremen, C., 2019. Proximity of restored hedgerows interacts with local floral diversity and species' traits to shape long-term pollinator metacommunity dynamics. *Ecology Letters*, *22*(7), pp.1048–1060.

Ponisio, L.C., M'Gonigle, L.K. and Kremen, C., 2016. On-farm habitat restoration counters biotic homogenization in intensively managed agriculture. *Global Change Biology*, *22*(2), pp.704–715.

Porter, S. 2017. *Assessing the efficacy of hedgerow management policies in farmland bird conservation*. PhD Thesis, Queen's Univ. Belfast, Belfast, UK.

Précigout, P.A. and Robert, C., 2022. Effects of hedgerows on the preservation of spontaneous biodiversity and the promotion of biotic regulation services in agriculture: towards a more constructive relationships between agriculture and biodiversity. *Botany Letters*, *169*(2), pp.176–204.

Prendergast-Miller, M.T., Jones, D.T., Berdeni, D., Bird, S., Chapman, P.J., Firbank, L., Grayson, R., Helgason, T., Holden, J., Lappage, M. and Leake, J., 2021. Arable fields as potential reservoirs of biodiversity: Earthworm populations increase in new leys. *Science of the Total Environment*, *789*, p.147880.

Puech, C., Poggi, S., Baudry, J. and Aviron, S., 2015. Do farming practices affect natural enemies at the landscape scale? *Landscape Ecology*, *30*, pp.125–140.

Pywell, R.F., James, K.L., Herbert, I., Meek, W.R., Carvell, C., Bell, D. and Sparks, T.H., 2005. Determinants of overwintering habitat quality for beetles and spiders on arable farmland. *Biological Conservation*, *123*(1), pp.79–90.

Raatz, L., Pirhofer Walzl, K., Müller, M.E., Scherber, C. and Joshi, J., 2021. Who is the culprit: is pest infestation responsible for crop yield losses close to semi-natural habitats? *Ecology and Evolution*, *11*(19), pp.13232–13246.

Rajbanshi, J., Das, S. and Paul, R., 2023. Quantification of the effects of conservation practices on surface runoff and soil erosion in croplands and their trade-off: A meta-analysis. *Science of The Total Environment*, *864*, p.161015.

Raven, P.H. and Wagner, D.L., 2021. Agricultural intensification and climate change are rapidly decreasing insect biodiversity. *Proceedings of the National Academy of Sciences*, *118*(2), p.e2002548117.

Ravetto Enri, S., Nucera, E., Lonati, M., Alberto, P.F. and Probo, M., 2020. The biodiversity promotion areas: effectiveness of agricultural direct payments on plant diversity conservation in the semi-natural grasslands of the Southern Swiss Alps. *Biodiversity and Conservation*, *29*(14), pp.4155–4172.

Rieger, M.R., Mailänder, S., Stier, L., Santon, M., Staggenborg, J. and Anthes, N., 2022. Optimizing flower fields as an effective farmland eco-scheme also during non-breeding. *Journal of Applied Ecology*, *59*(2), pp.514–525.

Reid N., McDonald R.A. & Montgomery W.I. (2007) Mammals and agri-environment schemes: hare haven or pest paradise? *Journal of Applied Ecology, 44*, 1200–1208.

Requier, F. and Leonhardt, S.D., 2020. Beyond flowers: including non-floral resources in bee conservation schemes. *Journal of Insect Conservation*, *24*(1), pp.5–16.

Rey Benayas, J.M., Altamirano, A., Miranda, A., Catalán, G., Prado, M., Lisón, F. and Bullock, J.M., 2020. Landscape restoration in a mixed agricultural-forest catchment: Planning a buffer strip and hedgerow network in a Chilean biodiversity hotspot. *Ambio*, *49*(1), pp.310–323.

Ricciardi, V., Mehrabi, Z., Wittman, H., James, D. and Ramankutty, N., 2021. Higher yields and more biodiversity on smaller farms. *Nature Sustainability*, *4*(7), pp.651–657.

Rieger, M.R., Mailänder, S., Stier, L., Santon, M., Staggenborg, J. and Anthes, N., 2022. Optimizing flower fields as an effective farmland eco-scheme also during non-breeding. *Journal of Applied Ecology*, *59*(2), pp.514–525.

Rigal, S., Dakos, V., Alonso, H., Auniņš, A., Benkő, Z., Brotons, L., Chodkiewicz, T., Chylarecki, P., de Carli, E., Del Moral, J.C. and Domşa, C., 2023. Farmland practices are driving bird population decline across Europe. *Proceedings of the National Academy of Sciences*, *120*(21), p.e2216573120.

Rime, Y., Luisier, C., Arlettaz, R. and Jacot, A., 2020. Landscape heterogeneity and management practices drive habitat preferences of wintering and breeding birds in intensively-managed fruit-tree plantations. *Agriculture, Ecosystems & Environment*, *295*, p.106890.

Rivest, M., Whalen, J.K. and Rivest, D., 2020. Variation of soil microbial and earthworm communities along an agricultural transect with tree windbreak. *Agroforestry Systems*, *94*, pp.1639–1649.

Rodríguez, E., Clemente-Orta, G., Crisol-Martínez, E., Gutiérrez, I., van der Blom, J. and González, M., 2023. Aphid suppression by natural enemies in hedgerows surrounding greenhouses in southern Spain. *Biological Control*, *177*, p.105126.

Roy, V. and de Blois, S., 2008. Evaluating hedgerow corridors for the conservation of native forest herb diversity. *Biological Conservation*, *141*(1), pp.298–307.

Ruas, S., Rotchés-Ribalta, R., Ahmed, K.D., Gormally, M., Stout, J.C., White, B. and Moran, J., 2021. Selecting appropriate plant indicator species for Result-Based Agri-Environment Payments schemes. *Ecological Indicators*, *126*, p.107679.

Russ, J.M. and Montgomery, W.I., 2002. Habitat associations of bats in Northern Ireland: implications for conservation. *Biological Conservation*, *108*(1), pp.49–58.

Šálek, M., Kalinová, K. and Reif, J., 2022. Conservation potential of semi-natural habitats for birds in intensively-used agricultural landscapes. *Journal for Nature Conservation*, *66*, p.126124.

Salat-Moltó, A., Caballero-López, B., Pérez-Hidalgo, N., Michelena, J.M., Ferrer Suay, M., Guerrieri, E. and Blanco-Moreno, J.M., 2023. Not all field margins are equally useful: effects of the vegetation structure of margins on cereal aphids and their natural enemies. *Insects*, *14*(2), p.156.

Sardiñas, H.S., Ponisio, L.C. and Kremen, C., 2016. Hedgerow presence does not enhance indicators of nest-site habitat quality or nesting rates of ground-nesting bees. *Restoration Ecology*, *24*(4), pp.499–505.

Sauser, C., Commagnac, L., Eraud, C., Guillemain, M., Morin, S., Powolny, T., Villers, A. and Lormée, H., 2022. Habitats, agricultural practices, and population dynamics of a threatened species: The European turtle dove in France. *Biological Conservation*, *274*, p.109730.

Schlinkert, H., Ludwig, M., Batáry, P., Holzschuh, A., Kovács-Hostyánszki, A., Tscharntke, T. and Fischer, C., 2016. Forest specialist and generalist small mammals in forest edges and hedges. *Wildlife Biology*, *22*(3), pp.86–94.

Seppelt, R., Arndt, C., Beckmann, M., Martin, E.A. and Hertel, T.W., 2020. Deciphering the biodiversity–production mutualism in the global food security debate. *Trends in Ecology & Evolution*, *35*(11), pp.1011–1020.

Sih, A., Barbosa, P. and Castellanos, I., 2005. Ecology of predator–prey interactions. *Predator–Prey Space Use as Emergent Outcome of a Response Race*, *256*, pp.240–255.

Smigaj, M. and Gaulton, R., 2021. Capturing hedgerow structure and flowering abundance with UAV remote sensing. *Remote Sensing in Ecology and Conservation*, *7*(3), pp.521–533.

Smith, J., Westaway, S., Mullender, S., Giannitsopoulos, M. and Graves, A., 2023. Making hedgerows pay their way: the economics of harvesting field boundary hedges for bioenergy. *Agroforestry Systems*, 97, pp. 291–303.

Spaans, F., Caruso, T., Hammer, E.C. and Montgomery, I., 2019. Trees in trimmed hedgerows but not tree health increase diversity of oribatid mite communities in intensively managed agricultural land. *Soil Biology and Biochemistry*, *138*, p.107568.

Spaans, F., Caruso, T. and Montgomery, I., 2018. The abundance and condition of hedgerow tree standards in Northern Ireland. In *Biology and Environment: Proceedings of the Royal Irish Academy* (Vol. 118(3), pp. 129–145). Royal Irish Academy.

Staley, J.T., Amy, S.R., Adams, N.P., Chapman, R.E., Peyton, J.M. and Pywell, R.F., 2015. Re-structuring hedges: rejuvenation management can improve the long term quality of hedgerow habitats for wildlife in the UK. *Biological Conservation*, *186*, pp.187–196.

Staley, J.T., Bullock, J.M., Baldock, K.C., Redhead, J.W., Hooftman, D.A., Button, N. and Pywell, R.F., 2013. Changes in hedgerow floral diversity over 70 years in an English rural landscape, and the impacts of management. *Biological Conservation*, *167*, pp.97–105.

Staley, J.T., Redhead, J.W., O'Connor, R.S., Jarvis, S.G., Siriwardena, G.M., Henderson, I.G., Botham, M.S., Carvell, C., Smart, S.M., Phillips, S. and Jones, N., 2021. Designing a survey to monitor multi-scale impacts of agri-environment schemes on mobile taxa. *Journal of Environmental Management*, *290*, p.112589.

Staley, J.T., Wolton, R. and Norton, L.R., 2023. Improving and expanding hedgerows–Recommendations for a semi-natural habitat in agricultural landscapes. *Ecological Solutions and Evidence*, *4*(1), p.e12209.

Stanbury, A., Eaton, M., Aebischer, N., Balmer, D., Brown, A., Douse, A., Lindley, P., McCulloch, N., Noble, D. and Win, I., 2021. The status of our bird. *British Birds*, *114*, pp.723–747.

Stanbury, D.P., Pescott, O.L. and Staley, J.T., 2020. Hedgerow management experiment relevant to agri-environment schemes: cutting regime impacts species richness of basal flora and Ellenberg indicator profiles. *Biodiversity and Conservation*, *29*(8), pp.2575–2587.

Stašiov, S., Diviaková, A., Svitok, M., Novikmec, M. and Dovciak, M., 2020. Hedgerows support rich communities of harvestmen (Opiliones) in upland agricultural landscape. *Basic and Applied Ecology*, *47*, pp.73–82.

Stathakis, T., Economou, L., Barda, M., Angelioudakis, T., Kati, V. and Karamaouna, F., 2023. Potential of hedgerows with āromatic plants as reservoirs of natural enemies of pests in Orange Orchards. *Insects*, *14*(4), p.391.

Sullivan, T.P. and Sullivan, D.S., 2009. Are linear habitats in agrarian landscapes source areas of beneficial or pest rodents? *Agriculture, Ecosystems & Environment*, *129*(1-3), pp.52-56.

Sybertz, J. et al. 2019. Biodiversity modelling in practice - predicting bird and woody plant species richness on farmlands, *Ecosystems and People*, *16*(1), pp. 19–34. doi: 10.1080/26395916.2019.1697900.

Tarjuelo, R., Concepcion, E.D., Guerrero, I., Carricondo, A., Cortes, Y. and Diaz, M., 2021. Agri-environment scheme prescriptions and landscape features affect taxonomic and functional diversity of farmland birds. *Agriculture, Ecosystems & Environment*, *315*, p.107444.

Tew, T.E., Todd, I.A. and Macdonald, D.W., 2000. Arable habitat use by wood mice (Apodemus sylvaticus). 2. Microhabitat. *Journal of Zoology*, *250*(3), pp.305–311.

Theves, F. and Zebitz, C.P.W., 2012. Biodiversity of carabid beetles (Carabidae) in field hedgerows-alternative approaches. *Mitteilungen der Deutschen Gesellschaft für allgemeine und angewandte Entomologie*, *18*, pp.173–176.

Thiele, J., Kellner, S., Buchholz, S. and Schirmel, J., 2018. Connectivity or area: what drives plant species richness in habitat corridors? *Landscape Ecology*, *33*, pp.173–181.

Thomas, Z., Ghazavi, R., Merot, P. and Granier, A., 2012. Modelling and observation of hedgerow transpiration effect on water balance components at the hillslope scale in Brittany. *Hydrological Processes*, *26*(26), pp.4001-4014.

Thompson, J.B., Symonds, J., Carlisle, L., Iles, A., Karp, D.S., Ory, J. and Bowles, T.M., 2023. Remote sensing of hedgerows, windbreaks, and winter cover crops in California's Central Coast reveals low adoption but hotspots of use. *Frontiers in Sustainable Food Systems*, *7*, p.1052029.

Tilman, D., Isbell, F. and Cowles, J.M., 2014. Biodiversity and ecosystem functioning. *Annual Review of Ecology, Evolution, and Systematics*, *45*, pp.471–493.

Tilzey, M., 2021. The political ecology of hedgerows and their relationship to agroecology and food sovereignty in the UK. *Frontiers in Sustainable Food Systems*, *5*, p.752293.

Timberlake, T.P., Vaughan, I.P. and Memmott, J., 2019. Phenology of farmland floral resources reveals seasonal gaps in nectar availability for bumblebees. *Journal of Applied Ecology*, *56*(7), pp.1585–1596.

Tougeron, K., Couthouis, E., Marrec, R., Barascou, L., Baudry, J., Boussard, H., Burel, F., Couty, A., Doury, G., Francis, C. and Hecq, F., 2022. Multi-scale approach to biodiversity proxies of biological control service in European farmlands. *Science of the Total Environment*, *822*, p.153569.

Tresise, M.E., Biffi, S., Field, R.H. and Firbank, L.G., 2021. Drivers of songbird territory density in the boundaries of a lowland arable farm. *Acta Oecologica*, *111*, p.103720.

Um, M.J., Kim, Y., Park, D., Jung, K., Wang, Z., Kim, M.M. and Shin, H., 2020. Impacts of potential evapotranspiration on drought phenomena in different regions and climate zones. *Science of the Total Environment*, *703*, p.135590.

Van Den Berge, S., Tessens, S., Baeten, L., Vanderschaeve, C. and Verheyen, K., 2019. Contrasting vegetation change (1974–2015) in hedgerows and forests in an intensively used agricultural landscape. *Applied Vegetation Science*, *22*(2), pp.269–281.

Vanneste, T., Govaert, S., De Kesel, W., Van Den Berge, S., Vangansbeke, P., Meeussen, C., Brunet, J., Cousins, S.A., Decocq, G., Diekmann, M. and Graae, B.J., 2020. Plant diversity in hedgerows and road verges across Europe. *Journal of Applied Ecology*, *57*(7), pp.1244–1257.

Vanneste, T., Govaert, S., Spicher, F., Brunet, J., Cousins, S.A., Decocq, G., Diekmann, M., Graae, B.J., Hedwall, P.O., Kapás, R.E. and Lenoir, J., 2020a. Contrasting microclimates among hedgerows and woodlands across temperate Europe. *Agricultural and Forest Meteorology*, *281*, p.107818.

Vanneste, T., Van Den Berge, S., Brunet, J., Hedwall, P.O., Verheyen, K. and De Frenne, P., 2021. Temperature effects on forest understorey plants in hedgerows: a combined warming and transplant experiment. *Annals of Botany*, *128*(3), pp.315–327.

Vanneste, T., Van Den Berge, S., Riské, E., Brunet, J., Decocq, G., Diekmann, M., Graae, B.J., Hedwall, P.O., Lenoir, J., Liira, J. and Lindmo, S., 2020b. Hedging against biodiversity loss: forest herbs' performance in hedgerows across temperate Europe. *Journal of Vegetation Science*, *31*(5), pp.817–829.

Veste, M., Littmann, T., Kunneke, A., Du Toit, B. and Seifert, T., 2020. Windbreaks as part of climate-smart landscapes reduce evapotranspiration in vineyards, Western Cape Province, South Africa. *Plant, Soil and Environment*, *66*(3), pp.119–127.

Volpato, A., Ahmed, K.S., Williams, C.D., Day, M.F., O'Hanlon, A., Ruas, S., Rotchés-Ribalta, R., Mulkeen, C., Huallachain, D.O. and Gormally, M.J., 2020. Using Malaise traps to assess aculeate Hymenoptera associated with farmland linear habitats across a range of farming intensities. *Insect Conservation and Diversity*, *13*(3), pp.229–238.

von Königslöw, V., Fornoff, F. and Klein, A.M., 2022. Pollinator enhancement in agriculture: Comparing sown flower strips, hedges and sown hedge herb layers in apple orchards. *Biodiversity and Conservation*, *31*(2), pp.433–451.

Wagner, D.L., 2020. Insect declines in the Anthropocene. *Annual Review of Entomology*, *65*, pp.457–480.

Walker, M.P., Dover, J.W., Sparks, T.H. and Hinsley, S.A., 2006. Hedges and green lanes: vegetation composition and structure. *Biodiversity & Conservation*, *15*, pp.2595–2610.

Wallace, E.E., McShane, G., Tych, W., Kretzschmar, A., McCann, T. and Chappell, N.A., 2021. The effect of hedgerow wild-margins on topsoil hydraulic properties, and overland-flow incidence, magnitude and water-quality. *Hydrological Processes*, *35*(3), p.e14098.

Weninger, T., Scheper, S., Lackóová, L., Kitzler, B., Gartner, K., King, N.W., Cornelis, W., Strauss, P. and Michel, K., 2021. Ecosystem services of tree windbreaks in rural landscapes–a systematic review. *Environmental Research Letters*, *16*(10), p.103002.

Wratten, S.D., Gillespie, M., Decourtye, A., Mader, E. and Desneux, N., 2012. Pollinator habitat enhancement: benefits to other ecosystem services. *Agriculture, Ecosystems & Environment*, *159*, pp.112–122.

Wright, J., 2016. *A natural history of the hedgerow: and ditches, dykes and dry stone walls*. Profile Books.

Yang, K.F., Gergel, S.E. and Baudron, F., 2020. Forest restoration scenarios produce synergies for agricultural production in southern Ethiopia. *Agriculture, Ecosystems & Environment*, *295*, p.106888.

Zamberletti, P., Sabir, K., Opitz, T., Bonnefon, O., Gabriel, E. and Papaïx, J., 2021. More pests but less pesticide applications: Ambivalent effect of landscape complexity on conservation biological control. *PLoS Computational Biology*, *17*(11), p.e1009559.

Zamorano, J., Bartomeus, I., Grez, A.A. and Garibaldi, L.A., 2020. Field margin floral enhancements increase pollinator diversity at the field edge but show no consistent spillover into the crop field: a meta-analysis. *Insect Conservation and Diversity*, *13*(6), pp.519–531.

Part 3

The role of government and the private sector in promoting on-farm conservation practices

Chapter 8

Stick your wellies on: messy development and co-design processes with England's new Environmental Land Management (ELM) policy

Jennifer Dodsworth and Rachel Lasko, University of Oxford, UK; and Ruth Little, University of Sheffield, UK

1 Introduction

Following the referendum vote to leave in June 2016, the United Kingdom officially left the European Union (EU) in January 2020. Agriculture has always played a key role in the history of the European Union and accounted for almost 40% of the total budget of the Union when the vote occurred (Helm, 2017). Post-referendum, the UK government declared an ambition to pursue a 'Green Brexit', marking a 'new era for farming, fishing and the environment' (Gove, 2018). The European Union exit has been viewed as both a challenge and an opportunity for the United Kingdom to develop a new vision for agriculture and the environment, combining targets for biodiversity and net zero with a framework for sustainable and productive farming. Brexit removed the United Kingdom from the Common Agricultural Policy (CAP), and the British

http://dx.doi.org/10.19103/AS.2024.0134.13

Government's Department for Environment, Food and Rural Affairs (Defra) has since moved England onto a new path in agriculture policy.

The Health and Harmony consultation, published by Defra (2018), was the first attempt to outline a blueprint for the future of UK agriculture outside of the European Union. It sought public views on how to balance food and farming with the need to protect and enhance the environment and garnered over 53 000 responses. The results of the consultation underpinned a new Agriculture Bill (2018) which set out the principle of paying 'public money for public goods'. These goods include but are not limited to, enriching wildlife habitats, preventing flooding, improving the quality of air, soil and peat, and planting trees with the overarching goal of managing and mitigating the effects of climate change (Bateman and Balmford, 2018).

Defra (2020) went on to publish 'The Path to Sustainable Farming: An Agricultural Transition Plan 2021–2024', setting out the government's ambitious approach to transforming the agricultural sector in England. The Environmental Land Management (ELM) schemes are central to the government's plans, encouraging farms to produce healthy food whilst being profitable and economically sustainable, and contributing significantly to global environmental goals, including addressing biodiversity loss and climate change.

In the first part of this chapter, we outline the context of ELM within England's contemporary agri-environment policy development so far, offering a conceptual overview of the process through which the policy is being rolled out. We describe the programme's 'messy' development, in comparison to the more structured and cohesive development of other transition policies, such as the Animal Health and Welfare Pathway. In Part II, we examine the Sustainable Farming Incentive (SFI) National Pilot as an example of the 'Innovation Process' in practice within ELM development.

2 Contextualising the new Environmental Land Management policy

ELM is the latest in a relatively long history of agri-environment schemes for England, whose first agri-environment schemes were developed in the late 1980s (Fig. 1). The Environmentally Sensitive Areas (ESA) scheme was established in 1987 and was the first scheme of its kind in the European Union to support farmers in delivering environmental benefits in targeted parts of the country (Batáry et al., 2015). In 1991, England's first version of the Countryside Stewardship scheme was introduced to operate in areas that were not covered by the ESA scheme and to target different environmental priorities in each county (Carey et al., 2000). In 2005, both the ESA scheme and the original Countryside Stewardship scheme were superseded by the Environmental

Stewardship (ES) scheme, which was open for applications from 2005 until 2014. The ES scheme had 3 main elements:

- Entry Level Stewardship,
- Organic Entry Level Stewardship, and
- Higher Level Stewardship.

Entry Level Stewardship was characterised as a 'broad and shallow' scheme, aiming to encourage high levels of farmer uptake through its non-competitive application process and simple administrative design (Cusworth and Dodsworth, 2021) over a 5-year agreement period. Higher Level Stewardship involved competitively assigned 10-year ES contracts, which required higher levels of delivery for environmental benefits according to priority areas and situations (Natural England, 2012). Despite the scheme's replacement with the second iteration of Countryside Stewardship in 2014, some Higher-Level Environmental Stewardship agreements are still running, and many have been offered short extensions as ELM is still under development.

Countryside Stewardship, as it was constructed in 2014, is England's current main agri-environment scheme. Countryside Stewardship, similar to Environmental Stewardship, has two categories of differing intensity:

- Mid-Tier Stewardship (supporting individual measures to improve wildlife diversity, water and air quality and flood management).
- Higher-Tier Stewardship (involving complex management of environmentally important sites such as commons (areas of land which may be privately owned but where others (commoners) have rights to exploit some natural resources) or woodland, e.g. by creating/restoring wildlife habitat or improving woodland).

Countryside Stewardship was England's agri-environment scheme in the country's rural development programme under Pillar Two of CAP until Britain's exit from the European Union. It has continued as the main source of agri-environment funding for farmers during the transition period as the new post-Brexit schemes are developed.

Defra's Agricultural Transition plan, published in late 2020, initially outlined the development of three strands within the ELM programme. First, the SFI is a universal scheme which is open to all farmers and aims to support more sustainable farming actions applying to a wide range of farm types. This is achieved through a suite of standards with relevant management actions. SFI is currently in its second iteration, 'SFI23', and includes a broad collection of actions which farmers can choose from. Options such as flower-rich margins, winter bird food provision, and herbal leys encourage direct actions that enhance or produce benefits for biodiversity. Other current standards include a variety of management plans for soils, nutrients and integrated pest management. These indirect actions will create baselines for specific target areas and arguably promote more environmentally friendly farming practices by providing farmers with additional capacity to better understand the current status of a farm's environmental quality. In this way, plan-based and assessment actions may also provide potential longer term biodiversity-related benefits by increasing farmer knowledge and readiness for on-farm improvements over time.

Secondly, the Local Nature Recovery (LNR) scheme was originally planned to broadly replace England's existing Countryside Stewardship scheme with a suite of options for larger farm-scale protection of farm habitats and target species, with a stronger focus on biodiversity (e.g. managing and creating habitats for wildlife or restoring peatland and wetland areas on farms).[1] Finally, Landscape Recovery will be a series of very large-scale, longer-term, one-off collaborative projects which bring together a wide range of stakeholders to tackle a particular environmental issue (e.g. wildlife conservation or flood management) in a particular catchment or geographic area.

More recently, however, the LNR scheme has been replaced by a third rendition of Countryside Stewardship, in a move framed as 'building on the success' (Hughes, 2020) of the existing Countryside Stewardship (2.0) programme, citing the increased uptake in recent years as an indicator of the scheme's qualities. In her announcement of Countryside Stewardship's continuation, the Defra Secretary at the time, Therese Coffey, said 'over 30 000 farmers are now involved in our simplified, streamlined, and I hope overall, enhanced Countryside Stewardship scheme – that's a 94% increase in 3 years. Something must be working.' (Coffey, 2022; emphasis added). This narrative, however, did not take into account the considerable uncertainty that farmers have faced in recent years since withdrawal from the European Union, where reductions in basic payments have already begun. With additional income streams from ELM still under development, farmers have turned to the only current offer available to them: Countryside Stewardship.

This significant adjustment to Defra's original (2020) outline of the ELMs in the Transition Plan is perhaps understandable given the significant upheaval within the department throughout 2022. In the ongoing evolution of Countryside Stewardship, which the government committed in 2023, some innovative elements outlined in the previous ambitions for the abandoned LNR scheme are still being developed. For example, in outlining their next steps for the programme, Defra commits to 'continuing to develop and test our approach to incentivising and rewarding those who join up across larger areas, targeting

1 Target species as defined under the LNR guidance include predators, invasive non-native species, and specific species such as deer, grey squirrel, rhododendron and bracken.

funding towards the right actions in the right places, and potentially including an element of payment by results' (Defra, 2023a, p. 14). These themes reflect the broader ambitions of the original LNR scheme to address collaboration, spatial prioritisation and outcomes-based approaches respectively.[2] As with Landscape Recovery and the SFI, the new Countryside Stewardship scheme will continue to evolve over the coming years, indicated by the ELM development phase shown in Fig. 1.

3 ELM development in the United Kingdom

As a group of three female researchers with varied experiences in academia, farming and policy, we argue that this notion of continuing co-designed evolution is central to how academics, practitioners and policymakers should understand ELM development. Over the transition period and, indeed, beyond, an outlook that considers how the schemes can be constantly improved and enhanced to address continuing challenges faced by farmers and the environment offers scope for a much more dynamic, attractive and potentially effective approach to broader agri-environment scheme development than the periodic publication of a static programme.

To outline and examine the development of ELM policy, we utilise a brief interpretation of de Boon's et al.'s (2022) descriptive framework for understanding innovations in agri-environment policy transitions. The framework invites us to consider four 'nested dimensions' of the context for agricultural policy transition:

- the 'macro' context as a setting or 'driver of change' in which the transition is occurring;
- the governance system in which decisions about the agricultural transition are being made and implemented;
- the 'foundation' – the immediate context and stakeholder's capacity to adapt to the transition; and
- the innovation processes of the agricultural transition itself, namely ELM development.

First, we outline the 'macro-context' in which ELM was formed. Brexit, and thus the UK's departure from the EU's CAP, was the setting or driver of change that catalysed the development of ELM. In contrast to the two pillars broadly composed of direct payments and agri-environment payments that form CAP, England is now broadly abandoning the hectarage-based direct payments

2 It is too soon to see tell how these elements will be incorporated into the evolving Countryside Stewardship scheme. The ongoing tests and trials continue to explore these elements to improve delivery mechanisms.

associated with Pillar One of CAP and, instead, emphasising a move towards paying farmers for the provision of environmental goods. The UK government opted to overhaul the EU policy due to the CAP's focus on land ownership rather than environmental considerations. Instead, ELM policy provides a shift to compensate farmers for providing environmental benefits alongside food production, while also offering grants to boost farm productivity and innovation. These changes aim to ensure a more resilient UK agriculture sector while aligning with national environmental and climate targets.

Under CAP (for a detailed overview of lessons learnt from CAP, see Little et al., 2022), farmers were paid principally through two mechanisms of support. Pillar One payments, such as the basic payment scheme (BPS), provide 'decoupled' income support to farmers (meaning the payment is not tied to productivity), which helps the European Union to meet CAP's aims of 'safeguarding farmers to make a reasonable living' (European Commission, n.d.) and maintaining rural areas and economies. The 'decoupled' nature of these payments is important to ensure that funding for farmers is not seen as unduly 'trade distorting', which could affect countries' abilities to export and import goods under the World Trade Organisation (WTO)'s rules in its Agreement for Agriculture.

Similarly, in Pillar Two, payments are not tied to productivity but rather delivered as payments for rural development through initiatives such as agri-environment schemes. Rural development embodied six different priorities under Pillar Two, which include knowledge sharing in agriculture and rural areas, boosting agricultural competitiveness and sustainability, organising food chains and managing risks, preserving ecosystems linked to agriculture and forestry, promoting resource efficiency for a low-carbon economy, and fostering social inclusion, poverty reduction and economic growth in rural areas (Nègre, 2022).

The multi-layered and dynamic governance system in which the agricultural transition is taking place is extremely complex. The uncertain and at times turbulent context of Brexit has increased the pressure on the British Government to deliver a great many policy changes at pace and at scale. The structures and processes of decision making for food, agriculture, and the environment is one of many pressing areas of public policy which is being developed as it is being delivered. These challenges are exacerbated by multiple political dynamics. For example, the brief and troubled tenure of Liz Truss as Britain's Prime Minister, and her appointment of Ranil Jayawardena as Secretary of State for Defra, left ELM in a state of precarity for a time as Jayawardena put the programme into a 'rapid review' (Marshall and Prior, 2022). This move further exacerbated the uncertainty and frustration amongst farming and environmental sectors (Hird, 2022).

Further, the United Kingdom's devolved governance system has led to a marked divergence between each nation's approach to post-Brexit agri-environment policy. The Scottish Government has explicitly and consistently stated (for instance, see 2018, p.8) that its domestic agriculture policy will be shaped by, amongst other areas of legislation, the CAP Greening Report (ECoA, 2017), thus continuing to align with key principles of agricultural support that will not be available to farmers just over the border in England. In moving to a policy focused primarily on agri-environment schemes and removing the notion of income support, according to WTO rules, England's ELM schemes can only compensate farmers through payments based on income foregone plus costs incurred (Mennig and Sauer, 2020). This will mean neighbouring farmers potentially receive vastly different levels of agri-environmental and income support from their respective governments and will face different pressures and expectations for public good delivery, despite new domestic policy for both being catalysed by Brexit, which both (or neither) may have voted for.

The lack of clarity, parity and consistency which these and other instances demonstrate in the governance of ELM are illustrative of an ad-hoc approach to agri-environment policy development. The unprecedented circumstances of Brexit have led to an entirely new agri-environment scheme development, wherein the existing rules and norms of governance based on CAP were set aside, giving way to the notion of 'public money for public goods'.

The concept of public money for public goods was first detailed by the government in relation to environmental enhancement in their post-Brexit paper presented to Parliament (Defra, 2018). In this document, Defra outlined the different possible public goods that a new ELM system could support, ranging from improved soil, water and air quality, increased biodiversity, and climate change mitigation, to enhanced animal health and welfare, heritage protection and the resilience of rural communities (Defra, 2018). The attitudinal shift alone towards public money for public goods in many ways marked a significant move towards innovative processes in British environmental policy development. Jordan et al. (2003, p. 185) emphasised that the 'traditional structure, style and content of British policy militated strongly against innovation.' The agricultural transition has, in many ways, dissolved these traditional structures and provided the opportunity to depart from legacy policies and the need to 'muddle through' (Lindblom, 1979).

The sentiment of public money for public goods is underlined by some important assumptions. First, this approach acknowledges that farming systems simultaneously deliver both private and public goods (Meuwissen et al., 2019). For example, food is considered a private good (paid for by the market) and broader, non-fungible benefits for society and the environment are seen as non-diminishing public goods, such as access to the countryside and biodiversity. A second, related assumption of the public money for public goods outlook is that the market adequately pays the production costs of private goods (i.e. food) and this assumption justifies the government's public goods-only approach to ELM. Again, these assumptions make a marked change from the farming support under CAP, which provided domestic support to farmers for both the private and public functions of farming.

A key 'foundation' element of ELM, central to the Agricultural Transition Plan and the first of Defra's 'guiding principles' for the transition, was a commitment to 'co-design' their new agricultural policy with farmers. Shortly after the Agricultural Transition Plan was published, Janet Hughes, the Director of Defra's Farming and Countryside Programme, outlined more specifically what they meant by co-design as an approach. Hughes (2020) defined co-design as:

> 'a design approach that **actively involves users** and stakeholders **from the beginning of a project, right through to roll-out**. It means we collaborate with everyone who has an interest to solve real problems with them. We actively seek their input and feedback, based on their lived experience, as we iterate and improve services. There's a lot of overlap with [other] approaches... ...but **co-design puts more emphasis on what we learn from the aggregated *opinions* of participants**.' (our emphasis in bold)

This definition aligns broadly with wider co-design literature, which defines the approach as 'a process to rapidly develop technologies better matched to customer needs [which] focusses [on] understanding the end-user context before designing and testing potential solutions', and also emphasises a focus on longer-term goals (Eastwood et al., 2022, p. 2). In principle, the co-design approach should have utilised farmers' innovative and adaptive capacity and resourced them to bring their ideas, knowledge and own social capital (de Boon et al., 2022, p. 412) to the fore. Moreover, the promise of co-design may have improved farmer willingness to adapt to policy changes where they felt acknowledged and invested in the process.

To put co-design into practice, Defra embarked on an ambitious programme of tests, trials and pilots (T&Ts). The learning from across these ongoing co-design initiatives has been utilised in various ways to support the development of ELM. They vary from short-term, hypothetical 'tests' which explore one specific element of the proposed new schemes in a short period, such as perspectives on paid advice, to longer-term and larger-scale pilot projects which, for instance exploring Landscape Recovery models of collaboration at a catchment scale.

However, the farmers and facilitators involved in these T&Ts have expressed frustration about the lack of meaningful and sustained engagement with Defra. Our research has found that for farmers: 'Co-design quite literally means ***the land manager on the left and the government authority on the right converging into the middle and designing something that works for***

everybody and delivers environmental objectives whilst, at the same time, the process of delivery is considered to be achievable by he or she who manages the land' (farmer interview). Co-design is understood as a fundamental element of the process.

On the other hand, Defra's approach to co-design, whilst undertaken with admirable ethos, has been more disparate and a matter of 'doing co-design' in particular stages, or in particular areas of the programme, a practice or method employed at particular times, but much less understood as embedded in the process. Moreover, the continued delays and changes in policy development (most evident in the reversion from LNR to CS and ministerial variations such as Jayawardena's brief term as Secretary) have left both farmers and policymakers in ELM at best uncertain about the scheme's future and, on occasion, considering the need for complete policy overhaul or a return to evolving a legacy policy.

In contrast to ELM, the Animal Health and Welfare Pathway has followed a more structured path of innovative policy development. The Agriculture Act 2020, whilst providing limited specific details on animal health and welfare improvements, established the legislative framework to enable the government to introduce regulations and schemes related to these areas. The key delivery mechanism for these enhancements is the Animal Health and Welfare Pathway. Defra's (2023b) policy paper outlines the pathway's three 'mutually reinforcing' strands:

- 'Financially rewarding farmers who deliver public goods' via an annually funded veterinary visit, capital grant funding, disease eradication and control programmes focusing on preventing and reducing priority endemic diseases and conditions and scoping the future role of 'payment by results' schemes.
- 'Stimulating market demand for higher welfare products' via market mechanisms such as labelling and enhanced product transparency for consumers.
- 'Strengthening the regulatory baseline' via, for example, the introduction of The Animals (Penalty Notices) Act 2022 to aid enforcement.

The initiation of the Pathway predates Brexit, with the impetus provided by the Animal Health and Welfare Board for England, an independent group of sector experts and senior Defra officials, with responsibility for strategic animal health and welfare policy and implementation oversight in England. Through an extended period of co-design with agricultural sector groups (principally cattle, sheep, pigs and poultry), the pathway has become part of the support mechanisms to transition the industry towards a model of public goods provision, focusing on incentives, regulations and enhanced market opportunities for higher welfare products.

There are clearly multiple conflicts and trade-offs when considering the role of animal agriculture in the agricultural transition, particularly with reference to achieving net zero. However, there is a clear indication from the Agriculture Act and subsequent transition documents that enhanced animal health and welfare is considered critical to reducing the environmental disbenefits of current agricultural practices, focusing on healthier and more productive animals. Questions remain on how payment for animal health and welfare as public goods will be sustained in the longer term and the extent to which these goods will need to be tied more closely with whole farm approaches towards achieving net zero. Improving, sustaining and rewarding these enhancements will likely rely on both market and government interventions.

In this first part of the chapter, we have sought to emphasise the complex and messy factors of agri-environment policy development which are evident in England's agricultural transition at varying scales and by reference to multiple socio-political factors. In the second part of the chapter, we explore the most sustained example of Defra's co-designed ELM schemes, the SFI National Pilot, which we examine in further detail as an innovation process.

4 Innovation in focus: the Sustainable Farming Incentive national pilot

Here, we describe the structure and rationale of the SFI national pilot, before examining the extent to which it can be understood as an innovative co-design process. De Boon et al. (2022, p. 415) identified five interrelated stages within the innovation process: problem identification, idea generation, prototype development, prototype testing and implementation.

The SFI was the first of the three new ELM schemes introduced to the public through a national-scale, co-designed pilot. Defra wanted SFI to appeal to the majority of farmers, setting itself the ambitious target of having 70% of farms and farmland in the scheme by 2028 (Defra, 2021). The pilot itself therefore focuses on developing a programme that equips farmers to incorporate environmentally friendly farming practices into their everyday land management activities in a relatively simple manner.

The SFI pilot welcomed the first cohort of farmers to participate in pilot studies in October 2021, and two more cohorts were selected to participate in the pilot to co-design the policy and help gather insights for SFI 2022 and the full roll-out of the programme. An SFI pilot agreement is three years in length, so the pilot will continue to run until mid-2025. The purpose of the pilot is to gather feedback from farmers about how the proposed system for SFI works in practice and investigate areas of weakness in the draft scheme. Pilot participants were told that penalties would be limited and that the pilot was a learning experience, in hopes that this reassurance would encourage farmers to be more ambitious in their decisions to produce greater environmental benefits.

The SFI pilot was structured as a limited series of environmental 'standards' with ambition levels. The 'standards' were a series of land management actions bundled together. Actions within each standard serve different roles in delivering public goods such as biodiversity restoration, water protection and other environmental goals in line with objectives of the Environment Act 2021 and subsequent Environmental Improvement Plan (2023a).

Farmers participating in the SFI pilot could choose from eight different standards. Figure 2 is an overview of the standards and the actions required for each of the different levels of ambition within the standard. Pilot participants were expected to select a bundle of different standards (at varying ambition levels) to form a combination of land management actions that worked best on their farm to deliver ELM goals. For example, an arable holding might have chosen the arable and horticultural land, arable and horticultural soils and hedgerow standards for their farm, whereas livestock farmers could have opted for the improved grassland soils, improved grassland and hedgerow standards. Farmers had the power to determine which standards were feasible and applicable to their holding. Farmers were then compensated for the delivery of public goods, by completing the actions of each selected standard, based on a set of predefined payment rates.

SFI pilot farmers could select from three ambition levels: Introductory, Intermediate and Advanced. The ambition levels are cumulative, building on one another to include additional actions farmers need to take to provide public goods. That is, the intermediate and advanced levels require all actions in the introductory, and introductory and intermediate levels, respectively, to be completed as well. The cumulative nature of ambition levels could relate either to the increased intensity of an action or an additional action. For instance, in Fig. 2, we see that for the arable land standard, ground cover for pollinator resources increases from 5% in the introductory, to 8% at the intermediate and then 10% at the advanced ambition level. An example of an additional action is the required use of efficient precision application equipment in the advanced grassland standard, in comparison to the low emission equipment requirement as necessitated in the intermediate ambition level. The advanced standard warrants the use of more expensive, less accessible machinery which is why that ambition level is compensated at a higher rate, compared to introductory or intermediate.

Figure 2 is a compilation of Defra's published guidance on the SFI Pilot action standards (Defra, n.d.). The *x*-axis in the table lists the different standards of the SFI Pilot, whilst the *y*-axis lists the different ambition levels which one can enter into each standard (i.e. introductory, intermediate and advanced). Underneath each standard are three boxes which define the actions that farmers

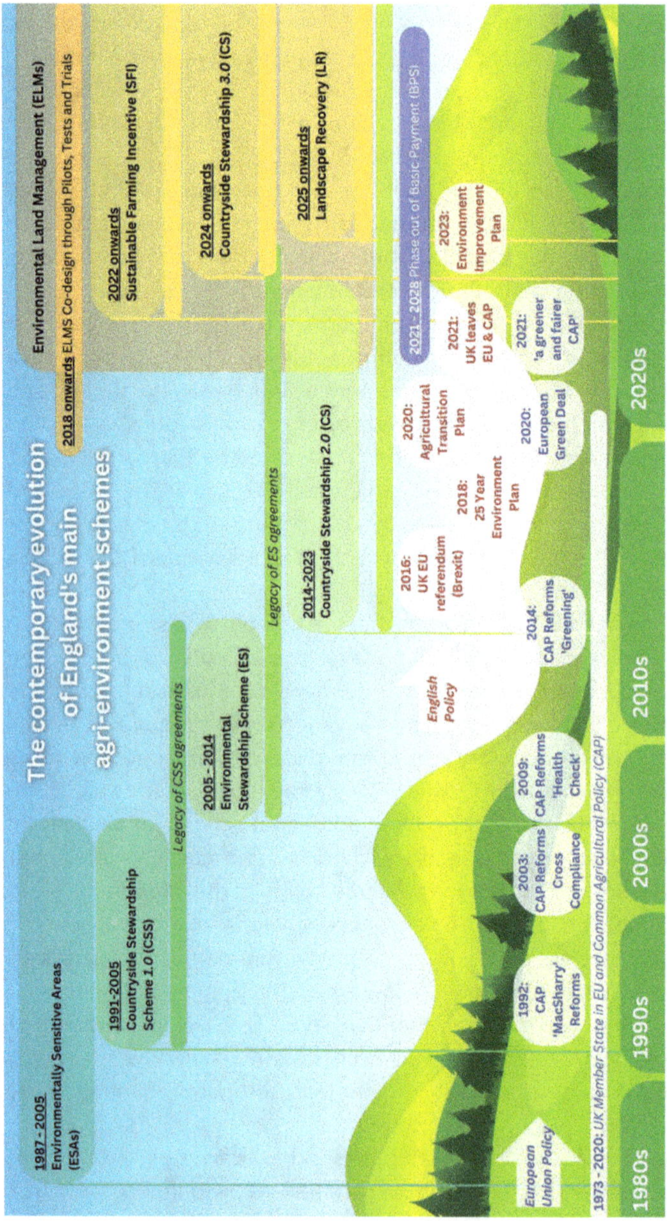

Figure 1 England's Agri-Environment Scheme Development

	Arable and horticultural land	Arable and horticultural soils	Farm woodland	Hedgerows	Improved grassland	Improved grassland soils	Low and no input grassland	Water body buffering
Introductory Level Actions	• Provide resources for birds and pollinators on 5% of eligible land • Create buffers around 50% of in-field trees • Use a nutrient management plan • Use low emission technologies	• Complete a soil assessment • Alleviate soil compaction • Establish green cover on at least 5% of land • Add soil organic matter to a minimum of 10% of land	• Complete a woodland condition assessment • Attend a woodland threats webinar • Create temporary open spaces and manage rides • Create and manage dead wood in woodland	• Leave at least 50% of the hedgerow uncut • Maintain hedgerow trees, or plant or tag new ones	• Manage fields for grazing • Manage historic features in grasslands • Leave margins uncut to encourage flowers and seed • Create buffers around in-field trees • Use a nutrient management plan • Allow tall vegetation and scrub to develop on 2% of grassland	• Complete a soil assessment • Alleviate soil compaction • Add or retain a mix of legumes, herbs and grasses on at least 5% of the grasslands • Maintain permanent grassland on land at risk of erosion or surface runoff on at least 5% of the land • Manage livestock density on grassland at risk of erosion or surface runoff on at least 5% of the land	• Manage grazed pasture with limited nutrient inputs • Manage weeds and invasive or competitive species • Create buffers around in-field trees • Leave margins uncut • Manage supplementary feeding • Manage ditches • Manage fields for grazing	• Do a runoff and soil erosion risk assessment • Establish grass buffer strips, 6 metres wide on at least 50% of the total length of water bodies in the agreement
Intermediate Level Actions	• Provide resources for birds and pollinators on 8% of eligible land • Support a wide range of farm wildlife • Complete a whole farm nutrient budget • Do soil mapping • Manage ditches	• Establish green cover on at least 10% of land • Add organic matter to a minimum of 15% of land • Use minimum tillage or no-tillage farming techniques		• Leave more hedgerows uncut or raise the cutting height • Leave the tops to grow uncut on short hedgerows • Maintain more hedgerow trees, or plant or tag new ones (There must be an average of one hedgerow tree every 200 metres across all the hedgerows) • Create buffer strips on both sides of the hedgerow on at least 50% of the hedgerows	• Allow tall vegetation and scrub to develop on 5% of grassland • Control cutting on silage fields • Manage ditches • Test your manures and soils • Complete a whole farm nutrient budget • Limit inputs of liquid manures • Use low emission technologies • Introduce clover and other legumes on 15% of improved grassland	• Add or retain a mix of legumes, herbs and grasses on at least 10% of the grasslands • Maintain permanent grassland on land at risk of erosion or surface runoff on at least 10% of the land • Manage livestock density on grassland at risk of erosion or surface runoff on at least 10% of the land	• Manage nutrient inputs on your grassland • Manage fields for grazing (increase the average sward height) • Manage rushes in grassland • Maintain areas of tall vegetation, scrub and wet features • Manage 50% of conserved forage area as hay meadow	• Establish grass buffer strips,6 metres wide on at least 20% of the total length of water bodies in the agreement
Advanced Level Actions	• Provide resources for birds and pollinators on 10% of eligible land • Create buffers around 75% of your in-field trees • Maintain areas of tall vegetation, scrub and wet features • Support crop-pest predators • Apply fertilisers and pesticides with precision equipment	• Establish green cover on at least 15% of land • Add organic matter to a minimum of 20% of land • Create a soil management plan		• Maintain more hedgerow trees, or plant or tag new ones (There must be an average of one hedgerow tree every 100 metres across all the hedgerows) • Create buffer strips on both sides of the hedgerow on at least 75% of the hedgerows	• Introduce diverse grass, herb and legumes on 20% of improved grassland • Manage grazing on improved grassland • Allow ryegrass to bear seed • Reduce nutrient inputs • Do soil mapping • Use efficient precision application equipment	• Add or retain a mix of legumes, herbs and grasses on at least 15% of the grasslands • Maintain permanent grassland on land at risk of erosion or surface runoff on at least 15% of the land • Manage livestock density on grassland at risk of erosion or surface runoff on at least 15% of the land • Create a soil management plan	• Manage 50% of conserved forage area as hay meadow • Provide habitat for wading birds	• Establish grass buffer strips, 10 metres wide on at least 30% of the total length of water bodies in the agreement

Figure 2 SFI pilot standards and actions. Table compilation of Defra's published guidance on the standards (Defra, n.d.)

need to complete for each respective ambition level. Farmers entered into the intermediate level or advanced levels of any given standard need to complete all of the actions of the superseding ambition level.[3] For example, farmers in the advanced level of a standard need to complete all introductory, intermediate level, and advanced ambition level actions. The summaries listed in the table reflect the wording of guidance provided by Defra, which was purposely left more open to interpretation for farmers' own land management choices, given previous commentary on guidance criticised its overly prescriptive nature.

The relatively narrow selection of standards offered for the pilot does not reflect the wider ambition of the SFI programme, which will offer many more actions to farmers across a much wider array of environmental targets and farm habitats. This narrow focus importantly highlights that Defra was holistically testing its approaches to agri-environment scheme design. Through the particular phrasing of the actions (as shown in Fig. 2), Defra were also testing the clarity of guidance, and exploring how farmers might interpret the actions differently. For instance, the wording of several SFI actions is less prescriptive than their prior Countryside Stewardship equivalents. In this way, Defra was able to ascertain both from the Pilot farmer's management and their reactions to the guidance where and how this change is appropriate. This strategy is an effective example of 'prototype testing' by exploring pilot farmers' 'current adaptive and innovative capacity' (de Boon et al., 2022, p. 415) to understand and implement scheme actions for the development of national policy.

A degree of flexibility was incorporated in the pilot design to encourage agreement holders to feel they have enough options to apply on their farms. They can also adjust their agreement annually should it not work in practice. The pilot offers agreement holders the option to select the standards that are suitable for their holding, and their choice of ambition level to deliver the biodiversity goals set out in the standards. Farmers have the option of selecting how many plots of land they will enter into each standard and ambition level. Pilot participants have the option to apply multiple compatible standards to one plot of land, apply different standards across multiple plots of land, or mix and match the level of ambition in different standards. Once farmers have entered into an agreement with Defra, they are bound to it for 12 months and are expected to deliver the actions within the year. If for any reason the farmer is unable to complete the actions set out in the agreement, the farmer can amend their agreement at the end of the year during the annual change period. The concept of having an annual change period to make adjustments and amendments to an active agreement is a relatively novel aspect of agri-environmental schemes and is intended to add a layer of flexibility for farmers to make changes if the scheme does not work out as intended (e.g. their land is incompatible with the required actions of specific standards).

3 The Woodland standard is the only standard that only has one ambition level.

During the pilot, farmers engaged with policymakers through a series of different pilot-specific co-design activities and engagement events. Firstly, farmers were compensated for their engagement in co-design activities through a pilot Participation Payment amounting to £5000. The overarching goal of co-design was to ensure that the policy functions as intended for all relevant stakeholders and that engagement with stakeholders would encourage uptake and ensure that the scheme worked in practice, helping garner greater uptake later on (Tsouvalis & Little, 2019). The participation payment highlighted an important facet of co-design in the innovation process: pilot farmers were recognised as providing a valuable contribution to ELM policy development.

Farmers offered their innovative ideas and reflections in the SFI pilot process through a variety of co-design research activities, including webinars, on-farm events, interviews and other approaches. These activities were aimed at engaging directly with pilot participants to better understand how the policy was working for them and could be improved in the future. Co-design events and studies lent insight for policy teams into the limitations of SFI and the practicalities of how standards actually work together to produce biodiversity outcomes. Through these events, Defra was able to investigate areas of concern for farmers such as the appropriateness of payment rates, how drought impacts SFI biodiversity outcomes and any 'grey' areas in the guidance provided to interpret the standards. Learning, firsthand, about the dynamics of different issues in the policy provided policymakers with insight into how to rework and improve the SFI for the full rollout of the scheme.

Co-design learning from the SFI pilot has been applied in diverse ways to gain insights into specific policy areas and their real-world application, and there have been several key examples of where findings from the co-design learning activities have been instrumental in shaping the structure of the scheme's full rollout. For example, learning from these activities on payment rates has revealed that farmers do not believe the rates reflect income forgone and all of the costs associated with the implementation of a given standard. Farmers also shared the fact that the key financial incentive to sign up for the pilot was the £5000 Participation Payment – not the rate that they would be compensated for completing their standards. Defra has incorporated this feedback from farmers into the design of the full scheme rollout.

In January 2023, Defra unveiled enhancements to payment rates across different standards and introduced an additional funding component, referred to as the 'SFI Management Payment,' offering participants up to £1000 in additional support to cover administrative costs for scheme participation. The hope in implementing these changes is to make the full rollout of SFI attractive enough to keep the SFI pilot farmers engaged and increase interest in the main scheme amongst non-pilot farmers.

Another key learning where there has been a clear impact from co-design in the SFI pilot has been in the scheme structure. Through participatory on-farm workshops, it became evident that some of the actions within standards deter farmers from entering into either a higher ambition level or the standard as a whole. For example, horticultural farmers noted that they would need to take profitable ground out of production to establish cover crops. They intimated that they would rather enter into a lower ambition level, and grow crops on more of their land than enter into a higher ambition level to establish cover crops. Similarly, some farmers found particular actions too difficult, unworkable, or incompatible with their farming practices. Examples included the buffer strips required in hedgerow standards. Some farmers found great difficulty in establishing buffer strips and struggled to successfully grow any vegetation. Taking these learnings into consideration, Defra changed the design of the 2023 version of SFI, allowing farmers to pick and choose which actions they want to complete each standard. For example, in the full rollout of the scheme, a farmer can select the bundle of different actions, each standard depending on what works best for their farm. This 'pick n mix' change aims to equip farmers with the flexibility and discretion to make the right choices for their farms.

Co-design activities have also enabled Defra to learn about other aspects of policy design. For example, co-design activities for the pilot have led to a richer understanding of how farmers are seeking advice about joining the scheme. By better understanding where farmers go for advice and guidance, Defra has the opportunity to learn about who or what guides farmers through the decision-making processes of determining whether SFI is a good fit for them and how and what to apply for within the scheme. Strategic advice workshops and one-on-one interviews have been used to better understand how farmers seek guidance once in the pilot. The goal of these activities is to make the application process easier for farmers and for the government to better engage and provide advice and guidance, from the start of the application process.

Farmers' insights demonstrated that they either had to dedicate a lot of time to learn about the pilot or find someone or the resources to help them understand what combination of pilot standards worked best for their holding. Seeking guidance was considered time-consuming and frustrating by most farmers. DEFRA noted the threat that the administrative burden posed for future applicants to the scheme and decided to simplify the full rollout of the scheme to lessen the burden on farmers. The 2023 SFI scheme was amended to have fewer actions per standard, no ambition levels, and a single payment rate for each action. The simplification of SFI's offering was aimed at making the policy easier for farmers to understand and interpret so they do not have to seek additional resources or help. Through workshops and interviews, Defra also learned that farmers like to have easily accessible information, so all of the essential information was put in one handbook instead of spread across multiple URLs.

Other examples of the use of co-design in the pilot included working with pilot participants to better understand how the wording of policy guidance resonated with them and how variables that cannot be controlled for, like extreme weather conditions, might impact policy outcomes. These learnings can be used to clarify confusing aspects of the policy and explain what to do in the case of unforeseeable events. In these cases, co-design is used to allay farmers' concerns. Overall, the use of co-design has enabled a holistic review of the policy and identified any weaknesses in policy implementation. The hope is that the process of addressing the policy concerns of the pilot farmers will provide clarity for the farmers looking to join the full rollout of the scheme.

Though co-design helped shape the full rollout of SFI, some concerns with the SFI pilot have yet to be addressed and certain policy issues were not explored via a co-design approach. Farmers, especially in the uplands, are not yet fully recognised for the different biodiversity outcomes they produce because they do not fit within the remit of the current SFI standards. The same is the case for organic farmers in England, who arguably farm to a higher standard but – as yet – receive limited direct recognition for their practices under the SFI specifications. These are just two examples of cases where farmer-generated biodiversity outcomes, aligned with the goals of the 25-Year Plan and Net Zero by 2050, are not recognised under SFI. The threat is that if there is no incentive or compensation through SFI, farmers will turn away from these environmentally-sound practices and turn to an alternative solution that will compensate them for the financial losses incurred due to a phasing out of payments under the previous Basic Payment Scheme (e.g. looking to replace a loss in BPS income and turning to intensification or being paid to offset the carbon emissions of big polluters).

5 Getting mudd(l)y: limitations of ELM development

Navigating through an environmental policy transition has its limitations. Foremost in this case study were the time constraints from the start of the pilot programmes and tests and trials to the full rollout of the schemes. The limited time the UK government gave themselves to digest the policy findings from the pilot and make changes to the scheme did not allow for a thorough analysis of the data or implementation of all the findings. Defra was criticised by several pilot farmers regarding the rushed and overlapping timelines of the SFI pilot and the full rollout of the scheme. Pilot farmers felt as though their time in the pilot was undermined because the scheme that their insights are meant to help shape, was already launched and open to the public. Pilot farmers also considered SFI 2023 differed markedly from the design of SFI 2022, signalling that farmers can anticipate further drastic changes in subsequent versions of the scheme. The uncertainty continues.

The notion that the introduction of SFI was 'rushed' was also reflected in participation in the full scheme rollout. Some farmers adopted a wait-and-see approach to signing up to SFI, due to concern that the policy would be superseded by a better version within a few months. The condensed timelines were also evident in shelving the plan to develop a new LNR Scheme, replaced by 'copy and pasting' the existing Countryside Stewardship scheme, according to one farmer, albeit with some modifications. To manage timelines and expectations, the hope for an innovative new scheme was scrapped. The continuing changes to the design of the ELM scheme are signalling a rushed approach to farmers with the potential impact on farmer participation and confidence that the policy will be worthwhile and successful.

Another factor to highlight is the influence political instability has on the attractiveness and uptake of government schemes. In several co-design events and interviews, farmers shared their concerns about the impact of a change in government on the long-term security of the new agri-environmental schemes. They were concerned whether future administrations would continue to fund the new ELM schemes at the currently contracted level and whether the initial rollout, which was undersubscribed, would jeopardise future funding. Farmer uncertainty about whether the new ELM schemes will be sustainable and provide a consistent level of financial support is affecting farmer participation. Farmers argue that if they participate in SFI, for example, they should be guaranteed payments at an agreed rate for several years in order to secure a return on their investment. The absence of this reassurance and security was reflected in initial lower-than-anticipated levels of uptake and farmer hesitancy to embrace the new ELM policy.

In evaluating the potential of ELM to deliver for both biodiversity targets and for farmers, we highlight several key considerations. The first is that, as a programme still mid-way through development, it is, essentially, still too soon to tell what material impact ELMs will have in terms of the UK's ability to achieve its nature protection commitments, such as 30by30 (a commitment to protect 30% of land for biodiversity conservation by 2030) (Defra, 2023c), and continuing to support farmers in England for sustainable food production and enhanced delivery of public goods. A key issue here has been the messy transition from old to new systems of governance, for farmers, policymakers and land management more broadly. The step-change out of the European Union was a catalyst and enabler of dramatic changes in governance, where arguably the most consistent element has been the *inconsistency* in terms of scale and pace of scheme development.

At the scale of the pilot, this inconsistency is evident in various ways. First, the lack of a 'representative sample' of farms in the pilot (in terms of farm type and size) (NAO, 2021) makes extrapolating the findings to the wider farming sector in England difficult. Secondly, the lack of any 'control farms' in the pilot means that determining the environmental effectiveness of measures taken is also hard to establish. The limitations of the demographics of farms included in the pilot lead one to question how representative the developments from the SFI pilot are for non-pilot farmers across the country.

There are also other areas of concern for the pilot which are symptoms of the hasty development of new policy and establishment of the new system. For example, the impact of climate change on the environment and, therefore, the successful establishment or expansion of biodiversity and how it will be incorporated into monitoring decisions, is not addressed in new environmental policies. In the case of the SFI pilot, extreme drought and rain impacted farms across England differently, limiting some farms' ability to meet the requirements of specific standards despite doing everything in their power to achieve the standard requirements. There were no proactive guidelines included in the original SFI pilot guidance[4] on how to address monitoring decisions in cases of extreme weather. With the threat of climate change growing, future development of ELM policy should examine how to waive penalties incurred to farmers due to the negative impacts of climate change on biodiversity and environmental elements. It is important to acknowledge the impact of including easements in environmental policy to reassure farmers that they will not be penalised for factors out of their control and encourage them to commit to net zero.

Concerns over the intersection of food security, climate change, and public goods are also discussed by farmers when evaluating ELM. Increasing globalisation coupled with ongoing international conflict poses risks to food security globally. This and the need for versatility and resilient farming practices in the event of extreme weather necessitates more attention be paid to the protection of food production and the systems which support it.

6 Conclusion: ELM and future potential for co-design in agri-environment schemes

The development of England's ELM schemes, and the SFI National Pilot in particular, represents a pivotal moment in the country's agricultural transition towards sustainability and biodiversity conservation. The UK's departure from the European Union provided the opportunity to redefine agricultural policy,

4 During the SFI pilot, Defra ultimately decided to retrospectively implement an easement policy which would protect farmers who experienced extreme weather from receiving penalties on their monitoring decisions. For the farmers to be protected from penalties, they needed to provide proof of being impacted by extreme weather. The issue was this easement guidance was provided retrospectively after extreme weather was experienced. Moving forward, there should be an established methodology on how to document the impact of extreme weather on agri-environmental scheme outputs.

moving away from the CAP towards a system that prioritises the delivery of environmental 'public goods'. The concept of public money for public goods has underpinned this shift, acknowledging the dual role of farming in providing both private and public benefits. Whilst the new approach aligns with ambitious environmental goals, including biodiversity conservation and climate change mitigation, it also raises questions about the sustainability of payments for public goods in the longer term and the need for an approach that takes into account all aspects of England's farming communities and how to best work with them to achieve net-zero emissions.

Geopolitically, the UK's departure from the European Union has presented challenges and opportunities for developing sustainable farming and biodiversity conservation policies. Whilst ELM represents a significant step towards aligning agriculture with environmental objectives, its evolution, including the recent Countryside Stewardship adjustments, reflects the dynamic nature of policymaking. The changing political landscape presents challenges in creating a coherent policy that is implementable in practice. As the ELM schemes continue to evolve, addressing the challenges of recognising and rewarding biodiversity outcomes that fall outside the SFI standards, navigating the rushed policy timeline, and providing long-term policy stability will be essential. These considerations are crucial for creating a sustainable, resilient, and biodiverse future for UK agriculture.

The SFI Pilot offers a promising model for managing biodiversity in agricultural landscapes, emphasising flexibility, co-design and farmer engagement. Alongside building on the successes of the pilot, proactively addressing its limitations and ensuring policy stability will be essential to achieving the broader goals of the UK's 25-Year Plan and Net Zero by 2050 while maintaining the active involvement of farmers in sustainable land management practices.

The co-design process adopted by Defra in the development of ELM reflects both a step-change towards participatory policymaking and a commitment to stakeholder engagement and inclusion. However, challenges remain in ensuring that the scheme reaches less accessible stakeholders and involves a broader range of people in shaping the future of agriculture (Hurley et al., 2022). Without this, the policies and practices of the new agricultural transition may suit only a minority, ignoring those outside the predefined ELM boundaries (e.g. organic farmers), and excluding enterprises unfamiliar with agroecological approaches. In practice, when co-design is utilised as an action or discrete 'step' in policy development, rather than an outlook necessary to embed in the entire process, there is not sufficient scope to ensure both the shared sense of ownership and shared responsibility for the policy goals and public good delivery between policymakers and farmers. Nonetheless, the SFI pilot has been a demonstrable effort as an impact case of a co-design outlook, wherein farmers' views and experiences have been integrated into scheme design at multiple stages, feeding into the development of the main SFI scheme which continues to evolve and garner high levels of interest from the farming community. The track may be muddy and muddled, but the SFI pilot demonstrates that even in an imperfect policy development process, co-design can have beneficial results for farmers and for nature. So, stick your wellies on and get involved.

7 Future trends in research

There are several key areas in which research will continue to play a vital role in the development of co-designed agri-environmental schemes worldwide. One key area gaining increasing traction relates to innovative delivery and payment mechanisms. We have briefly mentioned a commitment to payment-by-results approaches here, but more research is needed to provide compelling evidence about how result-based schemes can be upscaled for a variety of farm types and targets, and how action-based schemes can continue to complement result-based approaches as needed.

Another issue likely to resurface in coming years is the public money for public goods mantra which defined Defra's early rationale for ELM. Whether food is a public good is contested in academia and industry (Little et al., 2022). From an economic perspective, 'Agriculture is a private enterprise and food is sold in markets, directly to consumers, and is therefore by definition (and according to official government guidance) a private, rather than a public, good' (Bateman and Balmford, 2018, p. 294). However, little consideration is given to the imperfect nature of the market in which food operates (how often, really, is food sold 'directly' to the consumer?) and the public value of food *access* alongside production is largely ignored. In contrast, biodiversity conservation has consistently been defined as a public good where no one can be excluded from its benefits (Perrings et al., 2009). However, the marketisation of biodiversity in England (Apostolopoulou and Adams, 2019) as seen most recently through the new biodiversity net gain obligations (Defra, 2023d) will undoubtedly invigorate debate about both food security and biodiversity, calling further into question in what cases it is appropriate for public goods (Timmermann, 2018) to be paid for by the market now that biodiversity is a good that the private sector can purchase.

Another area for further research includes evaluating the lessons learned from incorporating methods of stakeholder engagement into policy development, more specifically co-designing policy. Future research can look to explore the viability of co-design for developing national policy and its efficacy as a tool in this context. The SFI pilot serves as a case study for the application of policy co-design on a national stage and its successes and failures should be further researched to improve co-design methodologies in this context.

8 Where to look for further information

Further Reading

- The Defra Website is the most comprehensive and up-to-date source of information for updates and guidance about the Environmental Land Management Scheme and the Sustainable Farming Incentive. Search www.gov.uk/defra/sustainablefarmingincentive
- Papers, scheme handbooks and publications from Natural England provide the most comprehensive instructions and rationale for legacy scheme actions.
- Little, R., Lyon, J. & Tsouvalis, J. (2022). The co-design of post-Brexit agri-environmental policy focusing on environmental land management in England. In A. Attorp, S. Heron, & R. McAreavey (Eds.), Rural Governance in the UK: Towards a Sustainable and Equitable Society (1st ed., pp. 54–77). Routledge. https://doi.org/10.4324/9781003200208-4
- For a review of the co-design process and issues with farmer engagement see Hurley, P., Lyon, J., Hall, J., Little, R., Tsouvalis, J., White, V. & Rose, D. C. (2022). Co-designing the environmental land management scheme in England: The why, who and how of engaging 'harder to reach' stakeholders. People and Nature, 4(3), 744–757.

Key journals/conferences

- Oxford Real Farming Conference.
- Oxford Farming Conferences (and correspondence/briefs coming out of the event).
- British Ecological Society Annual Meeting.
- Groundswell: The Regenerative Agriculture Festival.

Major research projects related to agriculture and biodiversity

- Contracts 2.0 was a 4-year Horizon2020 study which explored innovative agri-environment scheme delivery mechanisms across 11 European countries including the UK.
- Leverhulme Centre for Nature Recovery (Oxford).
- Leverhulme Centre for Anthropocene Biodiversity (York).
- Agri-Environmental Governance Post-Brexit (Sheffield & Reading).

9 List of acronyms

BPS	Basic Payments Scheme
CAP	Common Agricultural Policy
CS	Countryside Stewardship
Defra	Department for Environment, Food and Rural Affairs
ELM(s)	Environmental Land Management (Schemes)
ES	Environmental Stewardship
ESA	Environmentally Sensitive Areas
EU	European Union
LNR	Local Nature Recovery
SFI	Sustainable Farming Incentive
T&Ts	Tests and Trials
UK	United Kingdom

10 References

Agriculture Bill. (2018). *UK Parliament Parliamentary Bills.* https://bills.parliament.uk/bills/2292/publications (Accessed 9 January 2024).

Apostolopoulou, E. and Adams, W. M. (2019) 'Cutting nature to fit: Urbanization, neoliberalism and biodiversity offsetting in England', *Geoforum*, 98 (May 2017), pp. 214–225. doi: 10.1016/j.geoforum.2017.05.013.

Bateman, I. J. & Balmford, B. (2018). Public funding for public goods: a post-Brexit perspective on principles for agricultural policy. *Land Use Policy*, 79 (August 2018), 293–300. https://doi.org/10.1016/j.landusepol.2018.08.022

Carey, P., Finch, C., Hunt, J., Morris, C., Parkin, A., Priscott, A., Routh, C. & Short, C. (2000). *Monitoring and Evaluation of the Countryside Stewardship Scheme.* Natural England. https://publications.naturalengland.org.uk/publication/5629534368432128

Coffey, T. (2022). Speech. Environment Secretary speech at CLA Business Conference. Department for Environment, Food & Rural Affairs. https://www.gov.uk/government/speeches/environment-secretary-speech-at-cla-business-conference

Cusworth, G., & Dodsworth, J. (2021). Using the 'good farmer' concept to explore agricultural attitudes to the provision of public goods. A case study of participants in an English agri-environment scheme. *Agric Hum Values* 38, 929–941. https://doi.org/10.1007/s10460-021-10215-z

de Boon, A., Sandström, C. & Rose, D. C. (2022). Governing agricultural innovation: a comprehensive framework to underpin sustainable transitions. *Journal of Rural Studies, 89*, 407–422. https://doi.org/10.1016/j.jrurstud.2021.07.019

Department for Environment Food and Rural Affairs. (2018). Health and Harmony: the future for food, farming and the environment in a Green Brexit. In *Gov.Uk* (Issue February 2018).

Department for Environment Food and Rural Affairs. (2020). *The Path to Sustainable Farming: An Agricultural Transition Plan 2021 to 2024*. gov.uk/government/publications

Department for Environment Food and Rural Affairs. (2023d). Understanding Biodiversity Net Gain, Accessed January 9 2024, from https://www.gov.uk/guidance/understanding-biodiversity-net-gain

Department for Environment Food and Rural Affairs. (n.d.). *Sustainable Farming Incentive pilot guidance*. Defra Farming Blog. https://defrafarming.blog.gov.uk/sustainable-farming-incentive-pilot-guidance-2/

Eastwood, C. R., Turner, F. J. & Romera, A. J. (2022). Farmer-centred design: an affordances-based framework for identifying processes that facilitate farmers as co-designers in addressing complex agricultural challenges. *Agricultural Systems*, *195*(January 2021), 103314. https://doi.org/10.1016/j.agsy.2021.103314

Environment Act 2021, c. 30. Available at: https://www.legislation.gov.uk/ukpga/2021/30/contents

European Commission. (n.d.). *CAP at a glance - European Commission*. Accessed January 9, 2024, from https://agriculture.ec.europa.eu/common-agricultural-policy/cap-overview/cap-glance_en

European Court of Auditors. (2017). Special Report. *Greening: a more complex income support scheme, not yet environmentally effective*. European Union. https://www.arc2020.eu/wp-content/uploads/2018/03/SR_GREENING_EN.pdf

Gove, M. (2018). Speech. *Green Brexit: a new era for farming, fishing and the environment*. Department for Environment, Food & Rural Affairs. https://www.gov.uk/government/speeches/green-brexit-a-new-era-for-farming-fishing-and-the-environment

Helm, D. (2017) 'Agriculture after Brexit', *Oxford Review of Economic Policy*, *33*, S124–S133. https://doi.org/10.1093/oxrep/grx010.

Hird, V. (2022). *Farm policy drama and what really happened*. Sustain Blog: Sustainable Farming Campaign. https://www.sustainweb.org/blogs/oct22-elms-and-nature-in-peril/

Hughes, J. (2020). *What we mean by "co-design."* Defra Farming Blog. https://defrafarming.blog.gov.uk/2020/12/11/what-we-mean-by-co-design/

Hurley, P., Lyon, J., Hall, J., Little, R., Tsouvalis, J., White, V. & Rose, D. C. (2022) 'Co-designing the environmental land management scheme in England: The why, who and how of engaging "harder to reach" stakeholders', *People and Nature*, *4*(3), 744–757. https://doi.org/10.1002/pan3.10313.

Jordan, A., Wurzel, R. K. W., Zito, A. R. & Brückner, L. (2003). Policy innovation or "muddling through"? "New" environmental policy instruments in the United Kingdom. *Environmental Politics*, *12*(1), 179–200. https://doi.org/10.1080/09644010412331308344a

Lindblom, C. E. (1979). Still Muddling, Not Yet Through. *Public Administration Review*, 39(6), 517–526. https://doi.org/10.2307/976178.

Little, R., Lyon, J. & Tsouvalis, J. (2022). The co-design of post-Brexit agri-environmental policy focusing on environmental land management in England. In A. Attorp, S. Heron, & R. McAreavey (Eds.), *Rural Governance in the UK: Towards a Sustainable and Equitable Society* (1st ed., pp. 54–77). Routledge. https://doi.org/10.4324/9781003200208-4

Marshall, C. & Prior, M. (2022). Government review of ELMS farming subsidies stokes anger. *BBC News*. https://www.bbc.co.uk/news/science-environment-63029266

Mennig, P. & Sauer, J. (2020). The impact of agri-environment schemes on farm productivity: a DID-matching approach. *European Review of Agricultural Economics*, *47*(3), 1045–1053. https://doi.org/10.1093/erae/jbz006

Meuwissen, M. P. M., Feindt, P. H., Spiegel, A., Termeer, C. J. A. M., Mathijs, E., de Mey, Y., Finger, R., Balmann, A., Wauters, E., Urquhart, J., Vigani, M., Zawalińska, K., Herrera, H., Nicholas-Davies, P., Hansson, H., Paas, W., Slijper, T., Coopmans, I., Vroege, W., ... Reidsma, P. (2019). A framework to assess the resilience of farming systems. *Agricultural Systems, 176*(January), 102656. https://doi.org/10.1016/j.agsy.2019.102656

National Audit Office. (2021). *The Environmental Land Management scheme* (H.C. 664, Vol. SESSION 2021-22, pp. 1–56). National Audit Office Report by the Comptroller and Auditor General.

Natural England. (2012). *Higher Level Stewardship Handbook* (4th ed., Issue January). Department for Environment Food and Rural Affairs (DEFRA). https://publications.naturalengland.org.uk/publication/2827091

Nègre, F. (2022). Second Pillar of the Cap: Rural Development Policy: Fact Sheets on the European Union: European Parliament, www.europarl.europa.eu/factsheets/en/sheet/110/second-pillar-of-the-cap-rural-development-policy

Perrings, C., et al. (2009). The economics of Biodiversity and Ecosystem Services, Oxford. In S. Naeem, D. E. Bunker, A. Hector, M. Loreau, & C. Perrings (Eds.), *The Economics of Biodiversity and Ecosystem Services. In Biodiversity, Ecosystem Functioning, and Human Wellbeing: An Ecological and Economic Perspective*. Oxford University Press.

Scottish Government. (2018). *Stability and Simplicity: Proposals for a Rural Funding Transition Period*. https://consult.gov.scot/agriculture-and-rural-communities/economy-post-brexit-transition/user_uploads/00537221.pdf

Timmermann, C. (2018). Food security as a global public good. In J. R. V. Pol, T. Ferrando, O. de Schutter, and U. Mattei (Eds.), *Routledge Handbook of Food as a Commons* (pp. 85–99). Routledge.

Tsouvalis, J. & Little, R. (2019). *Factors Influencing Farmer Participation In Agri-Environment Schemes (AES) – Evidence From The Social Sciences*. The University of Sheffield.

Understanding biodiversity net gain

Chapter 9

Developments in agri-environment schemes (AES): North America

Gordon Rausser and David Zilberman[1], University of California, Berkeley, USA

1 Introduction

Government policies have played a major role in U.S. agriculture. Early policies include land grants, homesteading, investment in public research, extension, and land reclamation. The twentieth century saw the emergence of policies that supported farm income and enhanced demand for agricultural food domestically and internationally (Cochran 1993). The Dust Bowl of the 1930s initiated soil conservation programs in the United States and Canada. Increased environmental awareness and concerns in the 1970s led to the formation of the Environmental Protection Agency and legislation, including the Clean Air Act; Clean Water Act; Federal Insecticide, Fungicide, and Rodenticide Act (FIFRA); and Endangered Species Act. Individual states developed their own environmental legislation, and this new environment affected how agriculture was practiced. Public projects in agriculture have been subjected to rigorous environmental evaluation, with traditional agricultural policies that aim to protect farm income transforming to also target environmental conservation objectives. Thus, agricultural development has become directed, in part, by environmental objectives.

1 David would like to thank Sadie Shoemaker for editorial assistance.

http://dx.doi.org/10.19103/AS.2024.0134.14

This chapter presents an economic perspective on significant agri-environmental policies that affect U.S. agriculture and assesses some of the environmental outcomes of these policies. The following section will present a conceptual framework for designing and assessing environmental intervention in agriculture, emphasizing payment for environmental services programs, which became a major agri-environmental intervention. It will be followed by a presentation of the major agri-environmental programs affecting U.S. agriculture and other North American countries. Results of studies assessing the economic and some of the environmental outcomes of agri-environmental programs in North America will be presented next, followed by a concluding section.

2 Conceptual frameworks for assessing agri-environmental schemes

Economic theory suggests that government intervention may be needed to address negative externalities (e.g. pollution or other damages inflicted unintentionally on third parties as part of economic activities) and public goods (goods or services benefitting all members of society without access barriers (Laffont 1988)). Multiple policy interventions address externality problems, including incentive-based mechanisms such as taxes, subsidies, direct regulation, public provision, voluntary agreement, and educational programs.

There are various limitations to public interventions to address externalities. While taxes on externalities are efficient mechanisms provided there is sufficient information, they face political objections from affected parties (i.e. major corporations, some consumers – objecting to energy taxes – or small farmers) who may use the political process to avert these policies. Subsidies are costly to the government, and thus their application is limited. Tradable permits may introduce markets for externalities that may decrease costs for polluters. Direct regulations are blunt and tend to be inefficient when there is much heterogeneity among the affected population.[2] Voluntary agreements may have limited capacity to lead to change because of limited enforcement and are more effective when there is a threat of regulation. Education, which internalizes environmental values among the population, is very effective but limited in scope (Claassen et al. 2014). Consequently, most of our attention will be directed to payment for ecosystem services (PES): policies that were introduced and designed, to a large extent, to address agri-environmental problems. PES are voluntary agreements between at least one seller (service provider) and one buyer (service buyer) covering a well-defined environmental service with conditional payment on service delivery. Usually, the buyer is a

2 However, they tend to be helpful when costly research is needed to design new solutions or assess new technologies.

government agency but can be a private entity when incentivized with tax benefits.

There are three categories of PES programs (Baylis et al. 2008):

- **Land diversion programs:** They incentivize the diversion of land periodically or permanently from agricultural production to provide environmental services, such as carbon sequestration, habitat protection, or soil erosion and air pollution reduction.
- **Working land programs:** Farmers are subsidized to adopt environmentally friendly practices, such as integrated pest management (IPM) or improved animal production facilities to reduce pollution.
- **Conditional PES programs in livestock and other non-field crop production activities:** Participants are entitled to benefit from an agricultural support program depending on the adoption of environmental practices.

There is significant and ongoing literature on the design of PES. The literature suggests several principles for effective design (Zilberman and Segerson 2012):

- The first is using information and measurement tools to track performance.
- The second principle is to predicate payment based on improving performance rather than on practices.
- The third principle is to recognize biophysical and socioeconomic heterogeneity.
- The fourth principle is to build flexibility for adaptation to economic and environmental conditions.

Most agricultural PES programs have targeted land use. A key element in designing PES is determining the selection of land and its respective remuneration, given the heterogeneity in the provision of environmental amenities and land cost (reflecting rent from agricultural activities).

PES programs like the Conservation Reserve Program have targeted the cheapest land providing environmental services for a given budget. This cost-targeting approach may be inefficient if less productive land also has the lowest ecological amenities. Another targeting approach is benefit targeting, where the land with the greatest environmental benefits per acre is selected. However, this approach is suboptimal if land with the highest environmental benefits per acre is also the most agriculturally productive; greater aggregate environmental benefits can be obtained by buying cheaper lands. Accordingly, the desired approach (which maximizes the benefits given a budget) is benefit–cost targeting, where lands are ranked according to their benefits per acre minus the costs to compensate for potential agricultural productivity per acre. The program targets the land with the highest benefit–cost ratio.

In contrast to the government as a buyer for biodiversity services, Rausser and Small (2000) provided a market-based determination of biodiversity sites via biodiversity prospecting. The methodology is designed to implement statistical decision theory to evaluate various prospecting options. Unique genetic materials that can only be found by preserving endangered and threatened species hold great potential for commercial research and development with potential for innovative products and subsequent intellectual property rights. Products such as Paclitaxel, a chemotherapy medication that stems from the Pacific yew tree (*Taxus brevifolia*), and Tebanicline, a non-opioid analgesic drug derived from the secretion of an Ecuadorian poison dart frog (Dendrobatidae), display the value of preserving wildlife.[3] For lands with sufficient biodiversity, an opportunity exists for pharmaceutical companies to make payments in conjunction with land diversion PES programs to augment the amount of compensation that is offered to landowners for conservation.

Wu et al. (2001) argued that the benefit–cost ratio maximizes the overall social benefit of a PES program for a given budget if the affected industry is competitive and price-taking (i.e. can accommodate prevailing market prices). However, if the industry faces falling demand, the PES program may increase output prices, leading to slippage. Furthermore, farming on previously idle land may result in negative environmental outcomes from recultivation. In this case, PES programs may pay idle land not to enter production. In addition, the PES benefit–cost targeting formula may need to be adjusted to prevent environmental agencies from using their monopoly power to underpay for the environmental services they purchase. Wu et al. (2001) also showed that stakeholders may prefer different targeting schemes. Under plausible assumptions, landowners prefer cost targeting because it leaves them with the most productive land. The increased political power for environmental interests contributes to the transition from cost-targeting to benefit–cost targeting.

PES programs affect multiple groups. Zilberman et al. (2008) considered differences in terms of farm size and differences between landowners and farmworkers. They argued that land diversion PES programs tend to benefit landowners (who obtain a new source of income) but may hurt consumers because of higher food prices and farmworkers, who may lose their jobs. However, working land programs may benefit farmworkers if the new practices are labor-intensive and may benefit producers if yields increase. Baylis et al.'s (2008) comparison of the EU versus U.S. policy suggested that the EU

3 A valuation method under uncertainty for land that is a candidate for conservation must add its bioprospecting value as a function of success probabilities taking into account revenues per success, cost per test, a hit rate based on a certain number of species and time discounting. For various biodiversity sites for which differential information is available, rents will emerge that are much higher than scarcity rents. Under such circumstances, the values associated with high quality and diverse sites can be large enough to motivate conservation activities (Rausser and Small 2000).

emphasizes working land programs where farmers are paid to modify existing practices. In the United States much of the emphasis is still on land diversion programs. Two reasons for the differences are a relatively greater abundance of land in the United States and the greater focus on using PES programs as a mechanism for rural development. In both the United States and the EU, cross-compliance programs require conformity with specific environmental standards to be entitled to agricultural support policies.

Benefits per unit of land may vary with scale, and there is evidence of significant agglomeration effects (i.e. when programs are concentrated in the same area). Wu and Boggess (1999) developed an extended model to design PES considering multiple attributes and agglomeration effects. In some situations, the marginal increment of environmental benefits may increase with the land utilized. Wildlife needs critical amounts of space to survive and flourish, and corridors may need to be established to allow movement. Furthermore, the same piece of land may provide multiple environmental amenities (e.g. wetland protection, air quality protection, or wildlife shelter).

PES programs may have multiple objectives, so one of the challenges of design is the weighting and aggregation of multiple benefits (Engel et al. 2008). They suggest that the design of these policies requires an assessment of preferences for different environmental and relevant economic objectives. Willingness-to-pay studies may be used to estimate the weights given to different objectives, or policymakers may determine them. In other words, the design of PES programs is challenged by the vast amount of data[4] that are subjected to a high degree of randomness (i.e. weather and policy), in which PES program design is likely to change over time in order to take advantage of increased knowledge and improved technology. Furthermore, the design of PES programs needs to incorporate mechanisms for monitoring and enforcement, which are not addressed heavily in the literature. Monitoring is also likely to improve as experience is accumulated, with monitoring technologies (e.g. remote sensing and drones) also improving.

The literature on PES emerged in the early 1990s, and much of the important conceptual contributions on the design and potential impact of PES programs were written between 1995 and 2010. Once PES programs were established, more emphasis was given to assessing their impact and learning from their actual outcomes. Naeem et al. (2015) argued that early assessments of the design of PES show that the incorporation of scientific and economic knowledge in the design of the program was weak, limiting their effectiveness, and this is an area for improvement. Wunder et al. (2020)'s review of studies assessing PES programs suggests that the design and implementation are

4 These may include data on the distribution of environmental characteristics, rents over space and the likely impact of participating in the program under different environmental amenities.

lacking. In particular, targeting and enforcement are inadequate, and there are significant cases of adverse self-selection. While this study has global coverage, the results rely on studies from North America. Since PES programs are so young, it is no wonder that their execution leaves room for improvement and suggests the importance of studies that assess the results on the ground.

The dynamics of the PES literature has been analyzed by Salzman et al. (2018). The number of refereed articles on PES grew from a few articles in 1995 to more than nineteen hundred in 2016. However, despite the growth of these programs, there is very scant evidence of their effectiveness, which should be an area of future research, and requires appropriate sampling, benchmarking, and continuous data collection.

Watershed PES is the most widely used form of PES financed by governments with programs as part of compliance with water quality regulations. Biodiversity and habitat are a second group, mostly driven by government regulation. Mitigation credit is an emerging category. Regulatory PES aiming to maintain biodiversity and habitat objectives are quite prominent in the United States and other North American countries, as shown below.

3 Agri-environmental schemes in North America

Multiple policy tools are used to achieve environmental objectives in agriculture: a detailed regulatory process for the introduction of pesticides to U.S. agriculture (regulated by FIFRA) establishes mandatory testing before new material is introduced and develops mechanisms to ban materials that are found to be carcinogenic (Claassen et al. 2014). There is a complex regulatory system for introducing genetically engineered organisms, which tends to slow the development of new traits and the adoption of technology (Just et al. 2006). The U.S. government and several states, especially California, established policies to introduce and expand biofuel cultivation to strengthen the renewable energy sector (Sorda et al. 2010). The EPA introduced regulatory standards to control the effluence and disposal of large, confined animal feeding operations (CAFOs). There is a patchwork of state regulations regarding the disposal of animal waste (Claassen et al. 2014).

The Clean Water Act (CWA) established an upper bound on the concentration of chemicals in water. It introduced procedures (for both point and nonpoint pollution sources) to monitor and enforce these standards (Copeland 2006). Keiser and Shapiro (2019) assessed the performance and impact of the CWA, finding that it significantly increased the share of safe water for drinking. However, the cost of cleanup of streams is high ($1.5 million per mile), and more than half of U.S. rivers and streams still do not meet pollution standards. Some water quality regulations gave rise to water quality trading in several regions in the United States (Shortle 2013).

While environmental taxation is not frequently used, monetary penalties enforce standards. Weersink et al. (1998) reviewed environmental policies in Canadian agriculture. They suggested that there is limited use of taxation or other 'first-best' policies to achieve environmental goals because of heterogeneity, informational challenges, and political-economic constraints. They argued that regulatory pressure might lead to technological development and business initiatives that will make it feasible to introduce financial incentives to achieve environmental goals in agriculture. This is consistent with Nordhaus's (2021) *The Spirit of Green,* which suggests that the buildup of awareness leads to regulatory pressure and the emergence of financial incentives that lead to adoption of greener technologies.

Since the 1980s, there has been a growing use of PES programs in U.S. agriculture. Most of the spending has been allocated to the Conservation Reserve Program (CRP), which incentivized agricultural producers to take land out of production and instead adopt conservation practices. It was first introduced in the Farm Bill of 1985 and has been modified since. The CRP is a competitive program; landowners offer eligible land for enrollment. After specifying the targeted environmental objective (i.e. reduced soil erosion, improved water quality through wetlands and field buffers, reduced fertilizer use, and increased wildlife habitat), the Farm Service Agency utilizes a benefit–cost criterion to assess eligible land for CRP contracts. CRP payments come in multiple forms, including annual rental payments, cost-share payments for establishing eligible practices, and reimbursement payments for maintaining practices. While most contracts have been for a limited period of 10 years, environmentally important lands (such as riparian land) are entitled to continuous sign-ups. The top conservation practices include close to 5.7 million ha for establishing vegetative cover and native grass and about 1.6 million ha for wildlife habitat (Stubbs 2014).

A historical review of the CRP appears in USDA (2020). Between 1988 and 2010, the enrollment in the CRP varied in response to commodity program prices and other factors, reaching peaks of 14.6 million ha in 2007, while in 2020, only 8.9 million ha were enrolled in the program. Expecting higher commodity prices, the Agricultural Act of 2014 reduced enrollment in the program from 12.9 million ha to 9.7 million ha by 2018. This farm bill also permitted some agricultural activities on CRP land (e.g. grazing and other use of forage) and allowed withdrawal from the program after 5 years. About 70% of the USDA land was allocated to address negative environmental side effects such as soil erosion, nitrate emissions, and preserving native plants. Seventy per cent of the PES land that is retired from agricultural production under the CRP program is from land that was formerly used to produce corn, soybean, wheat, and other major crops. Other CRP land retired from production included 7% for land previously allocated for trees, 11% for wetland restoration, and 7% for buffer crops. The agri-environmental program of the Farm Bill received about 6.8% of the Farm Bill, which is about $5.4 billion. The CRP expenditures were about $2 billion annually in 2020.

Preserving wetlands has been one of the major environmental policy objectives in the United States. The North American Wetlands Conservation Act (NAWCA), passed in 1989, aims to curtail wetland loss. The agri-environmental component of the farm bill explicitly targeted wetland restoration and protection by supporting the Wetland Reserve Program, which enrolled about 1.2 million ha and restored approximately 44500 wetland ha in 2007 alone. Between 1985 and 2006, the U.S. Farm Bill included the cross-compliance program, the 'Swampbuster', that excluded individuals who farm on environmentally sensitive land from eligibility to participate in most of the USDA's agricultural support programs.

The Environmental Quality Incentives Program (EQIP), a voluntary working land program introduced in the 1966 Farm Bill, provided farmers with subsidies and technical support to plan and implement conservation practices (Khanna 2001). Its support dropped from $1.75 billion in 2002 to $1.3 billion in 2007. One program component is to assist farmers in adopting best management practices. Payments are based on the estimated value of the environmental benefits accrued to society, including the farmers. The program strongly emphasizes water quality improvement, including reducing non-point-source pollution in agriculture to meet water quality standards, conserving ground and surface water resources, reducing air pollution to meet water quality standards, reducing soil erosion, and protecting wildlife. While most of the CRP is allocated to crop farming, about 60% of the EQIP program is allocated to livestock agriculture, especially targeting support of programs that will reduce groundwater contamination by livestock. The four largest recipients of EQIP resources are Texas, California, Colorado, and Nebraska. However, there is concern that much of the resources benefit larger operators (Stubbs 2010). Seventeen per cent of approved applicants to the EQIP program withdrew from the program. Cattaneo (2003) suggested that the cost share of the program may be low, and although farmers' conservation plans meet regulations, they are often not profitable.

The United States has seen rapid growth in private conservation through land trusts and conservation easements that are incentivized by tax incentives. Large land trusts, like the Nature Conservancy, conserve land using private donations and public money; the gifts are exempt from federal, state, and local taxation. Conservation easements preclude development activities; the value of donations for the easement is calculated by the difference between the value of the land in the best commercial activities and its value under the constraints of the easements. More than 6.5 million ha of land are in land trusts and easements, which is more than half the area of the CRP (Parker and Thurman 2019).

Canada and Mexico have their own agri-environmental policies, which differ from the United States. Baylis et al. (2022) suggested that like in the United States, agri-environmental programs in Canada were started in the 1930s in response to dust storms. The Canadian government subsequently invested in water projects and working land programs, providing financial and technical support to enhance agricultural productivity, improve grasslands and pastures, and reduce soil erosion. Recent programs, particularly the Cover Crop Protection program, encouraged the conversion of erodible land to pasture and cover crops and diverted land from production. Land diversion programs are more prominent in the United States than Canada because the United States had the land set-aside program as part of its supply control policy. This program was converted to become more environmentally oriented. Canada has very few set-aside programs if any. Therefore agri-environmental policies emphasize the working land program, and diversion occurs only when it makes sense from an environmental perspective.

Mexican agriculture is different from the United States and Canada. It is not dominated by large volumes of land dedicated to the modern production of field crops and does not have the tradition of set-aside programs. It has significant amounts of land allocated to fruits and vegetables, and there is demand for the expansion of agricultural land by deforestation. The PES program in Mexico combines market mechanisms with regulations and subsidies. It aims to improve environmental quality and enhance efficiency, and at the same time, improve the well-being of Campesinos in rural communities (McAfee and Shapiro 2010). Satellite data of land coverage suggest that deforestation was substantially lower on properties that received PES than those that did not. Still, interviews on the ground suggest that the PES was only one cause of the reduced deforestation, and there is a potential for increased efficiency of PES funds through improved targeting (Scullion et al. 2011).

4 Assessment the impact of payment for ecosystem services programs

There is growing research assessing the impact of PES programs and suggesting mechanisms to improve them. Newburn et al. (2006) argued that current targeting methods, including benefit–cost targeting, are based on static analysis and information based on the present, and thus are inefficient because they ignore the likelihood that certain land will be converted, for example, to urban use, or environmental amenities that are subjected to future risks. In other words, the use of expected benefit–cost targeting must consider future land use patterns. This approach is necessary for developing PES programs in the urban fringe, where the conversion of agricultural land or precious ecosystems to commercial uses is very likely. Newburn et al. (2006) demonstrated the gain from expected benefit cost targeting by analyzing data on land use change and conservation opportunities under different targeting strategies in Sonoma County in California.

Miao et al. (2016) identified significant flaws in the targeting of the CRP program. They showed that many of the lands that enter the CRP program could also be enrolled in crop insurance programs. Because such lands tend to be marginal and high risk, they would have received significant crop insurance payments if they were not enrolled in the CRP program; for example, in North Dakota, average crop insurance subsidies were 70% of CRP rents. Their analysis suggests that if targeting had considered saving and insurance subsidies, the government could have obtained 50% of the offered acres in 2014 at zero nominal cost, acreage and payment would have increased in the Great Plains and Southeastern states, and overall environmental amenities obtained would have increased. If this saving was considered both in targeting and in the program's design, the targeting would have slightly changed, and the amount of land enrolled in the program would have increased significantly.

Assessing the costs and benefits of agri-environmental programs is challenging and may improve with better data. Nolte (2020) presented results that use a high-resolution map of the estimated value of private land in the United States to predict the cost of conservation interventions. Reassessing recent U.S. wildland protection and species conservation studies suggests that traditional estimation techniques underestimate the cost of achieving environmental goals by at least 50%. This underestimation is likely to lead to the misallocation of the budget for conservation programs and biases in the location of expenditure across locations. Duke et al. (2014) suggested the importance of using non-market benefit assessment techniques (hedonic prices, sometimes contingent valuation), to assess the benefit and costs of agri-environmental programs. Assessing targeting criteria using Delaware data, they found that benefit–cost targeting approximates more precise optimization targeting and is preferable to cost or benefit targeting. Lewis et al. (2011) suggested that designing habitat conservation programs to protect vulnerable species requires understanding species biology and considering spatial heterogeneity. Incentive-based policies not adjusted to these considerations are likely to be suboptimal and can gain by encouraging agglomeration of conserved habitat based on understanding the conserved species' behavior.

Baylis et al. (2022) surveyed some of the findings on the impact of agri-ecological programs while recognizing the limitations of measurement techniques. Studies found that the benefits of land diversion programs tended to exceed the costs of reducing soil erosion and improving wildlife habitat, water quality, carbon sequestration, and wetland acreage. Because of the heterogeneity of working land programs, it is difficult to generalize their economic impact. However, Atwood et al. (2003) estimated that EQIP had a benefit-to-cost ratio of 1.5 after its expansion in 2002. Baylis et al. (2022) proved that the benefits of the CRP and other set-aside programs are additive because they would not have been achieved otherwise. While some Canadian working land programs tend to enhance efficiency, their practices might have been adopted without government support. Baylis et al.'s survey also found a significant degree of additional inputs applied to non-idle land that has the indirect effects of generating additional negative environmental effects resulting from overuse of chemical fertilizers and other inputs to increase yields. These effects are directly affected by market conditions, namely, if prices are very favorable, there is more incentive for farmers to apply additional variable inputs that expand the yields of utilized land. Wu (2000) found that about 10% of the potential benefits of various aspects of the CRP was lost because of higher commodity prices and expanding acreage resulting from the program. Baylis et al. found mixed results regarding the permanent impacts of land diversion programs. In many cases, land returns to production once the programs ended. However, in some cases, the diversion may result in establishing recreational or other profitable activities that lead to permanent change.

The recent paper by Taylor and Druckenmiller (2022) provides the benchmark for quantitative assessments of the value of the PES programs that protect wetlands. They found that the average value of a hectare of wetland lost in the United States between 2001 and 2016 was about $1840 annually. By contrast, the environmental benefits vary and can reach $3200 per hectare in highly developed areas, mostly from flood protections. This research suggests that the value, for example, of PES to protect wetlands depends on other government policies. For example, subsidized homeowner insurance in vulnerable areas increases the value of flood protection and PES to achieve it.[5]

Wallander and Hand (2011) assessed the impact of the EQIP program that provided subsidies for adopting conservation technologies. They found that the program significantly reduced the application rates of water and chemicals per acre and increased water use efficiency while, at the same time, increasing total irrigated acreage. This is consistent with other findings on the impact of economic incentives to adopt modern irrigation technologies, particularly drip irrigation. Government support policies, in terms of extension and direct support, contributed to the increased adoption of modern irrigation technologies throughout the United States. This adoption tends to significantly increase yield, reduce water use per acre, and reduce drainage, ultimately generating a large net-economic benefit. But in some regions, it led to increased water use (Taylor and Zilberman 2017).[6] Kovacs et al. (2013) estimated the return on

5 Hannah Druckenmiller, personal communication, seminar at UC Berkeley 2022.

6 Similarly, several studies, including Zilberman et al. (1991), suggest that despite the significant environmental costs, pesticide use, and regulation produced net economic benefits. A few studies estimated the economic return from agri-environmental programs.

investment from public land conservation programs in Minnesota, discovering that the return on investment varies depending on assumptions, but benefits can reach $2–3 per dollar of investment, as long as the land is in the program. Many benefits come from carbon sequestration, water quality improvement, and other amenities.

Porto et al. (2020) surveyed multiple studies the results of multiple studies estimating the immense value of pollination services globally and in North America. The value of pollination services in the United States is estimated to be between half a billion to $52 billion annually. However, pollination services are negatively affected by pesticide application and other human activities, and there is a case for support. Research, regulations, and some incentives protect pollination, but they suggest that more needs to be done. The survey by Baylis et al. (2021) suggested that we lack reliable economic estimates of the value of pollination services. Organized markets have provided pollination services thus far in the United States, but diseases and externalities have challenged pollination, and there is a need to assess the interventions, including PES, for a sustainable future for the industry.

One of the objectives of PES programs is restoration of native plants. Gibson-Roy (2018) documented how the CRP leads to large-scale native plant restoration in the United States. The incentive led to the emergence of commercial seed and restoration sectors that enabled farmers to join the program profitably.

Not all the programs that aim to provide agri-environmental benefits are effective in achieving environmental objectives and may achieve other objectives; one example is biofuel policies. The Renewable Fuel Standard, at best, reduces greenhouse gas emissions from fuel production modestly compared to gasoline but reduces fuel cost in the United States and saves the government significantly in exchange. The California Low-Carbon Fuel standard discriminates between alternative biofuel compensating low-carbon fuels such as sugarcane, which has lower greenhouse gas emissions per gallon than corn ethanol. However, because of learning by doing, the impact of biofuel policies is improving over time (Khanna et al. 2021; Hochman and Zilberman 2016). Government subsidies to support organic farming[7] may actually have a negative environmental impact. The lower yield of most organic practices and the extra land they require tend to increase agriculture's footprint and greenhouse gas emissions (Meemken and Qaim 2018; Rausser et al. 2019). Another area where environmental concerns lead to policies that are likely to harm the environment is excessive regulation of agricultural biotechnology, mostly in Europe but also in the United States. There is evidence that agricultural biotechnology improves yields and leads to significant reductions in greenhouse gas emissions from

7 https://www.usda.gov/media/press-releases/2022/08/22/usda-invest-300-million-new-organic-transition-initiative.

agriculture (Barrows et al. 2014; Klümper and Qaim 2014). Furthermore, agricultural biotechnology, by accelerating crop breeding cycles and identifying new traits in plants (including nitrogen fixation and acceleration of photosynthesis), has the potential to accelerate decarbonization significantly. Policies that over-regulate technologies decrease investment in technology and learning by doing and thus can harm the environment.

Alix-Garcia et al. (2015) econometrically assessed an important PES program (Mexican Payments for Hydrological Services Program) to reduce deforestation and soil erosion and improve rural well-being. They found that land cover loss was reduced by about 50%, and there was a small positive poverty alleviation. The environmental gains were higher in areas where income was higher. In a more recent article, Alix-Garcia et al. (2018) have shown that PES programs in Mexico that are based on outcomes of resources controlled by a community improve land management and enhance different measures of social capital.

5 Conclusion and future trends

The twentieth century has seen the emergence and proliferation of AES aiming to control the excess of agricultural production and its negative externalities to provide enhanced environmental amenities while sustaining agricultural productivity. While conventional policy interventions exist, much attention has been directed toward PES agreements, which incentivize agricultural producers to divert land for conservation purposes, utilize the land in a more environmentally friendly manner, and/or adopt environmental practices using conditional subsidies. PES programs such as the Conservation Reserve Program, the Environmental Quality Incentives Program, and the Wetland Reserve Programs have successfully installed conservation practices. AES programs vary among different North American countries; the United States has emphasized land diversion programs, Canada has emphasized working land programs, and Mexico has highlighted forest protection programs.

PES and other environmental programs in agriculture are works in progress. PES programs were introduced before the new millennium, and policymakers, farmers, and the general public are learning how to utilize them effectively. The study and design of PES agreements and other AES programs require the consideration of dynamic factors in order to ensure efficiency and effectiveness in the future. Technological advancements in information and measurement technologies, including GPS, new remote sensing tools, and artificial intelligence, have led to consequential improvements in measuring environmental benefits as well as migration behaviors that shift the rural-urban fringe. These advancements have resulted in shifts in the linkages between agricultural production, environmental objectives, and the impact of alternative policies on people's behavior. Environmental objectives are evolving; thus, agri-environmental programs must be reevaluated to consider the difference between actual and predicted outcomes from *ex ante* conceptual formulations. We realize that we do not have a sufficient analysis of the past performance of PES programs. Since future programs must build on lessons from past ones, it is important to expand the research quantifying the impact of PES programs in terms of environmental, economic, and social outcomes.

With growing concerns about climate change, PES may play a much more important role in introducing activities that mitigate emissions, sequester carbon, or lead to adaptation or resilience to climate change. Some PES activities have already been established as part of international agreements to address climate change; for example, UN-REDD is a PES program resulting from the Kyoto Protocol. But it represents a start and we need to identify the potential of using PES to help make agriculture a part of the solution to climate change rather than a problem.

PES can be a major tool used to introduce new practices or crops capable of reducing greenhouse gas emissions in agriculture, sequestering carbon by agricultural activities, reducing emissions from livestock, and transitioning to activities that will be resilient to climate shocks. Some of these directions are already identified (Rausser and Zilberman 2022). This will require ongoing research to identify new technical solutions that can meet the challenge of climate change and mechanisms to assess compliance.

However, it is not sufficient to identify new potential PES activities. It is as challenging to identify sources of demand for these activities. The demand for PES activities originated with NGOs, environmental groups, regional, and federal governments. The NGOs pursue objectives that are important to their supporters. As long as we expect private individuals to have environmental objectives that cannot be paid fully by the government, we expect them to initiate PES activities. As income increases, we expect this type of support to continue and even increase. Local governments are developing PES that can improve quality of life within their jurisdiction; for example, by reducing disaster risk and improving water quality, and greater concern for these issues may increase demand for PES, especially as people become more concerned about their use. We expect more federal government involvement in climate-related PES activities over time. However, the design of policy intervention is a political-economic exercise, and we need to pursue further political-economic research to understand the circumstances leading to using PES in agriculture.

Preliminary efforts in this respect can be found in Jack et al. (2008), who showed that political feasibility depends on the distribution of political power in the setting of PES policies; Muradian et al. (2010), who showed that political economic relationships determine the ultimate setting in which PES operate; Farrell (2018), who showed the mechanisms by which PES schemes can impinge upon the political and economic autonomy of various Indigenous peoples in tropical countries; and Baylis et al. (2008), who demonstrated that the political economic setting and inherent setting differ between the EU and the United States, the former treating rural landscapes as a positive externality while the latter emphasize the associated negative externalities.

Nevertheless, environmental policymakers, as well as environmental groups or others interested in improved environmental objectives, need to be educated about the value of PES, how to design them effectively, and how to ensure their effectiveness. The increased awareness and literature about the use of PES, and the development of technologies that enable better monitoring of their effectiveness, suggest to us that they will play a major role in environmental policy design in agriculture in the future.

6 Where to look for further information

- Bulte, E. H., Lipper, L., Stringer, R. and Zilberman, D. Payments for ecosystem services and poverty reduction: concepts, issues, and empirical perspectives. *Environment and Development Economics* 13(3) (2008) 245–254.
- Engel, S., Pagiola, S. and Wunder, S. Designing payments for environmental services in theory and practice: An overview of the issues. *Ecological Economics* 65(4) (2008): 663–674.
- Salzman, J., Bennett, G., Carroll, N., Goldstein, A. and Jenkins, M. The global status and trends of payments for ecosystem services. *Nature Sustainability* 1(3) (2018): 136–144.
- Wunder, S. When payments for environmental services will work for conservation. *Conservation Letters* 6(4) (2013): 230–237.
- Wunder, S., Börner, J., Ezzine-de-Blas, D., Feder, S. and Pagiola, S. Payments for environmental services: Past performance and pending potentials. *Annual Review of Resource Economics* 12 (2020): 209–234.

7 References

Alix-Garcia, J. M., Sims, K. R. E., Orozco-Olvera, V. H., Costica, L. E., Medina, J. D. F. and Romo Monroy, S. Payments for environmental services supported social capital while increasing land management. *Proceedings of the National Academy of Sciences of the United States of America* 115(27) (2018): 7016–7021.

Alix-Garcia, J. M., Sims, K. R. E. and Yañez-Pagans, P. Only one tree from each seed? Environmental effectiveness and poverty alleviation in Mexico's payments for ecosystem services program. *American Economic Journal: Economic Policy* 7(4) (2015): 1–40.

Atwood, J. D., Knight, L., Cattaneo, A. and Smith, P. F. Benefit cost analysis of the 2002 EQIP farm bill provisions. No. 376-2016-20566 (2003).

Barrows, G., Sexton, S. and Zilberman, D. Agricultural biotechnology: the promise and prospects of genetically modified crops. *Journal of Economic Perspectives* 28(1) (2014): 99–120.

Baylis, K., Coppess, J., Gramig, B. M. and Sachdeva, P. Agri-environmental programs in the United States and Canada. *Review of Environmental Economics and Policy* 16(1) (2022): 83–104.

Baylis, K., Lichtenberg, E. M. and Lichtenberg, E. Economics of pollination. *Annual Review of Resource Economics* 13(1) (2021): 335–354.

Baylis, K., Peplow, S., Rausser, G. and Simon, L. Agri-environmental policies in the EU and United States: a comparison. *Ecological Economics* 65(4) (2008): 753–764.

Cattaneo, A. The pursuit of efficiency and its unintended consequences: contract withdrawals in the environmental quality incentives program. *Review of Agricultural Economics* 25(2) (2003): 449–469.

Claassen, R., Cooper, J., Salvioni, C. and Veronesi, M. Agri-environmental policies: a comparison of US and EU experiences. In: Duke, J. M. and Wu, J. *The Oxford Handbook of Land Economics* (2014).

Cochran, W. W. *The Development of American Agriculture*, pp. 183–184 (1993). University of Minnesota Press.

Copeland, C. *Water Quality: Implementing the Clean Water Act*. Congressional research service report, The Library of Congress.

Duke, J. M., Dundas, S. J., Johnston, R. J. and Messer, K. D. Prioritizing payment for environmental services: using nonmarket benefits and costs for optimal selection. *Ecological Economics* 105 (2014): 319–329.

Engel, S., Pagiola, S. and Wunder, S. Designing payments for environmental services in theory and practice: an overview of the issues. *Ecological Economics* 65(4) (2008): 663–674.

Farrell, K. N. Intellectual mercantilism and franchise equity: a critical study of the ecological political economy of international payments for ecosystem services. *Ecological Economics* 152 (2018): 378–387.

Hochman, G. and Zilberman, D. Corn ethanol and US biofuel policy ten years later: a systematic review and meta-analysis. No. 333-2016-14652 (2016).

Jack, B. K., Kousky, C. and Sims, K. R. Designing payments for ecosystem services: lessons from previous experience with incentive-based mechanisms. *Proceedings of the National Academy of Sciences of the United States of America* 105(28) (2008): 9465–9470.

Just, R. E., Alston, J. M. and Zilberman, D. (Eds). *Regulating Agricultural Biotechnology: Economics and Policy* (Vol. 30). New York, USA: Springer Science & Business Media (2006).

Keiser, D. A. and Shapiro, J. S. Consequences of the Clean Water Act and the demand for water quality. *The Quarterly Journal of Economics* 134(1) (2019): 349–396.

Khanna, M. Non-mandatory approaches to environmental protection. *Journal of Economic Surveys* 15(3) (2001): 291–324.

Khanna, M., Rajagopal, D. and Zilberman, D. Lessons learned from US experience with biofuels: comparing the hype with the evidence. *Review of Environmental Economics and Policy* 15(1) (2021): 67–86.

Klümper, W. and Qaim, M.. A meta-analysis of the impacts of genetically modified crops. *PLoS ONE* 9(11) (2014): e111629.

Kovacs, K., Polasky, S., Nelson, E., Keeler, B. L., Pennington, D., Plantinga, A. J. and Taff, S. J. Evaluating the return in ecosystem services from investment in public land acquisitions. *PLoS ONE* 8(6) (2013): e62202.

Laffont, J.-J. *Fundamentals of Public Economics*. Cambridge, Massachusetts: MIT Press Books, p. 1 (1988).

Lewis, D. J., Plantinga, A. J., Nelson, E. and Polasky, S. The efficiency of voluntary incentive policies for preventing biodiversity loss. *Resource and Energy Economics* 33(1) (2011): 192–211.

McAfee, K. and Shapiro, E. N. Payments for ecosystem services in Mexico: nature, neoliberalism, social movements, and the state. *Annals of the Association of American Geographers* 100(3) (2010): 579–599.

Meemken, E.-M. and Qaim, M. Organic agriculture, food security, and the environment. *Annual Review of Resource Economics* 10(1) (2018): 39–63.

Miao, R., Feng, H., Hennessy, D. A. and Du, X. Assessing cost-effectiveness of the Conservation Reserve Program (CRP) and interactions between the CRP and crop insurance. *Land Economics* 92(4) (2016): 593–617.

Muradian, R., Corbera, E., Pascual, U., Kosoy, N. and May, P. H. Reconciling theory and practice: an alternative conceptual framework for understanding payments for environmental services. *Ecological Economics* 69(6) (2010): 1202–1208.

Naeem, S., Ingram, J. C., Varga, A., Agardy, T., Barten, P., Bennett, G., Bloomgarden, E., Bremer, L. L., Burkill, P., Cattau, M., Ching, C., Colby, M., Cook, D. C., Costanza, R., DeClerck, F., Freund, C., Gartner, T., Goldman-Benner, R., Gunderson, J., Jarrett, D., Kinzig, A. P., Kiss, A., Koontz, A., Kumar, P., Lasky, J. R., Masozera, M., Meyers, D., Milano, F., Naughton-Treves, L., Nichols, E., Olander, L., Olmsted, P., Perge, E., Perrings, C., Polasky, S., Potent, J., Prager, C., Quétier, F., Redford, K., Saterson, K., Thoumi, G., Vargas, M. T., Vickerman, S., Weisser, W., Wilkie, D. and Wunder, S. Get the science right when paying for nature's services. *Science* 347(6227) (2015): 1206–1207.

Newburn, D. A., Berck, P. and Merenlender, A. M. Habitat and open space at risk of land-use conversion: targeting strategies for land conservation. *American Journal of Agricultural Economics* 88(1) (2006): 28–42.

Nolte, C. High-resolution land value maps reveal underestimation of conservation costs in the United States. *Proceedings of the National Academy of Sciences of the United States of America* 117(47) (2020): 29577–29583.

Nordhaus, W. D. *Spirit of Green: The Economics of Collisions and Contagions in a Crowded World*. Princeton, New Jersey: Princeton University Press (2021).

Parker, D. P. and Thurman, W. N. Private Land conservation and public policy: land trusts, land owners, and conservation easements. *Annual Review of Resource Economics* 11(1) (2019): 337–354.

Porto, R. G., de Almeida, R. F., Cruz-Neto, O., Tabarelli, M., Viana, B. F., Peres, C. A. and Lopes, A. V. Pollination ecosystem services: a comprehensive review of economic values, research funding and policy actions. *Food Security* 12(6) (2020): 1425–1442.

Rausser, G., Sexton, S. and Zilberman, D. The economics of the naturalist food paradigm. *Annual Review of Resource Economics* 11(1) (2019): 217–236.

Rausser, G. and Small, A. Valuing research leads: bioprospecting and the conservation of genetic resources. *Journal of Political Economy* 108(1): (February 2000): 173–206. Reprinted in The economics of biodiversity conservation. In: Polasky S. (Ed.), International Library of Environmental Economics and Policy. Aldershot, UK: Ashgate, 2002.

Rausser, G. and Zilberman, D. Resource economics and modern science to the rescue. *Annual Review of Resource Economics* 14(1) (2022): v–xvi.

Salzman, J., Bennett, G., Carroll, N., Goldstein, A. and Jenkins, M. The global status and trends of payments for ecosystem services. *Nature Sustainability* 1(3) (2018): 136–144. https://doi.org/10.1038/s41893-018-0033-0.

Scullion, J., Thomas, C. W., Vogt, K. A., Perez-Maqueo, O. and Logsdon, M. G. Evaluating the environmental impact of payments for ecosystem services in Coatepec (Mexico) using remote sensing and on-site interviews. *Environmental Conservation* 38(4) (2011): 426–434.

Shortle, J. Economics and environmental markets: lessons from water-quality trading. *Agricultural and Resource Economics Review* 42(1) (2013): 57–74.

Sorda, G., Banse, M. and Kemfert, C. An overview of biofuel policies across the world. *Energy Policy* 38(11) (2010): 6977–6988.

Stubbs, M. *Environmental Quality Incentives Program (EQIP): Status and Issues*. Washington, DC: Congressional Research Service (2010).

Stubbs, M. *Conservation Reserve Program (CRP): Status and Issues*. Washington, DC: Library of Congress, Congressional Research Service (2014).

Taylor, C. A. and Druckenmiller, H. Wetlands, flooding, and the clean water act. *American Economic Review* 112(4) (2022): 1334–1363.

Taylor, R. and Zilberman, D. Diffusion of drip irrigation: the case of California. *Applied Economic Perspectives and Policy* 39(1) (2017): 16–40.

USDA. *Wetland Reserve Program, Final Programmatic Environmental Assessment*. Natural Resources Conservation Services. Available at: https://www.nrcs.usda.gov/sites/default/files/2022-09/WRP-Programmatic-EA-January-2009.pdf (January 2009).

USDA. The conservation reserve program, a 35-year history. Available at: https://www.fsa.usda.gov/Assets/USDA-FSA-Public/usdafiles/Conservation/PDF/35_YEARS_CRP_B.pdf(2020).

Wallander, S. and Hand, M. S. Measuring the impact of the Environmental Quality Incentives Program (EQIP) on irrigation efficiency and water conservation. No. 321-2016-10734 (2011).

Weersink, A., Livernois, J., Shogren, J. F. and Shortle, J. S. Economic instruments and environmental policy in agriculture. *Canadian Public Policy/Analyse de Politiques* 24(3) (1998): 309–327.

Wu, J. Slippage effects of the conservation reserve program. *American Journal of Agricultural Economics* 82(4) (2000): 979–992.

Wu, J. and Boggess, W. G. The optimal allocation of conservation funds. *Journal of Environmental Economics and Management* 38(3) (1999): 302–321.

Wu, J., JunJie, D. Z. and Babcock, B. A. Environmental and distributional impacts of conservation targeting strategies. *Journal of Environmental Economics and Management* 41(3) (2001): 333–350.

Zilberman, D., Lipper, L. and McCarthy, N. When could payments for environmental services benefit the poor? *Environment and Development Economics* 13(3) (2008): 255–278.

Zilberman, D., Schmitz, A., Casterline, G., Lichtenberg, E. and Siebert, J. B. The economics of pesticide use and regulation. *Science* 253(5019) (1991): 518–522.

Zilberman, D. and Segerson, K. Top ten design elements to achieve more efficient conservation programs. No. 156623. Council on Food, Agricultural, and Resource Economics (C-FARE) (2012).

Chapter 10

Developments in agri-environment schemes (AES): Australia

Dean Ansell, Andrew Macintosh, Don Butler and Marie Waschka, Australian National University, Australia

1 Introduction

Australia is facing a biodiversity crisis, with marked declines in the extent and condition of many native ecosystems, increases in the number of listed threatened species, and forecasts that significant numbers of native species could go extinct in the coming decades, beyond the more than 100 endemic species that have gone extinct since colonisation (Woinarski et al., 2019; Murphy and van Leeuwen, 2021). The crisis in outcomes is a product of a failure of federal, state and local biodiversity policy. The nation's primary environmental protection legislation, the *Environment Protection and Biodiversity Conservation Act 1999* (Cth) (EPBC Act), has been found to be environmentally ineffective, or have considerable scope for improvement, in numerous government and independent reviews since the late 2000s (Australian National Audit Office, 2003, 2014, 2020; Macintosh, 2004, 2009; Macintosh and Wilkinson, 2005; Hawke, 2009; Milder et al., 2015; Ward et al., 2019; Samuel, 2020). Similar findings have been made on the operation of state regulatory frameworks (Fowler, 1986; Bartel, 2003; New South Wales Auditor General, 2006; Macintosh, 2012; Simmons et al., 2018; Evans et al., 2019; Audit Office of New South

http://dx.doi.org/10.19103/AS.2024.0134.15

Wales, 2022). At the same time, public investment in biodiversity conservation is widely considered to be inadequate (Samuel, 2020; Ritchie, 2022).

Agriculture, which covers more than half of Australia's 7.67 million km^2 land mass (Australian Bureau of Statistics, 2018), represents both a major cause of this biodiversity decline and a key part of the solution. Agricultural activity affects more than half of the country's approximately 1533 threatened species (Kearney et al., 2019). Privately owned (e.g. freehold) land disproportionately overlaps the distributions of these threatened species (Kearney et al., 2022). Effective conservation in Australia therefore requires significant investment in conservation management on agricultural land to address threats to biodiversity.

Agri-environment schemes have existed in various forms in Australia for the past two to three decades.[1] Hajkowicz (2009) and Salt (2016) provide comprehensive syntheses of the history and emergence of agri-environment schemes in Australia. In essence, it was only in the 1980s that Australian governments (state and federal) began to prioritise and invest in outcomes other than production in agricultural land. Prior to this, land clearing for the purpose of expansion was heavily incentivised. Growing recognition of the extent of environmental degradation, including impacts on water quality and soil health, led to a shift towards policies that encouraged sustainable production. This primarily took the form of research and extension activities as well as the formation of networks of community-based groups advocating for sustainable land use practices (e.g. Landcare). Funding for on-ground activities was primarily small, short-term grants, undertaken voluntarily by participating landholders. Though this focus on education and extension, coupled with small grants widely distributed through community-based groups, was arguably effective in shifting thinking around the impacts of agriculture and more sustainable ways of production, the level of expenditure was considered inadequate to address the scale of environmental degradation, including in a global context (Hajkowicz, 2009). This led to the establishment in 1996 of the National Heritage Trust (NHT), a 5-year $1.3 billion initiative involving the disbursement of funds by the federal government to regional natural resource management bodies for on-ground activities, planning and capacity-building. It was in the later stages of the NHT that governments began to use market-based instruments as a means of more directly financing improved land management practices on agricultural land (Hajkowicz, 2009). Increasingly, these programmes were at least partly targeted at retaining or enhancing native vegetation in order to address biodiversity declines. Since their introduction, market-based schemes have increasingly challenged the place of grants

1 It is worth noting that the term 'agri-environment scheme' is not widely used in Australia, including within the academic community. Terms such as 'natural resource management' programmes tend to be favoured. We use the term agri-environment schemes in this chapter for consistency with the theme of this book.

and community-based approaches as the preferred tool of governments for promoting natural resource management outcomes on farms, though not all agree with the change in focus (e.g. Royal, 2021).

In contrast to the broad-ranging and comparatively generic agri-environment schemes employed in Europe under the common agricultural policy framework, and in the USA under the Conservation Reserve Program, schemes in Australia have tended to be smaller in geographic scope and more targeted, both in terms of the environmental objective and in the eligible activities they promote (Pannell and Rogers, 2022). Early examples of Australian market-based schemes include the Victorian BushTender programme, a voluntary competitive scheme that ran between 2001 and 2003 that aimed to protect and enhance native vegetation on agricultural land. The programme used a reverse auction approach where landholders nominated a price for undertaking prescribed management actions designed to generate biodiversity benefits in areas of remnant native vegetation. Projects were selected by the state government (the funding body) based on value for money, where the conservation benefit of the project was estimated on the basis of the current ecological condition of the site and the predicted biodiversity gains generated through the chosen management actions (Stoneham et al., 2003; Department of Sustainability and Environment, 2006). Participating landholders were offered 3- or 6-year contracts, with 10-year extension options as well as the opportunity to establish in-perpetuity conservation covenants over their project site, receiving periodic payments for management over the term of the contract.

The BushTender programme introduced a number of key features that were to heavily influence the design of many similar schemes to follow, namely:

- the use of conservation tenders, which would come to represent the preferred delivery mechanism for the procurement of conservation outcomes on private land;
- the assessment of a project's conservation benefit using a combination of the value of the conservation assets encompassed by the project (e.g. endangered species or ecological communities) and the predicted change or gain in ecological condition through management; and
- the selection of projects for funding based on their cost-effectiveness.

The subsequent Environmental Stewardship scheme, established in 2007 by the federal government, was the first large scale agri-environment scheme that used a market-based approach to biodiversity conservation in agricultural landscapes, and it shared many features of the BushTender programme. The scheme provided financial incentives for the restoration and enhancement of remnant native vegetation on farmland, using a reverse auction mechanism that resulted in 15-year contracts for successful landholders (Burns et al., 2016). It targeted threatened ecological communities listed under federal environmental protection legislation through a series of tenders across the sheep wheat belt in New South Wales, Queensland, and South Australia. The sheep wheat belt, which contains some of Australia's most productive agricultural land, has been extensively cleared since European settlement in the late eighteenth century, and as a result contains a large number of endangered species and communities. Over the life of the scheme, 297 landholders were contracted to manage 53 458 ha of land for conservation purposes, for approximately AU$152 million (Burns et al., 2016). The programme marked a progression in the evolution of agri-environment schemes in Australia by incorporating a long-term ecological monitoring programme in its design and budget, a feature often omitted from such programmes and expenditure on environmental programmes in general (Lindenmayer et al., 2012).

2 Agri-environment schemes and emerging environmental markets in Australia

There are four notable trends in agri-environment schemes and environmental markets in Australia:

1 The rise and dominance of Australia's carbon offset market, and how the market has shaped interest in other agri-environment schemes;
2 Related to the first trend, a willingness to trial blended approaches, where incentives are provided for activities that generate both climate and biodiversity benefits;
3 An increasing interest and eagerness from federal and state governments to utilise market mechanisms as a way of incentivising changes in land management practices; and
4 Moves by governments to try to shift the burden of funding environment and conservation initiatives onto the private sector.

In terms of scope and volume of trade, the environmental market landscape in Australia is currently dominated by carbon offsets. The primary carbon offset market is the Australian Government's Australian Carbon Credit Unit (ACCU) scheme (previously known as the 'Emissions Reduction Fund'), a statutory-based offset scheme administered by the federal government that provides for the registration of carbon offset projects. Under the ACCU scheme, landholders, and other eligible entities, can register projects that avoid emissions or sequester carbon dioxide (CO_2) in trees, soils, or geological formations. Proponents who undertake offset projects in accordance with a set of rules, known as methods, are granted ACCUs, a form of tradeable credit that is supposed to represent one tonne of real and additional abatement of greenhouse gas emissions from registered offset projects.

The scheme has been subject to intense public scrutiny, including by the authors of this chapter, due to concerns about the integrity of the ACCU scheme and its governance (Butler et al., 2022). However, despite these issues, the market has flourished. There are now approximately 1500 projects registered under the ACCU scheme and, since its introduction in 2011, these projects have received over 130 million ACCUs. Almost 60% of the ACCUs issued to date have gone to three types of land-based projects: human-induced regeneration of even-aged native forests, avoided deforestation in western New South Wales, and early dry season burning to reduce late dry season wildfires in Australia's northern savannas (Clean Energy Regulator, 2023).

The dominance of the environmental markets landscape by the ACCUs scheme has had a profound influence on the direction and development of agri-environment schemes at the federal and state levels. Most notably, it has resulted in attempts to leverage the carbon market to drive biodiversity and other environmental outcomes. This is evidenced by the rise in agri-environment schemes that integrate carbon, biodiversity and other incentives. Such schemes have been introduced in a number of Australia's state and territories. For example, the Queensland government under its Land Restoration Fund, buys carbon credits – at a premium to typical carbon market prices – from farmers undertaking carbon farming projects that generate First Nations, environmental and/or socio-economic co-benefits (Queensland Government, 2022). A similar programme, the Carbon Farming and Land Restoration Program, was established in 2022 in Western Australia (Department of Primary Industries and Regional Development, 2023). Similarly, the New South Wales government agency responsible for private land conservation recently introduced carbon sequestration objectives into the design of its conservation tenders, allowing farmers to generate carbon credits through revegetation, and collect additional payments for biodiversity activities (Biodiversity Conservation Trust, 2022).

At the national scale, the federal government's Agriculture Biodiversity Stewardship Package, established in 2021, included a number of mechanisms aimed at trialling new market arrangements to enhance biodiversity and provide carbon abatement outcomes in agricultural landscapes while providing farmers with diversified income and increasing private investment. In the following section, we provide a case study of one of these mechanisms, to demonstrate the recent shift in agri-environment scheme design in Australia, and to demonstrate the risks and opportunities they present, from the perspectives of the farmer and other market participants, as well as from the government as market makers seeking to achieve environmental outcomes.

Partly due to the financial (if not environmental) success of the carbon market, over the past decade, federal and state governments have shown an increasing willingness to utilise market mechanisms as a way of incentivising changes in land management practices. Biodiversity offset markets have existed in some form for roughly 20 years in Australia. These markets have been predominantly compliance based, for example the biodiversity offset markets employed in state and territories to mitigate the impacts of development on biodiversity value. These schemes place an obligation on the proponent of the development to offset the biodiversity impacts of their activities. In New South Wales and Victoria, the biodiversity offset schemes are unitised – meaning proponents of offset projects receive tradeable biodiversity offset credits that can be sold to developers, who then use them to meet their statutory obligations. In other jurisdictions, the schemes work via approval conditions imposed on developers and the developers then directly sourcing the provision of relevant ecosystem services from landholders, or buying relevant land and providing the services themselves.

In recent times, federal and state governments have sought to go beyond biodiversity offset schemes by creating voluntary nature markets. The most prominent of these is the New South Wales Conservation Management Program, which is a private protected areas programme run by the NSW Biodiversity Conservation Trust. Under the programme, landholders can enter into a contract with the state government that provides in-perpetuity annual payments for placing their land under a permanent conservation covenant and implementing an agreed biodiversity management plan (State of New South Wales Biodiversity Conservation Trust, 2023).

The Australian Government has recently announced it will create a Nature Repair Market, a voluntary market for the trade of biodiversity certificates generated through the undertaking of projects by landholders consistent with specified methods. While compliance-based or compensatory markets aimed at mitigating or offsetting impacts of development on biodiversity exist in some form in many countries (Madsen et al., 2010), the Nature Repair Market is the world's first statutory unitised biodiversity market in which a significant proportion of demand is envisaged to come from voluntary buyers. At the time of writing, the legislation for the market was still before Parliament and it was scheduled to commence in late 2024 (DEECCW, 2023). The announcement of the Nature Repair Market has received a mixed reception, with concerns raised about its likely integrity and contribution to conservation outcomes (e.g. Chee, 2023). In addition, questions have been raised about the source of demand for the proposed project certificates. The government is claiming that voluntary private buyers will provide the primary source of demand. However, this seems unlikely, at least in the short to medium term. There is limited private sector funding currently available for nature-related initiatives, and nature-related commitments within the private sector remain low (Claes et al., 2022). Despite government statements to the contrary, the only foreseeable voluntary private sector demand for the certificates is where they are attached to carbon credits (ACCUs). Some ACCU buyers are likely to be willing to pay a premium for ACCUs that come from projects that provide certified biodiversity benefits. However, aside from the carbon market, the main source of demand for Nature Repair Market certificates in the near-term is likely to be compliance buyers and the Australian Government, via proposed changes to biodiversity offsets and the EPBC Act.

These emerging markets present both significant opportunities and risks to both market participants, and the environmental outcomes that they seek to achieve. The increasing number of schemes offer farmers with a more diverse choice of programmes to suit their individual needs and business planning, including different project lengths, from short-term grants to long-term and even in-perpetuity agreements registered on land title with associated payments.

The final notable trend in the environmental market landscape is the attempts by governments to shift the burden of funding environment and conservation initiatives onto the private sector. This appears to be a response, at least partly, to the growing recognition of the significant shortfall in public expenditure on effective biodiversity conservation in Australia (Wintle et al., 2019). The other driver appears to be the enthusiasm for markets and financialisation. Examples of this trend include the Nature Repair Market and the attempts at the federal and state levels to leverage the carbon market, and its private funding, to achieve biodiversity outcomes. The trend mirrors that seen in other countries as governments seek to ease pressure on public budgets and push the cost of conservation onto the private sector (Kedward et al., 2022).

3 Case study: the Australian Government's Carbon + Biodiversity Pilot

The Carbon + Biodiversity Pilot ('the Pilot'), a core component of the Australian Government's AUD$66 million Agriculture Biodiversity Stewardship Package, trialled a mechanism to provide biodiversity co-payments to carbon projects that generated both carbon and biodiversity benefits. The ultimate aim was to improve climate and biodiversity outcomes and help facilitate landholder participation in environmental markets. The scheme was co-designed by authors of this chapter for the Australian Government.

The pilot targeted intensive agriculture regions, which corresponds to the areas where terrestrial biodiversity is typically at greatest risk (Bennett et al., 2022). Two pilot rounds were run across 12 natural resource management (NRM) regions, the primary zoning system used by the federal government for natural resources policies and funding. Seventy-two environmental planting projects were contracted through the scheme.

To participate in the scheme, landholders had to be willing to undertake an environmental planting project of between 5 and 200 ha and register it under the ACCU scheme. In addition to meeting the ACCU scheme's eligibility and participation requirements, the plantings had to be established and managed in accordance with prescribed planting protocols that were created for the Pilot.

The planting protocols were tailored to each participating NRM region and were designed to ensure the projects generated benefits for biodiversity, with specific eligibility requirements regarding land use history, project dimensions, planting composition, establishment methods, and ongoing maintenance (e.g. weed and pest control). Applicants were given the choice of two types of plantings: the first aimed at restoring the composition and structure of the pre-clearing vegetation community (local vegetation community plantings); and the second aimed at establishing similar function but with less stringent requirements for species composition (simple mixed native planting). To aid comprehension and facilitate uptake, the planting protocols were designed in a user-friendly table format that presented the requirements of the ACCU scheme and Pilot side-by-side, and grouped the requirements thematically under four headings: land and project eligibility (e.g. based on land use history, consents and legal approvals), planting design (e.g. concerning species composition and structure, dimensions and stem density), establishment (e.g. soil preparation, use of herbicides, and tree protection), and maintenance (e.g. longevity, remedial plantings, grazing pressure, and prescribed burning).

Successful applicants were provided with upfront biodiversity payments under the Pilot over the first 2–3 years of the project. In addition, because the projects had to be registered under the ACCU scheme, participants will also be eligible to receive carbon credits for the carbon sequestered in the plantings over at least 25 years, which they can sell in the ACCU market.

To lower transaction costs and facilitate uptake, an online platform was developed that allowed applicants to map their desired planting areas within their property and complete and submit their application details (DCCEEW, 2023). Now known as the National Stewardship Trading Platform (www.agsteward.com.au), the web-based tool allowed landholders to generate instant estimates of the amount of carbon that is likely to be sequestered via environmental plantings in the planting areas over 25 years (the minimum 'permanence' period under the ACCU scheme) under average climate conditions. The tool also allowed applicants to test alternative project scenarios and optimise their application, while streamlining the paperwork involved in applications.

Three noteworthy innovations were developed as part of the Pilots:

- A biodiversity benefit scoring system was developed to rank and select projects. The system used an ecosystem condition scale (0–-100) to estimate likely change from proposed plantings based on the project area's current condition and the proponent's choice of planting type (local vegetation community plantings or simple mixed native planting). The magnitude of likely condition change was estimated for end of the 25-year project life, and included consideration of planting shape and connectivity to adjacent habitats. Benefits for biodiversity from projects were the weighted sum of changes in separate indices for regional biodiversity persistence based on three different asset classes: the region's native ecosystems (80% weighting), potential habitat for selected threatened birds (15% weighting), and the condition of vegetation in watercourse buffers across the region's catchments (5% weighting). Finally, modifiers were used to promote projects based on inclusion of remnant mature native trees within the planting footprint, and to discount benefits if proponents opted for 25-year permanence rather than 100 years on their carbon project. The Pilots applied a consistent conceptual approach across the 12 NRM regions, demonstrating how consistent approaches can be delivered despite variation in data available in each state.

- A hybrid pricing system was developed to determine the quantum of the biodiversity payment offers made to successful applicants. It is often argued that conservation tenders are more cost-effective than alternative approaches to pricing (Rolfe et al., 2017). However, the cost-effectiveness of tenders depends on how they are designed and the context in which they are used. Relevant factors that can undermine the cost-effectiveness of tenders, either by deterring participation or promoting overbidding, include small numbers of participants, running tenders in areas where landholders are well connected (i.e. creates opportunities for collusion and strategic bidding), running tenders in regions where landholders have too little or too much experience with tender processes, and using a discriminatory auction design (i.e. where successful landholders receive their bid price) (Rolfe et al., 2017; Lundberg et al., 2018). With the Pilots, there was a risk that using a standard conservation tender could reduce participation, due to landholders' unfamiliarity with the process and the short lead times (the application windows were less than 3 months) (Zammit, 2013). A standard tender process also carried material risks of overpayment – due to the likely small pool of applicants and opportunities for strategic bidding.[2] To improve accessibility for applicants, and guard against overpaying, a hybrid approach was developed, whereby the biodiversity payment offers were based on the lower of the bid submitted by the applicant (if any) and a pricing offer determined using

2 Relatively low participation was anticipated because of the short lead times, complexity of the rules governing participation in the ACCU scheme, and the limitation of each round to six NRM regions.

a build block-style pricing model. The pricing model generated offers by determining the biodiversity payment that was likely to be necessary to provide the project with a prescribed (and undisclosed) internal rate of return, having regard to the project costs and the revenues likely to be generated through the sale of ACCUs. Proponents were required to submit costing estimates in designated cost categories as part of the application process (i.e. site preparation, soil preparation, seed and seedlings, planting, tree protection, freight, watering, and ongoing pest and weed control). The project costs were calculated as the lower of the proponent's submitted cost estimates or prescribed (undisclosed) cost caps that applied to each cost category. Pricing guides were developed for each NRM region to help with project planning and the application process. The system was designed to rapidly generate payment offers that provide a robust approximation of the opportunity cost of supply, having regard to the risks and uncertainties associated with participation.

- A rapid cost-effectiveness rating system was used to rank and select projects. The system combined the biodiversity benefit scoring and hybrid pricing systems described earlier. Combining these systems allowed for projects to be ranked based on a cost-effectiveness index (biodiversity payment offer per unit of projected biodiversity benefit). The system allowed for project evaluations and rankings to be completed within 4 weeks.

Contracted landholders are required to report annually on the implementation of their project plans, including level of effort and expenditure on agreed project activities such as weed and pest control. The contract with the government also requires participating farmers to undertake basic photo point monitoring. This reporting is undertaken using the National Stewardship Trading Platform. Compliance monitoring will involve review of landholder reports combined with regular site assessments undertaken by ecologists as part of the ecological monitoring programme. This ecological monitoring, which will focus on assessing the contribution of the projects to regional biodiversity condition, was in the process of being initiated at the time of writing. Due to this, it is not possible to make any assessment of the effectiveness of the programme with respect to the Pilot's biodiversity objectives. However, our experience in the design and implementation of the Pilot, and other similar schemes, yields some key transferrable lessons, particularly for those seeking to merge carbon and biodiversity outcomes amid emerging markets:

1 The complexity of these nascent environmental markets, particularly carbon markets, represents a significant barrier to participation for many farmers. Complexity has been identified as a factor impacting

participation in several agri-environmental schemes (Comerford, 2014; Smith et al., 2015; Lute et al., 2018). The ACCU scheme that dominates the Australian carbon market is necessarily complex, with exhaustive rules, regulations and requirements regarding eligibility, project design, establishment and maintenance, and auditing and reporting. These features are needed to ensure the integrity of the scheme (see Macintosh et al., 2022), but in turn this can deter potential participants who may already be time-poor and face scepticism from peers, business partners and family members (pers. obs.) over their involvement in a non-traditional agricultural venture. In our experience with the Carbon + Biodiversity Pilot, many applicants withdrew following acceptance of their application, despite receiving a lucrative contract offer from the government, as a result of the time and effort required to progress through the formal registration of the project under the ACCU scheme. Similarly, many farmers encountered critical issues with ownership of land parcels that precluded their participation at an advanced stage. Understanding the complexities of these schemes, and rectifying issues as they arise, will further drive up transaction costs for participants, already a significant barrier to participation in environmental markets (Coggan et al., 2013), and create an environment that is easily exploited by intermediaries and selects for less resource-constrained landholders. For these schemes to be successful and achieve their environmental objectives, significant resources must be dedicated to extension services, awareness campaigns and education resources to ensure farmers have quick access to accurate, reliable, and easily understood information.

2 These new schemes are characterised by a higher degree of risk and uncertainty than traditional programmes. While the latter were characterised by fewer management activity options, shorter term contracts with stable, regular payments and little if any market involvement, the new schemes introduce greater financial complexity and risk exposure. In the case of carbon markets, and with emerging biodiversity markets, there is uncertainty in the future revenue stream that must be weighed against the cost of participation. Such uncertainty is not unique to these markets but adds another layer of risk to agricultural businesses in addition to commodity prices, domestic and international agri-environmental policy settings, and climatic variability. This uncertainty, coupled with the relative infancy of carbon and nature markets, can translate to hesitancy of financial institutions – who often play a critical role in agri-environment schemes as the eligible interest holder in the land under mortgage and therefore necessary signatory to the types of agreements often employed – to provide the necessary

consent for a farmer to enter into the scheme. This uncertainty also interacts with contract length, a key design feature of this and all other agri-environment schemes. While in-perpetuity contracts may be desirable from an environmental effectiveness perspective, notwithstanding the inherent administrative and financial challenges of securing and managing a perpetual fund, many studies have shown contract length as a significant factor in a landholder's decision to participate in AES, with a typical preference for shorter term contracts (e.g. Blackmore and Doole, 2013; Sheremet et al., 2018; Cortés-Capano et al., 2021).

3 The integration of markets in agri-environment schemes intensifies the risks associated with natural disturbances, including wildfire, drought, and flooding. These events can have significant impacts on projects using the vegetation methods often used in schemes involving carbon and nature markets, such as revegetation and assisted regeneration. Depending on the scheme in question, and the rules of the market involved, the landholder often wears the majority of the risk of these events impacting on a project. In our experience with the design of schemes in Australia, these risks, or a lack of understanding of the participant's obligations and appropriate mitigation strategies, impact on the type and the number of landholders that enrol. As with the issue of scheme complexity, the design of schemes that integrate carbon and nature markets should incorporate sufficient resources to educate and advise landholders on the risks and uncertainties of participation. This is harder to achieve when the operation of the market is left primarily to the private sector as is increasingly reflected in the agri-environmental policy rhetoric in Australia.

4 Future trends in research

Though significant research has been undertaken into the factors that influence landholders' willingness to adopt agri-environment schemes (Yasué and Kirkpatrick, 2020), the introduction of markets into scheme design warrants more targeted research on barriers and motivations for participation in this evolving policy space. The outcomes of this research should inform not only scheme design, including the development of appropriate tools and materials to assist farmers and other market participants to make informed decisions, but also would likely improve the effectiveness and efficiency of the markets.

There is an ongoing need for rigorous long-term ecological and socio-economic monitoring of schemes. Despite significant public investment in large agri-environment schemes in Australia over the past two decades, there remains a relative paucity of published research into their effectiveness. This includes a continued lack of data on the costs of designing and implementation, at the scheme and individual project level. A lack of knowledge of the cost-effectiveness of alternative strategies for achieving environmental outcomes in agricultural landscapes will hamper the efficiency of markets going forward.

Further research is also needed into the benefits of agri-environment schemes for farm businesses. Impacts on farm production remains a significant barrier to participation in schemes in Australia and internationally (Lute et al., 2018; Simmons et al., 2020; Vos and Davies, 2021). This includes the range of attitudes and perceptions of sustainable land management and biodiversity management activities on farms, from the 'loss' of productive land to conservation, to the perceived impacts of native vegetation on production (e.g. reduced pasture and crop growth) to reduced property values associated with land under conservation. There remains a serious lack of fundamental research into both the veracity of these perceived negative impacts (positive or negative), as well as that of the positive socio-economic benefits often cited, including improved production associated with sustainable land management practices, and improved access to new domestic and international markets. Until this research is undertaken, prevailing attitudes will persist and continue to impact enrolment.

The final missing research component is on optimal governance structures to promote environmental integrity and efficiency. The environmental markets that have been created in Australia have tended to have two defining governance characteristics: a high level of centralised control by government agencies over scheme rules, compliance and enforcement, and information; and heavy reliance on proponents, and private contractors engaged and paid by proponents, for project reporting and the verification of environmental outcomes. Reflecting these characteristics, the schemes have often been designed to largely exclude financially disinterested third parties and limit external scrutiny. These structures can lead to governance problems and, in extreme cases, institutional corruption. Further research is warranted on the role governance structures play in scheme performance and how greater involvement of third parties can help improve outcomes and community confidence in environmental markets.

5 Conclusion

The recent trend in the design of agri-environment schemes presents more opportunities for Australia's agricultural sector. Integration with carbon markets, and the emerging nature or biodiversity markets, provide for further income diversification and offsetting uncertainty in domestic and international commodity markets and mitigate against climate-driven impacts on production. They also allow farm businesses to offset their own production impacts, for example greenhouse gas emissions and environmental impacts, and in doing so potentially access new markets for sustainable agricultural produce. Shifting the funding of these schemes onto the private sector potentially allows governments to reallocate scare resources to other environmental problems.

At the same time, however, the nascency of these markets presents significant barriers to participants on both sides of the transaction. Sellers (i.e. landholders) must navigate the complexities of the rules and regulations associated, incurring significant transaction costs in the process. Participation can also require significant upfront capital for project establishment, while financial returns may not materialise for some years, if at all. In addition, the risk of project failure is borne primarily by the landholder, and this can impact on the willingness of financial institutions to support their involvement. Ecological outcomes, particularly biodiversity response, are extremely difficult to accurately predict and are often influenced by extraneous factors. Buyers (i.e. investors) also face risks, including financial and reputational impacts where projects are found to fail or where the integrity of a project, or the scheme itself, is questioned.

Despite the government's enthusiasm for the use of carbon and nature markets, few believe their rapid expansion and development to be the panacea for addressing biodiversity decline in Australia's agricultural landscapes (e.g. Evans and Maron, 2022). There will always be a continued role for effective and enforced regulation in the management of environmental values on farms. Similarly, there is an important role for engaging, educating, and supporting farmers to improve their environmental sustainability without the need for financial incentive.

In our experience, there is a substantial number of landholders, from a diverse range of agricultural businesses and demographics, who are motivated and willing to protect and enhance the biodiversity on their land with little or no financial support. Indeed many undertake such activities routinely across their properties. Nurturing and expanding this section of the agricultural community should be an ongoing priority in any agri-environmental policy framework. Often what is needed is support in the form of information and advice and, where possible, financial assistance to help cover the cost of relevant materials. The existing natural resource management infrastructure in Australia, including regional NRM bodies and community-driven land management networks and groups, should play a vital role in delivering this support.

A key risk associated with an over-reliance on market-based approaches is that they can crowd out the altruistic initiatives of landholders. This can lead to cost-shifting and, in the worse cases, a net decline in the provision of biodiversity services by landholders as they reduce their altruistic efforts because of the prospect of being paid or resentment that others are being paid for what they have provided voluntarily. Policy makers should be mindful of this risk when designing and administering market-based schemes.

6 Where to look for further information

For a detailed review of the history and design of agri-environment schemes and agri-environmental policy in Australia, see Hajkowicz (2009) and Salt (2016). Detailed discussion on aspects of the design and evaluation of Australian schemes can be found in chapters in Ansell et al. (2016). Royal (2021) explores the socio-ecological implications of the shift in approaches to biodiversity conservation on private land toward market-based approaches. For further information on the development of the federal government's Nature Repair Market discussed earlier, and for additional detail on the Agriculture Biodiversity Stewardship Package and the pilot programme referred in the case study, visit the government's website at https://www.dcceew.gov.au/. The National Stewardship Trading Platform can be found at https://agsteward.com.au/.

7 References

Ansell, D., Gibson, F. and Salt, D. (2016). In: Ansell, D., Gibson, F. and Salt, D. (Eds) *Learning from Agri-Environment Schemes in Australia: Investing in Biodiversity and Other Ecosystem Services on Farms*. Canberra: ANU Press. doi: 10.22459/LFAESA.05.2016.

Audit Office of New South Wales (2022). *Effectiveness of the Biodiversity Offsets Scheme - NSW Auditor-General's Report*. Sydney. Available at: https://www.audit.nsw.gov.au/our-work/reports/effectiveness-of-the-biodiversity-offsets-scheme.

Australian Bureau of Statistics (2018). *Land Management and Farming in Australia*. Available at: https://www.abs.gov.au/statistics/industry/agriculture/land-management-and-farming-australia/latest-release (accessed: 10 March 2023).

Australian National Audit Office (2003). *Referrals, Assessments and Approvals under the Environment Protection and Biodiversity Conservation Act 1999*. Canberra, Commonwealth of Australia.

Australian National Audit Office (2014). *Monitoring Compliance with Environment Protection and Biodiversity Conservation Act: Conditions of Approval*. Canberra.

Australian National Audit Office (2020). *Referrals, Assessments and Approvals of Controlled Actions under the Environment Protection and Biodiversity Conservation Act 1999*. Canberra. Available at: http://www.anao.gov.au/uploads/documents/Audit%5CnReport%5Cn38-2003.pdf.

Bartel, R. (2003). Compliance and complicity: an assessment of the success of land clearance legislation in New South Wales. *Environmental and Planning Law Journal* 20(2), 116–141.

Bennett, A. F., Holland, G. J., Haslem, A., Stewart, A., Radford, J. Q. and Clarke, R. H.(2022). Restoration promotes recovery of woodland birds in agricultural environments: a comparison of "revegetation" and "remnant" landscapes. *Journal of Applied Ecology* 59(5), 1334–1346. doi: 10.1111/1365-2664.14148.

Biodiversity Conservation Trust (2022). *Restoring Murray Woodlands Biodiversity and Carbon Conservation Tender - Landholder Guide 2022–2023*. Sydney. Available at: https://www.bct.nsw.gov.au/sites/default/files/2023-01/restoring-murray-woodlands-biodiversity-and-carbon-conservation-tender-landholder-guide.pdf.

Blackmore, L. and Doole, G. J. (2013). Drivers of landholder participation in tender programs for Australian biodiversity conservation. *Environmental Science and Policy* 33, 143–153. doi: 10.1016/j.envsci.2013.05.010.

Burns, E., Burns, E., Zammit, C., Attwood, S. and Lindenmayer, D. (2016). The Environmental Stewardship Program: lessons on creating long-term agri-environment schemes. In: Ansell, D., Gibson, F. and Salt, D. (Eds) *Learning from agri-Environment Schemes in Australia: Investing in Biodiversity and Other Ecosystem Services on Farms*. Canberra: ANU Press, pp. 33–52.

Butler, D., Evans, M. and Macintosh, A. (2022). Australian National University (ANU) - University of New South Wales (UNSW) ERF research team submission to the Chubb review. Available at: https://consult.dcceew.gov.au/independent-review-of-accu/submission/view/150.

Chee, Y. E. (2023). Would a nature repair market really work? Evidence suggests it's highly unlikely. Available at: https://theconversation.com/would-a-nature-repair-market-really-work-evidence-suggests-its-highly-unlikely-199975 (accessed: 19 May 2023).

Claes, J., Erben, I., Hopman, D., Jayaram, K., Katz, J. and Van Aken, T. (2022). Where the world's largest companies stand on nature. Available at: https://www.mckinsey.com/capabilities/sustainability/our-insights/where-the-worlds-largest-companies-stand-on-nature.

Clean Energy Regulator (2023). Emissions Reduction Fund project register. Available at: https://www.cleanenergyregulator.gov.au/ERF/project-and-contracts-registers/project-register (accessed 18 July 2023).

Coggan, A., Buitelaar, E., Whitten, S. and Bennett, J. (2013). Factors that influence transaction costs in development offsets: who bears what and why? *Ecological Economics* 88, 222–231. doi: 10.1016/J.ECOLECON.2012.12.007.

Comerford, E. (2014). Understanding why landholders choose to participate or withdraw from conservation programs: a case study from a Queensland conservation auction. *Journal of Environmental Management* 141, 169–176. doi: 10.1016/j.jenvman.2014.01.049.

Cortés-Capano, G., Hanley, N., Sheremet, O., Hausmann, A., Toivonen, T., Garibotto-Carton, G., Soutullo, A. and Di Minin, E. (2021). Assessing landowners' preferences to inform voluntary private land conservation: the role of non-monetary incentives. *Land Use Policy* 109 (August 2020). doi: 10.1016/j.landusepol.2021.105626.

DCCEEW (2023). National stewardship trading platform. Available at: https://agsteward.com.au/.

DEECCW (2023). Nature repair market. Available at: https://www.dcceew.gov.au/environment/environmental-markets/biodiversity-market (accessed: 19 May 2023).

Department of Primary Industries and Regional Development (2023). Western Australian carbon farming and land restoration program. Available at: https://www.agric.wa.gov.au/carbon-farming/western-australian-carbon-farming-and-land-restoration-program (accessed: 17 May 2023).

Department of Sustainability and Environment (2006). *BushTender - The Landholder Perspective: A Report on Landholder Responses to the BushTender Trial*. East Melbourne.

Evans, M. C., Chiu, G., Gibbons, P. and Macintosh, A. (2019). Quantitative evaluation of regulatory policies for reducing deforestation using the bent-cable regression model. Available at: http://arxiv.org/abs/1906.09365.

Evans, M. C. and Maron, M. (2022). The government hopes private investors will help save nature. Here's how its scheme could fail. Available at: https://theconversation.com/the-government-hopes-private-investors-will-help-save-nature-heres-how-its-scheme-could-fail-193010 (accessed: 13 May 2023).

Fowler, R. (1986). Vegetation clearance controls in South Australia–a change of course. *Environmental and Planning Law Journal* 3(1), 48–66.

Hajkowicz, S. (2009). The evolution of Australia's natural resource management programs: towards improved targeting and evaluation of investments. *Land Use Policy* 26(2), 471–478. doi: 10.1016/j.landusepol.2008.06.004.

Hawke, A. (2009). *The Australian Environment Act. Report of the Independent Review of the Environment Protection and Biodiversity Conservation Act 1999: Final Report*. Canberra, Australian Government Department of the Environment, Water, Heritage and the Arts.

Kearney, S. G., Carwardine, J., Reside, A. E., Fisher, D. O., Maron, M., Doherty, T. S., Legge, S., Silcock, J., Woinarski, J. C. Z., Garnett, S. T., Wintle, B. A. and Watson, J. E. M. (2019). The threats to Australia's imperilled species and implications for a national conservation response. *Pacific Conservation Biology* 25(3), 328. doi: 10.1071/PC18024.

Kearney, S. G., Carwardine, J., Reside, A. E., Adams, V. M., Nelson, R., Coggan, A., Spindler, R. and Watson, J. E. M. (2022). Saving species beyond the protected area fence: threats must be managed across multiple land tenure types to secure Australia's endangered species. *Conservation Science and Practice* 4(3). doi: 10.1111/csp2.617.

Kedward, K., zu Ermgassen, S. O. S. E., Ryan-Collins, J. and Wunder, S. (2022). Nature as an asset class or public good? The economic case for increased public investment to achieve biodiversity targets. *SSRN Electronic Journal*. doi: 10.2139/ssrn.4306836.

Lindenmayer, D. B., Zammit, C., Attwood, S. J., Burns, E., Shepherd, C. L., Kay, G. and Wood, J. (2012). A novel and cost-effective monitoring approach for outcomes in an Australian biodiversity conservation incentive program. *PLoS ONE* 7(12). doi: 10.1371/journal.pone.0050872.

Lundberg, L., Persson, U. M., Alpizar, F. and Lindgren, K. (2018). Context matters: exploring the cost-effectiveness of fixed payments and procurement auctions for PES. *Ecological Economics* 146(February 2017), 347–358. doi: 10.1016/j.ecolecon.2017.11.021.

Lute, M. L., Gillespie, C. R., Martin, D. R. and Fontaine, J. J. (2018). Landowner and practitioner perspectives on private land conservation programs. *Society and Natural Resources* 31(2), 218–231. doi: 10.1080/08941920.2017.1376139.

Macintosh, A. (2004). Why the environment protection and biodiversity conservation act's referral, assessment and approval process is failing to achieve its environmental objectives. *Environmental and Planning Law Journal* 21(4), 288–311.

Macintosh, A. (2009). The Environment Protection and Biodiversity Conservation Act 1999 (Cth): an evaluation of its cost-effectiveness. *Environmental and Planning Law Journal* 26, 337–362.

Macintosh, A. (2012). The Australia clause and REDD: a cautionary tale. *Climatic Change* 112(2), 169–188. doi: 10.1007/s10584-011-0210-x.

Macintosh, A., Butler, D., Ansell, D. and Waschka, M.(2022). The emissions reduction Fund (ERF): problems and solutions. Australian National University: Canberra. Available at: https://law.anu.edu.au/research/publications?nid=52374.

Macintosh, A. and Wilkinson, D. (2005). EPBC act – the case for reform. *Australasian Journal of Natural Resources Law and Policy* 10(1), 141–179.

Madsen, B., Carroll, N. and Moore Brands, K. (2010). State of Biodiversity Markets Report: Offset and Compensation Programs Worldwide. Ecosystem Marketplace.

Milder, J. C., Arbuthnot, M., Blackman, A., Brooks, S. E., Giovannucci, D., Gross, L., Kennedy, E. T., Komives, K., Lambin, E. F., Lee, A., Meyer, D., Newton, P., Phalan, B., Schroth, G., Semroc, B., Van Rikxoort, H. and Zrust, M. (2015). An agenda for assessing and improving conservation impacts of sustainability standards in tropical agriculture. *Conservation Biology* 29(2), 309–320. doi: 10.1111/cobi.12411.

Murphy, H. and van Leeuwen, S. (2021). Australia State of the Environment 2021: biodiversity, independent report to the Australian Government Minister for the Environment. doi: 10.26194/ren9-3639.

New South Wales Auditor General (2006). *Regulating the Clearing of Native Vegetation*. Audit Office of New South Wales: Sydney.

Pannell, D. and Rogers, A. (2022). Agriculture and the environment: policy approaches in Australia and New Zealand, *Review of Environmental Economics and Policy* 16(1), 126–145. doi: 10.1086/718053.

Queensland Government (2022). The land restoration fund. Available at: https://www.qld.gov.au/environment/climate/climate-change/land-restoration-fund (accessed 6 April 2023).

Ritchie, E. G. (2022). Australia's biodiversity crisis and opportunity. *Science* 375(6578), 275.

Rolfe, J., Whitten, S. and Windle, J. (2017). The Australian experience in using tenders for conservation. *Land Use Policy* 63, 611–620. doi: 10.1016/j.landusepol.2015.01.037.

Royal, T. (2021). Private land conservation policy in Australia: minimising social-ecological trade-offs raised by market-based policy instruments. *Land Use Policy* 109(April), 105473. doi: 10.1016/j.landusepol.2021.105473.

Salt, D. (2016). A brief history of agri-environment policy in Australia: from community-based NRM to market-based instruments. In: Ansell, D., Gibson, F. and Salt, D. (Eds) *Learning from Agri-Environment Schemes in Australia: Investing in Biodiversity and Other Ecosystem Services on Farms*. Canberra: ANU Press, pp. 91–105. doi: 10.22459/lfaesa.05.2016.07.

Samuel, G. J. (2020). *Independent Review of the EPBC Act: Final Report*. Canberra. Available at: https://epbcactreview.environment.gov.au/resources/final-report.

Sheremet, O., Ruokamo, E., Juutinen, A., Svento, R. and Hanley, N. (2018). Incentivising participation and spatial coordination in payment for ecosystem service schemes: forest disease control programs in Finland. *Ecological Economics* 152(June), 260–272. doi: 10.1016/j.ecolecon.2018.06.004.

Simmons, B. A., Wilson, K. A., Marcos-Martinez, R., Bryan, B. A., Holland, O. and Law, E. A. (2018). Effectiveness of regulatory policy in curbing deforestation in a biodiversity hotspot. *Environmental Research Letters* 13(12). doi: 10.1088/1748-9326/aae7f9.

Simmons, B. A., Archibald, C. L., Wilson, K. A. and Dean, A. J. (2020). Program awareness, social capital, and perceptions of trees influence participation in private land conservation programs in Queensland, Australia. *Environmental Management* 66(3), 289–304. doi: 10.1007/s00267-020-01321-5.

Smith, H. F., Adams, T., Sullivan, C. A. and Licari, D. (2015). *Evaluating Incentives for Biodiversity Conservation on Private Land - Informing the Development of Lismore City Council's Rural Landholder Initiative*. Lismore.

State of New South Wales Biodiversity Conservation Trust (2023). Fixed price offers. Available at: https://www.bct.nsw.gov.au/fixed-price-offers (accessed 23 May 2023).

Stoneham, G., Chaudhri, V., Ha, A. and Strappazzon, L. (2003). Auctions for conservation contracts: an empirical examination of Victoria's BushTender trial. *Australian Journal of Agricultural and Resource Economics* 47(4), 477–500. doi: 10.1111/j.1467-8489.2003.t01-1-00224.x.

Vos, A. and Davies, K. (2021). Landholder perceptions of revegetation on the Atherton Tablelands, far North Queensland. *Ecological Management and Restoration* 22(1), 64–71. doi: 10.1111/emr.12438.

Ward, M. S., Simmonds, J. S., Reside, A. E., Watson, J. E. M., Rhodes, J. R., Possingham, H. P., Trezise, J., Fletcher, R., File, L. and Taylor, M. (2019). Lots of loss with little scrutiny: the attrition of habitat critical for threatened species in Australia. *Conservation Science and Practice* 1(11), 1–13. doi: 10.1111/csp2.117.

Wintle, B. A., Cadenhead, N. C. R., Morgain, R. A., Legge, S. M., Bekessy, S. A., Cantele, M., Possingham, H. P., Watson, J. E. M., Maron, M., Keith, D. A., Garnett, S. T., Woinarski, J. C. Z. and Lindenmayer, D. B. (2019). Spending to save: what will it cost to halt Australia's extinction crisis? *Conservation Letters* 12(6), 1–7. doi: 10.1111/conl.12682.

Woinarski, J. C. Z., Braby, M. F., Burbidge, A. A., Coates, D., Garnett, S. T., Fensham, R. J., Legge, S. M., McKenzie, N. L., Silcock, J. L. and Murphy, B. P. (2019). Reading the black book: the number, timing, distribution and causes of listed extinctions in Australia. *Biological Conservation* 239(May), 108261. doi: 10.1016/j.biocon.2019.108261.

Yasué, M. and Kirkpatrick, J. B. (2020). Do financial incentives motivate conservation on private land? *Oryx* 54(4), 499–510. doi: 10.1017/S0030605318000194.

Zammit, C. (2013). Landowners and conservation markets: social benefits from two Australian government programs. *Land Use Policy* 31, 11–16. doi: 10.1016/j.landusepol.2012.01.011.

Part 4

Habitat and animal rewilding

Chapter 11

Restoring peatlands in European landscapes

Rudy van Diggelen[1] and Tobias Ceulemans, University of Antwerp, Belgium; Camiel Aggenbach, KWR Watercycle Research Institute, The Netherlands; and Willem-Jan Emsens, Royal Zoological Society of Antwerp and University of Antwerp, Belgium

1 Introduction

A survey in the EU (European Environmental Agency 2020) showed that the conservation status of protected sites belonging to the habitat category 'Bogs, Mires and fens' is most frequently assessed as 'bad' (50%). Less than 10% is considered of 'good conservation status'. Even worse, when compared to the previous survey in 2012, more than 50% of the wetlands that already had a status of 'unfavourable' or 'bad' have continued to deteriorate. The main reasons for this negative trend are agricultural activities, land-use conversion and drainage.

Wetlands are '*areas that are inundated or saturated by water for all or part of the year to the extent that it supports soil microbes and rooted plants adapted for life in saturated soil conditions*' (Joosten 2016). Wet conditions lead to low oxygen availability in the soil and result in incomplete decomposition of dead plant material. If the water table remains sufficiently high and stable throughout the year, organic material accumulates. Organic wetland soils are called *peat*

[1] Email: Ruurd.vandiggelen@uantwerpen.be.

http://dx.doi.org/10.19103/AS.2024.0134.19

when the soil organic matter content is higher than 30% and the thickness of the organic layer exceeds 30 cm (Joosten et al. 2017).

The present chapter focuses on peatlands in temperate Europe. First we describe undisturbed lowland peatlands and the services they provide, which we compare to progressively degraded systems. Especially in north-western Europe, human population density and land-use intensity are so high that many of the once ubiquitous peatlands have disappeared or have severely degraded. For example, in Flanders, in the north of Belgium, approximately 95% of the peatlands that were still present in 1950 were drained for agriculture and did not survive into the twenty-first century (Decleer et al. 2016). For other European countries, estimated peatland losses for that same time period range from 60% to 84% (European Commission 2007). At the same time, we know that already in 1950, most peatlands had been lost completely due to drainage or peat-cutting in earlier periods (Leenders 1989). Exact numbers from the period before 1950 are not available but it seems reasonable to assume that north-western Europe has lost almost all of its peatlands. In less densely populated regions, for example in Northern Europe, most peatlands have been drained for forestry. For Europe as a whole, it is estimated that approximately 75 000 km^2 has completely disappeared, 275 000 km^2 has been drained and 320 000 km^2 is still in a natural or near-natural state (Joosten et al. 2017). The large majority of the latter are nowadays found in Russia, while undisturbed peatlands in the rest of Europe only cover a fraction of their past surface. Consequently, there is a huge need not only to conserve the few remaining peatlands but also to restore degraded ones, especially because they deliver important ecosystem services.

2 Ecosystem services of peatlands

Several studies (e.g. Bonn et al. 2016; Decleer et al. 2016) have shown that undisturbed peatlands deliver at least 15–17 of the services that are distinguished in the Common International Classification of Ecosystem Services CICES (EU 2014, 2016). These can be grouped into services to water flow, water purification, carbon cycling and biodiversity.

2.1 Services to water flow

Undrained peatlands dampen the effects of sudden precipitation peaks on the surface water system, as intact peat soil expands ('swells') under conditions of high water availability, thereby creating additional storage room. This process is called 'mire breathing' (Ingram 1982; Nijp et al. 2019). Water flow to the surface water system is further reduced by the topography of these peatlands, characterised by an intricated pattern of alternating hummocks, tussocks and hollows (Price et al. 2016). However, this service is often over-estimated, mainly because the peatlands' storage capacity for additional water is limited because peatlands are - by definition - water saturated. Also, the hydraulic permeability of most peatlands is low, meaning that they cannot release sufficient amounts of water to sustain the base flow in dry periods.

2.2 Services to water purification

Natural peatlands contain large amounts of nutrients, especially nitrogen (N). It is estimated that Boreal peatlands of the Northern hemisphere alone contain 8–15 Gt of N (Limpens et al. 2006). Like all green plants, peatland plants need nutrients, especially nitrogen. This implies that C-sequestration in peatlands automatically leads to N-sequestration as well. The flooded zones with highly productive tall sedges and reeds particularly serve as nutrient sinks, not in the least because they also trap sediments and the nutrients therein (Olde Venterink et al. 2002; Fisher and Acreman 2004).

2.3 Services to carbon cycling

According to present estimates, peatlands store globally at least 650 Gt of C on approximately 3% of the terrestrial surface. This equates to more than 20% of the global total organic C stock (Yu 2011, 2012; Scharlemann et al. 2014; Leifeld and Menichetti 2018). As such, they hold the most C per unit area of any terrestrial ecosystem (Zomer et al. 2016; Griscom et al. 2017). Globally restoring all drained peatlands, as well as preventing pristine peatlands from degradation, would result in the capture of almost 2 Pg CO_2 eq. per year. This is approximately 7% of the mitigation needed to keep global warming below 2°C (Griscom et al. 2017).

2.4 Services to biodiversity

The peatland environment is rather hostile to most organisms. Low soil oxygen levels, low redox potential and high concentrations of dissolved compounds such as NH_4^+, HS^-, Fe^{2+} and sometimes Al^{3+} are toxic to many biota. Therefore, most peatlands are home to true specialist species that are adapted and restricted to this extreme environment. Because of the current global scarcity of undisturbed peatlands, these species have often become rare and/or endangered. Some species of sedges (e.g. *Carex limosa, C. chordorrhiza*) and mosses (e.g. *Tomentypnum nitens, Paludella squarrosa*) are currently only found in undisturbed peatlands. The biodiversity value of undrained peatlands is therefore high, certainly in Europe (Hájek et al. 2006).

3 Types of peatland

Peatlands are highly diverse and complex systems due to large differences in the supporting hydrology. Various peatland classification systems have been developed, but unfortunately they are often incompatible. In the present chapter, we therefore restrict ourselves to a rather simple but useful classification (Du Rietz 1954):

- *Bogs* occur where precipitation is the primary source of water. Rainwater is mineral-poor and hardly buffered against acidification, while the dominant plants in bogs, peat mosses of the genus *Sphagnum*, excrete protons. Bogs are always acidic.
- *Fens*, on the other hand, are also fed by upwelling groundwater. Depending on the flow path, this groundwater can be relatively poor to extremely rich in dissolved minerals (Pietsch 1976; Shotyk 1988; Hajkova et al. 2004; Kharanzhevskaya et al. 2020). If the water has percolated through geological substrata with soluble minerals, water can be saturated with minerals. Important soil minerals are Ca^{2+} compounds, especially $CaCO_3$ and $CaSO_4$, but other minerals can also be found. If, on the other hand, the groundwater has travelled only short distances or through mineral-poor substrata, its composition is more similar to that of rainwater.
- A third major peatland type is *floodplains* that lie along streams or rivers and are, as the name suggests, strongly affected by regular flooding with surface water. Depending on its origin, surface water can vary largely in chemical composition, ranging from almost rainwater-like to extremely mineral rich. In densely populated areas, surface water is typically also loaded with nutrients and pollutants.

4 Characteristics of undisturbed peatlands

4.1. Water dynamics

Different peatland types are supplied with water from different origin. This not only affects water chemical composition but also the water regime (= pattern in water level fluctuations) and with that the dynamics of water saturation in the top peat layer (Wierda et al. 1997; Barber et al. 2004; Ahmad et al. 2021). Water level fluctuations in rainwater-fed systems depend on the variation of rainfall and evapotranspiration over the year. *Bogs* are normally found in areas with a significant precipitation surplus and a rainfall pattern that is more or less evenly distributed over the year, so that fluctuations in water levels are rather small. Fluctuation patterns in *fens* depend on the origin of the water. Upwelling groundwater that originates from deep soil layers often leads to very stable water levels because its main source is usually a large infiltration area, and water can take decades to reach the surface of the fen. Groundwater from shallow layers usually stems from smaller and nearby situated infiltration areas, with larger fluctuations in water levels as a result. Water levels in surface water-fed systems are generally very dynamic. *Floodplains* therefore typically exhibit the least stable water levels of all wetland types. This is certainly the case today: drainage systems have been optimised to maximise the export of excess water to the surface water system, resulting in large discharge peaks (Hirt et al. 2011).

4.2 Water chemistry

Water chemistry and water regime both exert a primary influence on soil conditions and nutrient availability. Firstly, the water itself may transport nutrients (N, P, K) to the site. This is typically the case with surface water that is naturally more nutrient-rich than rain- and groundwater. Under natural conditions, the latter two are generally more nutrient-poor. Moreover, internal nutrient release by peat decomposition is normally very low in fens and bogs due to the stable water levels and anaerobic conditions, making aerobic decomposition near impossible. This is much less the case in floodplains with more dynamic water levels or in sites where upwelling groundwater originates from small infiltration systems. Under such conditions oxygen can at least temporarily penetrate into the soil with decomposition rates that are not negligible. Here the release of nutrients from decomposing organic material may contribute significantly to nutrient loading.

4.3 Vegetation zonation

There is large variation in local topography, nutrient availability, water type and soil water saturation within one and the same peatland system. A comparison between vegetation zonation in undisturbed lowland peatlands (Wassen et al. 1996; Schipper et al. 2007) and the distribution of plant remnants in soils of now disturbed peatlands (van Diggelen et al. 1991) shows that a zonation in vegetation is very typical within lowland peatlands (Fig. 1). Along the edges, i.e. where the wetland borders the mineral soil, water supply fluctuates throughout the year and the water regime is rather dynamic. Here, nutrient availability is moderately high and this zone is typically covered by riparian alder forest. The second zone is fed by the upwelling of nutrient-poor groundwater from deeper layers, leading to very stable water levels and low vascular vegetation biomass. The vegetation is very open and is dominated by 'brown mosses' (e.g. from the *Amblystegiaceae* family) and small sedges (*Carex* sp.). The groundwater impact diminishes closer to the river and flooding intensity increases, leading to larger water table fluctuations and higher nutrient availability. Consequently,

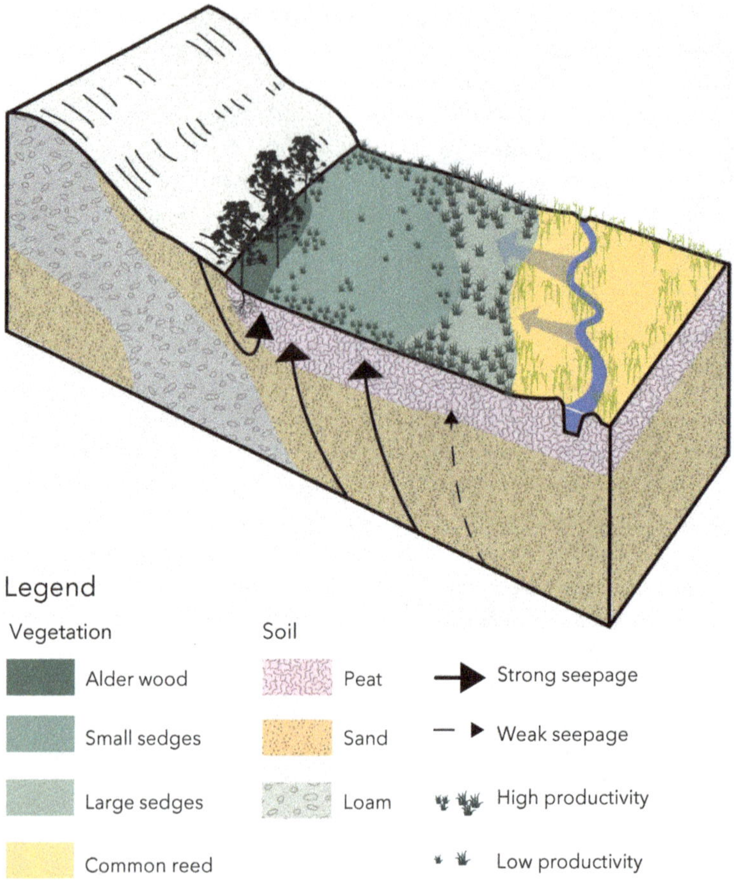

Figure 1 Typical vegetation zonation in an undisturbed riverine lowland peatland. The arrows indicate major water flows.

the vegetation is more productive there. The first zone is dominated by large sedges (*Carex* sp.), closer to the river these sedges are replaced by reed (*Phragmites australis*) and under very nutrient-rich conditions even by cattail (*Typha* sp.).

5 Effects of land-use changes on peatlands

5.1 Changes outside peatlands

5.1.1 Changes in the groundwater system

Increasing human interferences at the landscape scale have resulted in hydrological changes that directly and indirectly affect peatlands. A massive intervention is drainage, which often takes place at the level of the entire catchment. Drainage causes a reduction in groundwater recharge rates of the aquifer, lowers groundwater potential and subsequently reduces upward groundwater discharge into the peatland. In regions with a high precipitation surplus, such as western Europe, a decreased upward groundwater flow is generally replaced by infiltrating rainwater which leads to the acidification of the topsoil (Wassen et al. 1996; van Diggelen et al. 2006; Grootjans et al. 2006). Furthermore, local drainage of adjacent fields and groundwater abstraction also alters groundwater flows and lowers discharge rates (van Diggelen et al. 1994).

5.1.2 Changes in the surface water system

The majority of surface water systems have also been anthropogenically modified. Rivers have been straightened and embanked to prevent flooding, with large effects on hydrological conditions inside adjacent peatlands. Soil erosion and peak discharge of surface water started to increase in the Middle Ages because of the large-scale clearcutting of forests. In the peatlands adjacent to these rivers, this led to greatly enhanced deposition of mineral sediments (Notebaert et al. 2011).

5.1.3 Land-use intensification around the peatlands

Nowadays the remaining peatlands are typically situated in agricultural landscapes where they receive a large input of nutrients, pesticides and other pollutants via groundwater and surface water. Peatlands connected to an infiltration area that is intensively used by agriculture are typically subject to eutrophication. Often, however, increased nutrient availability is also the result of so-called internal eutrophication. This is caused by complex hydrogeochemical processes that are triggered by an altered chemical composition of groundwater (Hartog et al. 2004; Lamers et al. 2002; Smolders et al. 2010; Cirkel et al. 2014, see Box 1).

5.1.4 Air pollution

More recently, air pollution with sulphur and nitrogen further contributed to the eutrophication of peatlands (Robertson et al. 1989; Dise et al. 2009). Although atmospheric sulphur deposition strongly decreased due to measures taken by industry, nitrogen deposition is still very high in western Europe (European Environmental Agency 2022).

Box 1: Chemical processes in heavily fertilised landscapes

In western Europe, most groundwater-fed fens are hydrologically connected to infiltration areas that are covered by intensively used agricultural fields. This has consequences for the chemical composition of the infiltrating water, which is enriched with nitrate NO_3^- and sulphate SO_4^{2-}. Redox processes (Table 1) along the flow paths of groundwater (Fig. 2) further modify water chemistry, thus affecting hydrochemical conditions inside the peatlands wherever this water wells up.

Nitrate in the infiltrating water is usually removed through conversion to N_2 gas via redox reactions with pyrite (FeS_2) and reactive organic matter (reactions R1 to R3). Full oxidation of pyrite by nitrate (reaction R3) produces H^+, which in its turn leads to increased calcium and magnesium concentrations through dissolution of carbonates (reaction R6) and cation exchange processes (reaction R7). Depending on the specific groundwater flow paths and spatial geochemical variation, groundwater originating from agricultural infiltration areas is either nitrate-rich, nitrate + sulphate-rich or sulphate-rich. The combination of high concentrations with a high groundwater flux can lead to staggeringly high loads of nitrate (up to 1500 kg N/ha/year) and sulphate (up to 3000 kg S/ha/year) in discharge areas (Aggenbach et al. 2020). When such groundwater enters anaerobic organic soil layers, nitrate and sulphate are reduced via the reactions R1 and R4. This causes anaerobic decomposition of organic matter and thus losses of accumulated carbon from the peatland. These losses can be so high that in some peatlands the C-budget has changed from C-accumulation to C-release. The concomitant production of alkalinity elevates the pH and promotes decomposition processes inside the peatland. As a result, organically bound N and P are released. In a situation with high sulphate levels, in the absence of nitrate, sulphides precipitate with iron. This lowers the iron hydroxide pool in the soil and causes mobilisation of phosphate adsorbed to iron hydroxides. All reactions described previously lead to an increase in internal eutrophication.

Iron sulphides accumulate under waterlogged conditions but a strong drop in groundwater levels, for example in dry summers, causes aeration of the soil and oxidation of these sulphides. This leads to strong acidification by reaction R5 and causes leaching of calcium (reactions R6 and R7). Fens that are affected by the input of sulphate-rich groundwater are vulnerable to acidification, especially when the buildup of the acidification capacity exceeds the acid buffer capacity of the peat. When the peat is iron poor, a high sulphate load leads to high sulphide concentrations (HS^- and S^{2-}) in the porewater. High concentrations of sulphides are toxic to many plant species.

Table 1 Relevant chemical processes in heavily fertilised landscapes.

Reaction	Location	Chemical reaction
		Nitrate reduction
R1	2, 3, 5	by anaerobic decomposition: $4\ NO_3^- + 5\ CH_2O \rightarrow 2\ N_2\uparrow + CO_2 + 4\ HCO_3^- + 3\ H_2O$
R2	2, 3	by facultative pyrite oxidation: $5\ FeS_2 + 14\ NO_3^- + 4\ H^+ \rightarrow 7\ N_2\uparrow + 5\ Fe^{2+} + 10\ SO_4^{2-} + 2\ H_2O$
R3	2, 3	by full pyrite oxidation: $10\ FeS_2 + 30\ NO_3^- + 20\ H_2O \rightarrow 10\ Fe(OH)_3 + 15\ N_2\uparrow + 20\ SO_4^{2-} + 10\ H^+$
		Sulphate reduction
R4	2, 3, 5	by anaerobic decomposition: $2\ SO_4^{2-} + 3.5\ CH_2O + Fe^{2+} \rightarrow FeS_2 + 2\ HCO_3^- + 1.5\ CO_2 + 2.5\ H_2O$
		Pyrite oxidation
R5	5	by oxygen: $FeS_2 + 3.75\ O_2 + 3.5\ H_2O \rightarrow Fe(OH)_3 + 2\ SO_4^{2-} + 4\ H^+$
		Effects of acidification
R6	5	on carbonate dilution: $Ca_xMg_{(1-x)}CO_3 + 2\ H^+ \leftrightarrow x\ Ca^{2+} + (1-x)\ Mg^{2+} + CO_2 + H_2O$
R7	5	on cation exchange: $x\ H^+ + x\ Kat\text{-}CEC \rightarrow x\ Kat^+ + x\ H\text{-}CEC$

Numbers for locations refer to Fig. 2.

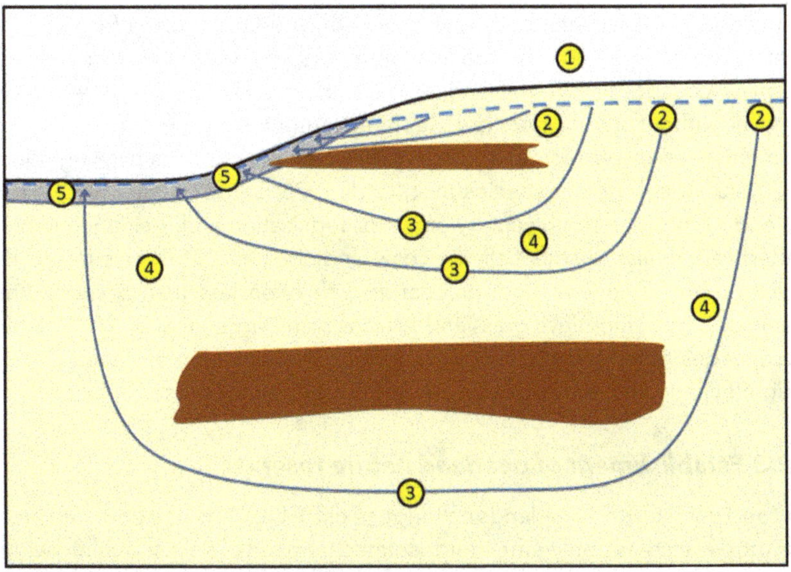

Figure 2 Cross section with important chemical processes along groundwater flow paths from infiltration area to discharge area in the peatland. Colors: yellow = aquifer, brown = aquitard, grey = peat; blue line = groundwater flow path; blue dotted line = groundwater level.

5.2 Changes inside peatlands

5.2.1 Peat extraction

A very drastic and irreversible change for peatlands is peat ('turf') extraction. In Europe, most peat extraction took place during the thirteenth to nineteenth century. Here, peat provided energy for the expanding cities (Unger 1984; Deforce et al. 2007; Jongepier et al. 2011). Most peat was extracted from bogs as bog turf has a higher carbon content than fen turf. This resulted in the complete loss of large bog complexes throughout Europe and with them the intermingled fen systems (Leenders 1989). Peat extraction for fuel was less common in brook valleys with fens, where extraction was often only a local phenomenon. Here, peat extraction deeper than the groundwater levels resulted in small pools, where peat growth started anew in the form of floating fens (van Diggelen et al. 1996). Yet many of these ponds, especially those in valleys, were kept open and used for fish farming or for hunting (Burny 1999). Due to these practices, large areas of fen habitat have been converted into pond systems.

5.2.2 Drainage of peatlands for agriculture

The most common land-use change inside peatlands is drainage for agriculture. In the past, many fens were converted into hay meadows by drainage. The drier conditions increased productivity and growth of edible grasses for livestock. Before the introduction of artificial fertilisers, conventional management normally consisted of haymaking at the end of summer. Centuries of hay removal without adding fertiliser depleted nutrients to the extent that low productive fen meadows developed, often with a very species-rich vegetation of 30–40 vascular plant species/m^2 (Klötzli 1978; Jansen et al. 2000). From the end of the nineteenth century onwards, fertilisation and drainage intensity started to increase substantially (Succow and Joosten 2012). This led to the development of species poor-grasslands with often less than 5 species/m^2, dominated by competitive grassland species (van Diggelen et al. 2005). In the last decades of the twentieth century, further intensification of drainage even enabled tillage and conversion to arable corn or potato fields.

5.2.3 Establishment of peatland nature reserves

Vast and undisturbed peatlands no longer exist in the densely populated areas of Europe. Instead, only small and isolated remnants persist, mainly within nature protection areas (see Box 2). They lost many of their characteristic species while most of the remaining species are endangered Red List species (IUCN 2021). Recolonisation by typical plant species is unlikely because these have a short-lived seed bank (Klimkowska et al. 2010) and dispersion is low or absent because the distance to source populations is too large (Ozinga et al. 2009).

Box 2. Case study of landscape changes in the Demer valley (Belgium) and future challenges for the conservation of its remaining peatlands

The peatlands of the Demer valley are a prime example of remnant peatlands in an agricultural landscape. Human impact here started in the Roman era when large-scale deforestation of the upstream catchment altered patterns of sedimentation (Notebaert et al. 2011; Verstraeten et al. 2018). This covered peat layers close to the river with alluvial deposits. Alluvial sedimentation expanded in the Middle Ages by digging canals to direct silt-laden river water over the peatlands as natural fertiliser. Inundation was also promoted by peat extraction in bogs upstream, thus reducing the capacity for water retention there. Furthermore, fens downstream were also subject to peat extraction, especially in the eighteenth and nineteenth centuries, lasting until the interbellum (Vervoort and Deneef, 1996). In the twentieth century, the river catchment underwent even more drastic changes. Large-scale industrial and household pollution in the 1960s and 1970s accelerated the straightening of the Demer that had started in the eighteenth century. This was done to remove the polluted water as fast as possible, causing a drop in groundwater and surface water levels. Cut-off meanders were subsequently used as landfills, converted into fishing ponds or levelled to be used for agriculture. Due to the drop in water levels, agriculture in the form of arable fields, particularly corn, was made possible. This subsequently replaced many of the traditional hay meadows. Areas that remained too wet were primarily planted with poplars. In addition, some of the peatlands were drained and built up as industrial area or also used as landfill up until the 1980s (Fig. 3a).

Despite these drastic changes, some small peatlands persisted. They were taken into nature management from the late 1970s onwards. Since then, restoration actions such as the removal of poplar plantations have managed to revive species such as *Carex diandra* (Fig. 3d). Furthermore, they are home to some of the last remaining populations of rare peatland mosses in the Northwest European lowlands such as *Hamatocaulis vernicosus*, *Philonotis marchica* and *Sphagnum affine*. Yet, the long-term persistence of these remnant peatlands faces many challenges on different scales. At a microscale, nature management is increasingly subject to mechanisation causing soil compaction and preventing the formation of the microtopography necessary for peat growth. At a mesoscale, the extant drainage network continues to cause peat oxidation. Furthermore these drainage networks act as 'inundation highways', spreading external surface water laden with pollutants from sewage and with nutrients and runoff from adjacent agriculture (Fig. 3b). This is exacerbated at a macroscale by urbanisation, expansion of arable fields and climate change. Urbanisation and agricultural expansion decrease water infiltration and cause faster run-off upstream, worsening inundation of the downstream peatlands. They also increase groundwater use, deplete aquifers and lower groundwater discharge. Climate change causes an increasing number of extreme events such as droughts (e.g. 2018, 2019, 2022) and long inundations (e.g. 1998, 2016, 2021). Consequently, several Red list species are now constricted to a limited number of buoyant patches on former peat excavation sites. The buoyancy prevents inundation during flooding and allows for waterlogged conditions during drought (Fig. 3c).

It is clear that the conservation of the last remaining peatlands of the Demer will require a multilevel approach that combines concerted efforts of all stakeholders, including nature and water managers, farmers and municipalities. At the micro level, nature management should take care not to compact soil and promote micro-topography for new peat growth. This will allow carbon and nutrient sequestration, on the one hand, and water retention in drought spells on the other. However, it will require more investment in skilled manual labourers working with adapted equipment. To halt peat oxidation and stimulate new peat growth, it will also be necessary to block drainage canals to allow for year-round water logging. At the meso-level, sufficient hydrological isolation is required to prevent inundation with surface water containing sewage and agricultural runoff. This will necessitate careful landscape planning as soil subsistence due to historical peat degradation has caused these peatlands to become among the lowest parts of the floodplain. Because the floodplain also serves as a water retention area to prevent inundation of urbanised areas, flood management will need to differentiate between alluvial areas and peatlands. Currently, peatland vegetation is limited to floating fens that float on open water bodies that go up and down with water table fluctuations (Arcadis Belgium, 2016). However, floating fens only comprise a few hundred square meters with limited room for expansion (Fig. 3b) Therefore, it is impossible to attain sustainable populations of endangered species here. Finally, at the watershed level, it is essential to continue water purification, protect peatland-feeding aquifers by limiting groundwater extraction and, importantly, to improve upstream water retention in urbanised and agricultural areas to reduce downstream flooding.

Only now the surface area under nature management has become large enough to start with the first two prerequisites: changing the internal management and blocking the drainage canals. Inundation prevention is not yet built into current landscape planning schemes. Flood protection of surviving peatlands will need to be put at a similar level as flood protection of urbanised areas. This remains an urgent task of water managers and municipalities. A similar challenge remains for mitigating water pollution and groundwater extraction. This requires that farmers and municipalities reduce agricultural and household use of water and prevent agricultural runoff and sewage discharge. Finally, upstream municipalities and farmers will need to adopt urban planning schemes and agricultural strategies to maximise local water retention. These management schemes, properly integrated, will ultimately benefit all ecosystem service goals, including adaptation to climate change.

Figure 3 (a) Remnant peatland, formerly with the endangered *Eriophorum gracile*, used as a landfill in 1982. (b) Flooding of a peatland in 2021 with brown-coloured silt-laden water characteristic of agricultural runoff. Patches with floating fens (fresh green color) escape inundation. (c) Small floating fens during natural flooding with local, non-polluted surface and groundwater. Notice the difference in water colour with (b). (d) Vegetation of the floating fens with red listed species *Carex diandra* and *Dactylorhiza majalis*. Photographs Luc Vervoort (a, c) and Kevin Feytons (b, d; https://www.kevinfeytons.be/).

5.3 Land-use legacies

Water table drawdown and concomitant peat oxidation triggers a myriad of cascading effects that alter peat properties. First, soil meso- and macrofauna such as earthworms find their way into the desiccated peat, where they fragment large particles of organic matter (Wu et al., 2017). Second, microbial community composition changes, which correlates with elevated activity of enzymes such as phenol oxidase (Fenner et al. 2005). Subsequent soil alterations can be rapid and include loss of organic matter as CO_2 or as dissolved organic carbon (DOC), soil subsidence, soil acidification, increase in bulk density, loss of hydraulic conductivity and decrease in water storage capacity (Zeitz and Velty 2002; Laiho 2006; Leifeld et al. 2011; Erkens et al. 2016; Liu et al. 2022). Drainage also leads to changes in the topography. Whereas the slope of the soil surface is usually small in undrained fens (Schipper et al. 2007; Succow and Joosten 2012), uneven subsidence leads to locally steep slopes in drained peatlands (Aggenbach et al. 2021). As a result of these processes, water table fluctuations are larger in drained peatlands (Price and Schlotzhauer 1999) thus enhancing further soil degradation. In addition, changes in peat quality and molecular composition take place in a sequential manner: easily degradable fractions of the peat such as cellulose are the first to be consumed upon oxidation, resulting in a relative accumulation of more recalcitrant material such as lignin (Emsens et al. 2020).

Drainage and subsequent increased decomposition rates trigger eutrophication by mobilisation of inorganic compounds from the organic soil (Lamers et al. 2015). This is evidenced by the concentration and accumulation of minerals and macronutrients such as ammonium (NH_4^+) followed by nitrate (NO_3^-), sulphates (SO_4^{2-}) or phosphates (PO_4^{3-}) in soil and pore water. Once large quantities of nutrients have been mobilised from the peat, this process cannot simply be reversed by peat rewetting. In fact, rewetting may even further enhance nutrient availability, particularly of phosphates, due to anaerobic reduction and desorption processes (Emsens et al. 2016; Zak and Gelbrecht 2007).

Encroachment of competitive vascular plants occurs at the expense of smaller and inherently slower-growing fen specialist species (Kotowski and van Diggelen 2004), which can no longer compete for light and consequently perish (Fig. 4). As a rule of thumb, light competition in fen communities of small sedges and brown mosses starts to become significant when less than 30% of the incoming solar radiation is able to reach the soil surface (Emsens et al. 2018).

The abovementioned profound shifts in peatland characteristics, functioning and resilience are not easily reversible within – at least – a decadal time span. This is why we refer to them as a 'land-use legacy' or 'degradation

Figure 4 Eutrophication leads to the encroachment of productive plant species such as reed (*Phragmites australis*) at the expense of slow-growing peatland specialists. Picture from the 'Zegge' in Belgium; a remnant peatland reserve that is completely surrounded and heavily affected by intensive agriculture (© KMDA/Jonas Verhulst.).

legacy'. Loss of soil carbon and soil structure due to drainage and degradation, for example, leads to profound changes in landscape topography (Fig. 5) and may remain clearly visible in the landscape for decades or centuries to come. Degradation legacies pose significant constraints on the restoration prospects of a peatland (Emsens et al. 2021). Tackling them always requires a combination of in situ restoration measures as well as measures that target the hydrology of the groundwater and surface water systems influencing the former peatland.

6 Restoration targets

The globally accepted definition of restoration says *Ecological restoration is the process of assisting the recovery of an ecosystem that has been degraded, damaged, or destroyed. It aims to move a degraded ecosystem*

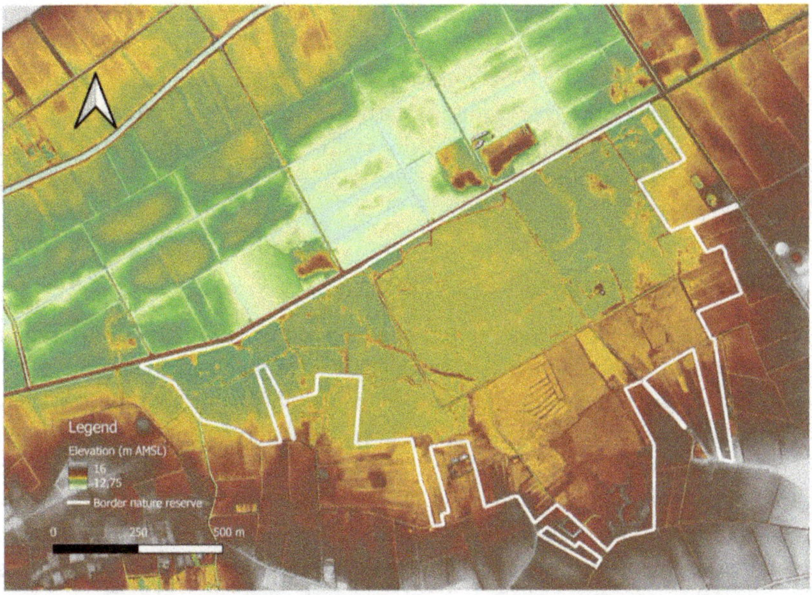

Figure 5 Elevation map of the peatland nature reserve 'De Zegge', Belgium, and its surroundings, showing clear signs of land-use legacies. Prior to large-scale land conversion and drainage, the area north of the reserve lay at approximately the same height as 'De Zegge'. In the beginning of the 1950s, however, the peatlands north of the reserve were drained and converted into agricultural land. This led to a soil subsidence of over 1 m and estimated losses of 31 ton CO_2/ha/year until all the peat there was gone. The groundwater that used to feed the reserve now drains directly into the subsided agricultural lands, causing ongoing soil desiccation, eutrophication and acidification within the borders of the reserve. Background: Digital Terrain Model (horizontal resolution: 1 m, vertical resolution: 1 cm).© Digitaal Vlaanderen

to a trajectory of recovery that allows adaptation to local and global changes, as well as persistence and evolution of its component species (Gann et al. 2019). What this definition does not elaborate on is where this recovery should lead to and what the final target should be. Harms et al. (1993) and Swart et al. (2001) depict several alternative targets to pursue, depending on the initial condition of the damaged ecosystem and the intended use of the restored system.

6.1 Natural target

Many people interpret restoration as a process that enables a return to a natural system that is, as much as possible, identical to the pre-degradation situation. The aim of such natural target is to restore a so-called *wilderness* landscape, in this case, peatlands that are identical to those under natural conditions. However, as discussed previously, long-term land-use legacies have irreversibly altered soil conditions and water flows. In such situations, restoration typically leads to a situation that differs considerably from the undisturbed one (Kreyling et al. 2021). Moreover, a complete return to a natural situation is often not wanted. A wilderness target for peatland restoration requires the restoration of large-scale hydrological systems and allows only very limited human activity inside the restored area. In other words, such a target requires a lot of space and is only feasible in sparsely populated areas.

6.2 Semi-natural target

An alternative target is the restoration of a *semi-natural* or *arcadian* landscape where the ecological conditions are modified by human activities but are still mainly determined by natural gradients. However, these gradients are typically no longer large-scale but instead split up in smaller, more local ones. Human impact on the landscape is larger and management activities are essential to keep the landscape in a certain state. The aforementioned fen meadows are a typical example of such semi-natural systems. They came into being after the superficial drainage of peatlands that were used for haymaking. Mowing is essential to keep these meadows open, but productivity is too low for modern farming standards. Their high biodiversity value can be preserved only with additional subsidies for nature conservation or landscape protection.

6.3 Functional target

A third target consists of the restoration of now degraded peatlands into a productive *functional* landscape in which intensive agricultural land use is combined with the preservation of important ecosystem services related to carbon cycling, hydrology and nutrients. This implies that fertilisers and pesticides are applied very sparingly and do not pollute groundwater or surface water. Tillage is avoided to prevent excess damage to the peat soils. The sites are not drained to avoid (further) carbon loss but are rewetted instead, with water tables that stay near the soil surface throughout the year. Instead of present-day 'classical' agriculture, the cultivation of 'new' crops needs to be developed. A promising initiative is what is called 'paludiculture' or wet agriculture (Wichtmann et al. 2016; Tanneberger et al. 2021), although most of the 'new' practices are in fact centuries old. Examples include the cultivation of crops like Cattail (*Typha*) for insulation material, reed (*Phragmites*) for thatching or the production of cellulose, willows (*Salix*) for cellulose or fuel pellets, or peat mosses (*Sphagnum*) for potting soil (Mulholland et al. 2020). In February 2023, an increasing number of paludiculture experiments is being carried out in countries like Germany, Poland, the Netherlands, the UK and Ireland and grant schemes to develop commercially viable paludiculture on lowland peat soils are starting to appear. The big challenge here is to develop cultivation types that yield enough income to live from but do not require so much fertilisation that there is a high risk of water pollution by nutrient runoff.

7 The way forward

Whatever restoration target is chosen, rewetting is a crucial first step to stop further degradation (Evans et al. 2021). At the same time not all rewetting is equal. The hydrochemical composition of the water and the rewetting technique applied strongly determine which targets can be achieved. A sudden rewetting by deep inundation with surface water generally leads to strong internal eutrophication and often also leads to high methane emissions (Zak and McInnes 2022). This option therefore hampers the achievement of more demanding restoration targets. On the other hand, 'slow' rewetting, i.e. gradually increasing water levels, preferably with unpolluted groundwater, leaves more options open and is therefore, wherever possible, always the preferred choice .

A wilderness restoration target has the most stringent requirements with respect to stability of water levels and hydrochemical conditions and will normally not be easy to attain in densely populated regions. However, where possible, this target should be the highest priority because it delivers most ecosystem services (Table 2). Wherever this target is pursued, the first requirement is to restore the natural hydrology. As long as water levels are not suitable for peat formation, it makes no sense to stop former management in order to rewild a drained peatland. Unsuitable hydrological conditions do not favour the development of a carbon-accumulating peatlands but instead promote the encroachment of eutrophic shrubs and bushes of low

Table 2 Ecosystem services delivered by different restoration targets

Ecosystem service	Wilderness	Semi-natural	Wet agriculture	Drained peatlands
Regulation of water flow and water purification	++	++	++	--
Carbon sequestration	++	-	0/+?	--
Biodiversity	++	++	0/-	-
Agricultural profits	--	0/+	+	++

++: highly positive effect; +: positive effect; 0: negligible effect; -: negative effect; --: highly negative effect. After van Diggelen and Verdonschot (2021).

conservation value (Opdekamp et al. 2012). Even when the hydrological system has been restored, prospects are not very hopeful in sites with a large degradation legacy. The wilderness target thus seems most appropriate for less disturbed landscapes. Human use of rewilded areas is limited and consists mainly of low-intensity tourism and possibly some hunting and fishing. Agriculture is not an option.

The target 'semi-natural restoration' has less strict requirements and is more easily achievable in landscapes where former land use has left its legacy. Contrary to the wilderness target, the semi-natural target does not necessarily require a conversion of the whole landscape, provided the activities around the site do not affect the hydrological conditions or nutrient levels within the site (Grootjans et al. 2006). In practice, this implies cessation of fertilisation anywhere near the peatland, as well as cessation of excessive fertilisation in or near infiltration areas and streams that feed the peatland. In addition, all drainage and groundwater abstraction in the vicinity should be halted and the natural morphology of streams restored to reduce the rates of catchment-scale water runoff. Regional canals and ditches should be closed or, if blocking is not possible, water levels should be raised, and the water of heavily eutrophied or polluted streams that flow into the wetland should be purified prior to entering.

In situ restoration measures typically include rewetting by closing drainage ditches, cessation of all direct fertilisation activities, suppression of competitive plants by regular mowing and biomass removal, or even species reintroductions (Lamers et al. 2015). In extreme cases, even the complete removal of a eutrophic and degraded topsoil can be considered, thereby exposing a less disturbed underlying peat layer (Klimkowska et al. 2015). The latter measure is very invasive and expensive, with potentially negative effects on the surrounding landscape and on the short-term carbon footprint (Zak and McInnes 2022). Therefore, it should always be treated with caution or as a last resort. Nonetheless, it may locally result in a complete 'reset' of the degraded topsoil layer, which lowers nutrient levels and may benefit the establishment of peat-forming specialist vegetation (Emsens et al. 2015; Zak et al. 2017).

Human use of semi-natural areas includes agriculture, hunting, fishing and a wide range of tourist activities, such as walking, cycling or – if climate permits – skiing or skating in winter. Agricultural activities are essential to manage the landscape but must be nature-friendly. Consequently, current intensive agriculture is not possible and farmers need an additional income besides agriculture. Recreation can serve as a good alternative. A recent study (Robinson et al. 2022) on the economy of nature revealed that tourists spend between 270 and 450 million Euros/year in the Dutch area of Zuid-Limburg alone, i.e. 6000–10 000 Euros/ha. In reality they spend most of their time in the semi-natural parts of the region, which is ca. 10 000 ha out of 45 000 for the total countryside, suggesting that the income per hectare for that part is even higher.

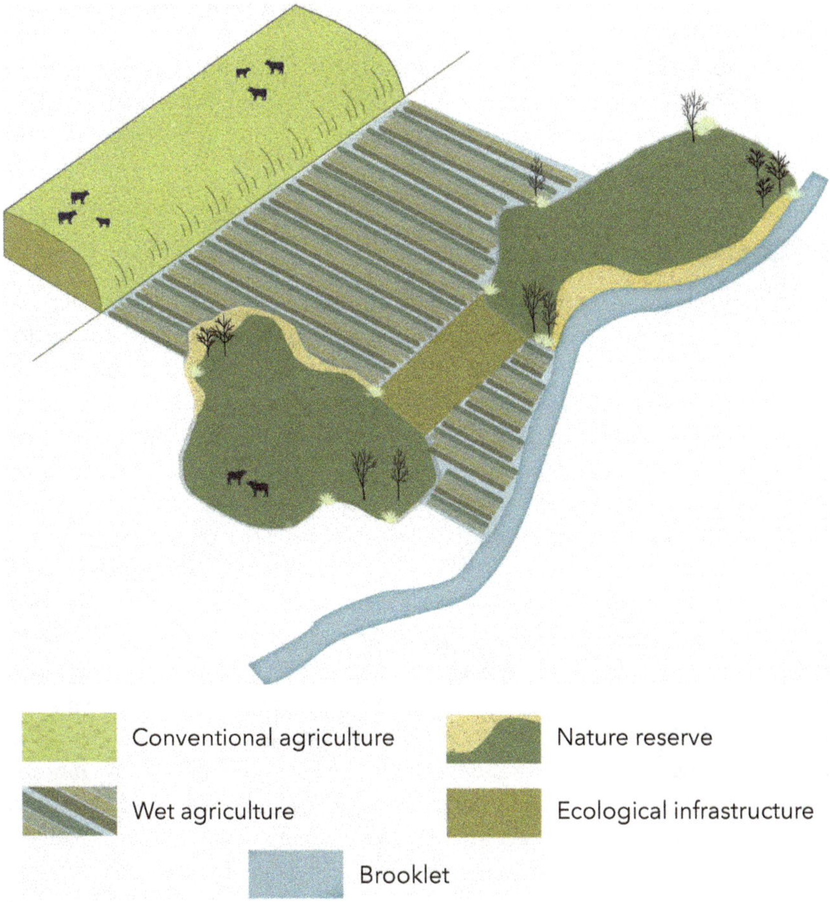

Figure 6 Landscape layout where wet agriculture buffers two nature reserves from influences of intensive land use and at the same time connects them in the form of a Green Infrastucture element.

The target 'wet agriculture' has few abiotic requirements and is perfectly suitable for sites with a persistent land-use legacy. In fact, a soil saturated with nutrients is to some degree beneficial for this type of land use because it increases productivity (Geurts et al. 2020). The only constraint for wet agriculture is that harvest is sustainable, i.e. that water levels are at the surface during the whole year and, depending on the crop, that nutrients are not depleted. At present, this farming technique is still in its infancy. Several crops are being tested and economic models being developed. Once established, this type of land use would be a perfect solution for buffer zones on former agricultural land located between sites with higher biodiversity value and areas of intensive and conventional land use (Fig. 6).

Setting sustainable restoration goals requires unambiguous choices, based on the legacies of the former land use, and physical and social conditions. A pitfall is to retain old land-use practices that hamper the development of feasible restoration targets. Or, worse, not make any choice at all. When clear and explicit choices have been made, peatlands generating multiple ecosystem services can be revived and provide new opportunities for both people and nature.

8 Conclusion and future trends

We distinguish at least two key trends in future research to look out and aim for.

The first trend we see is on the fundamental level: there is an urgent need for integrative and multidisciplinary research that takes into account multiple aspects of wetland functioning and restoration, which can be obtained by combining various scientific disciplines. This is particularly relevant for peatlands, which are highly diverse and complex ecosystems. A solid and profound understanding of key research topics, including ecosystem resilience (in response to change or stressors) or ecosystem restoration trajectories, can only be obtained using such approach. For example, there is a growing trend of combining molecular and soil microbial ecology with soil biogeochemistry, vegetation science and landscape ecology, which may unravel formerly unknown links or feedback loops between relevant actors within an ecosystem. Or, relevant biogeochemical processes that are studied on the local plot scale can be upscaled to the regional scale by GIS extrapolation and remote sensing. At the same time, there is a growing need for large-scale and cross-country studies with proper spatial replication, as extrapolation of data from only one site to the regional scale may produce biased or inaccurate results (Zalman et al. 2018). It is key that researchers with different scientific backgrounds form international consortia to jointly work on a topic or research question in a holistic manner. Multidisciplinary studies that cover a wide spatial gradient are thus likely to increase over the next few decades.

Second, the focus of applied peatland research is gradually moving towards a search for alternative and sustainable forms of land use in which multiple ecosystem services can be restored or conserved. This is relevant for degraded peatlands that are currently still used for conventional agriculture that includes fertilisation, drainage or peat soil tillage. As discussed in this chapter, paludiculture may provide an important part of the solution. Even though the concept of paludiculture is very old, scientific research on paludiculture is just emerging (Wichtmann et al. 2016; Ziegler et al. 2021). In the next few decades, it will therefore be important to initiate (more) large-scale trials on paludiculture and to subsequently investigate all relevant aspects ranging from economic feasibility and production processes to effects on the ecosystem carbon cycle, the water balance and biodiversity.

9 Where to look for further information

The following books provide a good overview of general wetland functioning, including relevant biogeochemical processes:

- Mitsch, W. J. and Gosselink, J. G. 2015. *Wetlands*. John Wiley & Sons, Hoboken.
- Maltby, E. and Barker, T. (Eds) 2009. *The Wetlands Handbook*. Wiley-Blackwell, Oxford.

The following books, journals or papers provide more detailed information specifically on peatlands, including peatland conservation, restoration and paludiculture:

- 'Mires & Peat' journal (http://www.mires-and-peat.net/)
- Wichtmann, W. et al. 2016. *Paludiculture: Productive Use of Wet Peatlands*. Schweizerbart Science Publishers, Stuttgart.
- Succow, M. and Joosten, H. 2012. *Landschaftsökologische Moorkunde* (in German). Schweizerbart Science Publishers, Stuttgart.
- Lamers, L. P. et al. 2015. Ecological restoration of rich fens in Europe and North America: from trial and error to an evidence-based approach. *Biological Reviews* 90(1), 182–203.

Key research and conservation efforts in this area can, amongst others, be found at the following organisations:

- International Mire Conservation Group (IMCG) (http://www.imcg.net/pages/home.php)
- Greifswald Mire Centre (https://greifswaldmoor.de/home.html)
- IUCN UK Peatland Programme (https://www.iucn-uk-peatlandprogramme.org/)

10 References

Aggenbach, C. J. S., Huyghe, P., Nijp, J. and van Diggelen, R. 2020. *Effects of nutrient polluted groundwater on seepage depended ecosystems*. Report number 2020/OBN242-BE. Vereniging van Bos- en Natuureigenaren, Driebergen.

Aggenbach, C. J. S., van Loon, A. H., Nijp, J. J., van Diggelen, R. and Ferrario, I. 2021. *Herstel van beekdalvenen door vernatting. Effecten na 30 jaar vernatting van het Gasterensche Diep* [Restoration of brook valleys by rewetting. Effects of 30 years rewetting]. *Landschap* 38(3): 175–183.

Ahmad, S., Liu, H., Alam, S., Günther, A., Jurasinski, G. and Lennartz, B. 2021. Meteorological controls on water table dynamics in fen peatlands depend on management regimes. *Frontiers in Earth Science* 9: 630469.

Arcadis Belgium 2016. PLAN-MER FASE II voor het Plan "Herinrichting Demervallei tussen Diest en Werchter" – Research into environmental effects for plan alternatives A, B, C en I. (dutch) Commissioned by W&Z.

Barber, K. R., Leeds-Harrison, P. B., Lawson, C. S. and Gowing, D. J. G. 2004. Soil aeration status in a lowland wet grassland. *Hydrological Processes* 18(2): 329–341.

Bonn, A., Allott, T., Evans, M., Joosten, H. and Stoneman, R. 2016. Peatland restoration and ecosystem services: an introduction. In: Bonn, A., Joosten, H., Evans, M., Stoneman, R. and Allott, T. (Eds). *Peatland Restoration and Ecosystem Services: Science, Policy and Practice*, pp. 1–16. *Ecological Reviews*. Cambridge University Press, Cambridge.

Burny, J. 1999. Bijdrage tot de historische ecologie van de Limburgse Kempen (1910–1950): tweehonderd gesprekken samengevat. *Natuurhist. Genootschap*, 211 p.

Cirkel, D. G., Van Beek, C. G. E. M., Witte, J. P. M. and Van der Zee, S. E. A. T. M. 2014. Sulphate reduction and calcite precipitation in relation to internal eutrophication of groundwater fed alkaline fens. *Biogeochemistry* 117(2–3): 375–393.

Decleer, K., Wouters, J., Jacobs, S., Staes, J., Spanhove, T., Meire, P. and van Diggelen, R. 2016. Mapping wetland loss and restoration potential in Flanders (Belgium): an ecosystem service perspective. *Ecology and Society*21(4): 46. https://doi.org/10.5751/ES-08964-210446.

Deforce, K., Bastiaens, J. and Ameels, V. 2007. Peat re-excavated at the Abbey of Ename (Belgium): archaeobotanical evidence for peat extraction and long-distance transport in Flanders around 1200 AD. *Environmental Archaeology* 12(1): 87–94.

Dise, N. B., Rothwell, J. J., Gauci, V., van der Salm, C. and de Vries, W. 2009. Predicting dissolved inorganic nitrogen leaching in European forests using two independent databases. *Science of the Total Environment* 407(5): 1798–1808.

Du Rietz, G. E. 1954. Die Mineralbodenwasserzeigergrenze als Grundlage einer natürlichen Zweigliederung der nord- und mitteleuropäischen Moore. *Vegetatio* 5/6: 571–585.

Emsens, W.-J., Aggenbach, C. J. S., Rydin, H., Smolders, A. J. P. and van Diggelen, R. 2018. Competition for light as a bottleneck for endangered fen species: an introduction experiment. *Biological Conservation* 220: 76–83.

Emsens, W.-J., Aggenbach, C. J. S., Schoutens, K., Smolders, A. J. P., Zak, D. and van Diggelen, R. 2016. Soil iron content as a predictor of carbon and nutrient mobilization in rewetted fens. *PLoS ONE* 11(4): e0153166.

Emsens, W.-J., Aggenbach, C. J. S., Smolders, A. J. P. and van Diggelen, R. 2015. Topsoil removal in degraded rich fens: can we force an ecosystem reset? *Ecological Engineering* 77: 225-232.

Emsens, W.-J., van Diggelen, R., Aggenbach, C. J. S., Cajthaml, T., Frouz, J., Klimkowska, A., Kotowski, W., Kozub, L., Liczner, Y., Seeber, E., Silvennoinen, H., Tanneberger, F., Vicena, J., Wilk, M. and Verbruggen, E. 2020. Recovery of fen peatland microbiomes and predicted functional profiles after rewetting. *The ISME Journal* 14(7): 1701-1712.

Emsens, W.-J., Verbruggen, E., Shenk, P., Liczner, Y., van Roie, M. and van Diggelen, R. 2021. Degradation legacy and current water levels as predictors of carbon emissions from two fen sites. *Mires and Peat* 27: article 14. DOI: 10.19189/MaP.2020.SNPG.StA.2149.

Erkens, G., van der Meulen, M. J. and Middelkoop, H. 2016. Double trouble: subsidence and CO_2 respiration due to 1,000 years of Dutch coastal peatlands cultivation. *Hydrogeology Journal* 24(3): 551-568.

European Commission 2007. *LIFE and Europe's wetlands. Restoring a vital ecosystem*. Report DG Environment, 65 pp. Publications Office of the European Union, Luxemburg.

European Environmental Agency 2020. *State of Nature in the EU - Results from Reporting Under the Nature Directives 2013-2018*. Publications Office.

European Environmental Agency 2022. *Eutrophication caused by atmospheric nitrogen deposition in Europe*. Available at: https://www.eea.europa.eu/ims/eutrophication-caused-by-atmospheric-nitrogen?utm_source=EEASubscriptions&utm_medium=RSSFeeds&utm_campaign=Generic.

Evans, C. D., Peacock, M., Baird, A. J., Artz, R. R. E., Burden, A., Callaghan, N., Chapman, P. J., Cooper, H. M., Coyle, M., Craig, E., Cumming, A., Dixon, S., Gauci, V., Grayson, R. P., Helfter, C., Heppell, C. M., Holden, J., Jones, D. L., Kaduk, J., Levy, P., Matthews, R., McNamara, N. P., Misselbrook, T., Oakley, S., Page, S. E., Rayment, M., Ridley, L. M., Stanley, K. M., Williamson, J. L., Worrall, F. and Morrison, R. 2021. Overriding water table control on managed peatland greenhouse gas emissions. *Nature* 593(7860): 548-552.

Fenner, N., Freeman, C. and Reynolds, B. 2005. Hydrological effects on the diversity of phenolic degrading bacteria in a peatland: implications for carbon cycling. *Soil Biology and Biochemistry* 37(7): 1277-1287.

Fisher, J. and Acreman, M. C. 2004. Wetland nutrient removal: a review of the evidence. *Hydrology and Earth System Sciences* 8(4): 673-685.

Gann, G. D., McDonald, T., Walder, B., Aronson, J., Nelson, C. R., Jonson, J., Hallett, J. G., Eisenberg, C., Guariguata, M. R., Liu, J., Hua, F., Echeverría, C., Gonzales, E., Shaw, N., Decleer, K. and Dixon, K. W. 2019. International principles and standards for the practice of ecological restoration. Second edition. *Restoration Ecology* 27S1: S1-S46.

Geurts, J. J. M., Oehmke, C., Lambertini, C., Eller, F., Sorrell, B. K., Mandiola, S. R., Grootjans, A. P., Brix, H., Wichtmann, W., Lamers, L. P. M. and Fritz, C. 2020. Nutrient removal potential and biomass production by *Phragmites australis* and *Typha latifolia* on European rewetted peat and mineral soils. *Science of the Total Environment* 747: 141102.

Griscom, B. W., Adams, J., Ellis, P. W., Houghton, R. A., Lomax, G., Miteva, D. A., Schlesinger, W. H., Shoch, D., Siikamäki, J. V., Smith, P., Woodbury, P., Zganjar, C., Blackman, A.,

Campari, J., Conant, R. T., Delgado, C., Elias, P., Gopalakrishna, T., Hamsik, M. R., Herrero, M., Kiesecker, J., Landis, E., Laestadius, L., Leavitt, S. M., Minnemeyer, S., Polasky, S., Potapov, P., Putz, F. E., Sanderman, J., Silvius, M., Wollenberg, E. and Fargione, J. 2017. Natural climate solutions. *Proceedings of the National Academy of Sciences of the United States of America* 114(44): 11645–11650.

Grootjans, A. P., Adema, E. B., Bleuten, W., Joosten, H., Madaras, M. and Janáková, M. 2006. Hydrological landscape settings of base-rich fen mires and fen meadows: an overview. *Applied Vegetation Science* 9(2): 175–184.

Hájek, M., Horsák, M., Hájková, P. and Dítě, D. 2006. Habitat diversity of central European fens in relation to environmental gradients and an effort to standardise fen terminology in ecological studies. *Perspectives in Plant Ecology, Evolution and Systematics* 8(2): 97–114.

Hajkova, P., Wolf, P. and Hajek, M. 2004. Environmental factors and Carpathian spring fen vegetation: the importance of scale and temporal variation. *Annales Botanici Fennici* 41: 249–262.

Harms, B., Knaapen, J. P. and Rademakers, J. G. 1993. Landscape planning for nature restoration: comparing regional scenarios. In: Vos, C. C. and Opdam, P. (Eds) *Landscape Ecology of a Stressed Environment*, pp. 197–218. Springer Netherlands, Dordrecht.

Hartog, N., Van Bergen, P. F., De Leeuw, J. W. and Griffioen, J. 2004. Reactivity of organic matter in aquifer sediments: geological and geochemical controls. *Geochimica et Cosmochimica Acta* 68(6): 1281–1292.

Hirt, U., Wetzig, A., Amatya, M. D. and Matranga, M. 2011. Impact of seasonality on artificial drainage discharge under temperate climate conditions. *International Review of Hydrobiology* 96(5): 561–577.

IUCN 2021. *The IUCN Red List of Threatened Species*. https://www.iucnredlist.org/.

Jansen, A. J. M., Grootjans, A. P. and Jalink, M. H. 2000. Hydrology of Dutch Cirsio-Molinietum meadows: prospects for restoration. *Applied Vegetation Science* 3: 51–64.

Jongepier, I., Soens, T., Thoen, E., Van Eetvelde, V., Crombé, P. and Bats, M. 2011. The brown gold: a reappraisal of medieval peat marshes in Northern Flanders (Belgium). *Water History* 3(2): 73.

Joosten, H. 2016. Peatlands across the globe. In: Bonn, A., Joosten, H., Evans, M., Stoneman, R. and Allott, T. (Eds) *Peatland Restoration and Ecosystem Services: Science, Policy and Practice*, pp. 19–43. *Ecological Reviews*. Cambridge University Press, Cambridge.

Joosten, H. F. Tanneberger and Moen, A. 2017. *Mires and Peatlands in Europe. Status, Distribution and Conservation*, 780 p. Scheizerbart Science PublishersStuttgart.

Kharanzhevskaya, Y. A., Voistinova, E. S. and Sinyutkina, A. A. 2020. Spatial and temporal variations in mire surface water chemistry as a function of geology, atmospheric circulation and zonal features in the south-eastern part of Western Siberia. *Science of the Total Environment* 733: 139343.

Klimkowska, A., Bekker, R. M., Van Diggelen, R. and Kotowski, W. 2010. Species trait shifts in vegetation and soil seed bank during fen degradation. *Plant Ecology* 206(1): 59–82.

Klimkowska, A., van der Elst, D. J. D. and Grootjans, A. P. 2015. Understanding long-term effects of topsoil removal in peatlands: overcoming thresholds for fen meadows restoration. *Applied Vegetation Science* 18(1): 110–120.

Klötzli, F. 1978. Ursachen für Verschwinden und Umwandlung von Molinion- Gesellschaften in der Schweiz. In: *Werden und Vergehen von Pflanzengesellschaften*. Berichte über das internationale Symposium der IVV in Rinteln, pp. 451–467. Cramer Verlag, Vaduz.

Kotowski, W. and van Diggelen, R. 2004. Light as an environmental filter in fen vegetation. *Journal of Vegetation Science* 15(5): 583–594.

Kreyling, J., Tanneberger, F., Jansen, F., van der Linden, S., Aggenbach, C., Blüml, V., Couwenberg, J., Emsens, W. J., Joosten, H., Klimkowska, A., Kotowski, W., Kozub, L., Lennartz, B., Liczner, Y., Liu, H., Michaelis, D., Oehmke, C., Parakenings, K., Pleyl, E., Poyda, A., Raabe, S., Röhl, M., Rücker, K., Schneider, A., Schrautzer, J., Schröder, C., Schug, F., Seeber, E., Thiel, F., Thiele, S., Tiemeyer, B., Timmermann, T., Urich, T., van Diggelen, R., Vegelin, K., Verbruggen, E., Wilmking, M., Wrage-Mönnig, N., Wołejko, L., Zak, D. and Jurasinski, G. 2021. Rewetting does not return drained fen peatlands to their old selves. *Nature Communications* 12(1): 5693.

Laiho, R. 2006. Decomposition in peatlands: reconciling seemingly contrasting results on the impacts of lowered water levels. *Soil Biology and Biochemistry* 38(8): 2011–2024.

Lamers, L. P. M., Falla, S.-J., Samborska, E. M., van Dulken, I. A. R., van Hengstum, G. and Roelofs, J. G. M. 2002. Factors controlling the extent of eutrophication and toxicity in sulfate-polluted freshwater wetlands. *Limnology and Oceanography* 47(2): 585–593.

Lamers, L. P. M., Vile, M. A., Grootjans, A. P., Acreman, M. C., van Diggelen, R., Evans, M. G., Richardson, C. J., Rochefort, L., Kooijman, A. M., Roelofs, J. G. M. and Smolders, A. J. 2015. Ecological restoration of rich fens in Europe and North America: from trial and error to an evidence-based approach. *Biological Reviews* 90(1): 182–203.

Leenders, K. A. H. W. 1989. *Verdwenen venen: een onderzoek naar de ligging en exploitatie van thans verdwenen venen in het gebied tussen Antwerpen, Turnhout, Geertruidenberg en Willemstad (1250–1750)*. Pudoc, Wageningen.

Leifeld, J. and Menichetti, L. 2018. The underappreciated potential of peatlands in global climate change mitigation strategies. *Nature Communications* 9(1): 1071.

Leifeld, J., Müller, M. and Fuhrer, J. 2011. Peatland subsidence and carbon loss from drained temperate fens. *Soil Use and Management* 27(2): 170–176.

Limpens, J., Heijmans, M. M. P. D. and Berendse, F. 2006. The nitrogen cycle in boreal peatlands. In: Wieder, R. K., Vitt, D. H. (Eds) *Boreal Peatland Ecosystems*, pp. 195–230. *Ecological Studies*. Springer, Berlin.

Liu, H., Rezanezhad, F. and Lennartz, B. 2022. Impact of land management on available water capacity and water storage of peatlands. *Geoderma* 406: 115521.

Liu, W., Fritz, C., Nonhebel, S., Everts, H. F. and Grootjans, A. P. 2022. Landscape-level vegetation conversion and biodiversity improvement after 33 years of restoration management in the Drentsche Aa brook valley. *Restoration Ecology* 30(7): 11.

Mulholland, B., Abdel-Aziz, I., Lindsay, R., McNamara, N., Keith, A., Page, S., Clough, J., Freeman, B. and Evans, C. 2020. *An Assessment of the Potential for Paludiculture in England and Wales*. Report to Defra for Project SP1218. Department for Environment, Food & Rural Affairs. London.

Nijp, J. J., Metselaar, K., Limpens, J., Bartholomeus, H. M., Nilsson, M. B., Berendse, F. and van der Zee, S. 2019. High-resolution peat volume change in a northern peatland: spatial variability, main drivers, and impact on ecohydrology. *Ecohydrology* 12(6): 17.

Notebaert, B., Verstraeten, G., Vandenberghe, D., Marinova, E., Poesen, J. and Govers, G. 2011. Changing hillslope and fluvial Holocene sediment dynamics in a Belgian loess catchment. *Journal of Quaternary Science* 26(1): 44–58.

Olde Venterink, H. O., Pieterse, N. M., Belgers, J. D. M., Wassen, M. J. and de Ruiter, P. C. 2002. N, P, and K budgets along nutrient availability and productivity gradients in wetlands. *Ecological Applications* 12(4): 1010–1026.

Opdekamp, W., Beauchard, O., Backx, H., Franken, F., Cox, T. J. S., van Diggelen, R. and Meire, P. 2012. Effects of mowing cessation and hydrology on plant trait distribution in natural fen meadows. *Acta Oecologica* 39: 117–127.

Ozinga, W. A., Römermann, C., Bekker, R. M., Prinzing, A., Tamis, W. L. M., Schaminée, J. H. J., Hennekens, S. M., Thompson, K., Poschlod, P., Kleyer, M., Bakker, J. P. and Groenendael, J. M. V. 2009. Dispersal failure contributes to plant losses in NW Europe. *Ecology Letters* 12(1): 66-74.

Pietsch, W. 1976. Vegetationsentwicklung und wasserchemische Faktoren in Moorgewässern verschiedener Naturschutzgebiete der DDR. *Archiv für Naturschutz und Landschaftsforschung* 16: 1-43.

Price, J., Evans, C., Evans, M., Allott, T. and Shuttleworth, E. 2016. Peatland restoration and hydrology. In: Bonn, A., Joosten, H., Evans, M., Stoneman, R. and Allott, T. (Eds) *Peatland Restoration and Ecosystem Services: Science, Policy and Practice*, pp. 77-94. *Ecological Reviews*. Cambridge University Press, Cambridge.

Price, J. S. and Schlotzhauer, S. M. 1999. Importance of shrinkage and compression in determining water storage changes in peat: the case of a mined peatland. *Hydrological Processes* 13(16): 2591-2601.

Robertson, W. D., Cherry, J. A. and Schiff, S. L. 1989. Atmospheric sulfur deposition 1950–1985 inferred from sulfate in groundwater. *Water Resources Research* 25(6): 1111–1123.

Robinson, P., van Schendel, M., Botzen, W., van Beukering, P., van den Heuvel, R., Koetse, M. and Aerts, J. 2022. *Economische waardering van Natuur en Landschap in Zuid-Limburg*. Insitute for Environmental Studies Free University of Amsterdam, Amsterdam.

Scharlemann, J. P., Tanner, E. V., Hiederer, R. and Kapos, V. 2014. Global soil carbon: understanding and managing the largest terrestrial carbon pool. *Carbon Management* 5(1): 81-91.

Schipper, A. M., Zeefat, R., Tanneberger, F., van Zuidam, J. P., Hahne, W., Schep, S. A., Loos, S., Bleuten, W., Joosten, H., Lapshina, E. D. and Wassen, M. J. 2007. Vegetation characteristics and eco-hydrological processes in a pristine mire in the Ob river valley (Western Siberia). *Plant Ecology* 193(1): 131–145.

Shotyk, W. 1988. Review of the inorganic geochemistry of peats and peatland waters. *Earth-Science Reviews* 25(2): 95–176.

Smolders, A. J. P., Lucassen, E. C. H. E. T., Bobbink, R., Roelofs, J. G. M. and Lamers, L. P. M. 2010. How nitrate leaching from agricultural lands provokes phosphate eutrophication in groundwater fed wetlands: the sulphur bridge. *Biogeochemistry* 98(1-3): 1-7.

Succow, M. and Joosten, H. 2012. *Landschaftsökologische Moorkunde*. Schweizerbart Science Publishers, Stuttgart.

Swart, J. A. A., Windt, H. J. V. D. and Keulartz, J. 2001. Valuation of nature in conservation and restoration. *Restoration Ecology* 9(2): 230–238.

Tanneberger, F., Appulo, L., Ewert, S., Lakner, S., Ó Brolcháin, N., Peters, J. and Wichtmann, W. 2021. The power of nature-based solutions: how peatlands can help us to achieve key EU sustainability objectives. *Advanced Sustainable Systems* 5(1): 2000146.

Unger, R. W. 1984. Energy sources for the Dutch golden age: peat, wind, and coal. *Research in Economic History* 9: 221-253.

van Diggelen, R., Grootjans, A. and Burkunk, R. 1994. Assessing restoration perspectives of disturbed brook valleys: the Gorecht area, The Netherlands. *Restoration Ecology* 2(2): 87-96.

van Diggelen, R., Middleton, B., Bakker, J., Grootjans, A. and Wassen, M. 2006. Fens and floodplains of the temperate zone: present status, threats, conservation and restoration. *Applied Vegetation Science* 9(2): 157-162.

van Diggelen, R., Molenaar, W. J., Casparie, W. A. and Grootjans, A. P. 1991. Paläoökologische Untersuchungen als Hilfe in der Landschaftsanalyse im Gorecht-Gebiet (Niederlande). *Telma* 21: 57-73.

van Diggelen, R., Molenaar, W. J. and Kooijman, A. M. 1996. Vegetation succession in a floating mire in relation to management and hydrology. *Journal of Vegetation Science* 7(6): 809-820.

van Diggelen, R., Sijtsma, F. J., Strijker, D. and van den Burg, J. 2005. Relating land-use intensity and biodiversity at the regional scale. *Basic and Applied Ecology* 6(2): 145-159.

van Diggelen, R. and Verdonschot, P. F. M. 2021. Beekdallandschappen in beweging. De Weg vooruit. *Landschap* 2021: 195-199.

Verstraeten, G., Notebaert, B., Broothaerts, N., Vandenberghe, J. and De Smedt, P. 2018. River landscapes in the Dijle catchment: from natural to anthropogenic meandering rivers. In: Demoulin, A. (Ed.) *Landscapes and Landforms of Belgium and Luxemburg*, Chapter 16, pp. 269-280. Springer International Publishing, Cham.

Vervoort, L. and Deneef, R. 1996. The meander of Vorsdonkbos-Turfputten in Gelrode (Aarschot, Belgium) - a historical land ecological review (Dutch). *Natuur en Landschap* 15(5): 39-62.

Wassen, M. J., van Diggelen, R., Wolejko, L. and Verhoeven, J. T. A. 1996. A comparison of fens in natural and artificial landscapes. *Vegetatio* 126(1): 5-26.

Wichtmann, W., Schröder, C. and Joosten, H. (Eds.) 2016. *Paludiculture - Productive Use of Wet Peatlands*. Schweizerbart Science Publishers, Stuttgart.

Wierda, A., Fresco, L. F. M., Grootjans, A. P. and Diggelen, R. 1997. Numerical assessment of plant species as indicators of the groundwater regime. *Journal of Vegetation Science* 8(5): 707-716.

Wu, X., Cao, R., Wei, X., Xi, X., Shi, P., Eisenhauer, N. and Sun, S. 2017. Soil drainage facilitates earthworm invasion and subsequent carbon loss from peatland soil. *Journal of Applied Ecology* 54(5): 1291-1300.

Yu, Z. 2011. Holocene carbon flux histories of the world's peatlands: global carbon-cycle implications (W. F. Ruddiman, M. C. Crucifix, & F. A. Oldfield, Eds.). *The Holocene* 21(5): 761-774.

Yu, Z. C. 2012. Northern peatland carbon stocks and dynamics: a review. *Biogeosciences* 9(10): 4071-4085.

Zak, D. and Gelbrecht, J. 2007. The mobilisation of phosphorus, organic carbon and ammonium in the initial stage of fen rewetting (a case study from NE Germany). *Biogeochemistry* 85(2): 141-151.

Zak, D. and McInnes, R. J. 2022. A call for refining the peatland restoration strategy in Europe. *Journal of Applied Ecology* 59(11): 2698-2704.

Zak, D., Meyer, N., Cabezas, A., Gelbrecht, J., Mauersberger, R., Tiemeyer, B., Wagner, C. and McInnes, R. 2017. Topsoil removal to minimize internal eutrophication in rewetted peatlands and to protect downstream systems against phosphorus pollution: a case study from NE Germany. *Ecological Engineering* 103: 488-496.

Zalman, C., Keller, J. K., Tfaily, M., Kolton, M., Pfeifer-Meister, L., Wilson, R. M., Lin, X., Chanton, J., Kostka, J. E., Gill, A., Finzi, A., Hopple, A. M., Bohannan, B. J. M. and Bridgham, S. D. 2018. Small differences in ombrotrophy control regional-scale variation in methane cycling among Sphagnum-dominated peatlands. *Biogeochemistry* 139(2): 155-177.

Zeitz, J. and Velty, S. 2002. Soil properties of drained and rewetted fen soils. *Journal of Plant Nutrition and Soil Science* 165(5): 618-626.

Ziegler, R., Wichtmann, W., Abel, S., Kemp, R., Simard, M. and Joosten, H. 2021. Wet peatland utilisation for climate protection - an international survey of paludiculture innovation. *Cleaner Engineering and Technology* 5: 100305.

Zomer, R. J., Neufeldt, H., Xu, J., Ahrends, A., Bossio, D., Trabucco, A., van Noordwijk, M. and Wang, M. 2016. Global tree cover and biomass carbon on agricultural land: the contribution of agroforestry to global and national carbon budgets. *Scientific Reports* 6: 29987.

Published by Burleigh Dodds Science Publishing Limited, 2024.

Chapter 12

Rewilding grasslands and rangelands

Thomas A. Jones, Forage and Range Research Laboratory – U.S. Department of Agriculture – Agricultural Research Service, USA

1 Introduction

Both grasslands and rangelands support grasses and other herbaceous plants (Allen et al., 2011). Together, these lands cover about 40% of the Earth's area (Buisson et al., 2022; Bardgett et al., 2021). Grasslands include prairies, grassy shrublands, and savannas (Bardgett et al., 2021). They are typically found on relatively fertile soils in reasonably humid climates and therefore can produce more forage and support a higher livestock-carrying capacity than rangelands.

http://dx.doi.org/10.19103/AS.2024.0134.20

Grasslands may be described as 'natural,' occupied primarily by native species, or 'tame' or 'improved,' primarily occupied by introduced species. Grasslands may be managed intensively for agricultural production, which arrests successional processes (Allen et al., 2011).

Rangelands usually support indigenous vegetation that is managed as a natural ecosystem to support grazing livestock and wildlife (Holechek et al., 2010; Stoddart et al., 1975). Grasslands may be described as tame, primarily occupied by introduced species, or natural, occupied primarily by native species. While grasslands have the potential to be managed intensively with irrigation and fertilization, rangelands are necessarily managed extensively as minimal-input natural ecosystems (Holechek et al., 2010). Hence, management intensity is a distinction between grasslands and rangelands. Relative to grasslands, rangelands typically have greater abiotic limitations than grasslands, e.g. low or erratic precipitation, rough or steep topography, poor or excessive drainage, extremely high or low temperatures, or saline or sandy soils (Briske, 2017; Holechek et al., 2010; Stoddart et al., 1975). Thus, rangelands display a high degree of spatial or temporal variability (Herrick and Whitford, 1995), while grasslands tend to be more uniform. While grasslands may be cultivated, most rangelands cannot sustain cultivation (Holechek et al., 2010). Rangelands may display a significant woody-plant component, shrubs and/or trees, while grasslands support primarily grasses and other herbaceous plants, with perhaps a small frequency of woody plants (Allen et al., 2011).

Since the agricultural revolution of the Neolithic Period, humans have expanded grasslands to support livestock agriculture (Slodowicz et al., 2023) through livestock management, controlled burns, and clearing and thinning timber (Staude et al., 2023). These ecosystems provide ecosystem services, including biodiversity, forage, water, erosion control, pollinator and wildlife habitat, carbon (C) sequestration, and meat, milk, and fiber products (Buisson et al., 2022). Relative to forests, grasslands and rangelands are especially valuable for theoretical and applied ecological studies due to small plant size, ease of manipulation, and short time scales (Wilsey, 2021).

Natural and seminatural grasslands, as well as rangelands, are disturbance-driven ecosystems (Török et al., 2021). Disturbance events curtail the growth of dominant species, thus facilitating light penetration that supports the growth of subordinate species (Sturchio et al., 2022). When these lands are abandoned, however, reversion to woody vegetation may ensue following cessation of grazing (Knapp and Seastedt, 1986), at least partly due to the expansion of scrub, i.e. shrub-dominated vegetation (Bardgett et al., 2021). The impacts of fire may be either favorable (e.g. suppression of woody vegetation, stimulation of growth and diversity) or unfavorable (e.g. altered fire regimes, providing a niche opening for invasive species, displacement by cropland, loss of soil C)

(Bardgett et al., 2021; Wagle and Gowda, 2018; Chambers et al., 2014; Rowe, 2010; Copeland et al., 2002).

Grasslands and rangelands are largely underappreciated and misunderstood by both the public and much of the scientific community, leading to limited recognition and priority in policy-making despite ongoing degradation (Staude et al., 2023; Bardgett et al., 2021). Tropical grasslands are ignored to an even greater extent than temperate grasslands (Buisson et al., 2021b). Misconceptions about grasslands include the presumptions that they are necessarily of recent origin, they may be reassembled quickly and easily, and they are merely early successional stages of forests (Buisson et al., 2022). In fact, ancient (i.e. 'old-growth') grasslands require centuries to assemble their high species and functional diversity. Their subterranean structure drives recovery from aboveground disturbances such as grazing and fire. However, once these lands are damaged below ground (Buisson et al., 2022), ecological thresholds may be crossed that are impractical to reverse, especially on a broad scale (Hobbs et al., 2009; Briske et al., 2006), though notable exceptions do exist (Gibson-Roy and McDonald, 2014).

The seven objectives for this chapter are as follows:

- Spotlight threats to the integrity of grasslands and rangelands worldwide;
- Delineate ecological principles that drive grasslands and rangeland functioning and upon which their rewilding is based;
- Highlight the diversity of best practices used to rewild grasslands and rangelands along with their supporting science. Such practices include those to restore biological soil crusts, apply prescribed fire, increase stand establishment and persistence, overcome soil limitations, rectify nutrient pollution, and sequester soil C;
- Describe concepts and emerging technologies with great potential to rewild grasslands and rangelands;
- Showcase the role of native seed industries in the USA, Europe, and Australia in supporting the rewilding of grasslands and rangelands;
- Provide numerous science-based examples from around the world supporting the above objectives; and
- Offer a case study of the restoration of sagebrush-steppe rangelands in the state of Utah (Intermountain West, USA), including the scientific documentation of the successes and limitations of these efforts.

2 Past and current threats to grasslands and rangelands

Grasslands and rangelands are vulnerable to conversion to cropland (tillage), mining, overgrazing by livestock, invasive species, and afforestation (Buisson et al., 2022; Bardgett et al., 2021). Because unprotected grasslands and rangelands are often considered expendable, land-use change is a major cause of their loss. In Rio Grande do Sul, Brazil, grasslands are being lost due to afforestation and conversion to cropland for maize and soybean cultivation (Staude et al., 2023) as the agricultural frontier is extended into undeveloped areas (Rey Benayas and Bullock, 2012). Likewise, large expanses of temperate grasslands have been lost in southeastern Australia due to diversion to agriculture, leaving only small and largely disconnected relict areas (Williams and Morgan, 2015).

Iowa, USA, was opened for agricultural development in 1833, but once settlers arrived in a particular area, tallgrass prairie disappeared within 10–20 years (Smith, 1990). John Deere's moldboard plow was designed in 1837, and by 1857, his Iowa factory was manufacturing 10 000 plows annually. Large numbers of former Civil War soldiers moved to the state to farm after 1865, and beginning in 1888, large-scale tile drainage opened lands to cultivation that were previously inaccessible. By the end of the century, 97% of the vast Iowa tallgrass prairie was no more.

Semi-natural grasslands were once widespread throughout Europe (Ridding et al., 2015; Hedberg and Kotowski, 2010). They were primarily occupied by native flora without fertilizer or pesticide application. However, semi-natural grasslands were utilized for grazing and haying. An example is the Salisbury Plain of southern England, home of Stonehenge, once an extensive area of species-rich semi-natural grasslands (Wedin, 1992). These 'chalk grasslands' occurred on thin infertile soils over calcareous bedrock. They were maintained by sheep, whose diet includes woody species, for centuries. However, in the 1870s, sheep grazing declined for economic reasons, which led to an invasion of tree and shrub species, though grazing by wild rabbits continued.

During the Great War, much of the chalk grasslands was diverted into cropland, a process that continued through World War II (Wedin, 1992). Subsequently, these vast areas were largely compromised by the advent of technology that increased agricultural productivity but led to the demise of small-scale farming. These novel inputs included mineral fertilizers, herbicides, pesticides, and drainage infrastructure (Lawson et al., 2004). These, along with widespread seedings of perennial ryegrass (*Lolium perenne*), increased forage production but, at the same time, sharply curtailed species diversity due to N fertilizer-induced eutrophication (Wedin, 1992). In the Salisbury Plain, species composition and critical functional traits may take up to 100 years to reappear after the abandonment of arable agriculture, as many of the ancient grassland species are limited by low seed dispersal (Redhead et al., 2014).

Agricultural lands, particularly in Europe but more recently in developing countries, are being abandoned due to migration to urban areas (Rey Benayas and Bullock, 2012). In other cases, land is abandoned due to agricultural policy addressing commodity surpluses, water-quality problems, and biodiversity

Published by Burleigh Dodds Science Publishing Limited, 2024.

losses (Wedin, 1992). Regardless, abandonment is followed by forest and scrub encroachment (Navarro and Pereira, 2015) and often an expansion of exotic plant populations, along with a concomitant decline in native plant biodiversity (Wedin, 1992). While fires and herbivores may function to check woody-plant invasions of grasslands (Staude et al., 2023), fire suppression and afforestation policies may lead to grassland and rangeland degradation (Bardgett et al., 2021).

Certain agricultural practices may also trigger degradation. Improperly timed or excessive livestock grazing primarily impacts herbaceous understory plants, but woody-plant expansion may be difficult to reverse once the grassy understory is insufficient to support grazing or fuel a fire (Buisson et al., 2022). Conversion to cropland, beginning with tillage, leads to a rapid loss of grassland biodiversity, a negative impact on wildlife, and dramatic losses in sequestered soil C (Bardgett et al., 2021). Intensification of agriculture typically reduces native-plant diversity (Bardgett et al., 2021, McIntyre and Lavorel, 1994; Wedin, 1992), though it often increases the diversity of exotic species (McIntyre and Lavorel, 1994).

Climate change has led to more frequent and intense fires and droughts (Bardgett et al., 2021). As 67% of the world's grasslands are found in semiarid, arid, or cold climates, climate change is expected to have a negative impact on their ability to sequester soil organic C (Bai and Cotrufo, 2022). In addition, desertification is expected to expand the ranges of drylands (Buisson et al., 2021a). Because of climate's overarching influence, it may be necessary to modify restoration plans to accommodate shifting restoration trajectories in response to labile future climates (Buisson et al., 2022).

3 Application of ecological models to restoration management

The traditional model of rangeland management was based on Clementsian succession, where vegetation moves along a trajectory toward a single climax state (Westoby et al., 1989). In this model, stocking rate and drought are presumed to work in a similar fashion, but in opposite directions. Therefore, rangeland management sought to balance the two opposing forces to maintain an equilibrium, such that animal weight gains could be achieved without damaging the rangeland. This equilibrium is maintained by reducing the stocking rate under conditions of drought and increasing the stocking rate in years of higher precipitation.

Despite the intuitive appeal of the Clementsian model, however, empirical evidence supported by accompanying theory has suggested that this model generally fails to hold, particularly in drier climates (Prach and Walker, 2020; Stringham et al., 2003; Westoby et al., 1989). For such situations, state-and-transition models have been developed with multiple stable states, both desirable and undesirable. Stable states are separated by thresholds, either disturbances or management decisions, that trigger a non-reversible conversion to an alternate stable state. Within a stable state, multiple community phases may also be present. The difference between a stable state and a community phase is that the latter is reversible, while to former is not (Briske et al., 2008). Transitions between states often occur in patches, but as the process continues, patches may coalesce, consolidating the transition (Bestelmeyer et al., 2003).

Once a threshold has been crossed, management should focus on restoring the damaged ecological process rather than establishing a plant community that cannot be sustained in the new state (Stringham et al., 2003). In other words, for all practical purposes, restoration to the previous state may not be feasible (Monaco et al., 2012; Bestelmeyer et al., 2003). For example, using long-term vegetation-monitoring data, Bagchi et al. (2013) compared a higher-elevation wetter rangeland (1923–1973) to a lower-elevation drier rangeland (1950–2006), both in eastern Idaho. Monitoring occurred before and following invasion by downy brome (*Bromus tectorum*). When the sagebrush-perennial bunchgrass community of the higher-elevation rangeland crossesa threshold to a downy brome state, it may reverse itself, indicating a phase shift. However, when the same occurrs on the lower-elevation rangeland, the transition is irreversible, indicating a threshold.

Thresholds are central to the 'novel ecosystem' concept. Novel ecosystems are systems of 'abiotic, biotic and social components (and their interactions) that, by virtue of human influence, differ from those that prevailed historically, having a tendency to self-organize and manifest novel qualities without intensive human management' (Hobbs et al., 2013). Novel ecosystems are derived from natural historic ecosystems through a disturbance-induced threshold crossing. However, once the threshold is crossed, the novel ecosystem is stable, self-perpetuating, and irreversible. An example is the degraded stable state known as 'Intermountain West annual grasslands' in the Great Basin of the western USA (Davies et al., 2021b). This state has resulted from crossing a threshold from the native perennial sagebrush-bunchgrass ecosystem. Smith et al. (2022) documented an eight-fold increase in annual grass-dominated areas in the Great Basin over a 30-year period (1990–2020), corresponding to a 0.6% annual increase in area. This includes invasion into higher elevations and north-facing aspects. Climate change is expected to favor annuals due to warmer winters, altered precipitation patterns, and an earlier and more-active fire season (Davies et al., 2021b).

4 Approaches to grassland and rangeland rewilding: passive restoration

Rewilding has been described as assisting the natural regeneration of natural habitats through the passive management of ecological succession to restore natural ecosystem processes and reduce human control of formerly agricultural landscapes (Navarro and Pereira, 2015; Gillson et al., 2011). According to this plant-based definition, former crop- or grazing lands are allowed to assemble vegetation without seeding (Navarro and Pereira, 2015), thus success depends on the quality of the native seed bank (Rey Benayas et al., 2012) and seed dispersal from adjacent wildland tracts if present. Active intervention is minimal, though it may be required in the early stages of rewilding (Navarro and Pereira, 2015). Succession often leads to scrubland and possibly later to forest (Conti and Fagarazzi, 2005). While rewilding has been disregarded by restoration practitioners until recently in Europe, it does provide a low-cost opportunity to increase biodiversity (Navarro and Pereira, 2015).

In central and eastern Europe, large areas of marginal cropland located on poor soils are being abandoned, but researchers posed the question as to whether these grassland dunes of calcareous sand could experience spontaneous recovery (i.e. respond to passive restoration) (Török et al., 2018). On reference sites in central Hungary, the composition of seed banks correlated poorly with extant vegetation, and seeds of characteristic species were mostly lacking in seed banks of both reference and restoration sites. However, over time, both vegetation and seed banks progressed toward the target, while weedy seeds decreased. This pattern suggests that succession was driven by seeds that dispersed across space by vectors rather than via the seed bank.

Invasion by exotic trees and shrubs is the greatest threat to the last remaining natural grasslands of South America (Cuevas and Zalba, 2010). The Pampas grasslands of central Argentina were originally free of trees, but by the early twentieth century, they were populated by woody exotics. Cuevas and Zalba (2010) sought to restore low-altitude (550 m) and high-altitude (700 m) montane grasslands suffering from a dense invasion of Aleppo pine (*Pinus halepensis*) that was reducing light penetration and lowering soil pH. Felling of the trees launched a slow recovery of native grasses that displaced exotic herbs (Cuevas and Zalba, 2010). Monitored over 4 years, recovery of the low-altitude clearing was fueled by its native seed bank alone, as it was surrounded by pines. The high-altitude site, on the other hand, benefitted from both its native seed bank and dispersal from adjacent natural grassland upslope. Due to early intervention (i.e. trees were less than 12 years old when felled), native recruitment was not seed limited, enabling spontaneous recovery. Despite a substantial pine seed bank, seedling recruitment was low. The authors suggested low pine-seedling recruitment might be a consequence of winter felling followed by regeneration of native vegetation in spring, which possibly interfered with pine seedling recruitment in fall.

It may be uncertain whether a native seed bank is able to regenerate a plant community, for example, when restoring a formerly cultivated field. Growing out a soil sample can determine the frequency and identity of native species present on the site. If the native-plant community was removed some time ago, seeds of species with seed dormancy are more likely to be present than those of readily germinating species. Alternatively, seed dispersal from nearby intact sites may facilitate recolonization. Both recruitment from the seed bank and dispersal from nearby sites qualify as passive restoration. It simply requires a source of propagules and dispersal vectors plus moderate soil fertility (Török et al., 2018). While passive restoration may be successful under certain circumstances, many grassland species lack good seed-dispersal mechanisms (Redhead et al., 2014). Moreover, this problem is greatly exacerbated by a high degree of grassland fragmentation (Hedberg and Kotowski, 2010). When a site is seed limited, active restoration (i.e. seeding) will be required to increase biodiversity, but this may not be feasible if appropriate seed is unavailable.

5 Approaches to grassland and rangeland rewilding: active restoration

Compared to passive restoration, active restoration has a longer history and is much more widely practiced in Europe (Navarro and Pereira, 2015). Low levels of agricultural practice are maintained, which tend to counter successional processes. Wet meadow is a species-diverse grassland ecosystem that occurs throughout northern Europe and is maintained by haying and sometimes grazing (Wedin, 1992). In the Netherlands, wet meadows have become highly productive by applying high levels of N fertilizer. This has supported dairying on a small land base but has also led to declines in species diversity. For policy reasons, intensive agriculture has been largely halted in these areas, providing an opportunity for restoration research.

Reintroducing cattle grazing and eliminating fertilization initially failed to restore the native-plant communities. However, after mowing was applied in place of grazing animals and clippings were removed over 45 years, historical species composition returned to near-historical levels as soil nutrients dwindled. Nevertheless, N levels became elevated due to the atmospheric deposition of N from agricultural and industrial nutrient pollution. Under this scenario, it became possible to reintroduce cattle. However, modern breeds selected for utilizing high-quality forage were unable to subsist on the low-quality rushes, orchids, and forbs of the wet meadows. Nonetheless, imported Scottish Highland cattle have been able to thrive on this low-quality diet.

Persistent seed banks that better support passive restoration occur frequently in wetlands and at sites where historical disturbances are frequent and unpredictable, as they are better able to buffer disturbance (Kiss et al., 2018). However, active restoration is likely to be required where seed banks are transient and associated with the lack of historical disturbance. These systems are most likely to suffer severe changes in species composition upon disturbance or climate change. A changing climate is also expected to drive species with good spatial dispersal ability to higher altitudes or toward either pole (Walther et al., 2002).

The Cerrado, in the highlands of central Brazil, is the world's most biodiverse savanna region, equaling the richness of the Brazilian Amazon, with almost 37% of Cerrado species being endemic (Lewis et al., 2022). Species distribution models were used to identify restoration hotspots, areas expected to provide the greatest economic return on restoration investment (Lewis et al., 2022). Four hotspots were identified, the central Cerrado and three transitional areas between biomes. Each hotspot displayed high species diversity and high potential C sequestration with a relatively low degree of fragmentation.

Exotic annual grasses, especially downy brome and medusahead wildrye (*Taeniatherum caput-medusae*), are occupying increasing portions of the Great Basin, USA, a vast semi-arid to arid region stretching from Nevada to western Utah, southern Idaho, and southeastern Oregon (Davies et al., 2021b). Downy brome alone occupies 21 million ha, almost a third of the region (Bradley et al., 2018), and perennial plant communities are converting to exotic annual grasslands at a rate of nearly 200 000 ha annually (Smith et al., 2022). These annual grasslands are now recognized as a stable state that is very difficult to reclaim with perennial species (Davies et al., 2021b; Clements et al., 2017). Furthermore, the litter from these annual species fuels recurring wildfires that favor the annuals, exacerbating the problem (Davies et al., 2021b).

Downy brome is a prolific seed producer that builds persistent seed banks at or near the soil surface. In northern Nevada, USA, Clements et al. (2017) implemented a seed-bank strategy to control the highly flammable invasive annual grass, downy brome. When wildfire burns a big sagebrush community, the high fire intensity consumes most of the downy brome seed. But once sagebrush has been removed by a previous wildfire, the fire intensity is often too low to deplete the downy brome seed bank, triggering recurring wildfire. Spring disking removed downy brome plants before the seed set and buried the seed bank, reducing seed density by 82%. In another tactic, the use of pre-emergent herbicides achieved over 95% reduction in seed density.

Once this competitive weed was controlled, the objective of establishing perennial grasses for long-term suppression was attainable (Clements et al., 2017). Seeding native perennial grasses was subject to a high failure rate due to the challenging environment. An alternative may be a reclamation seeding of crested wheatgrass (*Agropyron* spp.). This exotic is easier to establish than the native grasses due to its high seedling vigor. A crested wheatgrass seeding achieved 92% control of downy brome and stabilized the site.

5.1 Green hay as a seed source

A common source of seed used in Europe for grassland restoration is green hay (Wagner et al., 2021; Kiehl et al., 2010), a practice that dates from the Roman era (Kiehl et al., 2010). A hay cut is timed to maximize its seed content and is then spread on a prepared seedbed within 24 h to avoid seed-shattering losses (Wagner et al., 2021; Kiehl et al., 2010). Baasch et al. (2016) reported similar results for green hay and threshing material for seeding a lowland hay meadow in Saxony-Anhalt, Germany.

When a regional commercial seed mix was added to either green hay or threshing material, Baasch et al. (2016) achieved greater success in terms of the number of target species that were established. Over the next few years, species spread into adjacent undisturbed grassland. Depending on its source, green hay may be a viable option to deliver local native seeds to the restoration site, particularly if those are unavailable commercially (Kiehl et al., 2010). It may also provide propagules of non-vascular plants, such as mosses and lichens (i.e. biocrust species).

5.2 Biological soil crusts

Biological soil crusts, or biocrusts, are communities of mosses, lichens, cyanobacteria, eukaryotic algae, liverworts, and other organisms that form a thin horizontal layer on the soil surface of many rangeland ecosystems (Bowker, 2007; Bowker et al., 2018). Biocrusts are found across a diversity of biomes and on all continents (Reed et al., 2019). They require high light, hence an open canopy, thus they predominate in arid and polar ecosystems (Reed et al., 2019). Biocrust species are regarded as both ecosystem engineers (Muñoz-Rojas et al., 2021) and indicators of rangeland health (Bowker, 2007).

Cyanobacteria are fast-growing pioneer biocrust organisms (Rubio and Lázaro, 2023) that can survive under extreme temperatures, saline conditions, long-term drought, and high UV irradiance (Muñoz-Ramos et al., 2021). Biocrust succession begins with cyanobacteria, which tolerate high irradiance, and proceeds as shade increases. In southern Spain, succession may continue with cosmopolitan generalist moss species, sometimes referred to as 'fire mosses' (Muñoz-Ramos et al., 2021). Succession then proceeds with fast-growing lichen species, and finally, slow-growing lichens and mosses that prefer shade. Rubio and Lázaro (2023) described the microbial crust as an all-encompassing matrix, with a lichen crust developing upon or replacing that matrix.

Biocrusts provide numerous ecosystem services, including soil aggregation and stabilization, N-fixation, nutrient cycling, dust-trapping, hydrological functioning, and thermal-energy balance (Bowker et al., 2018). They are also believed to facilitate secondary succession (Bowker, 2007). Interestingly, seeds of certain native species 'bio-primed' with cyanobacterial strains from the Pilbara region of Western Australia demonstrated superior radicle or shoot growth, though other species did not (Chua et al., 2020). In Gansu, northwestern China, biocrusts play an ecological filtering role by promoting one vascular-plant species but limiting two others, a process called sieving (Song et al., 2022). Furthermore, biocrusts may preclude seed penetration, which may explain observations that biocrusts are associated with reduced exotic-plant cover (Antoninka et al., 2020). This may be a mechanism by which biocrusts limit the crossing of ecological thresholds to less desirable states (Bowker, 2007).

In some cases, biocrusts may be restored passively, which requires varying lengths of time (Condon et al., 2020; Hilty et al., 2004). They may or may not be sensitive to livestock grazing, depending on the biocrust organism (Condon et al., 2020). Installation of a physical structure, such as jute netting, offers wind protection, creates some shade, and captures wind-blown propagules (Rosentreter, 2020).

Active restoration of biocrusts may be initiated by inoculation with salvaged dry material from the restoration site or with a slurry that has been reared ex situ (Young et al., 2016). Inoculants salvaged from the restoration site may be presumed to be adapted to that site, and they avoid the problem that some species are difficult to culture (Chiquoine et al., 2016). However, salvage efforts may be challenging and expensive, particularly if large quantities of inoculum are required for extensive projects.

Culturing biocrust species ex situ requires facilities, equipment, and time (Chiquoine et al., 2016). Cyanobacterial inoculants have been cultured for large-scale application in Almeria, southern Spain (Roncero-Ramos et al., 2019). Chiquoine et al. (2016) suggested using the salvage technique for species difficult to culture, such as lichens, and the slurry technique for producing cyanobacteria inoculum. Incorporation of topsoil from the restoration site proved to be less effective for biocrust restoration than either salvaging or culturing (Chiquoine et al., 2016).

5.3 Seeding native forbs

Perennial grass species are widely seeded for grassland restoration, but perennial forbs are used to a much lesser extent due to high cost and low availability of seed, as well as the difficulty of field collection (Jones, 2019). Many forb species can be present in the tallgrass prairie ecosystem of North America, so they often make a large contribution to species richness (Grman et al., 2021).

However, forb seedings often establish poorly or succumb to competition from grasses. Thus, species richness and diversity can be compromised in restoration seedings, especially because grasses tend to be more competitive than forbs.

A tallgrass prairie restoration experiment tested the impact of four C_4 grass species, big bluestem (*Andropogon gerardii*), little bluestem (*Schizachyrium scoparium*), switchgrass (*Panicum virgatum*), and Indiangrass (*Sorghastrum nutans*), on forbs in 78 ex-arable field seedings in Illinois, Indiana, and Michigan, USA (Grman et al., 2021). These native grasses are widely used in tallgrass prairie seedings because they dominate reference sites, generate economical seed crops, produce sufficient biomass to fuel burns, and compete against weeds. Big bluestem and, to a lesser extent, little bluestem had a direct negative impact on forbs. To rectify this problem, Grman et al. (2021) recommended decreasing seeding rates of warm-season (C_4) grasses, increasing seeding rates of forbs, and burning more frequently to limit grass competition.

5.4 Strip seeding and restoration islands

Strip seeding is a technique designed to reduce costs yet enhance the probability of seeding success on a limited scale (Rayburn and Laca, 2013). Strip seeding concentrates resources, such as seed and weed control, on a linear strip of land without seeding adjacent sections, where weeds may be easily controlled. Segregation of seeded species within the strip allows less competitive species to avoid competition from more vigorous species. While only a minority of the land is seeded, colonization of the adjacent vacant areas can begin once the seeded plants mature.

Restoration islands can be used to concentrate high-value plant material on relatively small sites that can then expand or 'nucleate' to the vicinity (Hulvey et al., 2017). The small island area allows expensive resources to be concentrated to ultimately maximize impact on the surrounding area. Such resources include creation of favorable microsites, high seeding rates, supplemental water, and weed control. Islands may be strategically located for microclimate, topography, soil quality and moisture, minimal weed competition, and fire avoidance. Islands should be designed to minimize edge-to-area ratio to limit perimeter encroachment, and seeded species may be chosen for good seed-dispersal ability. Islands provide an opportunity to sow high-value seeds that could not be seeded across the entire restoration site due to prohibitive cost.

To evaluate a seeding of 35 target species and to examine their dispersal into adjacent areas, Kiss et al. (2021) tilled three sizes (1, 4, and 16 m^2) of 'establishment gaps' into a species-poor 8-year-old grass seeding in the Great Hungarian Plain, Hungary. They then seeded the high-diversity seed mix into the gaps. Within 5 years, colonization of the species-poor surrounding area had begun despite a 1-year drought. The intermediate-sized gaps displayed the best dispersal. Their smaller edge-to-area ratio offered more protection from recolonization by the matrix species than the small-sized gaps. Yet the intermediate-sized gaps were more inconspicuous to grazing cattle than the large-sized gaps.

6 Approaches to grassland and rangeland rewilding: prescribed burning

Prescribed burning is widely employed in the USA, especially in the tallgrass prairie ecosystem of the Midwest (Rowe, 2010). Practitioners aim to burn a tallgrass prairie seeding as soon as it produces sufficient fuel to carry a fire, averaging 2 years and ranging from one to six (Rowe, 2010). The management goals of burning tallgrass prairie are to increase forb diversity (Rowe, 2010), maintain dominant C_4 grasses, remove litter to warm the soil, and stimulate early spring green-up (Wagle and Gowda, 2018). To favor a particular functional group, the site is burned when that functional group is dormant, but when a group needs to be controlled, the burn is implemented in its peak growth season. For example, forbs may be increased by early spring or fall burning, while cool-season (C_3) and C_4 grasses may be controlled by frequent spring and summer burning, respectively (Wagle and Gowda, 2018). Early-September burns remove the growth of dominant C_4 grasses without damaging them, thus favoring subdominant grasses and forbs and increasing species richness (Copeland et al., 2002). Frequent burning favors C_4 grasses but decreases C_3 grasses, while forbs are favored by infrequent burning or grazing (Wagle and Gowda, 2018).

Prescribed burning can be an effective tool to manage restored tracts of land when fire is a component of the ecosystem being restored. A selling point of prescribed burning is its low expense. However, a minimum fuel load is required to carry a fire, and specific weather conditions are necessary to safely complete the plan (Toledo et al., 2012). While prescribed burning has the potential to be a useful tool for many ecosystems, perceived safety and legal concerns have hindered its adoption (Toledo et al., 2012). In parts of Texas, USA, the landowner can join a regional prescribed burn association, a cooperative organization that promotes safe and effective burning on a landscape scale through training, pooled equipment and labor, and insurance policies. Members join because, while they favor prescribed burning due to its low cost, ease of implementation, and high efficacy, they lack the personal financial resources to implement a burn on their own. A common realization is that it is better to consume fuels through a prescribed burn than to experience a catastrophic event with unforeseeable consequences.

In the high-altitude Himalayan meadows of Bhutan, invasion by rhododendrons has reduced yak carrying capacity (Wangchuk et al., 2013).

Prescribed burning controlled rhododendron, stimulating the growth of grasses, sedges, and broadleaf herbaceous plants, thereby restoring grassland. While the traditional practice had been to burn annually, a 6-year burn interval informed by this research is now recommended to generate a fuel load of 1 ton of dry-matter yield per hectare per year.

Burning was once a traditional management tool in Europe to improve forage quality, counter woody encroachment, and reduce litter accumulation (Valkó et al., 2014). However, its use disappeared with the adoption of intensive agricultural practices. Furthermore, burning is prohibited in many European countries due to the potential for damage to life and property, as well as air pollution. In an additional complication, warmer and drier climates are expected in the Mediterranean Region, where abandoned lands accumulate fuels that propel wildfires. Europe has historically lacked extensive fire-adapted C_4 grasslands as were once present in the midwestern USA. Consequently, for prescribed burning to be implemented in Europe, a fire-return interval of at least 3 years may be required because of the fire susceptibility of European vegetation.

While fire is no longer a frequently used restoration tool in Europe, interest in its use has been stimulated by the declines in the extent and species richness of European grasslands (Valkó et al., 2014). These declines have been attributed to reductions in livestock in rural areas and the accompanying disappearance of traditional grazing regimes. The question is whether fire can substitute for grazing and mowing of hay to reverse these richness declines.

7 Species establishment and persistence

7.1 Fire-related germination

Thirty years ago, Brown (1993) reported that smoke could stimulate the germination of fynbos species in the Cape Region, South Africa. About 10 years later, the active ingredient karrikinolide (KAR_1) was identified (Flematti et al., 2004). Effective at levels as low as parts per trillion, KAR_1 links with proteins to generate hydrolytic enzymes that trigger germination once hydrothermal conditions are favorable (Lamont, 2023). KAR_1 operates in the post-dormancy phase, so it stimulates germination rather than breaks dormancy (Dixon et al., 2009; Ramos et al., 2019).

Physiological seed dormancy is common in grasses and is the most common dormancy mechanism in ecosystems with seasonal rainfall, including many fire-prone systems (Ramos et al., 2019), particularly in Mediterranean climates (Lamont, 2023). Species that display physiological dormancy and water-permeable seed coats are more likely to respond to either smoke alone or smoke plus heat than species that do not (Hodges et al., 2021; Cuello et al., 2020). Some species respond to smoke water or fire but not to KAR_1, which suggests other factors may be operating (Ramos et al., 2019). Germination response to fire is much less consistent in temperate grasslands than in Mediterranean climates (Hodges et al., 2021). However, in the temperate grasslands of southeastern Australia, grass species were the most responsive to smoke (Hodges et al., 2021). In Uruguay's Rio de la Plata temperate sub-humid grassland, however, a field burn favored monocots and dicots similarly (Cuello et al., 2020).

7.2 Seed treatments for establishment

While several pre-emergent herbicides are available for use in restoration seedings, they can negatively impact germinating seedlings following application, delaying seeding a year or more. During this delay, weeds may rebound, defeating the purpose of the herbicide. Holfus et al. (2021) described a 'vortexing' method to coat seeds with activated charcoal that adsorbs the pre-emergent herbicide, rendering the herbicide inactive in the vicinity of the seed. When a pre-emergent herbicide was applied just prior to a greenhouse seeding, no damage was sustained by seedlings germinating from coated bluebunch wheatgrass seeds.

Seed enhancement technologies are being developed for rangeland seeds that increase soil-water availability, enhance seedling emergence from soil crusts, delay germination until conditions become more favorable, improve planting ability and emergence of small-seeded species, and enhance seed distribution of aerial seedings without impacting existing native vegetation (Madsen et al., 2016). While application of these technologies may significantly increase costs, they may offer great savings on cost per established seedling in harsh rangeland environments.

7.3 Stand establishment and persistence

To be useful, restored populations of long-lived perennial species should be able to persist through a variety of environmental stresses, both biotic and abiotic. After 2 years of spraying herbicide and discing, Cox and Anderson (2004) established a thriving native grass and forb community in western Utah, USA, on land previously occupied by crested wheatgrass, an exotic widely seeded in the cold-desert valleys of the Intermountain West. While early results were impressive, native species had been mostly extirpated 20 years later. By this time, crested wheatgrass was in decline, and the site had become colonized by downy brome (Anderson and Johnson, 2020). Contrasting results were seen with a native-plant seeding in a dry montane meadow in central Utah where another perennial exotic, smooth bromegrass (*Bromus inermis*) had been seeded for forage (Anderson and Johnson, 2020). After 2 years, the restoration seeding was regarded as a failure, being covered by weedy annuals. After 10 years, annual species continued to dominate with only a few seeded native species present. Nevertheless, after 20 years, many seeded grasses, forbs, and shrubs had formed a diverse plant community, along with smooth bromegrass as a subdominant. Most surprising was that 15% of the ground was bare despite the presence of the highly rhizomatous smooth bromegrass. But what will the future hold?

High-diversity seed mixes can be recommended as a general practice to enhance the probability of persistence of at least some species in the mix (Wilsey, 2021). Genetic diversity within populations can be increased by mixing populations of a species from across a targeted area (Broadhurst et al., 2008; Larson et al., 2000). Functional diversity, controlled by specific traits, should also prove useful for seed mixes. Such traits can be identified and selected for in restoration populations, including direct selection for persistence (Jones et al., 2022; Chivers et al., 2016).

Testing for persistence requires long-term trials, ideally in many different environments. With patience, field experiments designed to collect establishment data can also be extremely useful sources of persistence data. Alternatively, persistence data at the species level may be collected from long-term monitoring of restoration seedings though priority effects may come in to play if species are introduced sequentially (Wilsey, 2021).

Robins et al. (2013) reported testing results for 18 cool-season grass plant materials in 34 trials conducted in eight western states, USA. Sites receiving less than 310 mm of average annual precipitation displayed relatively poor establishment at 1-year post-seeding and relatively poor persistence at 3 years post-seeding. Rhizomes were able to increase the stand percentage of the native, western wheatgrass (*Pascopyrum smithii*), from 1 to 3 years but not for another rhizomatous native, thickspike wheatgrass (*Elymus lanceolatus*), though this species displayed the greatest establishment at low-precipitation sites. Rigby et al. (2018) tested 32 plant materials of seven native and two introduced C_3 grass species in the western USA states of Idaho, Utah, and Wyoming. The native grasses bottlebrush squirreltail (*E. elymoides*), bluebunch wheatgrass (*Pseudoroegneria spicata*), slender wheatgrass (*E. trachycaulus*), and Snake River wheatgrass (*E. wawawaiensis*) displayed establishment percentages as good as the successful exotic, 'Vavilov II' Siberian wheatgrass (*Agropyron fragile*). However, by 5 years post-seeding, western wheatgrass, thickspike wheatgrass, and Snake River wheatgrass were the only native grasses equaling Vavilov II.

In central Utah, USA, Clifford (2023) evaluated 52 plant materials of 10 native and three exotic grass species for establishment and persistence traits. Based on a multivariate analysis of data, 25 plant materials qualified for Group 2, which scored higher for both establishment and persistence than three other groups. These included all plant materials of the three exotic species, all plant materials of natives basin wildrye (*Leymus cinereus*) and Snake River wheatgrass, and some plant materials of the natives, bluebunch wheatgrass, salina wildrye (*Leymus salina*), and thickspike wheatgrass. These kinds of evaluation data are invaluable for informing seed-mix design.

8 The native seed industry

8.1 United States of America

Due to high demand for native seed and the ability of the native seed industry in the Intermountain West to supply that demand, the magnitude of this industry is greater than anywhere else in the world (Jones, 2019). Demand is driven by state and federal government response to wildfire, government nonemergency needs, private landowners participating in government programs, and corporate procurement to fulfill government mandates. Unfortunately, increasing disturbance, particularly increasing wildfire combined with exotic annual grass invasions, is driving increasing demand.

Aridity in the Intermountain West is a challenge to restoration with native species, but success has been increasing. Native seeds are supplied by an industry of private companies. Some specialize in wildland-seed collection, mostly of shrub seed, particularly Wyoming big sagebrush (*Artemisia tridentata* ssp. *wyomingensis*) (Stevenson, 2016). Other companies specialize in the field production of herbaceous species, primarily a diverse array of native cool-season grasses, along with several native forbs (Jones, 2019), while many companies offer both. In the more arid areas, success has been much greater with species introduced from central Asia (Rigby et al., 2018; Clements et al., 2017; Robins et al., 2013). Thus, most companies produce or offer seeds of exotic cool-season grasses and forbs as well as natives (Jones, 2019).

The wildland seed collection industry originally developed in central Utah and now dominates supply for the western USA. Two hundred species may be collected in a typical year (Stevenson, 2016). While shrub species are the primary objective of wildland collection, forb and grass species are collected as well. Most shrubs do not lend themselves to field production, as they may take several years to reach reproductive maturity, which makes them less profitable in a cultivated setting than herbaceous species (Jones, 2019). Wildland-collected seed constitutes about a third of the seed sold to the Bureau of Land Management, the primary federal agency responsible for rangeland management and the largest seed-buying customer in the Intermountain West (Stevenson, 2016). Roughly similar quantities of this seed are harvested from state and federal lands versus privately owned lands. Annual harvests of big sagebrush seed may net over 450 000 kg, with totals of all species collected ranging from 567 000–800 000 kg annually.

While land ownership in the Intermountain West is predominantly public (Jones 2019), land ownership in the Midwest, USA it is mostly private, with the predominant use being cropland. Thus, the biggest driver of restoration acreage in the Midwest has been the Conservation Reserve Program. This federal cost-share program encourages agricultural producers to establish habitat-friendly seedings for wildlife that sequester more soil C on existing farmland than annual crops (Zinnen and Matthews, 2022).

Like the Intermountain West, the Midwest also has a well-developed native-plant seed industry. Zinnen and Matthews (2022) examined how this industry is serving the needs of restoration in the region with native-seed mixes. Their database included 1031 seed mixes from 32 vendors, encompassing 486 native species. The seed mixes were divided into six categories: pollinator habitat, tallgrass prairie, wetland, wet prairie, dry prairie, and woodland/ savanna. Species richness was greatest for pollinator habitat and least for tallgrass prairie, while the percentage of graminoids in the mix displayed a reverse pattern. Zinnen and Matthews (2022) concluded that seed-mix diversity is lacking. For example, the most diverse relict areas display over 100 native species, but most mixes had fewer than 30 species with a mean species richness of 22.5, leading to simplified plantings. An important component, Cyperaceae species (sedges), was either absent or underrepresented. Mixes had few, if any, early blooming forbs, which may be due to their slow growth and/or poor seed-production potential. Zinnen and Matthews (2022) identified 'popular species,' based on the percentage of mixes that included the species and its percentage contribution by weight. The most popular species, in this case warm- and cool-season grasses and early seral showy forbs, are known in the seed industry as 'workhorse species.' They establish easily, display broad adaptation, produce prolific quantities of seed, dominate the seed trade, and sell for a reasonable price. While low seed cost is advantageous to practitioners' budgets, it may lead to overuse of these species, thus reducing diversity within and among seedings.

The USA has a well-developed system for native-seed certification termed 'pre-variety germplasm' (PVG; Jones and Young, 2005). PVGs may be released under two tracks (natural or manipulated) and four classes (source identified, selected, tested, or variety/cultivar). Manipulated track indicates hybridization and/or artificial selection, the intentional selection for specific traits, while natural track indicates their absence. The four PVG classes refer to the amount of research-testing data provided at the time of release, i.e. comparisons of the given pre-variety germplasm with other plant materials of the same species. This ranges from no testing (source identified) to a single year of testing (selected) to testing across generations to establish heritability (tested) to cultivar (multiple years). Along with track and class, the generation number (downstream from the original source) is included on the seed tag.

8.2 Europe

Compared to the USA, the scale of restoration in Europe is small. Wildland-harvested seed is limited by both availability and demand (Tischew et al., 2011). The European native-seed industry primarily consists of small- to medium-sized companies, which favor the development of seed-transfer zones, geographical regions within which seed from a particular point-of-origin can be used (De Vitis et al., 2017). In 2017, 100 native-seed production organizations were identified in Europe, 93% of which were private companies and nearly half of which were concentrated in Germany, Spain, Switzerland, and the UK. This industry focuses on herbaceous species for grassland restoration but is limited by a lack of experience with many individual native species and high production costs.

Prior to 2010, trade in wild seeds of commercial forage species was prohibited in Germany, but since 2020, legislation has allowed only native seeds to be used in the wild (Mainz and Wieden, 2019). Native seed companies have been hampered, on one hand, by seed legislation that limits wild seed production of forage species to protect the forage seed industry, and on the other, by the lack of EU Nature Conservation standards requiring proof of provenance, zones of seed origin, and seed propagation within these zones. For example, seed companies must compete with inexpensive foreign wildflower-seed mixes, which are currently legal if they don't include any of the 44 regulated forage species. Some European countries have regulations like Germany (e.g. Austria, France, Italy, Switzerland, UK, and Spain), while others have no regulations. Specific recommendations for native seed policy were made by Abbandonato et al. (2018).

8.3 Australia

In Australia, over 90% of restoration seed is wildland collected (Broadhurst et al., 2016). Wildland-collected seed is generally much more expensive than field-produced seed, but the quality tends to be poorer due to sub-optimal environments, such as drought conditions (Broadhurst et al., 2016). Contamination with invasive-species seed is also a problem. Small, fragmented native-species remnants used for wildland collection may be genetically depauperate (Broadhurst et al., 2008). Researchers have supported broader seed-transfer zones to expand genetic diversity and to accommodate changing climates (Prober et al., 2015; Breed et al., 2013).

Australia's acute need for much greater volumes of restoration seed has long been recognized (Broadhurst et al., 2015, 2016; Mortlock et al., 2000). Seed of many species needed for restoration are unavailable in the marketplace (Broadhurst et al., 2016). Meeting restoration targets, especially given climate change, will be challenging. Much more infrastructure, specifically strategically located seed production areas with warehousing space for seed storage, is critically needed (Broadhurst et al., 2015).

9 Weed management in restoration projects

Examples presented next describe successes and failures of efforts to control annual and perennial weedy species in restoration projects. Some weeds are pernicious despite multiple control techniques, including mowing herbaceous plants, cutting down trees, mowing shrubs, seeding competitive mixtures, and applying herbicides. Monitoring weed populations on the ground and sometimes in the seed bank is critical for guiding weed-control efforts.

9.1 Chemical control of rangeland weeds

Wilson and Pärtel (2003) applied glyphosate to initiate replacement of a 50-year-old rangeland stand of the exotic forage, crested wheatgrass, with a C_3/C_4 grassland in Grasslands National Park in southwestern Saskatchewan, Canada. An early May herbicide application was timed to control crested wheatgrass when it was actively growing but before blue grama (*Bouteloua gracilis*), the desired native C_4 grass, began to grow. Blue grama was broadcast-seeded on slightly tilled soil shortly after the herbicide was applied. By fall, crested wheatgrass control was only about 50% (Wilson and Pärtel, 2003). Therefore, glyphosate was applied with a tractor-mounted wick applicator multiple times in each of the four following years to control the taller crested wheatgrass amongst the shorter native grasses. This was followed by herbicide application with a hand-held wick for two more years. For additional control, a 1-cm clipping treatment was applied four times annually for 3 years. However, no combination of seeding, glyphosate, and mowing controlled crested wheatgrass, suggesting it will be very difficult to remove over the long term. Anderson and Johnson (2020) came to a similar conclusion for this species in Utah, USA.

Spotted knapweed (*Centaurea stoebe*) and downy brome were targeted for control in western Montana, USA, using treatments of broadleaf, grass, and glyphosate herbicides in combination with seedings of five perennial cool-season grass species, four being native (Mangold et al., 2015). Good weed control was achieved with herbicides, while a successful seeding helped to suppress knapweed but had little impact on downy brome. One of the native species, bluebunch wheatgrass, and the exotic, tall wheatgrass (*Thinopyrum ponticum*) were successful 4 years post-seeding, though initially they appeared to have only negligible establishment. This result affirms the conventional wisdom that one should avoid premature judgments regarding seeding success.

A fall application of pre-emergent herbicide successfully controlled two invasive annual grasses, downy brome and medusahead wildrye, on

Wyoming big sagebrush (*Artemisia tridentata* ssp. *wyomingensis*) rangeland in northeastern California, USA (Kyser et al., 2013). Control of the invasive grasses increased cover of native forbs.

9.2 Mechanical control of rangeland shrubs

Degraded Wyoming big sagebrush plant communities, featuring high shrub cover and a depleted herbaceous perennial understory, are widespread in western North America (Davies et al., 2021a). Mowing big sagebrush, followed by drill-seeding large native bunchgrasses, has been recommended to rejuvenate the understory, thus limiting invasion by exotic annual grasses. However, slow-to-recover biocrusts may sustain damage from the mechanical disturbances of mowing and seeding, and exposed microsites offer an opportunity for invasive annuals to gain a foothold, particularly with sagebrush dominance curtailed. Furthermore, big sagebrush cover rebounded over an 11-year recovery period. Similar results were previously reported by Wilder et al. (2019), who suggested that this shrub-reduction approach offers limited long-term success, particularly on sites with low resilience to disturbance.

9.3 Mechanical control of grassland weeds

In eastern Hungary, Török et al. (2012) addressed weed-control issues for grassland restoration of former cropland by seeding low-diversity, native cool-season grass seed mixtures. Basic grass diversity was recovered within 3 years using the seed mixtures, combined with annual mowing. These treatments successfully controlled annual weeds. However, control of perennial weeds, such as *Elymus repens*, *Thinopyrum intermedium*, and *Cirsium arvense*, proved to be much more difficult. No herbicides were used in this restoration experiment.

In the Great Hungarian Plain in eastern Hungary, Valkó et al. (2021) compared post-seeding weed response to 8 years of mowing vs. mowing for only 3 years, followed by abandonment. This study, in one of the largest (760 ha) grassland restoration projects in Europe, compared treatment responses in lower-elevation alkaline grasslands on nutrient-poor soils with species-poor vegetation to higher-elevation loess grasslands on highly fertile 'black-earth' soils. Abandonment after 3 years led to a decline of sown grasses, an increase of weeds, and a large weed-dominated seed bank, especially in the alkaline grasslands. The loess grasslands, on the other hand, were able to maintain dominance of native perennial grasses due to the ability of the rhizomatous perennial, smooth bromegrass, to control annual weeds. These results suggest that long-term monitoring, seed-bank analysis, along with long-term mowing, may be required to assure eventual restoration success.

In the Pannonian sand grassland of central Hungary, a plantation of black locust (*Robinia pseudo-acacia*) was cut down, and stumps were treated with herbicide to prevent resprouting (Reis et al., 2021). This leguminous tree is considered the most problematic invasive tree in Europe. Its ability to fix atmospheric N enriches the surrounding soil, impacting the vegetation. Mowing plus hay removal was conducted twice annually for 7 years to control weed and shrub establishment and to reduce N inputs into the soil. Herbicide plus mowing successfully eliminated black locust. However, mowing also encouraged secondary invasions of alien species in areas with low amounts of grassland in the vicinity. Without mowing, however, dense cover of native shrubs returned.

10 Dealing with the legacy of degraded soils

Restoration of plant communities may require restored soils as a prerequisite, particularly when unnaturally high N and/or phosphorus (P) levels are present. Enriched soils drive growth of ruderal and exotic plant species, thus interfering with reestablishment of native-species plant communities. In Flanders, northern Belgium, species-rich semi-natural grasslands are some of the most diverse in the world. This plant community assembled due to a long history of low soil nutrients, resulting from grazing and haying without application of chemical fertilizers (Goossens et al., 2022). After widespread adoption of fertilizers post WWII, however, high levels of soil P and sometimes N were sustained, and most native species were lost.

Soil P can be depleted by mowing and clipping removal, or if that is insufficient, through a slow and labor-intensive process called P mining. This involves N and potassium fertilization to stimulate high P regrowth followed by mowing and clipping removal (Goossens et al., 2022). If the problem persists, removal of the P-enriched topsoil may be required. Once the abiotic soil component is addressed, seeding is needed to enrich the depauperate native-species seed bank, while depleting the ruderal seed bank through weed-control measures.

In Victoria, Australia, most successful restorations have been accomplished on land where N and P levels fall below pre-determined thresholds (Gibson-Roy and McDonald, 2014). Low P favors a greater diversity and quantity of native forbs, while high P favors broad-leaved weeds. Low N favors native grasses, while high N favors exotic grasses. Soil is first tested for N and P levels, as well as seed of ruderal species. If problems are detected, the site is scalped by removing the top 10 cm of soil, as a first step to restoring a grassy ecosystem.

Psammophilous (sand-loving) grasslands of central Europe are nutrient-poor habitats that are threatened by aerial deposition of industrial N and consequent domination by competitive grasses, such as *Calamagrostis epigejos*

(Řehounková et al., 2021). In the southeastern Czech Republic, 10–30 cm of contaminated soil was removed to expose soil low in nutrients and organic matter. This expensive process allowed psammophilous vegetation, collected in the vicinity from intact parcels, to quickly recolonize the impacted area. After 6 years, vegetation of the targeted area approached that of reference sites.

Downy brome is a pernicious, exotic annual weed of Eurasian origin that fuels ecosystem-altering wildfires in the Intermountain West, USA. In this region, downy brome converts soils from a tightly connected nutrient-cycling system, with rapid nutrient uptake by plants and microbes, to a leaky system featuring high nutrient availability courtesy of soil mineralization (Norton et al., 2007). A recurring question is, why is this species invasive in this alien New-World environment yet non-invasive in its home Old-World environment? In the Intermountain West, downy brome can increasingly favor itself over native perennials by increasing soil-nutrient availability via soil mineralization. This response is akin to that of cultivation and wildfire, both of which also release nutrients to the soil (Norton et al., 2007). Thus, not only is downy brome invasive because of its growth traits, but also because of its ability to perpetuate its favored environment. On the other hand, in downy brome's native Old-World environments, nutrient depletion long ago deprived this grassy annual of its ability to spread as an invasive weed.

11 Atmospheric nitrogen deposition

Atmospheric N deposition can dramatically impact low-nutrient grasslands. Rather than being a legacy of the past, however, N deposition from agricultural, industrial, and exhaust emissions is typically an ongoing problem (Fenn et al., 2003). Moreover, it exerts its influence at the landscape level rather being restricted to a locality (Fenn et al., 2003). Nutrient additions can reduce species richness of grasslands by extirpating less-competitive species and by limiting colonization by novel ones (Muehleisen et al., 2023).

Thirteen years of N application reduced the size and richness of the seed bank (forbs, sedges, and C_3 grasses) of an acidic (pH = 3.5–4.0) *Festuca-Agrostis* grassland in the Peak District National Park, UK (Basto et al., 2015). This site, proximal to the historically industrial cities of Manchester, Sheffield, and Leeds in northern England, had received one of the UK's highest rates of twentieth-century N deposition. The impact on the seed bank was greater than on the aboveground vegetation, thereby degrading the seed bank's functions as a buffer against local extinctions and a reservoir of genetic diversity.

In northern California, USA, serpentine grasslands host 10% of the state's endemic plant species, ten times the number expected based on the state's land area occupied by this ecosystem (Hernández et al., 2021). Serpentine soils are stony, shallow, low in macronutrients, and high in heavy metals. Native-species diversity is high, but under N-enriched conditions, this ecosystem is susceptible to exotic, annual grass invasion, particularly annual ryegrass (*Lolium multiflorum*) (Vallano et al., 2012). A combination of prescribed fire and grazing can counter this threat, but it must be applied on an ongoing basis (Hernández et al., 2021).

12 Soil-carbon sequestration

While prolonged cultivation decomposes soil organic matter, stimulating soil-C emissions to the atmosphere, abandoned agricultural lands can sequester C as succession proceeds (Yang et al., 2019). However, this can be a slow process with limited C sequestration, particularly in the early years of succession. In a tallgrass prairie ecosystem in Minnesota, USA, Yang et al. (2019) reported results from the initial establishment of a high diversity of late-seral species, including grasses and legumes. In this experiment, high levels of C sequestration, comparable to that seen in years 13–22 of a natural succession, were achieved early in the restoration process.

Grassland plants allocate about 60% of their net primary productivity below ground as roots and store about 90% of their total C below ground as root biomass and soil organic C, accounting for about 34% of the global terrestrial stock of C (Bai and Cotrufo, 2022). Though soil is the largest terrestrial reservoir of C globally, soil C is being lost through erosion, respiration, and leaching, both in agricultural and natural ecosystems (De Deyn et al., 2011). In a 16-year experiment in northern England, cessation of NPK fertilizer and seeding of the N_2-fixing native, red clover (*Trifolium pratense*), increased C relative to N storage, reduced ecosystem respiration, increased soil organic matter, and improved soil structure. Soil aggregation likely plays a role in the physical protection of soil organic matter, thus facilitating sequestration of C and N.

Restoration activities that ameliorate degradation will also increase soil-C storage (Bai and Cotrufo, 2022). Because about half of global grasslands are degraded, soil-C sequestration can be favored by improved plant and livestock management. Cessation of tillage upon conversion of cropland to grassland improves C sequestration, while improved vegetation management, particularly seedings of N_2-fixing legumes, is also helpful. In addition, light or rotational grazing favoring plant and soil health can also stimulate soil-C stocks, particularly on grazing-susceptible grasslands dominated by cool-season species.

13 Case study: Utah's Watershed Restoration Initiative and the restoration of sagebrush-steppe rangelands

13.1 Background

In the western USA, piñon–juniper (P–J) woodlands, dominated by piñon pine (*Pinus edulis*) and juniper (*Juniperus monophylla*, *J. osteosperma*), have expanded from less than 3 million ha prior to European settlement to 50 million ha today (Miller et al., 2019; Fig. 1). Over the last 160 years, these woodlands have encroached into big sagebrush (*Artemisia tridentata*) shrub-steppe communities (Fig. 2). This has resulted from natural range expansion, fire (Fig. 3) suppression, livestock grazing that reduced fuels, and favorable wetter climates in past centuries. In turn, this has led to serious reductions in herbaceous vegetation and species diversity; increases in flammable, exotic annual grasses; and instability, erosion, and impaired hydrology of impacted soils.

Figure 1 Dense piñon–juniper woodland. Source: Photo courtesy of Watershed Restoration Initiative.

Figure 2 Mule deer in big sagebrush. Source: Photo courtesy of Watershed Restoration Initiative.

Figure 3 Wildfire burning on a piñon–juniper site. Source: Photo courtesy of Watershed Restoration Initiative.

Since the 1940s, woody plants, both trees and shrubs, have been removed from the sagebrush steppe to release soil moisture and nutrients for the herbaceous understory, often in combination with seedings of perennial herbaceous species, particularly grasses (Munson et al., 2020; Miller et al., 2019). However, today shrub reduction is practiced much less frequently than in the past (Riginos et al., 2023). This is due to a 2015 moratorium on sagebrush reduction instated by the U.S. Department of the Interior to protect Greater sage-grouse (*Centrocercus urophasianus*) habitat on federal lands. While the moratorium has greatly reduced the loss of sagebrush habitat, losses to wildfire, largely driven by downy brome (Fig. 4), have continued (Riginos et al., 2023).

Utah's Watershed Restoration Initiative (WRI; https://watershed.utah.gov and wri.utah.gov/wri/) is a public–private partnership-driven effort dedicated to improving high-priority watersheds in Utah. WRI focuses on (1) watershed health and biological diversity, (2) water quality and yield, and (3) opportunities for sustainable uses of natural resources. Each of five regions in the state has a team that elects its own leaders, evaluates project proposals, and assists its members in project implementation. Hence, the WRI is designed to be a bottom-up hierarchy. Centralized efforts are administered by the Utah Department of Natural Resources (UDNR) in Salt Lake City but are limited to the project database, administration (contracting, accounting support), fundraising, marketing, and training. In short, the UDNR administration supports local decisions for projects and their implementation.

The WRI assists its local partners in applying restoration treatments, while monitoring data are collected by the Big Game Range Trend Studies program, directed by the Great Basin Research Center in Ephraim, Utah, a unit of the

Figure 4 Downy brome, also known as cheatgrass, occupying a stand of Wyoming big sagebrush (*Artemisia tridentata* ssp. *wyomingensis*) in Idaho's Snake River Plain. Source: Photo courtesy of Tom Monaco, USDA-ARS.

UDNR's Division of Wildlife Resources (UDWR). This is where seed is bought in bulk (6.53 million kg through 2022), stored in a seed warehouse (Fig. 5), and mixed before application. Since its inception in 2006, the WRI has completed 2672 projects encompassing 1 million ha (Fig. 6) with US$370 million of state funding plus US$43 million of in-kind funding. Currently, 274 projects are in progress covering 255 000 ha, with more proposed for the future. Proactive restoration accounts for nearly two-thirds of the projects, with the remainder being fire rehabilitation. Most range trend (monitoring) studies are conducted on winter range for mule deer (*Odocoileus hemionus*) and Rocky Mountain elk (*Cervus canadensis*).

13.2 Restoration treatments

The UDWR applies treatments to modify the vegetation of woodlands and shrublands to improve habitat for wildlife. These treatments are described here. For *herbicide/drill seeding*, a pre-emergent herbicide is sprayed from an aircraft to reduce downy brome (Fig. 7). Then a tractor-drawn rangeland drill is used to plant seed at the specific desired depth (Fig. 8). Woody-reduction treatments include harrowing, chaining, and mastication (i.e. grinding, crushing, or chewing) (Munson et al., 2020). The overall management objective

Figure 5 Interior of the seed warehouse at the Great Basin Research Center, Ephraim, Utah. Source: Photo courtesy of Watershed Restoration Initiative.

of these mechanical treatments is to release perennial grasses and forbs from woody competition, while minimizing increase of flammable annual grasses (Riginos et al., 2019b). Harrowing involves pulling an Ely (toothed) chain or pipe harrow. *Harrowing/broadcast seeding* involves a broadcast seeder pulled behind the harrow (Fig. 9), which may be pulled either *one-way* or *two-way* (in opposite directions), depending on the management goal. *Chaining/aerial seeding* consists of dragging a heavy anchor chain (Fig. 10) in a U-formation between two bulldozers moving in parallel (Fig. 11). Seed is applied aerially (Fig. 12) following the first pass of the chain, which is then covered by a second pass moving in the opposite direction. Like harrowing, chaining may be either one- or two-way. *Mastication/aerial seeding* involves aerial seeding followed by shredding with a mulcher rotor (toothed rotating drum) attached to an excavator or tractor.

Treatments with high disturbance intensity, such as harrowing and chaining, are applied to sites with more advanced woodland invasion (Pyke et al., 2015). These manipulations reduce woody vegetation, but they can also have the undesirable effect of increasing infestations of downy brome due to soil disturbance. While harrowing is either one- or two-way, the latter is more disruptive, effective, and expensive. Seeding often accompanies woody reduction, but the decision whether to seed depends on the extent of herbaceous perennial cover present (Monaco and Gunnell, 2020). For example, sites selected for chaining have a degraded understory and nearly always need to be seeded.

Published by Burleigh Dodds Science Publishing Limited, 2024.

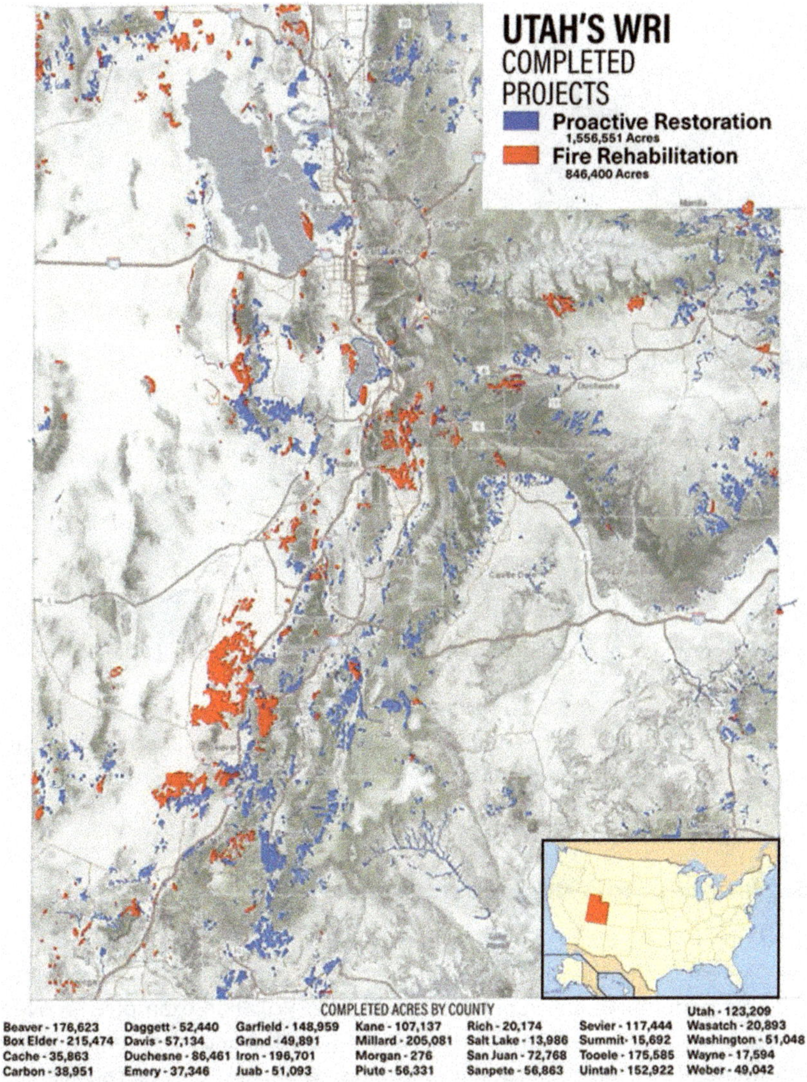

Figure 6 A map of restoration and rehabilitation projects completed by Utah's Watershed Restoration Initiative. In the lower right-hand corner, see an inset USA map with Utah shaded. Source: Photo courtesy of Watershed Restoration Initiative.

Mastication works well for removing trees in an advanced invasion without impacting desirable shrub cover. The mulched residue provides erosion control, but it may also favor establishment of downy brome or reduce emergence of seeded species. The use of mastication is growing due to its selectivity (i.e. greater control over the species and size classes of trees being removed), lower soil disturbance and erosion, and reduced fuel loads on the soil surface)

Figure 7 Spraying pre-emergent herbicide from a helicopter. Source: Photo courtesy of Watershed Restoration Initiative.

(Munson et al., 2020). Consequently, indiscriminate methods of woody-plant removal, such as harrowing and chaining, are being used less frequently than before.

13.3 Response to restoration treatments

Based on the data collected in the Big Game Range Trend Studies program, even 10 years of monitoring data may give false impressions concerning the post-treatment trajectory (Copeland et al., 2019b). For example, recovery of perennials, particularly native perennials, can be underestimated in this time frame. Thus, their efficacy in suppressing downy brome can also be underestimated. The findings presented here have been made possible by long-term collection of monitoring data over an extensive geographic region. Nevertheless, the impact of restoration treatments cannot be determined until a steady state is reached or a trajectory is clear. Thus, it is advisable to continue monitoring until data are analyzed, at which point an informed decision can be made whether to continue or terminate monitoring. Extended monitoring efforts obviously require dedicated funds and personnel.

Riginos et al. (2019b) reported responses of Wyoming big sagebrush (*A. tridentata* ssp. *wyomingensis*; drier lower elevations) and mountain big sagebrush

Figure 8 Drill-seeding with a ten-row seeder. Furrows are opened with discs, planted with seed from two planter boxes, and followed by chains that cover up the furrows. Source: Photo courtesy of Watershed Restoration Initiative.

(*A. tridentata* ssp. *vaseyana*; moister higher elevations) communities to fire and mechanical shrub-reduction treatments (with or without seeding) across 94 sites in Utah. All four mechanical treatments plus fire reduced sagebrush cover over the short term (1–4 years), but fire was more effective than all but the two-way harrow in the short term. However, this difference was less pronounced over the long term. All treatments increased perennial grass cover in Wyoming big sagebrush communities, probably resulting from the release of a degraded herbaceous understory from sagebrush competition. However, in mountain big sagebrush communities, perennial grass cover only increased following fire, to which big sagebrush is highly susceptible. Forb cover increased with seeding but only in the short term. Over the long term, the fire/seeding treatment resulted in the greatest increase in perennial grass cover, suggesting shrub reduction combined with seeding is necessary to achieve desired outcomes.

Monaco and Gunnell (2020) monitored woodland encroachment over 15 years at 129 sagebrush sites following P–J reduction. P–J reduction increased shrubs and herbaceous vegetation, including downy brome, to varying extents depending on the treatment and sagebrush plant community. However, the longevity of reduction may not exceed a decade, depending on tree density and the native vegetation, two variables that are often inversely related.

Figure 9 A Dixie harrow, used to remove shrubby vegetation and scarify the soil for seeding. Photo courtesy of Watershed Restoration Initiative.

Chaining and mastication are used when tree invasion is advanced. Both treatments resulted in greater herbaceous perennial recovery but greater downy brome issues due to soil disturbance. This may lead to achievement of short-term goals for ecosystem services (e.g. reduced competition for water, stimulation of herbaceous perennials, or predisposition to ecological risks), but also a long-term exacerbation of downy brome. In contrast to these high disturbance-intensity treatments, tree-cutting can be employed in more recent invasions. Cutting, also referred to as 'lop and scatter' (Fig. 13), is more appropriate when tree density is low and seeding is unnecessary (Monaco and Gunnell, 2020; Mller et al., 2019). This method maintains shrub and herbaceous cover with little ground disturbance, but only minimal enhancement of the perennial herbaceous understory is expected, as it was already in good condition.

Recovery of herbaceous perennials was greater on seeded sites, suggesting that these species are often seed limited and seeding is beneficial (Monaco and Gunnell, 2020). However, sagebrush seeding did not enhance sagebrush recovery. Sagebrush establishment may be limited by the infrequency of germination episodes, dry spring weather, grazing by big game, and competition from seeded herbaceous perennials. In summary, seeding strongly improves recovery of the perennial herbaceous understory and is essential for reversal of

Figure 10 A large chain, pulled by two bulldozers moving in parallel, being used to remove piñon and juniper trees. Photo courtesy of Watershed Restoration Initiative.

Figure 11 A tractor pulling both ends of a chain for treating small areas. Photo courtesy of Watershed Restoration Initiative.

Figure 12 Aerial seeding by helicopter. Photo courtesy of Watershed Restoration Initiative.

P–J encroachment. Though mountain big sagebrush communities have been regarded as more resilient than Wyoming big sagebrush communities due to the former's higher elevation and more favorable environment, their treatment responses were similar.

Due to a dearth of cost-tracking and monitoring data, accountability for restoration and rehabilitation projects has been limited (Munson et al., 2020). However, cost-effectiveness is expected to become increasingly important as project sizes, inflation-adjusted costs, and complexities increase, yet resources fail to keep pace (Munson et al., 2020; Copeland et al., 2018). Munson et al. (2020) examined monitoring results of 74 sites from range trend studies. In this meta-analysis, woody plants regrew similarly regardless of reduction method, either harrowing, chaining, or mastication. All three treatments increased downy brome cover due to soil disturbance caused by heavy equipment. Chaining was more effective at increasing perennial grass cover than harrowing, and a second pass of harrowing led to increased cover over the first pass. Seeding of perennial grasses increased perennial cover relative to downy brome up to 12 years post-treatment. However, while seeding of forbs increased species richness in the short term, forbs began to decline after two to three years as woody plants recovered and perennial grasses established. At 9% of the sites, seeding reduced perennial cover due to the inability to

Figure 13 Lop and scatter (tree-cutting) in a piñon–juniper community. Photo courtesy of Watershed Restoration Initiative.

compensate for reductions caused by the disturbance of the woody-removal operation.

13.4 Climate and weather impacts on restoration efficacy

Recovery of both woody and herbaceous species depends on climate and extreme-weather events, with recovery being limited by low precipitation and high temperatures (Copeland et al., 2019b). Consequently, recovery is expected to be lower in the future, as climate change is expected to result in drier and warmer conditions (Copeland et al., 2019a). As weather may alter competition during establishment, weather anomalies during this time can greatly alter long-term community composition. While positive effects of high precipitation from April to September did not persist after 5 years, extreme annual post-treatment temperatures had long-lasting negative effects. Balazs et al. (2020) also concluded that temperature is a stronger driver of community assembly than precipitation.

13.5 Natives and exotics

Seeding exotic perennial grasses, such as crested wheatgrass, after woody-plant removal can accelerate perennial cover and reduce downy brome but also inhibit native perennial cover (Copeland et al., 2019b). Nevertheless, exotics continue to be included in seed mixes though at lesser rates. Their inclusion is due to their high establishment, their high productivity, and their ability to stabilize soils and suppress downy brome. Every 10% decrease in downy brome cover resulted in a 1.5% increase in native cover.

13.6 Functional traits of plant materials

Balazs et al. (2020) compared restoration species with varying levels of seed mass, plant height, and specific leaf area (SLA) for their seeding establishment on 150 sites in the Colorado Plateau ecoregion of southeastern Utah, USA. These functional traits were chosen for their relevance to important ecological strategies or trade-offs. For example, seed mass reflects plant strategy for reproduction and colonization. Whereas small seeds are more persistent in seed banks, large seeds promote persistence under limited resources. Plant height represents a trade-off between light competition (tall plants) and stress tolerance (short plants). SLA, leaf mass per unit area, reflects a trade-off between resource capture (high SLA) and resource-use efficiency (low SLA). Balazs et al. (2020) reasoned that choice of the best trait for the best site corresponds to the nature of the trait × environment interaction. They attempted to predict a species' restoration outcome at a site (dependent variable) by fitting the species' trait values, six climatic variables, and species × climate interactions (all independent variables) into a model. Height and seed mass interacted with a particular climate variable, temperature seasonality, which quantifies the divergence between summer and winter temperatures of a site. Thus, Balazs et al. (2020) concluded that both tall species and large-seeded species are favored by high temperature seasonality. In the Colorado Plateau, the sites that display the greatest temperature seasonality (hot summers and cold winters) are valley bottoms. Interestingly, this is where many UDWR rangeland restoration sites are located. Unlike height and seed mass, SLA did not display such an interaction, so no such pattern could be suggested for this trait. These patterns held for both native and exotic species.

13.7 Economic decision-making

In an economic analysis, Munson et al. (2020) evaluated the cost-effectiveness of woody-plant destruction, seeding treatments, and control of annual invasive grasses. Costs of increased perennial grass cover were driven more by woody-plant destruction costs than seeding costs. Increased spending on woody destruction effectively reduced woody-plant cover and favored perennial grass cover due to competitive release. Costs of woody-plant removal per unit removed are similar regardless of treatment. However, due to selective removal, the cost of mastication/aerial seeding is two to three times greater than other treatments, though this treatment yielded only small improvements in perennial grass cover.

Seeding costs were driven up by the greater expense of native seeds, particularly forbs, relative to crested wheatgrass and other exotic grasses. Seed mixes with more forbs, when applied aerially after mastication or chaining resulted in only a 2–3-year gain in perennial forb cover and species richness, with no measurable benefit thereafter. However, seed-mix costs were unrelated to the composition of perennial grass cover. While more-expensive native seeds increased seed-mix costs, this spending did not lead to greater native cover. Instead, due to its high growth rate and stress tolerance, the use of inexpensive crested wheatgrass seed led to increased exotic grass cover relative to native grass cover.

Herbicide/drill seeding was the only treatment that effectively reduced downy brome. To increase cost-effectiveness, Munson et al. (2020) recommended spending more on a pre-emergent herbicide to suppress downy brome and more on drill-seeding to increase perennial grass cover. Using an herbicide is relatively inexpensive, thus its use is cost effective. However, the herbicide negatively impacted perennial forb cover and species richness.

13.8 Tools

Monitoring on the ground is highly labor intensive and cannot always be achieved across vast rangeland areas. Copeland et al. (2019a) investigated the value of Landsat satellite imagery to alleviate these issues using SATVI (Soil-Adjusted Total Vegetation Index) to estimate vegetation cover, where 'total' refers to green plus senescent cover. They reported a high correlation between SATVI and ground-based data, suggesting this tool has potential for this application. An advantage of SATVI is a continuous data stream instead of infrequent records, as for traditional ground-based monitoring. As SATVI also collects data across geographical space, it avoids the high labor requirements that would limit the number of restoration sites that could be monitored.

Resistance and resilience (R&R) is a concept based on two measures of ecosystem integrity: (1) resistance, the ability to preclude invasion, in this case by downy brome, and (2) resilience, the ability to recover from disturbance (Riginos et al., 2023). High R&R sites display low soil temperature and high soil moisture, and in Utah, these variables are closely related to elevation. Researchers used data from 51 Utah sites to relate R&R classes (high, medium, and low) to changes in plant cover following disturbance. Seeding resulted in increased perennial grass cover following fire or mechanical treatment for all three R&R classes (Riginos et al., 2023). While over the short term (0–4 years), seeding increased perennial grass cover most at the low R&R sites, over the long term (5–12 years), cover increased only at the high R&R sites. Like grass cover, perennial forb cover increased in all R&R classes in the short term, but over the long term, an increase was only seen at high R&R sites, with an actual decline at low R&R sites. In response to fire or mechanical treatment, over the short term, big sagebrush decreased most at the high R&R sites and least at the low R&R sites. However, recovery appeared to be greater at medium and high R&R sites but lesser at low sites. Over the long term, annual grass cover increased slightly at high and medium R&R sites but declined at low sites. The researchers concluded that the R&R framework is useful for making broad relative predictions of restoration outcomes, but they cautioned that it should be used in concert with other decision-making tools rather than alone.

13.9 Restoration for Greater sage-grouse management

At least 10% cover of big sagebrush is considered necessary for suitability for Greater sage-grouse (Riginos et al., 2019a; Fig. 14), This shrub is in decline over vast areas of the Intermountain West due to its high mortality from wildfire, which is increasing in scope and frequency (Beck et al., 2009). Lower-elevation sagebrush communities are more prone to annual-grass invasion and wildfire than at high elevations (Chambers et al., 2014). The fire-recovery interval of big sagebrush can range from 30 to 120 years, exceeding the time frame of conservation planning (Lesica et al., 2007; Baker, 2006). Thus, the conventional wisdom of wildlife managers is that both wildfire and prescribed burns are detrimental to management of sage-grouse and other sagebrush obligates, as fire rotations in big sagebrush range from 70 to 240 years (Baker, 2006). Thus, the perception is that any fire essentially results in

Figure 14 Greater sage-grouse. Source: Photo courtesy of Watershed Restoration Initiative.

a semi-permanent loss of sage-grouse habitat (Riginos et al., 2019a; Baker, 2006).

Riginos et al. (2019a) found that when pre-fire sagebrush cover fell below 10%, recovery after fire was insufficient to meet habitat standards. However, when pre-burn sagebrush cover exceeded 10%, a recovery trajectory materialized by 6–10 years post fire, suggesting habitat standards could be met. This favorable trajectory was indicated by increasing perennial-grass and forb cover, increasing sagebrush cover, and declining downy brome.

14 Conclusion

Rewilding of grasslands and rangelands is challenged by changing climates, ecosystems, and landscapes that are associated with losses in biodiversity. At the same time, however, opportunities have arisen due to land-use change. Marginal lands being diverted from production agriculture offer a land base for rewilding in many locations around the world. This trend is also favored by increasing populations in urban areas that may provide political and economic support for rewilding. Three components are necessary to systematically move forward in this effort. First, strong ecological monitoring efforts are imperative. Second, newly developed plant materials that are more effective in modified environments must be developed. Third, a reliable demand must exist to support and incentivize a seed-production industry to supply that demand.

15 Future research needs

Our world is subject to increasing ecological modification, which is likely to be exacerbated in the foreseeable future by an increasing human population. Until population growth stabilizes and then begins to decline, research to identify soil and water conservation strategies will be necessary to maintain these resources and prevent large-scale social catastrophes. Prophylactic approaches to wildfire management and wildfire mitigation via restoration are also critical for sustaining ecosystems in a warming world. Control of invasive species must also be a research imperative, given an increasing array of novel ecosystems. Restoration of biocrusts, a burgeoning research area, may play an outsized role in limiting the niche space available to invasive species. A shortage of restoration seeds is commonly cited as a limiting factor for restoration, but research leading to improved seed-production practices is necessary to support a thriving seed industry that can respond to this need. Finally, the greatest challenge may be to apply scientific information in a real-world context that recognizes economic and political realities and limitations.

16 Where to look for further information

Grasslands

- 1927-present. Proceedings, International Grassland Congress (www.internationalgrasslands.org)
- 2005. Grasslands of the world. Suttie, Reynolds, Batello
- 2009. Grasses and grassland ecology. Gibson
- 2015. Land of sweeping plains: managing and restoring the native grasslands of south-eastern Australia. Williams, Marshall, Morgan, eds.

Rangelands

- 1975. Range management, 3rd ed. Stoddart, Smith, Box
- 1978-present. Proceedings, International Rangeland Congress (https://rangelandcongress.org)
- 2010. Range management: principles and practices, 6th ed. Holechek, Pieper, Herbel
- 2011. Conservation benefits of rangeland practices: assessment, recommendations, and knowledge gaps. Briske, ed.
- 2017. Rangeland systems: processes, management and challenges. Briske, ed.

Restoration ecology

- 2004. Assembly rules and restoration ecology: bridging the gap between theory and practice. Temperton, Hobbs, Nuttle, Halle, eds.
- 2006. Restoration ecology. van Andel, Aronson
- 2013. Novel ecosystems. Hobbs, Higgs, Hall, eds.

Ecological restoration

- 1999. Repairing damaged wildlands. Whisenant
- 2010. Linking restoration and ecological succession. Walker, Walker, Hobbs, eds.
- 2011. Restoring disturbed landscapes: putting principles into practice. Tongway, Ludwig
- 2015. Land of sweeping plains: managing and restoring the native grasslands of south-eastern Australia. Williams, Marshall, Morgan, eds.
- 2023. Ecological restoration: moving forward using lessons learned. Florentine, Broadhurst, Dixon, Gibson-Roy, eds.

Fire

- 1982. Fire ecology: United States and southern Canada. Wright, Bailey

Published by Burleigh Dodds Science Publishing Limited, 2024.

- 2014. Ecology of plant-derived smoke. Jefferson, Pennacchio, Havens-Young
- 2022. Global application of prescribed fire. Weir, Scasta

Biological soil crusts

- 2003. Biological soil crusts: structure, function and management. Belnap, Lange, eds.
- 2016. Biological soil crusts: an organizing principle in drylands. Weber, Büdel, Belnap, eds.
- 2020. Restoration Ecology (volume 28, special issue S2: Synthesis of emerging research and perspectives: using biological soil crusts for restoration. Antoninka, Faist, Chaudhary, Condon, Rodriguez-Caballero, Pyke, eds.)

17 Acknowledgments

Thanks to my colleague, Tom Monaco, who peer-reviewed the manuscript multiple times and made many helpful suggestions. Thanks also to Kevin Gunnell, who directed me to information about the Watershed Restoration Initiative for the case study. Thanks, too, to the Watershed Restoration Initiative for making images from their website available for this chapter. Finally, thanks to Rhiannon Smith and Nick Reid (University of New England, NSW) for their expert and thorough reviews of the submitted manuscript.

18 References

Abbandonato, H., Pedrini, S., Pritchard, H. W., et al. 2018. Native seed trade of herbaceous species for restoration: a European policy perspective with global implications. *Restoration Ecology*, 26, 820–826.

Allen, V. G., Batello, C., Berretta, E. J., et al. 2011. An international terminology for grazing lands and grazing animals. *Grass and Forage Science*, 66, 2–28.

Anderson, V. J. and Johnson, R. L. 2020. Roughing up smooth brome and dethroning crested wheatgrass with native plants: dominant to subordinate on Utah Rangeland. *Native Plants Journal*, 21, 281–289.

Antoninka, A., Faist, A., Rodriguez-Caballero, E., et al. 2020. Biological soil crusts in ecological restoration: emerging research and perspectives. *Restoration Ecology*, 28, S3–S8.

Baasch, A., Engst, K., Schmiede, R., et al. 2016. Enhancing success in grassland restoration by adding regionally propagated target species. *Ecological Engineering*, 94, 583–591.

Bagchi, S., Briske, D. D., Bestelmeyer, B. T., et al. 2013. Assessing resilience and state-transition models with historical records of cheatgrass *Bromus tectorum* invasion in North American sagebrush-steppe. *Journal of Applied Ecology*, 50, 1131–1141.

Bai, Y. and Cotrufo, M. F. 2022. Grassland soil carbon sequestration: current understanding, challenges, and solutions. *Science*, 377, 603–608.

Baker, W. L. 2006. Fire and restoration of sagebrush ecosystems. *Wildlife Society Bulletin*, 34, 177–185.

Balazs, K. R., Kramer, A. T., Munson, S. M., et al. 2020. The right trait in the right place at the right time: matching traits to environment improves restoration outcomes. *Ecological Applications*, 30, e2110.

Bardgett, R. D., Bullock, J. M., Lavorel, S., et al. 2021. Combatting global grassland degradation. *Nature Reviews Earth and Environment*, 2, 720–735.

Basto, S., Thompson, K., Phoenix, G., et al. 2015. Long-term nitrogen deposition depletes grassland seed banks. *Nature Communications*, 6, 6185.

Beck, J. L., Connelly, J. W. and Reese, K. P. 2009. Recovery of Greater sage-grouse habitat features in Wyoming big sagebrush following prescribed fire. *Restoration Ecology*, 17, 393–403.

Bestelmeyer, B. T., Brown, J. R., Havstad, K. M., et al. 2003. Development and use of state-and-transition models for rangelands. *Journal of Range Management*, 56, 114–126.

Bowker, M. A. 2007. Biological soil crust rehabilitation in theory and practice: an underexploited opportunity. *Restoration Ecology*, 15, 13–23.

Bowker, M. A., Reed, S. C., Maestre, F. T., et al. 2018. Biocrusts: the living skin of the earth. *Plant and Soil*, 429, 1–7.

Bradley, B. A., Curtis, C. A., Fusco, E. J., et al. 2018. Cheatgrass (*Bromus tectorum*) distribution in the intermountain Western United States and its relationship to fire frequency, seasonality, and ignitions. *Biological Invasions*, 20, 1493–1506.

Breed, M. F., Stead, M. G., Ottewell, K. M., et al. 2013. Which provenance and where? Seed sourcing strategies for revegetation in a changing environment. *Conservation Genetics*, 14, 1–10.

Briske, D. D. 2017. *Rangeland systems: processes, management, and challenges*. Cham: Springer Nature.

Briske, D. D., Bestelmeyer, B. T., Stringham, T. K., et al. 2008. Recommendations for development of resilience-based state-and-transition models. *Rangeland Ecology and Management*, 61, 359–367.

Briske, D. D., Fuhlendorf, S. D. and Smeins, F. E. 2006. A unified framework for assessment and application of ecological thresholds. *Rangeland Ecology and Management*, 59, 225–236.

Broadhurst, L., Driver, M., Guja, L., et al. 2015. Seeding the future - the issues of supply and demand in restoration in Australia. *Ecological Management and Restoration*, 16, 29–32.

Broadhurst, L. M., Jones, T. A., Smith, F. S., et al. 2016. Managing seed resources for restoration in an uncertain future. *BioScience*, 66, 73–79.

Broadhurst, L. M., Lowe, A., Coates, D. J., et al. 2008. Seed supply for broadscale restoration: maximizing evolutionary potential. *Evolutionary Applications*, 1, 587–597.

Brown, N. A. C. 1993. Promotion of germination of fynbos seeds by plant-derived smoke. *New Phytologist*, 123, 575–583.

Buisson, E., Archibald, S., Fidelis, A., et al. 2022. Ancient grasslands guide ambitious goals in grassland restoration. *Science*, 377, 594–598.

Buisson, E., De Almeida, T., Durbecq, A., et al. 2021a. Key issues in Northwestern Mediterranean dry grassland restoration. *Restoration Ecology*, 29, e13258.

Buisson, E., Fidelis, A., Overbeck, G. E., et al. 2021b. A research agenda for the restoration of tropical and subtropical grasslands and savannas. *Restoration Ecology*, 29, e13292.

Chambers, J. C., Bradley, B. A., Brown, C. A., et al. 2014. Resilience to stress and disturbance, and resistance to *Bromus tectorum* L. invasion in cold desert shrublands of western North America. *Ecosystems*, 17, 945–952.

Chiquoine, L. P., Abella, S. R. and Bowker, M. A. 2016. Rapidly restoring biological soil crusts and ecosystem functions in a severely disturbed desert ecosystem. *Ecological Applications*, 26, 1260–1272.

Chivers, I. H., Jones, T. A., Broadhurst, et al. 2016. The merits of artificial selection for the development of restoration-ready plant materials of native perennial grasses. *Restoration Ecology*, 24, 174–183.

Chua, M., Erickson, T. E., Merritt, D. J., et al. 2020. Bio-priming seeds with cyanobacteria: effects on native plant growth and soil properties. *Restoration Ecology*, 28, S168–S176.

Clements, C. D., Harmon, D. N., Blank, R. R., et al. 2017. Improving seeding success on cheatgrass-infested rangelands in northern Nevada. *Rangelands*, 39, 174–181.

Clifford, A. M. 2023. *An evaluation of stress tolerance in restoration plant species in response to fire, drought, or invasive plants through the lens of functional traits.* M.S. Thesis, Utah State University, Logan, Utah.

Condon, L. A., Pietrasiak, N., Rosentreter, R., et al. 2020. Passive restoration of vegetation and biological soil crusts following 80 years of exclusion from grazing across the Great Basin. *Restoration Ecology*, 28, S75–S85.

Conti, G. and Fagarazzi, L. 2005. Forest expansion in mountain ecosystems: "environmentalist's dream" or societal nightmare? *Plantum*, 11, 1–20.

Copeland, S. M., Munson, S. M., Bradford, J. B., et al. 2019a. Influence of climate, post-treatment weather extremes, and soil factors on vegetation recovery after restoration treatments in the southwestern US. *Applied Vegetation Science*, 22, 85–95.

Copeland, S. M., Munson, S. M., Bradford, J. B., et al. 2019b. Long-term planned non-native plant community trajectories suggest divergent responses of native and non-native perennials and annuals to vegetation removal and seeding treatments. *Restoration Ecology*, 27, 821–831.

Copeland, S. M., Munson, S., Pilliod, D. S., et al. 2018. Long-term trends in restoration and associated land treatments in the southwestern United States: vegetation treatment trends related to restoration. *Restoration Ecology*, 26, 311–322.

Copeland, T. E., Sluis, W. and Howe, H. F. 2002. Fire season and dominance in an Illinois Tallgrass Prairie Association. *Restoration Ecology*, 10, 315–323.

Cox, R. D. and Anderson, V. J. 2004. Increasing native diversity of cheatgrass-dominated rangeland through assisted succession. *Journal of Range Management*, 57, 203–210.

Cuello, N., López-Mársico, L. and Rodriquez, C. 2020. Field burn versus fire-related cues: germination from the soil seed bank of a South American temperate grassland. *Seed Science Research*, 30, 206–214.

Cuevas, Y. A. and Zalba, S. M. 2010. Recovery of native grasslands after removing invasive pines. *Restoration Ecology*, 18, 711–719.

Davies, K. W., Bates, J. D. and O'Connor, R. 2021a. Long-term evaluation of restoring understories in Wyoming big sagebrush communities with mowing and seeding native bunchgrasses. *Rangeland Ecology and Management*, 75, 81–90.

Davies, K. W., Leger, E. A., Boyd, C. S., et al. 2021b. Living with exotic annual grasses in the sagebrush ecosystem. *Journal of Environmental Management*, 288, 112417.

De Deyn, G. B., Shiel, R. S., Ostle, N. J., et al. 2011. Additional carbon sequestration benefits of grassland diversity restoration. *Journal of Applied Ecology*, 48, 600–608.

De Vitis, M., Abbandonato, H., Dixon, K. W., et al. 2017. The European native seed industry: characterization and perspectives in grassland restoration. *Sustainability*, 9, 1682.

Dixon, K. W., Merritt, D. J., Flematti, G. R., et al. 2009. Karrikolinide - a phytoreactive compound derived from smoke with applications in horticulture, ecological restoration and agriculture. *Acta Horticulturae*, 813, 155–170.

Fenn, M. E., Baron, J. S., Allen, E. B., et al. 2003. Ecological effects of nitrogen deposition in the western United States. *BioScience*, 53, 404–420.

Flematti, G. R., Ghisalberti, E. L., Dixon, K. W., et al. 2004. A compound from smoke that promotes seed germination. *Science*, 305, 977.

Gibson-Roy, P. and Mcdonald, T. 2014. Reconstructing grassy understories in South-Eastern Australia. *Ecological Management and Restoration*, 15, 111–122.

Gillson, L., Ladel, R. J., Araújo, M. B. 2011. Baselines, patterns and process. In R. J. Ladle and R. J. Whittaker (eds.) *Conservation biogeography* (pp. 31–44). Oxford: Wiley-Blackwell.

Goossens, E. P., De Schrijver, A., Schelfhout, S., et al. 2022. Phosphorus puts a mortgage on restoration of species-rich grasslands on former agricultural land. *Restoration Ecology*, 30, e13523.

Grman, E., Zirbel, C. R., Bauer, J. T., et al. 2021. Super-abundant C_4 grasses are a mixed blessing in restored prairies. *Restoration Ecology*, 29, e13281.

Hedberg, P. and Kotowski, W. 2010. New nature by sowing? The current state of species introduction in grassland restoration, and the road ahead. *Journal for Nature Conservation*, 18, 304–308.

Hernández, E., Shaw, E. A., Aoyama, L., et al. 2021. Fire versus grazing as tools to restore serpentine grasslands under global change. *Restoration Ecology*, 29, e13353.

Herrick, J. E. and Whitford, W. G. 1995. Assessing the quality of rangeland soils: challenges and opportunities. *Journal of Soil and Water Conservation*, 50, 237–242.

Hilty, J. H., Eldridge, D. J., Rosentreter, R., et al. 2004. Recovery of biological soil crusts following wildfire in Idaho. *Journal of Range Management*, 57, 89–96.

Hobbs, R. J., Higgs, E. S. and Hall, C. M. 2013. Defining novel ecosystems. In R. J. Hobbs, E. S. Higgs and C. M. Hall (eds.) *Novel ecosystems - intervening in the new ecological world order*. Chichester, West Sussex, UK: Wiley-Blackwell.

Hobbs, R. J., Higgs, E. and Harris, J. A. 2009. Novel ecosystems: implications for conservation and restoration. *Trends in Ecology and Evolution*, 24, 599–605.

Hodges, J. A., Price, J. N., Nicotra, A. B., et al. 2021. Smoke and heat accelerate and increase germination in fire-prone temperate grassy ecosystems. *Ecosphere*, 12, e03851.

Holechek, J. L., Pieper, R. D. and Herbel, C. H. 2010. *Range management: principles and practices*. 6th ed. New York, NY: Pearson.

Holfus, C. M., Rios, R. C., Boyd, C. S., et al. 2021. Preemergent herbicide protection seed coating: a promising new restoration tool. *Rangeland Ecology and Management*, 76, 95–99.

Hulvey, K. B., Leger, E. A., Porensky, L. M., et al. 2017. Restoration islands: a tool for efficiently restoring dryland ecosystems? *Restoration Ecology*, 25, S124–S134.

Published by Burleigh Dodds Science Publishing Limited, 2024.

Jones, T. A. 2019. Native seeds in the marketplace: meeting restoration needs in the Intermountain West, United States. *Rangeland Ecology and Management*, 72, 1017–1029.

Jones, T. A. and Young, S. A. 2005. Native seeds in commerce. *Native Plants Journal*, 6, 286–293.

Jones, T. A., Monaco, T. A., Larson, S. R., et al. 2022. Using genomic selection to develop performance-based restoration plant materials. *International Journal of Molecular Sciences*, 23, 8275.

Kiehl, K., Kirmer, A., Donath, T. W., et al. 2010. Species introduction in restoration projects - evaluation of different techniques for the establishment of semi-natural grasslands in Central and Northwestern Europe. *Basic and Applied Ecology*, 11, 285–299.

Kiss, R., Deák, B., Török, P., et al. 2018. Grassland seed bank and community resilience in a changing climate. *Restoration Ecology*, 26, S141–S150.

Kiss, R., Déak, B., Tóthmérész B, et al. 2021. Establishment gaps in species-poor grasslands: artificial biodiversity hotspots to support the colonization of target species. *Restoration Ecology*, 29, e13135.

Knapp, A. K. and Seastedt, T. R. 1986. Detritus accumulation limits productivity of tallgrass prairie. *BioScience*, 36, 662–668.

Kyser, G. B., Wilson, R. G., Zhang, J., et al. 2013. Herbicide-assisted restoration of Great Basin sagebrush steppe infested with medusahead and downy brome. *Rangeland Ecology and Management*, 66, 588–596.

Lamont, B. B. 2023. Grass species with smoke-released seed dormancy: a response to climate and fire regime but not photosynthetic pathway. *Plant Biology*, 25, 24–31.

Larson, S. R., Jones, T. A., Hu, Z.-M., et al. 2000. Genetic diversity of bluebunch wheatgrass cultivars and a multiple-origin polycross. *Crop Science*, 40, 1142–1147.

Lawson, C. S. J., Ford, M. A. and Mitchley, J. 2004. The influence of seed addition and cutting regime on the success of grassland restoration on former arable land. *Applied Vegetation Science*, 7, 259–266.

Lesica, P. Cooper, S. V. and Kudray, G. 2007. Recovery of big sagebrush following fire in Southwest Montana. *Rangeland Ecology and Management*, 60, 261–269.

Lewis, K., Barros, F. D. V., Moonlight, P. W., et al. 2022. Identifying hotspots for ecosystem restoration across heterogeneous tropical savannah-dominated regions. *Philosophical Transaction of the Royal Society B*, 378, 20210075.

Madsen, M. D., Davies, K. W., Boyd, C. S., et al. 2016. Emerging seed enhancement technologies for overcoming barriers to restoration. *Restoration Ecology*, 24(S2), S77–S84.

Mainz, A. K. and Wieden, M. 2019. Ten years of native seed certification - a summary. *Plant Biology*, 21, 383–388.

Mangold, J. M., Orloff, N., Parkinson, H., et al. 2015. Integrating herbicides and re-seeding to restore rangeland infested by an invasive forb-annual grass complex. *Ecological Restoration*, 33, 16–19.

Mcintyre, S. and Lavorel, S. 1994. Predicting richness of native, rare, and exotic plants in response to habitat and disturbance variables across a variegated landscape. *Conservation Biology*, 8, 521–531.

Miller, R. F., Chambers, J. C., Evers, L., et al. 2019. *The ecology, history, ecohydrology, and management of pinyon and juniper woodlands in the Great Basin and Northern Colorado Plateau of the western United States*. Fort Collins, CO, USA: Department of Agriculture, Forest Service, Rocky Mountain Research Station.

Monaco, T. A. and Gunnell, K. L. 2020. Understory vegetation change following woodland reduction varies by plant community type and seeding status: a region-wide assessment of ecological benefits and risks. *Plants*, 9, 1113.

Monaco, T. A., Jones, T. A. and Thurow, T. L. 2012. Identifying rangeland restoration targets: an appraisal of challenges and opportunities. *Rangeland Ecology and Management*, 65, 599–605.

Mortlock, W. 2000. Local seed for revegetation: where will all that seed come from? *Ecological Management and Restoration*, 1, 93–101.

Muehleisen, A. J., Watkins, C. R. E., Altmire, G. R., et al. 2023. Nutrient addition drives declines in species richness primarily via enhanced species loss. *Journal of Ecology*, 111, 552–563.

Muñoz-Rojas, M., Machado De Lima, N. M., et al. 2021. Restoring post-fire ecosystems with biocrusts: living, photosynthetic soil surfaces. *Current Opinion in Environmental Science & Health*, 23, 100273.

Munson, S. M., Yackulic, E. O., Bair, L. S., et al. 2020. The biggest bang for the buck: cost-effective vegetation management outcomes across drylands of the western United States. *Ecological Applications*, 30, e02151.

Navarro, L. M. and Pereira, H. M. 2015. Rewilding abandoned landscapes in Europe. In H. M. Pereira and L. M. Navarro (ed.) *Rewilding European landscapes*. New York, NY: Springer.

Norton, J. B., Monaco, T. A. and Norton, U. 2007. Mediterranean annual grasses in Western North America: kids in a candy store. *Plant and Soil*, 298, 1–5.

Prach, K. and Walker, L. R. 2020. *Comparative plant succession among terrestrial biomes of the world*. Cambridge, UK: Cambridge University Press.

Prober, S. M., Byrne, M., Mclean, E. H., et al. 2015. Climate-adjusted provenancing: a strategy for climate-resilient ecological restoration. *Frontiers in Ecology and Evolution*, 3, Article 65.

Pyke, D. A., Chambers, J. C., Pellant, M., et al. 2015. *Restoration handbook for sagebrush steppe ecosystems with emphasis on greater sage-grouse habitat*. Washington, DC, USA: Department of the Interior, U.S. Geological Survey.

Ramos, D. M., Valls, J. F. M., Borghetti, F., et al. 2019. Fire cues trigger germination and stimulate seedling growth of grass species from Brazilian Savannas. *American Journal of Botany*, 106, 1190–1201.

Rayburn, A. P. and Laca, E. A. 2013. Strip-seeding for grassland restoration: past successes and future potential. *Ecological Restoration*, 31, 147–153.

Redhead, J. W., Sheall, J., Bullock, J. M., et al. 2014. The natural regeneration of calcareous grassland at a landscape scale: 150 years of plant community re-assembly on Salisbury Plain, UK. *Applied Vegetation Science*, 17, 408–418.

Reed, S. C., Delgado-Baquerizo, M. and Ferrenberg, S. 2019. Biocrust science and global change. *New Phytologist*, 223, 1047–1051.

Řehounková, K., Jongepierová, I., Šebelíková, L., et al. 2021. Topsoil removal in degraded open sandy grasslands: can we restore threatened vegetation fast? *Restoration Ecology*, 29, e13188.

Reis, B., Kövendi-Jáko, A., Szitár, K., et al. 2021. Long-term effect of mowing on the restoration of Pannonian sand grassland to replace invasive black locust plantation. *Restoration Ecology*, 29, e13152.

Rey Benayas, J. M. and Bullock, J. M. 2012. Restoration of biodiversity and ecosystem services on agricultural land. *Ecosystems*, 15, 883–899.

Ridding, L. E., Redhead, J. W. and Pywell, R. F. 2015. Fate of semi-natural grassland in England between 1960 and 2013: a test of national conservation policy. *Global Ecology and Conservation*, 4, 516–525.

Rigby, C. W., Jensen, K. B., Creech, J. E., et al. 2018. Establishment and trends in persistence of selected perennial cool-season grasses in western United States. *Rangeland Ecology and Management*, 71, 681–690.

Riginos, C., Monaco, T. A., Veblen, K. E., et al. 2019a. Potential for post-fire recovery of Greater Sage-grouse habitat. *Ecosphere*, 10, e02870.

Riginos, C., Veblen, K. E., Thacker, E. T., et al. 2019b. Disturbance type and sagebrush community type affect plant community structure after shrub reduction. *Rangeland Ecology and Management*, 72, 619–631.

Riginos, C., Veblen, K. E., Thacker, E. T., et al. 2023. Resilience and resistance framework predicts regional vegetation responses to shrub reduction treatments in the sagebrush ecosystem. *Rangeland Ecology and Management*, 86, 35–43.

Robins, J. G., Jensen, K. B., Jones, T. A., et al. 2013. Stand establishment and persistence of perennial cool-season grasses in the Intermountain West and the Central and Northern Great Plains. *Rangeland Ecology and Management*, 66, 181–190.

Roncero-Ramos, B., Román, J. R., Gómez-Serrano, C., et al. 2019. Production of a biocrust-cyanobacteria strain (*Nostoc commune*) for large-scale restoration of dryland soils. *Journal of Applied Phycology*, 31, 2217–2230.

Rosentreter, R. 2020. Biocrust lichen and moss species most suitable for restoration projects. *Restoration Ecology*, 28, S67–S74.

Rowe, H. I. 2010. Tricks of the trade: techniques and opinions from 38 experts in tallgrass prairie restoration. *Restoration Ecology*, 18, 252–262.

Rubio, C. and Lázaro, R. 2023. Patterns in biocrust recovery over time in Semiarid Southeast Spain. *Frontiers in Microbiology*, 14, 1184065.

Slodowicz, D., Durbecq, A., Ladouceur, E., et al. 2023. The relative effectiveness of different grassland restoration methods: a systematic literature search and meta-analysis. *Ecological Solutions and Evidence*, 4, e12221.

Smith, D. D. 1990. Tallgrass prairie settlement: prelude to demise of the tallgrass ecosystem. In D. D. Smith and C. A. Jacobs (eds.) *Proceedings of the Twelfth North American Prairie Conference* (pp. 195–200). Cedar Falls, IA: University of Northern Iowa.

Smith, J. T., Allred, B. W., Boyd, C. S., et al. 2022. The elevational ascent and spread of exotic annual grass dominance in the Great Basin, USA. *Diversity and Distributions*, 28, 83–96.

Song, G., Hui, R., Yang, H., Wang, B. and Li, X. 2022. Biocrusts mediate the plant community composition of dryland restoration ecosystems. *Science of the Total Environment*, 844, 157135.

Staude, I. R., Segar, J., Temperton, V. M., et al. 2023. Prioritize grassland restoration to bend the curve of biodiversity loss. *Restoration Ecology*, 31, e13931.

Stevenson, R. M. 2016. Commercial wildland harvested seed and the Utah connection. *Rangelands*, 38, 273–277.

Stoddart, L. A., Smith, A. D., and Box, T. W. 1975. *Range management*, 3rd ed. New York, NY: McGraw-Hill Book Company.

Stringham, T. K., Krueger, W. C. and Shaver, P. L. 2003. State and transition modeling: an ecological process approach. *Journal of Range Management*, 56, 106–113.

Sturchio, M. A., Mackmick, J. E., Barron-Gafford, G. A., et al. 2022. Grassland productivity responds unexpectedly to dynamic light and soil water environments induced by photovoltaic arrays. *Ecosphere*, 13, e4334.

Published by Burleigh Dodds Science Publishing Limited, 2024.

Tischew, S., Youtie, B., Kirmer, A., et al. 2011. Farming for restoration: building bridges for native seeds. *Ecological Restoration*, 29, 219–222.

Toledo, D., Kreuter, U. P., Sorice, M. G., et al. 2012. To burn or not to burn: ecological restoration, liability concerns, and the role of prescribed burning associations. *Rangelands*, 34, 18–23.

Török, P., Brudvig, L. A., Kollmann, J., et al. 2021. The present and future of grassland restoration. *Restoration Ecology*, 29, e13378.

Török, P., Miglécz, T., Valkó, O., et al. 2012. Recovery of native grass biodiversity by sowing on former croplands: is week suppression a feasible goal for grassland restoration? *Journal for Nature Conservation*, 20, 41–48.

Török, P., Kelemen, A., Valkó, O., et al. 2018. Succession in soil seed banks and its implications for restoration of calcareous sand grasslands. *Restoration Ecology*, 26, S134–S140.

Valkó, O., Deák, B., Török, P., et al. 2021. Dynamics in vegetation and seed bank composition highlight the importance of post-restoration management in sown grasslands. *Restoration Ecology*, 29, e13192.

Valkó, O., Török, P., Deák, B., et al. 2014. Review: prospects and limitations of prescribed burning as a management tool in European grasslands. *Basic and Applied Ecology*, 15, 26–33.

Vallano, D. M., V., Selmants, P. C. and Zavaleta, E. S. 2012. Simulated nitrogen deposition enhances the performance of an exotic grass relative to native serpentine grassland competitors. *Plant Ecology*, 213, 1015–1026.

Wagle, P. and Gowda, P. H. 2018. Tallgrass prairie responses to management practices and disturbances: a review. *Agronomy*, 8, 300.

Wagner, M., Hulmes, S., Hulmes, L., et al. 2021. Green hay transfer for grassland restoration: species capture and establishment. *Restoration Ecology*, 29, e13259.

Walther, G. R., Post, E., Convey, P., et al. 2002. Ecological responses to recent climate change. *Nature*, 416, 389–395.

Wangchuk, K., Gyaltshen, T., Yonten, T., et al. 2013. Shrubland or pasture? Restoration of degraded meadows in the mountains of Bhutan. *Mountain Research and Development*, 33, 161–169.

Wedin, D. A. 1992. Grasslands: a common challenge. *Restoration and Management Notes*, 10, 137–143.

Westoby, M., Walker, B. and Noy-Meir, I. 1989. Opportunistic management for rangelands not at equilibrium. *Journal of Range Management*, 42, 266–274.

Wilder, L. E., Veblen, K. E., Gunnell, K. L., et al. 2019. Influence of fire and mechanical sagebrush reduction treatments on restoration seedings in Utah, United States. *Restoration Ecology*, 27, 308–319.

Williams, N. S. G. and Morgan, J. W. 2015. The ecology and dynamics of temperate native grasslands in south-eastern Australia. In A. Marshall, N. W. Williams and J. Morgan (eds.) *Land of sweeping plains: managing and restoring the native grasslands of south-Eastern Australia*. Clayton: CSIRO Publishing.

Wilsey, B. 2021. Restoration in the face of changing climate: importance of persistence, priority effects, and species diversity. *Restoration Ecology*, 29, e13132.

Wilson, S. D. and Pärtel, M. 2003. Extirpation or coexistence? Management of a persistent introduced grass in a prairie restoration. *Restoration Ecology*, 11, 410–416.

Yang, Y., Tilman, D., Furey, G., et al. 2019. Soil carbon sequestration accelerated by restoration of grassland biodiversity. *Nature Communications*, 10, 718.

Young, K. E., Grover, H. S. and Bowker, M. A. 2016. Altering biocrusts for an altered climate. *New Phytologist*, 210, 18–22.

Zinnen, J. and Matthews, J. W. 2022. Species composition and ecological characteristics of native seed mixes in the Midwest (USA). *Ecological Restoration*, 40, 247–258.

Chapter 13

Lessons from reforestation of agricultural landscapes in South-Eastern Australia

David Lindenmayer, David Smith, Daniel Florance, Clare Crane, Eleanor Lang, Angelina Siegrist, Michelle Young and Ben C. Scheele, The Australian National University, Australia

1 Introduction

Agriculture has had massive impacts on biodiversity and ecosystem function (IPBES, 2019; Gibbs and Salmon, 2015; Maxwell et al., 2016). The environmental effects of agriculture are reflected in many metrics: 60% of the world's mammal biomass is now livestock and only 4% of the remaining 40% is wild animals, with humans contributing 36% (Bar-On et al., 2018). An estimated 29% of the earth's arable land is degraded (Gibbs and Salmon, 2015), directly affecting 3.2 billion people (IPBES, 2019).

Restoration has been proposed as a way to overcome the challenge of land degradation, with a large number of restoration programs instigated around the world (e.g. see Palmer et al., 2016; Brancillion and Holl, 2020; Crouzeilles et al., 2020, 2016) and the United Nations declaring 2021-2030 to be the decade of restoration (United Nations, 2020). However, the scale of the restoration challenge is immense, with some workers estimating that the cost of remediating degraded land globally could exceed US$12 trillion (Crouzeilles et al., 2020) or approximately two-thirds of the GDP of the USA (Chapman and Lindenmayer, 2019).

http://dx.doi.org/10.19103/AS.2024.0134.21

The scale of restoration challenges in Australian agricultural landscapes is, like in most of the world, considerable (Gibbs and Salmon, 2015). There has been widespread transformation of Australia's natural landscapes for agriculture, including for livestock grazing, cropping, and mixed broadacre farming (Williams and Price, 2010; Walker et al., 1993; Norton and Reid, 2013; Dickman and Lindenmayer, 2022). As a result, ecosystems such as the temperate woodlands of south-eastern Australia are among the most heavily modified biomes globally (Fischer et al., 2009). For example, an estimated 12-15 billion trees have been cleared in the Murray-Darling Basin since European settlement (Walker et al., 1993), with much of this region formerly temperate woodlands (Prober and Thiele, 1995; Hobbs and Yates, 2000; Lindenmayer et al., 2022). Some temperate woodland ecosystem types have been 95-99% cleared (Lindenmayer et al., 2016b). Cleared pastures, small patches of remnant woodlands, and scattered paddock trees now characterize vast areas that were once extensive temperate woodlands (Lindenmayer et al., 2022; Gibbons and Boak, 2002; Manning et al., 2006). Likewise, within the broader temperate woodland region, other natural ecosystems such as temperate grasslands have experienced extensive losses (Department of the Environment and Energy, 2016).

In this chapter, we focus on restoration efforts in the temperate woodlands of south-eastern Australia. These ecosystems are not readily conserved by conventional approaches such as the establishment of large protected areas. This is for several key reasons. First, the vast majority of temperate woodlands is on the private property of tens of thousands of individual landholders. Second, most temperate woodlands occur on productive agricultural land unlikely to be transferred to a protected area network. Third, the vast majority of remaining patches of temperate woodlands are small, fragmented, and heavily degraded. Integration of agricultural production with biodiversity conservation therefore appears to be the best approach to achieve demonstrable conservation outcomes in these ecosystems. Major restoration programs to date have attempted to tackle a range of environmental problems directly or indirectly allied with biodiversity loss, including land degradation through soil loss, declining vegetation condition, and secondary salinity (e.g. Freudenberger, 2001; Freudenberger and Harvey, 2004; Gibbons and Freudenberger, 2006; Lindenmayer et al., 2022; Norton and Reid, 2013; Prober and Thiele, 2005; Stirzaker et al., 2002). Restoration programs that aimed to recover biodiversity have included the following: (1) the establishment of fences to protect patches of remnant vegetation from high-intensity livestock grazing (Spooner and Briggs, 2008; Fischer et al., 2009); (2) increasing the amount of native vegetation cover through plantings (Lindenmayer et al., 2010; Haslem et al., 2021; Mac Nally et al., 2010; Munro et al., 2007) and facilitating natural regeneration (Geddes et al., 2011); (3) the provision of artificial habitat elements such as

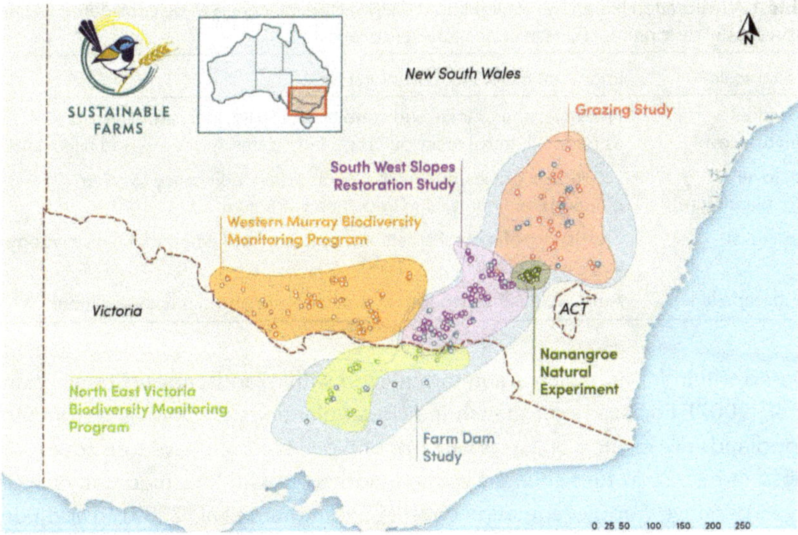

Figure 1 The spatial extent of long-term monitoring of vegetation management and restoration in temperate woodlands in the agricultural landscapes of south-eastern Australia. Restoration work in temperate woodlands encompasses a series of large-scale studies, all surveyed in broadly the same way, albeit answering different questions.

nest boxes (Lindenmayer et al., 2015; Goldingay et al., 2015); and (4) the control of the hyper-aggressive native bird species, the Noisy Miner (*Manorina melanocephala*; Melton et al., 2021; Crates et al., 2020; Grey et al., 1997), which engages in interference competition and drives away many other bird species (Beggs et al., 2020; Mac Nally et al., 2012; Maron et al., 2013).

The Sustainable Farms project at The Australian National University has, for the past 24 years (Fig. 1), examined a range of elements of restoration and revegetation programs, from individual trees and other structures (like dams and rocky outcrops) to plantings, whole farms, and across the broader landscape (Lindenmayer et al., 2022). This work has drawn on detailed empirical data gathered on over 800 long-term sites located on more than 300 farms; this chapter draws on insights for restoration programs derived from that project.

2 Reforestation at multiple spatial scales

A key lesson from the past 24 years of work in temperate woodlands of south-eastern Australia is that factors at multiple scales are critical for guiding thinking about effective reforestation of agricultural landscapes to maximize their value for biodiversity conservation (Table 1; Hingee et al., 2022a). Our long-term studies are well positioned to make recommendations in a structured hierarchy because some of the studies have a nested design; individual sites are

Table 1 A restoration hierarchy highlighting the key factors that need to be considered as part of successful restoration efforts in agricultural landscapes

Spatial scale	Important factors for consideration
Individual structure level	Presence, abundance, and condition of individual natural assets such as paddock trees, rocky outcrops, farm dams, mistletoe, and fallen logs
Site level (e.g. plantings)	Location in gullies, size, width, context (e.g. adjacency to other plantings, remnants), and vegetation structure
Farm level	The variety of natural assets across a farm and amount of native woody vegetation cover
Landscape level	Amount of native woody vegetation cover and climatic conditions

nested within farms, which are in turn nested within landscapes (Cunningham et al., 2007). For example, sites in individual plantings or patches of remnant woodlands are located on farms with low or high levels of native tree cover and these farms are in turn situated within landscapes with low, medium, or high levels of native woody vegetation cover (Cunningham et al., 2007). In addition, a hierarchically based statistical analysis has demonstrated the importance of factors at different scales influencing the occurrence of key elements of the biota such as temperate woodland birds (Hingee et al., 2022a; Hingee et al., 2022b). Consideration of scale effects in restored landscapes makes intuitive ecological sense given that different species typically respond to various factors at different scales (Allen and Hoekstra, 1992; Allen and Starr, 1988), and the same species can respond to factors at multiple scales (Mackey and Lindenmayer, 2001; Forman, 1964).

2.1 Individual structure level

The maintenance of individual structures within restored areas, such as replantings and patches of naturally regenerating woodland, can make a substantial contribution to their value for biodiversity. Our long-term data indicate that bird and reptile biodiversity in restored areas tends to be significantly higher when such areas support individual structures such as the following: (1) large old trees (which are the biological legacies of the previous natural woodland cover; Lindenmayer et al., 2010); (2) fallen trees and logs (Cunningham et al., 2007); (3) mistletoe (Lindenmayer et al., 2010); (4) farm dams (Smith et al., unpublished data); and (5) rocky outcrops (Michael and Lindenmayer, 2018; Michael et al., 2010). We have coined the term 'natural assets' for these structures (Lindenmayer et al., 2022). These structures provide microhabitats that are needed by particular species for shelter, foraging, or breeding. In some cases, individual structures can be maintained or recreated through targeted micro-restoration efforts. For example, micro-fencing programs can enable the establishment of new cohorts of scattered paddock trees (Lindenmayer et al., 2022) and, in the long term, restore the many values of such trees (such as providing stepping stones for animal movement through agricultural landscapes; Fischer and Lindenmayer, 2002); restoration of individual farm dams can have marked positive impacts, not only for biodiversity (Westgate et al., 2022) but also for reducing greenhouse gas emissions (Malerba et al., 2022).

2.2 Site level (e.g. whole-of-planting level)

Attributes of restoration plantings can have important positive impacts on their value for biodiversity. For example, plantings located in gully lines and around streams tend to support more species of birds than those located elsewhere in the landscape (Lindenmayer et al., 2010). Likewise, larger and wider plantings are superior to smaller and narrower plantings in terms of bird and reptile species richness (Lindenmayer et al., 2018b). Plantings that have an understorey and an intact leaf litter layer support more native bird species than where these structural features of plantings are absent (Lindenmayer et al., 2018b). An intact leaf litter also promotes reptile species richness (Lindenmayer et al., 2018b). Plantings that are dominated by exotic tree species are more likely to be used by exotic pest bird species like the European Starling (*Sturnus vulgaris*) (Lindenmayer et al., 2022). Finally, plantings connected to other plantings or patches of remnant native vegetation tend to support more bird species than restored areas that are isolated (Lindenmayer et al., 2007) (Figs 2 and 3).

2.3 Farm level

Work in the temperate woodlands has consistently shown that biodiversity is greatest on farms with high levels of native vegetation cover (Cunningham et al., 2014a, b, Hingee et al., 2022a). In addition, farm-level biodiversity is significantly increased when restored areas are added to the array of other natural assets on those farms, including patches of remnant native vegetation, scattered paddock trees, swards of native grasslands, and areas where woody debris remains on the ground (Cunningham et al., 2008). One of the reasons for this finding is that plantings, natural woodland regrowth, and old growth woodlands support different assemblages of bird species (Lindenmayer et al., 2012), and hence farms that support a range of vegetation types corresponding to these broad structural types tend to be more species rich. Similarly, there are differences in the mammal assemblages found in plantings, natural woodland regrowth, and old growth woodlands (Lindenmayer et al., 2020b) and again farms with different kinds of native vegetation types will be those where the greatest range of mammal species will most likely occur. One implication of these findings is that clearing old growth woodlands and

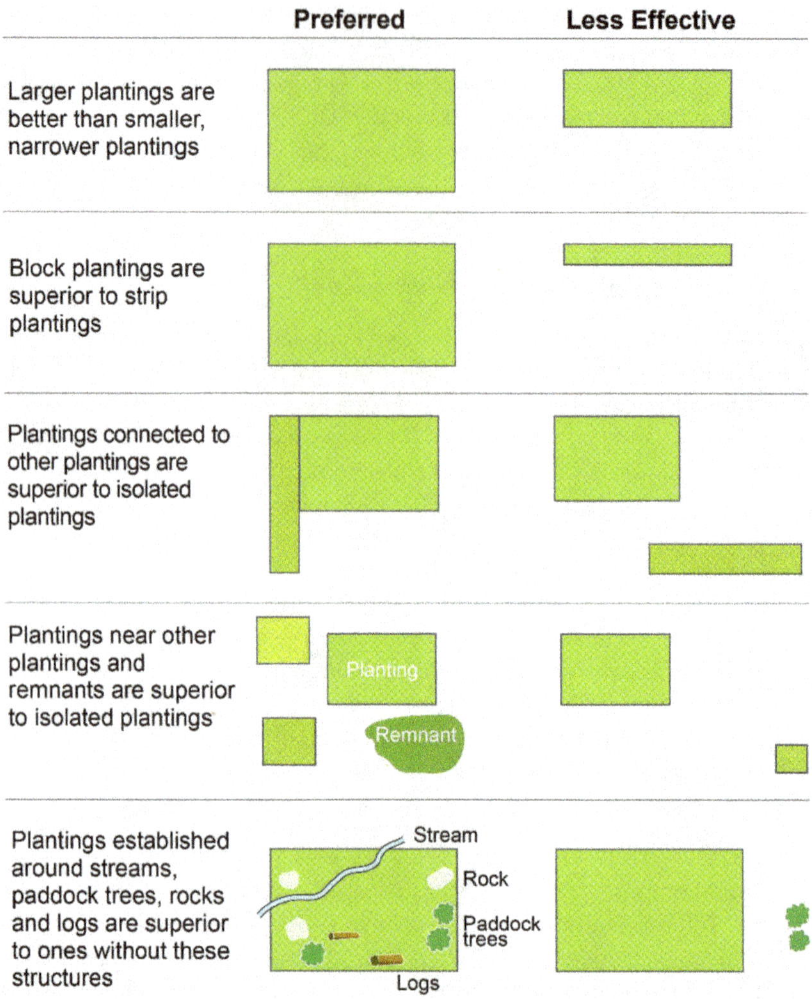

Figure 2 Key features of plantings that are better for wildlife. Redrawn from *Wildlife Conservation in Farm Landscapes* (Lindenmayer et al., 2016b).

then replacing them with replantings elsewhere on a farm is an invalid offset process. This is because they are markedly different habitats and the large old trees that characterize old growth woodlands can require many hundreds of years to be replaced. Similarly, removing large old trees that support many hollows and replacing them with nest boxes (that have a short effective lifespan for use by cavity-dependent fauna) is a flawed offset (Lindenmayer et al., 2017).

Figure 3 Intersected plantings in farming landscapes in south-eastern Australia support more bird species compared to isolated plantings. Note the planted area encompasses a large paddock tree in the foreground which can help boost the number of bird species that occur in restored areas (see Figure 2 and text; photograph by D. Florance).

2.4 Landscape and regional level

At a landscape scale (defined here as an area covering more than 10 000 ha), the total amount of native vegetation cover, including the amount of woodland that has been replanted, has a highly significant effect on bird biodiversity (Lindenmayer et al., 2020a; Hingee et al., 2022a). Indeed, increasing native woody vegetation cover is the single most effective action to promote biodiversity conservation in agricultural landscapes (Tulloch et al., 2016). Other factors at a landscape level can influence the effectiveness of restoration efforts. For example, the value of replanted areas for birds tends to be highest in high rainfall regions (Lindenmayer et al., 2018c). There also appear to be productivity effects, with some bird species better able to persist in the presence of interference competition from the hyper-aggressive Noisy Miner in landscapes with high levels of net primary productivity relative to where net primary productivity values are low (Lindenmayer et al., unpublished data).

2.5 Interactions between factors at different spatial scales

The preceding commentary has focused on a hierarchy of scale effects for restoration from individual structures to individual plantings, farms, and landscapes. However, factors at these different scales do not operate in isolation with, for example, efforts to increase the amount of vegetation at the individual planting level resulting in greater bird species richness at the farm and landscape levels (Cunningham et al., 2014a, b). In addition, different factors influence the suitability for birds of restored and naturally regenerated areas at different scales (Hingee et al., 2022a), with vegetation structure important at fine scales (individual planting levels), vegetation cover at farm and landscape scales, and climate variables at landscape and regional scales (Hingee et al., 2022a). Importantly, there is limited redundancy between drivers of bird occurrence at different scales. This means that fine-scale studies can benefit from consideration of climate context, and large-scale studies need to consider the importance of measures like vegetation structure, which are typically measured at relatively fine scales (Hingee et al., 2022a).

2.6 Connectivity issues

Connectivity of natural assets can have an important impact on the effectiveness of restoration efforts for biodiversity. At an individual structure level, scattered paddock trees can play an important stepping stone role for bird movements in agricultural landscapes (Fischer and Lindenmayer, 2002). Paddock trees are also valuable foraging habitats for species such as the Squirrel Glider (*Petaurus norfolcensis*) (Crane et al., 2014) and microbats (Lumsden and Bennett, 2005). At a site level (such as for individual plantings), intersecting plantings tend to support higher bird species richness than isolated plantings (Lindenmayer et al., 2007). In addition, adjacency effects, whereby plantings with better connectivity to other plantings and/or patches of remnant vegetation support more species than isolated plantings (Lindenmayer et al., 2010; Lindenmayer et al., 2007), are also likely to reflect the contribution of connectivity to species occurrence in agricultural landscapes where restoration programs have been implemented. Finally, greater bird diversity on farms and in landscapes characterized by higher levels of native woody vegetation cover (Cunningham et al., 2014a,b) is also likely to be related, at least in part, to greater levels of connectivity.

In some cases, reduced connectivity can be important for biodiversity conservation in restored landscapes. For example, disconnecting incised streamlines where continuous streamflow has developed can be critical to create chains of ponds, which are, in turn, important habitat for a range of amphibian species (Hazell et al., 2003). Similarly, coordinated poison baiting programs across contiguous farmland areas can be an effective way to reduce connectivity for populations of feral carnivores like the Red Fox (*Vulpes vulpes*) and reduce predation pressure on native prey species (e.g. see https://pestsmart.org.au/case_studies/bounceback-fox-control-in-the-flinders-ranges/).

3 Reforestation and temporal scales

So far, we have focused on spatial scale and context effects on the response of biodiversity to restoration efforts. However, there are also temporal considerations. For example, young plantings tend to support different biotic assemblages relative to older plantings, with distinct patterns of species turnover through time (Lindenmayer et al., 2016a). Hence, efforts to maximize measures like bird species richness on a farm require a well-planned and systematic approach to revegetation efforts with the regular addition of new areas of plantings to farms and landscapes over time to ensure that a range of planting age classes are represented at farm and landscape level. Similarly, some values of restored areas can take time to manifest, such as the decadal periods required for tree hollow formation (Vesk et al., 2008) and restoration efforts to suppress the occurrence of the Noisy Miner (Lindenmayer et al., 2018a) and in turn reduce its impacts on smaller bodied temperate woodland birds.

Other temporal perspectives on plantings include a decline over time in the effectiveness of infrastructure like fences that limit the impacts of grazing on vegetation in restored areas (Lindenmayer et al., 2018b). This highlights the need to repair or replace such infrastructure to ensure that the biodiversity values of restored areas are maintained (Lindenmayer et al., 2018b).

Overlaying these patterns is the observation that the conservation value of restored areas can vary over time depending on weather conditions and climate. For example, our long-term work, which has spanned drought and inter-drought periods, has shown that restored areas can be particularly important refuges for birds (especially small-bodied species of conservation concern) during prolonged dry periods, particularly plantings in areas with higher annual average rainfall (Lindenmayer et al., 2018c).

4 Improving the effectiveness of restoration programs

Many discussions about restoration programs focus on the need for landscape-level initiatives. We agree that this is important, but the empirical data from our work clearly indicate that factors at multiple scales from individual structures to farms and landscapes can influence the effectiveness of restoration programs for biodiversity in agricultural landscapes. That is, achieving effective landscape restoration requires consideration of factors at smaller spatial scales, like the location of restoration plantings on a farm, or the improvement in condition of natural assets like farm dams (Westgate et al., 2022). In the remainder of this chapter, we discuss some other considerations that are key to bolstering the effectiveness of restoration programs and to facilitating expansion of the extent of restored areas in agricultural landscapes.

4.1 Management of restored areas

Restored woodlands often need to be actively managed to maintain or enhance their biodiversity values. A key part of the management of woodlands, and especially replantings, is to limit grazing pressure. Grazing can directly reduce leaf litter and midstorey cover that in turn reduces bird and reptile diversity (Lindenmayer et al., 2018b) and indirectly through promoting tree dieback. The need to limit grazing pressure is increasingly important in many replanted woodlands as the fences erected when they were first established (often more than 25 years ago) are now degrading and need to be replaced (Lindenmayer et al., 2018b). Where fences need to be replaced or repaired, may provide an opportunity to reconfigure restored areas by, for example, expanding their width or adding an understorey to enhance their effectiveness for animal groups such as birds. Similarly, restored areas can be targets for pest animal control such as those for the Red Fox and the European Rabbit (*Oryctolagus cuniculus*).

In some cases, the biodiversity value of restored areas can be maximized where they are established around other natural assets like large old (paddock) trees, logs, farm dams, streams and creeklines, and rocky outcrops (see Figs 2 and 3). Structural augmentation of restored or planted areas can include towing logs into them from surrounding paddocks and erecting nest boxes where there is an absence of natural cavities in large old trees (Lindenmayer et al., 2015). However, the kinds of nest boxes added to restored areas can be critical to limit the risks of boosting populations of pest species like the Feral Honeybee (*Apis mellifera*; Cunningham et al., 2022) or the European Starling (Lindenmayer et al., 2017) or enhancing populations of species that are already common that do not need conservation actions (e.g. the Common Brushtail Possum [*Trichosurus vulpecula*]; Lindenmayer et al., 2015).

4.2 The importance of long-term monitoring

Almost everything that has been discussed in this chapter on restoration in temperate woodlands is based on the results of long-term monitoring and research. That work has produced the knowledge to inform evidence-based management (*sensu* Sutherland et al., 2004). Long-term monitoring has been critical because it has enabled the quantification of values of restoration that take time to emerge and has shown how those values vary across both extremely wet periods and droughts (Lindenmayer et al., 2018c). Such monitoring has, for example, highlighted the considerable value of plantings under dry conditions, when they act as refuges for small-bodied woodland birds (Lindenmayer et al., 2022). Indeed, temperate woodlands can be considered to be ecosystems where changes such as those resulting from restoration efforts can sometimes take prolonged periods to manifest. Hence the need to monitor changes over time, especially to determine the effectiveness of interventions such as grazing control or the establishment of plantings (such as underplantings in the understorey of otherwise degraded old growth woodlands). In the latter case, it took about 8 years before the effectiveness of enhancement plantings became apparent (Lindenmayer et al., 2018a). Such long-term work is particularly informative where the experimental design for investigations is set around testing ecological theory. Identifying robust, general patterns can allow extrapolation of results to other settings beyond the specific areas where research and monitoring has been conducted (Lindenmayer, 2020).

Our Sustainable Farms project is underpinned by a generalized program logic that has been broadly successful in ensuring the maintenance of long-term work. That model entails field-based ecologists who are part of farming communities, living and working permanently in different agricultural regions. They not only gather high-quality data in the long-term research and monitoring programs, but all are also involved in extensive outreach programs in the regions they work. This model approach helps facilitate knowledge transfer and accelerates the rate of uptake of new practices to improve restoration outcomes on sites, farms, and landscapes.

4.3 Catalyzing greater support for expanded restoration efforts

We argue that greater support for the expansion of existing and future restoration programs will be generated where it can be demonstrated that there are multiple co-benefits. An excellent case in point is the restoration of farm dams, an activity at the individual structure level (see Fig. 2). Restoration programs for farm dams entail controlling livestock access through fencing, creating high ground vegetation cover immediately surrounding dams to filter animal waste and sediment runoff (thereby preventing them from entering the water), establishing trees to shade the water body, and encouraging vegetation cover at the margins of the dam and in the water (Lindenmayer et al., 2022). The financial costs of these restoration actions can be recouped relatively quickly, primarily through the economic gains associated with improved livestock weight gain as a result of animals drinking better quality water (Dobes et al., 2021). There are also major biodiversity dividends associated with improving the water quality in farm dams (Westgate et al., 2022) and significant reductions in greenhouse gas emissions (Malerba et al., 2022). These benefits from renovating farm dams – for livestock production, biodiversity conservation, and climate change mitigation – all mean that restoration efforts are more likely to occur, either at a farmer's own expense or through co-investment programs with natural resource management agencies.

Another potentially important co-benefit lies with the use of carbon abatement and long-term carbon storage funding creating new streams of income for farmers that, if well designed, can also contribute to biodiversity conservation in agricultural landscapes. As an example, the areas of greatest biomass in agricultural landscapes are in riparian zones where trees and other plants grow largest and stand structural density can be highest (Soderquist and Mac Nally, 2000). These same areas are also where revegetation efforts result in the greatest biodiversity gains (Lindenmayer et al., 2010). Well-designed and implemented restoration efforts that target riparian areas can therefore yield carbon storage benefits and major biodiversity dividends (Lindenmayer et al., 2022).

Finally, a key part of engaging with landholders and natural resource managers is to be able to demonstrate directly the biodiversity dividends from programs that restore native vegetation cover. We have developed a novel computer tool that can be easily used by farmers and managers to predict changes in bird biodiversity likely to result from management actions such as establishing new areas of plantings (see Hingee et al., 2022b). The tool is not a simulation model but rather is based on statistical relationships constructed from detailed analyses of extensive field data gathered from hundreds of sites over almost 20 years (Hingee et al., 2022a). The model can make robust predictions based on the current amount of native vegetation cover, the character of the area in question (remnant vegetation or planting), the rainfall in the past year, and the presence or absence of the Noisy Miner (Hingee et al., 2022b). In essence, this provides a quantitative projection of the kinds of biodiversity dividends that can accrue from efforts to restore temperate woodlands in agricultural landscapes and increase the extent of native vegetation cover at different spatial scales - at a site level (for an individual planting or individual patch of remnant vegetation), at a farm level, and at a landscape level.

5 Conclusion

Revegetation is a major focus of restoration programs in agricultural landscapes around the world (Suding et al., 2015; Suding, 2011; Crouzeilles et al., 2016). Empirical evidence from our case study of temperate eucalypt woodlands in south-eastern Australia strongly suggests that effective restoration demands consideration of key factors at multiple spatial scales, from individual structures such as farm dams and isolated paddock trees to sites, whole farms, and landscapes. We present a restoration hierarchy (see Table 1) to highlight these considerations. We believe that the approach should be broadly applicable to restoration programs in other agriculturally dominated biomes outside the temperate woodlands where we have focused this chapter.

6 Acknowledgements

We thank the many farmers whom have allowed us access to their properties to monitor biodiversity and ecosystem condition over the past 20+ years. We thank the Ian Potter Foundation, the Riverina, Murray and Central Tablelands Local Land Services in NSW and the Goulburn-Broken and North East Catchment Management Authorities in Victoria for their support of the research and monitoring work that underpinned this chapter. Tabitha Boyer kindly assisted with a range of fiddly editorial tasks that ensured this article could be completed. We thank the Editors of this book for asking us to write this chapter. The work reported here is part of Sustainable Farms at The Australian National University.

7 Where to look for further information

Many key aspects of restoration ecology in temperate woodlands have been summarized in our book *Natural Asset Farming* (Lindenmayer et al. 2022). In addition, the Sustainable Farms website:

- (https://www.anu.edu.au/about/strategic-planning/sustainable-farms) supports a range of information brochures and other materials that can help private landholders access the latest science to guide woodland restoration efforts.

8 References

Allen, T. F. H. and Hoekstra, T. W. 1992. *Toward a Unified Ecology*, New York, Columbia University Press.

Allen, T. F. H. and Starr, T. B. 1988. *Hierarchy Perspectives for Ecological Complexity*, Chicago, The University of Chicago Press.

Bar-On, Y. M., Phillips, R. and Milo, R. 2018. The biomass distribution on Earth. *Proceedings of the National Academy of Sciences of the United States of America* 115(25), 6506–6511.

Beggs, R., Pierson, J., Tulloch, A. I. T., Blanchard, W., Westgate, M. and Lindenmayer, D. 2020. An empirical test of the mechanistic underpinnings of interference competition. *Oikos* 129(1), 93–105.

Brancillion, P. H. S. and Holl, K. D. 2020. Guidance for successful tree planting initiatives. *Journal of Applied Ecology* 57(12), 2349–2361.

Chapman, B. and Lindenmayer, D. B. 2019. A novel approach to the sustainable financing of the global restoration of degraded agricultural land. *Environmental Research Letters* 14(12), Art. 124084.

Crane, M. J., Lindenmayer, D. B. and Cunningham, R. B. 2014. The value of countryside elements in the conservation of a threatened arboreal marsupial Petaurus norfolcensis in agricultural landscapes of south-eastern Australia – the disproportional value of scattered trees. *PLoS ONE* 9(9), e107178.

Crates, R., Rayner, L., Webb, M., Stojanovic, D., Wilkie, C. and Heinsohn, R. 2020. Sustained and delayed noisy miner suppression at an avian hotspot. *Austral Ecology* 45(5), 636-643.

Crouzeilles, R., Beyer, H. L., Monteiro, L. M., Feltran-Barbieri, R., Pessoa, A. C. M., Barros, F. S. M., Lindenmayer, D. B., Lino, E. D. S. M., Grelle, C. E. V., Chazdon, R. L., Matsumoto, M., Rosa, M., Latawiec, A. E. and Strassburg, B. B. N. 2020. Achieving cost-effective landscape-scale forest restoration through targeted natural regeneration. *Conservation Letters* 13(3), e12709.

Crouzeilles, R., Curran, M., Ferreira, M. S., Lindenmayer, D. B., Grelle, C. E. V. and Rey Benayas, J. M. 2016. A global meta-analysis on the ecological drivers of forest restoration success. *Nature Communications* 7, art. no. 11666.

Cunningham, R. B., Lindenmayer, D. B., Crane, M., Michael, D. and MacGregor, C. 2007. Reptile and arboreal marsupial response to replanted vegetation in agricultural landscapes. *Ecological Applications* 17(2), 609-619.

Cunningham, R. B., Lindenmayer, D. B., Barton, P., Ikin, K., Crane, M., Michael, D., Okada, S., Gibbons, P. and Stein, J. 2014a. Cross-sectional and temporal relationships between bird occupancy and vegetation cover at multiple spatial scales. *Ecological Applications* 24(6), 1275-1288.

Cunningham, R. B., Lindenmayer, D. B., Crane, M., Michael, D. R., Barton, P. S., Gibbons, P., Okada, S., Ikin, K. and Stein, J. A. R. 2014b. The law of diminishing returns: woodland birds respond to native vegetation cover at multiple spatial scales and over time. *Diversity and Distributions* 20(1), 59-71.

Cunningham, R. B., Lindenmayer, D. B., MacGregor, C., Crane, M. and Michael, D. 2008. The combined effects of remnant vegetation and replanted vegetation on farmland birds. *Conservation Biology* 22, 742-752.

Cunningham, S. A., Crane, M. J., Evans, M. J., Hingee, K. L. and Lindenmayer, D. B. 2022. Density of feral western honey bee (Apis mellifera) colonies in fragmented woodland indicates potential for large ecosystem impacts. *Scientific Reports* 12(1), 3603.

Department of the Environment and Energy 2016. *Natural Temperate Grassland of the South Eastern Highlands: a Nationally Protected Ecological Community*, Canberra, Australia, Department of the Environment and Energy.

Dickman, C. R. and Lindenmayer, D. 2022. Australia's environment. In: Williams, S. (Ed.). *Paradigm Shift*, Berlin, Springer.

Dobes, L., Higgins, T., Crane, M., Van Dijk, A. and Lindenmayer, D. B. 2021. A cost-benefit analysis of improved farm dam water quality for livestock weight gain. *PLoS ONE* 16, e0256089.

Fischer, J. and Lindenmayer, D. B. 2002. The conservation value of paddock trees for birds in a variegated landscape in southern New South Wales. 2. Paddock trees as stepping stones. *Biodiversity and Conservation* 11, 833-849.

Fischer, J., Stott, J., Zerger, A., Warren, G., Sherren, K. and Forrester, R. I. 2009. Reversing a tree regeneration crisis in an endangered ecoregion. *Proceedings of the National Academy of Sciences of the United States of America* 106(25), 10386-10391.

Forman, R. T. T. 1964. Growth under controlled conditions to explain the hierarchical distributions of a moss, *Tetraphis pellucida*. *Ecological Monographs* 34(1), 1-25.

Freudenberger, D. 2001. Bush for birds: biodiversity enhancement guidelines for the Saltshaker Project, Boorowa, NSW, CSIRO Wildlife and Ecology / Greening Australia, Arts and Culture Trust and SE NSW Inc.

Freudenberger, D. and Harvey, J. 2004. Predicting biodiversity benefits of the Saltshaker Project, Boorowa, NSW. *Ecological Management and Restoration* 5, 5-14.

Geddes, L. S., Lunt, I. D., Smallbone, L. T. and Morgan, J. W. 2011. Old field colonization by native trees and shrubs following land use change: could this be Victoria's largest example of landscape recovery? *Ecological Management and Restoration* 12(1), 31-36.

Gibbons, P. and Boak, M. 2002. The value of paddock trees for regional conservation in an agricultural landscape. *Ecological Management and Restoration* 3(3), 205-210.

Gibbons, P. and Freudenberger, D. 2006. An overview of methods used to assess vegetation condition at the scale of the site. *Ecological Management and Restoration* 7(s1), S10-S17.

Gibbs, H. K. and Salmon, J. M. 2015. Mapping the world's degraded lands. *Applied Geography* 57, 12-21.

Goldingay, R. L., Rueegger, N. N., Grimson, M. J. and Taylor, B. D. 2015. Specific nest box designs can improve habitat restoration for cavity-dependent arboreal mammals. *Restoration Ecology* 23(4), 482-490.

Grey, M. J., Clarke, M. F. and Loyn, R. H. 1997. Initial changes in the avian communities of remnant eucalypt woodlands following a reduction in the abundance of noisy Miners, Manorina melanocephala. *Wildlife Research* 24(6), 631-648.

Haslem, A., Clarke, R. H., Holland, G. J., Radford, J. Q., Stewart, A. and Bennett, A. F. 2021. Local management or wider context: what determines the value of farm revegetation plantings for birds? *Journal of Applied Ecology* 58(11), 2552-2565.

Hazell, D., Osborne, W. and Lindenmayer, D. B. 2003. Impact of post-European stream change on frog habitat: south-eastern Australia. *Biodiversity and Conservation* 12(2), 301-320.

Hingee, K., Westgate, M. and Lindenmayer, D. B. 2022a. Complementary drivers of bird biodiversity across spatial scales. *Journal of Biogeography*.

Hingee, K. L., Lindenmayer, D. B., Florance, D. and Siegrist, A. 2022b. A bird occupancy estimator for land practitioners in the NSW South Western Slopes bioregion. *Ecological Management and Restoration* 23(2), 184-193.

Hobbs, R. J. and Yates, C. J. (Eds.) 2000. *Temperate Eucalypt Woodlands in Australia: Biology, Conservation, Management and Restoration*, Chipping Norton, Surrey Beatty and Sons.

IPBES 2019. *Intergovernmental Science-Policy Platform on Biodiversity and Ecosystem Services, Global Assessment Summary for Policymakers*, United Nations.

Lindenmayer, D. B. 2020. Improving restoration programs through greater connection with ecological theory and better monitoring. *Frontiers in Ecology and Evolution* 8, Art. 50.

Lindenmayer, D. B., Blanchard, W., Crane, M., Michael, D. and Florance, D. 2018a. Size or quality. What matters in vegetation restoration for bird biodiversity in endangered temperate woodlands? *Austral Ecology* 43(7), 798-806.

Lindenmayer, D. B., Blanchard, W., Crane, M., Michael, D. and Sato, C. 2018b. Biodiversity benefits of vegetation restoration are undermined by livestock grazing. *Restoration Ecology* 26(6), 1157-1164.

Lindenmayer, D. B., Crane, M., Blanchard, W., Okada, S. and Montague-Drake, R. 2016. Do nest boxes in restored woodlands promote the conservation of hollow-dependent fauna? *Restoration Ecology* 24(2), 244-251.

Lindenmayer, D. B., Crane, M., Evans, M. C., Maron, M., Gibbons, P., Bekessy, S. and Blanchard, W. 2017. The anatomy of a failed offset. *Biological Conservation* 210, 286–292.

Lindenmayer, D. B., Cunningham, R., Crane, M., Michael, D. and Montague-Drake, R. 2007. Farmland bird responses to intersecting replanted areas. *Landscape Ecology* 22(10), 1555–1562.

Lindenmayer, D. B., Knight, E. J., Crane, M. J., Montague-Drake, R., Michael, D. R. and MacGregor, C. I. 2010. What makes an effective restoration planting for woodland birds? *Biological Conservation* 143(2), 289–301.

Lindenmayer, D. B., Lane, P., Crane, M., Florance, D., Foster, C. N., Ikin, K., Michael, D., Sato, C. F., Scheele, B. C. and Westgate, M. J. 2019. Weather effects on birds of different size are mediated by long-term climate and vegetation type in endangered temperate woodlands. *Global Change Biology* 25(2), 675–685.

Lindenmayer, D. B., Blanchard, W., Foster, C. N., Scheele, B. C., Westgate, M. J., Stein, J., Crane, M. and Florance, D. 2020a. Habitat amount versus connectivity: an empirical study of bird responses. *Biological Conservation* 241, 108377.

Lindenmayer, D. B., Lane, P., Westgate, M. J., Scheele, B. C., Crane, M., Florance, D., Crane, C. and Smith, D. 2020b. Long-term mammal and nocturnal bird trends are influenced by vegetation type, weather and climate in temperate woodlands. *Austral Ecology* 45(6), 813–824.

Lindenmayer, D. B., Macbeth, S. M., Smith, D. G. and Young, M. L. 2022. *Natural Asset Farming, Creating Productive and Biodiverse Farms*, Melbourne, Australia, CSIRO Publishing.

Lindenmayer, D. B., Lane, P. W., Barton, P. S., Crane, M., Ikin, K., Michael, D. R. and Okada, S. 2016a. Long-term bird colonization and turnover in restored woodlands. *Biodiversity and Conservation* 25(8), 1587–1603.

Lindenmayer, D. B., Michael, D., Crane, M., Okada, S., Florance, D., Barton, P. K. and I. 2016b. *Wildlife Conservation in Farm Landscapes*, Melbourne, CSIRO Publishing.

Lindenmayer, D. B., Northrop-Mackie, A. R., Montague-Drake, R., Crane, M., Michael, D., Okada, S. and Gibbons, P. 2012. Not all kinds of revegetation are created equal: revegetation type influences bird assemblages in threatened Australian woodland ecosystems. *PLoS ONE* 7(4), e34527.

Lumsden, L. F. and Bennett, A. F. 2005. Scattered trees in rural landscapes: foraging habitat for insectivorous bats in south-eastern Australia. *Biological Conservation* 122(2), 205–222.

Mac Nally, R., Bowen, M., Howes, A., McAlpine, C. A. and Maron, M. 2012. Despotic, high-impact species and the subcontinental scale control of avian assemblage structure. *Ecology* 93(3), 668–678.

Mac Nally, De Vries, L. and Thomson, J. R. 2010. Are replanted floodplain forests in southeastern Australia providing bird biodiversity benefits? *Restoration Ecology* 18(1), 85–94.

Mackey, B. G. and Lindenmayer, D. B. 2001. Towards a hierarchical framework for modelling the spatial distribution of animals. *Journal of Biogeography* 28(9), 1147–1166.

Malerba, M. E., Lindenmayer, D. B., Scheele, B. C., Waryszak, P., Yilmaz , I. N., Schuster, L. and Macreadie, P. I. 2022. Fencing farm dams to exclude livestock improves water quality and halves methane emissions. *Global Change Biology* 28(15), 4701–4712.

Manning, A. D., Fischer, J. and Lindenmayer, D. B. 2006. Scattered trees are keystone structures - implications for conservation. *Biological Conservation* 132(3), 311–321.

Maron, M., Grey, M. J., Catterall, C. P., Major, R. E., Oliver, D. L., Clarke, M. F., Loyn, R. H., Mac Nally, R., Davidson, I. and Thomson, J. R. 2013. Avifaunal disarray due to a single despotic species. *Diversity and Distributions* 19(12), 1468-1479.

Maxwell, S. L., Fuller, R. A., Brooks, T. M. and Watson, J. E. 2016. Biodiversity: the ravages of guns, nets and bulldozers. *Nature* 536(7615), 143-145.

Melton, C. B., Reside, A. E., Simmonds, J. S., Mcdonald, P. G., Major, R. E., Crates, R., Catterall, C. P., Clarke, M. F., Grey, M. J., Davitt, G., Ingwersen, D., Robinson, D. and Maron, M. 2021. Evaluating the evidence of culling a native species for conservation benefits. *Conservation Science and Practice* 3(12), e549.

Michael, D. and Lindenmayer, D. B. 2018. *Rocky Outcrops in Australia*, Melbourne, CSIRO Publishing.

Michael, D. R., Lindenmayer, D. B. and Cunningham, R. B. 2010. Managing rock outcrops to improve biodiversity conservation in Australian agricultural landscapes. *Ecological Management and Restoration* 11(1), 43-50.

Munro, N. T., Lindenmayer, D. B. and Fischer, J. 2007. Faunal response to revegetation in agricultural areas of Australia: a review. *Ecological Management and Restoration* 8(3), 199-207.

Norton, D. and Reid, N. 2013. *Nature and Farming*, Melbourne, CSIRO Publishing.

Palmer, M. A., Zedler, J. B. and Falk, D. A. (Eds.) 2016. *Foundations of Restoration Ecology*, Washington, DC, Island Press.

Prober, S. and Thiele, K. R. 1995. Conservation of the Grassy White Box Woodlands: relative contributions of size and disturbance to floristic composition and diversity of remnants. *Australian Journal of Botany* 43(4), 349-366.

Prober, S. M. and Thiele, K. R. 2005. Restoring Australia's temperate grasslands and grassy woodlands: integrating function and diversity. *Ecological Management and Restoration* 6(1), 16-27.

Soderquist, T. R. and Mac Nally, R. 2000. The conservation value of mesic gullies in dry forest landscapes: mammal populations in the box-ironbark ecosystem of southern Australia. *Biological Conservation* 93(3), 281-291.

Spooner, P. G. and Briggs, S. V. 2008. Woodlands on farms in southern New South Wales: A longer-term assessment of vegetation changes after fencing. *Ecological Management and Restoration* 9(1), 33-41.

Stirzaker, R., Vertessey, R. and Sarre, A. (Eds.) 2002. *Trees, Water and Salt. An Australian Guide to Using Trees for Healthy Catchments and Productive Farms*, Canberra, Joint Venture Agroforestry Program.

Suding, K., Higgs, E., Palmer, M., Callicott, J. B., Anderson, C. B., Baker, M., Gutrich, J. J., Hondula, K. L., Lafevor, M. C., Larson, B. M., Randall, A., Ruhl, J. B. and Schwartz, K. Z. 2015. Conservation. Committing to ecological restoration. *Science* 348(6235), 638-640.

Suding, K. N. 2011. Toward an era of restoration in ecology: successes, failures and opportunities ahead. *Annual Review of Ecology, Evolution, and Systematics* 42(1), 465-487.

Sutherland, W. J., Pullin, A. S., Dolman, P. M. and Knight, T. M. 2004. The need for evidence-based conservation. *Trends in Ecology and Evolution* 19(6), 305-308.

Tulloch, A. I. T., Mortelliti, A., Kay, G. M., Florance, D. and Lindenmayer, D. B. 2016. Using empirical models of species colonization under multiple threatening processes to identify complementary threat-mitigation strategies. *Conservation Biology* 30(4), 867-882.

United Nations 2020. The UN decade on ecosystem restoration 2021–2030. *UNEP/FAO Factsheet*, Rome, United Nations Environment Programme and Food and Agriculture Organization of the United Nations.

Vesk, P., Nolan, R., Thomson, J. W., Dorrough, J. W. and Mac Nally, R. 2008. Time lags in the provision of habitat resources through revegetation. *Biological Conservation* 141, 174–186.

Walker, J., Bullen, F. and Williams, B. G. 1993. Ecohydrological changes in the Murray-Darling Basin. I. The number of trees cleared over two centuries. *Journal of Applied Ecology* 30(2), 265–273.

Westgate, M. J., Crane, C., Smith, D., O'Malley, C., Siegrist, A., Florance, D., Lang, E., Crane, M., Hingee, K., Scheele, B. C. and Lindenmayer, D. B. 2022. Improved management of farm dams increases vegetation cover, water quality, and macroinvertebrate biodiversity. *Ecology and Evolution* 12(3), e8636.

Williams, J. E. and Price, R. J. 2010. Impacts of red meat production on biodiversity in Australia: a review and comparison with alternative protein production industries. *Animal Production Science* 50(8), 723–747.

Chapter 14

The future of animal rewilding in agricultural landscapes[1]

Kiarrah J. Smith, Iain. J. Gordon[2], Belinda A. Wilson and Adrian D. Manning, The Australian National University, Australia

1 Introduction

On a continuum of human influence on ecosystems (Lesslie and Taylor, 1985; McIntyre and Hobbs, 1999), **rewilding** refers to the decisions implemented to produce *any* shift in an ecosystem from an anthropogenically modified state, towards total wilderness, where non-anthropogenic processes dominate (Carver, 2014; Carver *et al.*, 2021; Gordon *et al.*, 2021) (Fig. 1). The complex and transdisciplinary nature of rewilding is best captured in the summary from Carver *et al.* (2021):

> *rewilding exists on a continuum of scale, connectivity, and level of human influence and aims to restore ecosystem structure and functions to achieve a self-sustaining autonomous nature.*

1 Throughout this chapter, terms defined in the glossary (Table 1) are indicated in **boldface** on their first appearance.
2 In addition to the affiliation listed, the author is also affiliated to: The James Hutton Institute, United Kingdom; Central Queensland University, Australia; CSIRO, Australia; and James Cook University, Australia.

http://dx.doi.org/10.19103/AS.2024.0134.22

Animal rewilding refers to the process of shifting an ecosystem towards the 'wilder' end of the wilderness continuum (Fig. 1) by restoring any degree of the **ecosystem functions** (i.e. **abiotic and biotic interactions**) of invertebrate and vertebrate animals in an area where these have been lost. This process can involve **translocations** (including **reintroductions** and **ecological replacement**, as defined by IUCN/SSC, 2013, see Table 1, habitat enhancement, or other facilitation of natural recolonisation (e.g. replacing a portion of arable farmland with wildlife habitat, Pywell *et al.*, 2015). Animal rewilding aims to restore part of a food web or other ecosystem function and, in this way, is distinct from reintroductions and management undertaken for the sole purpose of conserving a species or population.

Animal rewilding has sometimes been viewed as incompatible with agricultural landscapes (i.e. **agroecosystems**). However, there are increasing precedents that show that animal rewilding can take place in agroecosystems – where a balance is struck between maximising the **provisioning, regulating, and cultural ecosystem services** of wildlife, while minimising their **disservices**.

In this chapter, we:

- provide an overview of the trade-offs related to animal rewilding in agroecosystems (i.e. ecosystem services versus disservices); and

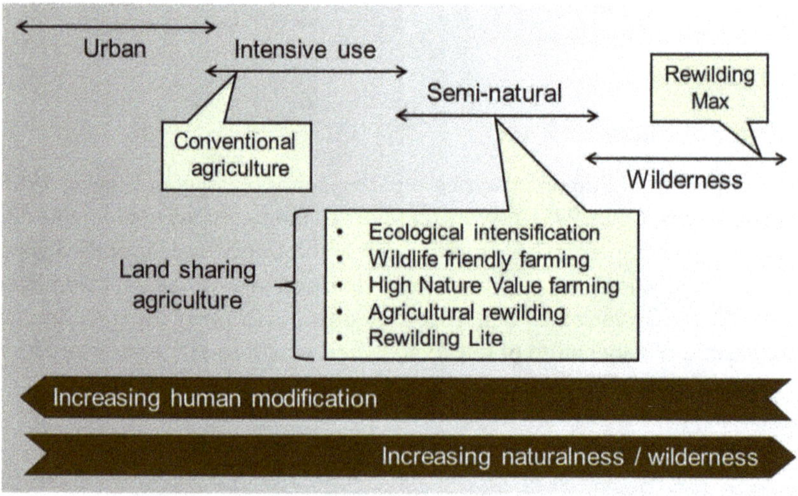

Figure 1 Concepts related to animal rewilding in agricultural landscapes fall around the middle of the wilderness continuum (*sensu* Lesslie and Taylor, 1985), between conventional agriculture and Rewilding Max (adapted from Carver, 2014; Gordon *et al.*, 2021). Each of these middle concepts (defined in Table 1) is a type of land-sharing strategy (*sensu* Gabriel *et al.*, 2009) and involves management that shifts agricultural land use towards the 'wilder' end of the wilderness continuum.

Table 1 Glossary

Term	Definition
Abiotic interactions	Relationships between non-living components (e.g. temperature, precipitation, soil, light, water, etc.) that shape the structure, function, and dynamics of environments.
Agricultural rewilding	Conservation and land management strategy aiming to restore ecological functions and enhance biodiversity within agricultural landscapes. This approach seeks to reconcile agricultural production with conservation goals by incorporating elements of wildness and natural ecosystem processes into farming practices. This differs from the other forms of rewilding by including more human interventions, specifically harvest of plants and/or animals, along with the restoration of ecological processes (Corson *et al.*, 2022).
Agroecosystem	An agricultural ecosystem that encompasses the interactions between agricultural organisms, their environment, and human activities within a managed landscape.
Animal rewilding	Process of shifting an ecosystem towards the 'wilder' end of the wilderness continuum (Figure 1) by restoring (through translocation, habitat enhancement, or other facilitation of natural recolonisation) any degree of the ecosystem functions of invertebrate and vertebrate animals in an area where these have been lost. Animal rewilding is distinct from reintroductions and management undertaken for the sole purpose of conserving a species or population (i.e. without a goal to restore part of a food web or other ecosystem function).
Biotic interactions	Relationships between living organisms that shape the structure, function, and dynamics of ecosystems.
Coexistence conservation	The long-term, iterative, and adaptive process of driving or enabling adaptation to facilitate the coexistence of species with the predators (including humans) that threaten their long-term persistence (Urlich, 2015; Pooley *et al.*, 2021; Evans *et al.*, 2022).
Cultural services	The non-material benefits that ecosystems provide to humans, contributing to cultural identity, well-being, and social values. These services are closely linked to human perceptions, beliefs, traditions, and practices, shaping the cultural and spiritual connections between people and their environment.
Dispersal barriers	Physical, ecological, or behavioural obstacles that impede the movement of individuals or propagules (such as seeds, spores, or larvae) between different areas within a landscape. Dispersal barriers can occur naturally or result from human activities, and they play a crucial role in shaping the distribution, genetic structure, and dynamics of populations and communities.
Ecological intensification	Sustainable approach to agricultural production that aims to increase crop yields and resource use efficiency while minimising negative environmental impacts and maintaining ecosystem services. Unlike conventional intensification methods, which often rely heavily on external inputs such as synthetic fertilizers, pesticides, and irrigation, ecological intensification emphasises ecological principles and regenerative practices to enhance productivity and resilience within agroecosystems (Bommarco *et al.*, 2013; Gaba *et al.*, 2014; Pywell *et al.*, 2015; Kremen, 2020).

(Continued)

Table 1 *(Continued)*

Term	Definition
Ecological replacement	Intentional human-mediated establishment of a species outside its indigenous range for the purpose of restoring a specific ecological function that was lost through the extinction of another species from that area (IUCN/SSC, 2013).
Ecosystem disservices	Adverse effects of ecosystem processes, structures, or functions on human well-being or societal interests.
Ecosystem function	Processes, interactions, and activities that occur within an ecosystem, contributing to its structure, stability, and dynamics. These functions encompass a wide range of biological, physical, and chemical processes that govern the flow of energy, cycling of nutrients, and maintenance of ecological balance within ecosystems.
Ecosystem services	Benefits that ecosystems provide to human societies, directly or indirectly, contributing to human well-being, livelihoods, and quality of life. Ecosystem services encompass a wide range of ecological processes, functions, and goods that are essential for supporting human survival, economic activities, and cultural values (e.g. healthy soils, nutrient cycling, and waste decomposition, Daily, 1997).
Genetic engineering	Manipulating the genetic material of an organism to introduce specific traits or characteristics (e.g. pest resistance in sugarcane, Iqbal *et al.*, 2021). This manipulation typically involves the insertion, deletion, or modification of genes within an organism's genome, allowing for the precise control and modification of heritable traits.
High Nature Value (HNV) farming	Agricultural land management characterised by its significant ecological and cultural importance (Strohbach *et al.*, 2015; Sutcliffe *et al.*, 2015; Plieninger *et al.*, 2019). Such systems are typically found in areas with rich biodiversity, unique habitats, and traditional land-use practices, and they provide multiple environmental benefits while supporting local livelihoods and cultural heritage.
Landsharing	Land use management approach that integrates agricultural production and biodiversity conservation within the same landscape (Gabriel *et al.*, 2009; Phalan *et al.*, 2011). Rather than segregating agricultural areas and conservation areas, land sharing achieves a semi-natural state with synergies between agricultural activities and biodiversity conservation through a range of strategies (examples are listed in Figure 1).
Land sparing	Land use management approach that seeks to separate areas for agricultural production from areas designated for biodiversity conservation (Phalan *et al.*, 2011). Under this paradigm, intensive agricultural production creates a relatively homogenous landscape, while separate, often island-like areas are set aside and managed for conservation purposes (Fischer *et al.*, 2008). The intent is that high yields minimise the conversion of natural areas to farmland (Green *et al.*, 2005).
Metapopulation	A spatially distributed population of interconnected subpopulations of the same species that interact through dispersal processes. Subpopulations usually occupy habitat patches that are fragmented and isolated from one another but are connected by occasional movement of individuals between patches.

Term	Definition
Provisioning services	Direct benefits that ecosystems provide to humans in the form of tangible resources that are essential for supporting human well-being. These services encompass a wide range of ecosystem products that are harvested or extracted for human consumption, use, or trade.
Regulating services	Ecological processes and functions that help regulate environmental conditions, maintain ecosystem stability, and mitigate natural hazards, thereby supporting human well-being and societal interests (e.g. water regulation, improved soil structure and fertility, pollination, waste decomposition, and control of pests and weeds, Millennium Ecosystem Assessment, 2003; Zhang *et al.*, 2007).
Reintroduction	The intentional release of an organism within its indigenous range from which it has been extirpated, with the aim of re-establishing a viable population (IUCN/SSC, 2013).
Rewilding	Conservation strategy that aims to promote ecological restoration and biodiversity conservation by implementing measures that produce any shift in an ecosystem from an anthropogenically modified state towards total wilderness, where non-anthropogenic processes dominate (Carver, 2014; Carver *et al.*, 2021; Gordon *et al.*, 2021).
Rewilding Lite	Form of rewilding that includes targeted, small-scale interventions within human-dominated landscapes, delivering wildness while producing some economic benefits (Carver, 2014; Gordon *et al.*, 2021). This approach acknowledges the limitations and complexities of implementing large-scale rewilding projects and seeks to achieve similar goals through more manageable actions.
Rewilding Max	Total abandonment of formerly managed or degraded landscapes following restoration activities (Carver, 2014; Gordon *et al.*, 2021).
Translocation	The human-mediated movement of living organisms from one area to another (IUCN/SSC, 2013).
Wildlife friendly farming	Agricultural practices and land management strategies that prioritise the conservation of biodiversity and habitat within agricultural landscapes. This strategy aims to minimise negative impacts on wildlife populations, enhance habitat quality and connectivity, and promote coexistence between agricultural activities and wildlife conservation goals.

- discuss three key management trends that we foresee will facilitate net positive outcomes in animal rewilding: (a) capturing wild provisioning services; (b) coordinating broad spatial heterogeneity in land use; and (c) employing **coexistence conservation** tactics.

A resurgence of appreciation for wild animal products would provide opportunities for landholders to obtain new or supplementary income. Further gains could be facilitated by creating heterogeneous landscapes with various **land-sharing** strategies that shift ecosystems away from intensive use towards higher quality wild animal habitat. Finally, increased implementation of tactics that may drive or enable the adaptation of livestock, crops, their predators, as well as human perceptions and management strategies could facilitate the coexistence of wild animals and agriculture that will help to produce net positive outcomes.

1.1 Agricultural context

For agroecosystems, rewilding aims to harness the ecological processes that operate in natural systems to promote both wildness and agricultural production. In other words, rewilding interventions are not implemented at the expense of all provisioning services (i.e. humans can still obtain food, fuel, and other products from their use of the land; Millennium Ecosystem Assessment, 2003). These circumstances encompass the concepts of '**ecological intensification**', '**wildlife friendly farming**', 'land sharing', '**High Nature Value farming**', '**agricultural rewilding**', and '**Rewilding Lite**' (Table 1), as opposed to intensive conventional agriculture, or the total abandonment of formerly managed or degraded landscapes following restoration activities (i.e. '**Rewilding Max**', Fig. 1; Beaufoy *et al.*, 1994; Green *et al.*, 2005; Gabriel *et al.*, 2009; Carver, 2014; Gaba *et al.*, 2014; Strohbach *et al.*, 2015; Kremen, 2020; Gordon *et al.*, 2021; Corson *et al.*, 2022; Thomas *et al.*, 2022).

In this chapter, we do not discuss the concept of '**land sparing**', which is where intensive agriculture is kept completely separate from wilderness areas (i.e. the opposite of land sharing, Green *et al.*, 2005; Phalan *et al.*, 2011; Edwards *et al.*, 2014). However, we acknowledge its importance in other contexts, such as the conservation of biodiversity that is sensitive to any human influence (Mattison and Norris, 2005), or where a different bundle of ecosystem services is preferred over that which could be achieved through land sharing interventions (Cerqueira *et al.*, 2015). A 'bundle' of ecosystem services refers to the quantity and type of ecosystem services typically delivered by a particular land management strategy (Raudsepp-Hearne *et al.*, 2010). For instance, the bundle of ecosystem services from feedlot agriculture, a rural village, and farmland converted with the goal of achieving Rewilding Max are likely to be very different (Raudsepp-Hearne *et al.*, 2010; Cerqueira *et al.*, 2015).

1.2 Notes on interpretation

Greater coexistence of wildlife and agriculture is an emerging field of research and endeavour that we present with an optimistic and ambitious view to achieving economic and environmental sustainability. Because there are currently limited examples of animal rewilding in agricultural landscapes, not all examples in this chapter are demonstrations of this *in toto*. Rather, many examples detail situations where landholders have always coexisted with wildlife in agricultural landscapes (i.e. intensive land use has never occurred in these locations), and others are drawn from the field of conservation biology. The purpose of these non-rewilding examples is to highlight the range of management strategies that landholders may choose to investigate or trial as part of their own animal rewilding path. The strategies most appropriate for each situation will vary, and any uncertainty around the more novel approaches should be interpreted as opportunities for research and experimentation, not a reason to maintain the status quo. The government schemes and incentives implemented to facilitate rewilding in agricultural landscapes are not discussed in this chapter.

2 Benefits of wild animals in agricultural landscapes

The balance between agriculture, biodiversity, and cultural values that can be achieved on an area of land is exemplified by the long-established, low-intensity, High Nature Value traditional farms that persist in parts of Europe (Beaufoy *et al.*, 1994; Bignal and McCracken, 1996; Plieninger *et al.*, 2019); for example, the wood pastures of Transylvania (Sutcliffe *et al.*, 2015). While these agroecosystems are often espoused for the benefits they provide *to* wildlife, how do they benefit *from* the wildlife they support? Semi-natural agricultural landscapes (Fig. 1) will facilitate more regulating and cultural ecosystem services than a conventional intensive agricultural system and not necessarily at the expense of high provisioning services (Raudsepp-Hearne *et al.*, 2010; Clough *et al.*, 2011; Bommarco *et al.*, 2013; Cerqueira *et al.*, 2015). For example, in experimental rotations of wheat (*Triticum aestivum*), rapeseed (*Brassica napus*), and broad beans (*Vicia faba*), fields with low-yielding arable edges replaced by wildlife habitat had no net loss of yield (and in fact 25–35% greater yield of beans) than control fields, where no wildlife habitat was added (Pywell *et al.*, 2015). The positive effect of wildlife habitat on crop yields became more pronounced over 6 years and was attributed to a combination of (a) ceasing to plant on the field edges, which are least optimal for crop growth, (b) a higher abundance and diversity of pollinators, and (c) attracting and supporting the natural enemies (i.e., species that exert control) of crop pests (Pywell *et al.*, 2015).

In addition to pollination and pest management, the regulating services provided by wildlife can also include water regulation, improved soil structure and fertility, waste decomposition, and weed control (Millennium Ecosystem Assessment, 2003; Zhang *et al.*, 2007). Where wild animals fill these functions, the economic and environmental costs of conventional agriculture inputs (e.g. pesticides and fences) may be reduced or eliminated (Power, 2010). It is estimated, for example, that the consumption of pest cotton bollworm moths

(*Helicoverpa armigera*) by Gould's wattled bat (*Chalinolobus gouldii*) has an annual value of around AU$64 million to the Australian cotton industry (Kolkert *et al*., 2021). In addition, wild animals may contribute to cultural ecosystem services, including spiritual, aesthetic, and inspirational values, recreation, eco- and agri-tourism, heritage, and education (Millennium Ecosystem Assessment, 2003). For example, the ecosystem engineering behaviour of Eurasian and American beavers (*Castor fiber*, *C. canadensis*, respectively) can moderate the impacts of droughts and floods on crops and livestock by redirecting water flow while attracting new streams of revenue associated with recreation and ecotourism (Auster *et al*., 2020; Puttock *et al*., 2021; Thompson *et al*., 2021).

3 Disservices of wild animals to agricultural landscapes

In agricultural settings, natural biotic interactions can impose disservices that reduce yields or increase production costs (Zhang *et al*., 2007). Returning to the previous example of beavers, while simultaneously providing the aforementioned ecosystem services, their activities can also impact agricultural and forestry areas (e.g. timber plantations and cereal crops) through feeding, bank erosion, and extensive flooding (Shwiff *et al*., 2011; Campbell-Palmer *et al*., 2016; Lodberg-Holm *et al*., 2022). In addition to the direct consumption and destruction of crops, wild animals may predate or compete with livestock for food and water and distribute weeds, pathogens, and other pests (Tisdell, 1982; Hill, 1997; Amador-Alcalá *et al*., 2013). So while broader ecosystem disservices may be produced by intensive conventional agriculture (e.g. water pollution from chemical, sediment, and animal effluent runoffs, and greenhouse gas emissions), there are also trade-offs associated with animal rewilding in agricultural landscapes (Daily, 1999; Power, 2010; Saunders *et al*., 2016; Tschumi *et al*., 2018). Such trade-offs have been the focus of much economic, social, and ecological study.

4 Future trends: capturing wild provisioning services

To achieve a balance between provisioning, regulating, and cultural services through animal rewilding, the beneficial services of wildlife need to be maximised while minimising their disservices. The compromises and interventions required to facilitate rewilding in the agricultural context are different from those in Rewilding Max situations because the goal is to harness ecological processes for human use, not to transition the land to wilderness (Fig. 1). Beyond an ecological research agenda (Kremen, 2005), and detailed studies of the net outcome from multiple animals' activities in agricultural landscapes (Saunders *et al*., 2016; Tschumi *et al*., 2018), we foresee that three key management trends will facilitate a balance of economic and environmental gains when undertaking animal rewilding. This section goes on to discuss the first of these: capturing wild provisioning services.

In addition to regulating and cultural services, opportunities to obtain new products that supplement or replace other income streams can be facilitated by the return of wild animals to agricultural landscapes. One example is the emerging market for stingless bee products (Jaffé *et al.*, 2015; Salatino *et al.*, 2019). A resurgence of appreciation for wild animal products would benefit both landholders and the landscapes within which they reside (Gordon, 2009; Gordon, 2019). Hunting and cooking guides are available for a huge range of wild animal products. In addition to the income or services, a landholder could gain by encouraging wild animals and recreational hunters on their farms (Lund and Jensen, 2017), there are opportunities for large-scale commercial operations (Menichetti *et al.*, 2019). These operations could reduce the impacts of the most abundant invasive or nuisance species in natural- and agroecosystems (e.g. by consuming wild pigs *Sus scrofa*, Tisdell, 1982; Parkes *et al.*, 1996; McKee *et al.*, 2020). The market for wild animal products could be grown by highlighting the wide range of environmental, health, and ethical advantages, and building trust in the safety of food products (Wilson and Edwards, 2019).

Running traditional and native livestock breeds (or species, e.g., llama *Lama glama*; Chillingham cattle *Bos taurus*, kangaroos family Macropodidae) rather than modern breeds is a strategy that would (a) return more wild provisioning services, while (b) attracting agritourism and (c) reducing the methane outputs that contribute to climate change (Carver, 2014; Cromsigt *et al.*, 2018; Wilson and Edwards, 2019; Gordon *et al.*, 2021; Thomas *et al.*, 2022). Known as 'Rewilding Lite' (Carver, 2014; Gordon *et al.*, 2021), this strategy is also referred to as a 'domestication' of rewilding and 'agricultural rewilding' (Corson *et al.*, 2022; Thomas *et al.*, 2022). It takes an agroecosystem a few steps closer to wilderness (i.e. further to the right on Fig. 1) than a wildlife-friendly conservation grazing strategy, where sheep or cattle stocking densities are simply reduced, or grazing management altered, to lessen adverse effects on wildlife habitat (Carver, 2014; Lyons *et al.*, 2017). Rewilding Lite creates a wilder landscape because it entails the running of traditional breeds or species that are best adapted to local conditions (including the threat of native predators), thereby requiring less-intensive husbandry (Gordon *et al.*, 2021; Thomas *et al.*, 2022). The local traditional knowledge of a rancher of a High Nature Value dehesa (i.e. an Iberian wood–pasture agroecosystem) highlights the value of this practice:

> we utilized native cattle breeds because they are better adapted and consequently less susceptible to diseases as well as they consume more efficiently the resources and need less external food supplies Garrido *et al.*, 2017

Rewilding Lite may also be a more socially, politically, and economically palatable alternative for balancing ecosystem services where Rewilding Max would not be feasible (Carver, 2014; Gordon *et al.*, 2021; Corson *et al.*, 2022; Thomas *et al.*, 2022). The principle being: *it is better to have some form of rewilding than no rewilding at all.*

5 Future trends: coordinating broad spatial heterogeneity in land use

Due to the mobility of animals, animal rewilding is ultimately a landscape-scale endeavour. Numerous authors have emphasised that the flow of ecosystem services and disservices depends not only on the management of an individual farm but also the state and condition of the surrounding landscape (Thies *et al.*, 2003; Fischer *et al.*, 2006; Zhang *et al.*, 2007; Jarnemo *et al.*, 2014; Saunders *et al.*, 2016; Senapathi *et al.*, 2017; Kolkert *et al.*, 2020). Even small invertebrates, which may perceive their own environment at less than a farm's scale (see 'Continua-Umwelt' in Manning *et al.*, 2004), still depend on immigration from a broader area to recolonise following local disturbance (Ponisio *et al.*, 2019) or avoid inbreeding (López-Uribe *et al.*, 2015). A farm that is not isolated by **dispersal barriers** can rely on the passive immigration of beneficial animals from the broader **metapopulation** rather than needing human-facilitated translocations (Sunderland and Samu, 2000; Viana *et al.*, 2014; Ponisio *et al.*, 2019).

Animal rewilding does not necessarily require that an entire agricultural region needs to shift towards the wilder end of the continuum to support a greater balance of ecosystem services (Cunningham *et al.*, 2013). Rather, the key is a coordinated approach among stakeholders (Zheng *et al.*, 2019) to facilitate broad-scale spatial heterogeneity in land use (Benton *et al.*, 2003; Lemaire *et al.*, 2017; Lund and Jensen, 2017). Highly heterogeneous landscapes provide a diversity of habitats within and between fields and are, therefore, likely to have greater animal diversity compared to a more homogenous landscape (Benton *et al.*, 2003; Weibull *et al.*, 2003; Lindenmayer and Fischer, 2006). This strategy can increase the presence of species that mitigate the disservices of other animals (e.g. parasitoids of pest beetles, Thies *et al.*, 2003) or enable synergistic (i.e. boosted) benefits from complementary species (e.g. wild bees have been shown to enhance pollination by honey bees *Apis mellifera*, through behavioural interactions, Greenleaf and Kremen, 2006). In addition, the availability of natural resources in the landscape provides alternative food that may reduce the disservices of otherwise beneficial herbivores, such as kangaroos (Ampt and Baumber, 2014) or red deer (*Cervus elaphus*, Jarnemo *et al.*, 2014), which can be appreciated for their provisioning services, ecosystem engineering, or role as prey species. The latter is exemplified in the Bokkeveld plateau in South Africa, where the retained landscape heterogeneity and habitat complexity underpins the diversity and abundance of small mammals, which are valued by landholders as an alternative food source for predators of sheep (O'Farrell *et al.*, 2008).

In those agricultural regions where animal rewilding is socially, economically, and geographically appropriate (Fischer *et al.*, 2008; Cunningham *et al.*, 2013; Norton *et al.*, 2013), we advocate for a mosaic of heterogeneous patches where the vast majority of land uses fall in the centre of the wilderness continuum (i.e. land sharing agricultural strategies, Fig. 1), forming a high-quality habitat matrix for wildlife, compared to the environment provided by conventional agriculture (Fischer *et al.*, 2006; Perfecto and Vandermeer, 2017; Grass *et al.*, 2019; Gordon *et al.*, 2021). To maximise heterogeneity in this landscape model, patches of intense conventional agriculture and Rewilding Max (i.e. land sparing) may still have a place embedded within that matrix (Robinson *et al.*, 2001; Fischer *et al.*, 2006; Manning and Lindenmayer, 2009; Grass *et al.*, 2019; Gordon *et al.*, 2021). The traditional shaded coffee plantations in Latin America are an oft-cited example of a heterogeneous agricultural landscape with a high-quality habitat matrix that supports bird conservation efforts (Leyequién *et al.*, 2010). Elsewhere, landholders are increasingly regaining an appreciation of the benefits of 'working with nature' by implementing rewilding actions (e.g. revegetation) and undertaking their own trials (e.g. ceasing insecticide use to realise the value of invertebrates as pest control agents, Smith and Watson, 2018). These trailblazers increase land use heterogeneity beyond their property boundaries by inspiring neighbours in their own region and further afield to trial animal rewilding strategies on their own farms (Smith and Watson, 2018).

6 Future trends: employing coexistence conservation tactics

'Coexistence conservation' was originally proposed for managing the predation of threatened species by invasive predators (Evans *et al.*, 2022; see also Urlich, 2015 for a similar concept) but is also applicable to animal rewilding in agricultural landscapes. Coexistence conservation seeks to facilitate the coexistence of species with the predators (including humans) that threaten their long-term persistence by enabling adaptation (Urlich, 2015; Pooley *et al.*, 2021; Evans *et al.*, 2022). In an expansion of this concept, we consider 'predators' to include both invasive (e.g. red foxes *Vulpes vulpes*, in Australia) and native (e.g. grey wolves *Canis lupus*, in North America) species. The primary principles (with expansion to include herbivores, herbivory, and human elements) are that (1) *predation* (i.e. consumption, including herbivory and carnivory) rather than the predator (i.e. consumer, herbivore, or carnivore) should be the focus of management to achieve desired outcomes, and (2) solutions should aim to drive or enable the adaptation of the predator, prey, and/or human perceptions and management strategies that will facilitate their coexistence over the long term. In a conservation context, the vulnerable organism is a threatened or rare plant or animal, while in an agricultural context, it is the crops or livestock that are at risk.

Agricultural science has already made great strides with respect to reducing the impact of invertebrate predators on crops through artificial selection (i.e. selectively breeding individuals that have desirable traits, Obermeier *et al.*, 2022) and **genetic engineering** (i.e. direct modification of genetic material, Iqbal *et al.*, 2021). Coexistence conservation provides a complementary approach to these practices and is expected to play an important role in facilitating a balance of economic and environmental gains when undertaking animal rewilding. In Table 2 we provide examples of the coexistence conservation tactics (derived from Evans *et al.*, 2022) that we expect could mitigate the impact of wildlife predation of crops and livestock in agricultural landscapes.

The social adaptation of human perceptions and management strategies (Pooley *et al.*, 2021) should be included alongside the evolutionary adaptation of predators and prey under the umbrella of coexistence conservation, because both are necessary for sustained ecosystem change. Evidence-based campaigns underpinned by studies of human–wildlife coexistence, where it exists (Pooley *et al.*, 2021), and experimental evaluation of coexistence tactics in varying contexts, will be important for driving the social adaptation of humans (van Eeden *et al.*, 2018; van Eeden *et al.*, 2021a). Landholders recognise that no landscape is static and, therefore, that adaptability is key. A particular land use strategy that was profitable in the past may not be so in the future, due to dynamics like climate change (Tahmasebi *et al.*, 2020), the emergence of insecticide resistance (Hansen, 2003), or changes in markets or the economy. Animal rewilding in agricultural landscapes will require that same adaptability to minimise human–wildlife conflicts. For example, landholders can look to others who already coexist with beavers and choose to respond proactively to a proposed reintroduction of this species by adapting their land use to a more beaver-compatible strategy, such as livestock grazing (Charnley *et al.*, 2020).

7 Conceptual case study: kangaroos and dingoes

As has typically been the case for large carnivores around the world, there is a complex debate about the value of the ecosystem services of dingoes (*Canis familiaris*), weighed against the costs of managing their predation of livestock and vulnerable wildlife (Allen *et al.*, 2013; Fleming *et al.*, 2021; Van Eeden *et al.*, 2021b). As the largest apex predator (besides humans) extant in Australia, dingoes have unique predator–prey and competitive interactions (Fleming *et al.*, 2022). However, in an effort to exclude them from large sheep-grazing

Table 2 Coexistence conservation tactics applicable to mitigating predation (i.e. consumption, including herbivory, and carnivory) by wildlife on crops and livestock. Choosing to apply these tactics, in concert with other strategies, may help to achieve 'wilder' agricultural landscapes where wildlife, humans, and their agricultural interests coexist.

Tactic	Description	Examples
Selective culling ('bettering the devil')	Imposing targeted selection for individuals that have lower impact on livestock and crops.	In contrast to conventional non-targeted culling, selectively culling only those individuals that display 'bold' traits, such as those that are not deterred by a novel stimulus, may drive the local population's dominant traits towards a more 'shy' personality that actively avoids livestock or crops (e.g. as proposed for sika deer *Cervus nippon*, Honda, 2021). See also Garvey *et al.* (2020) and Manning *et al.* (2021).
Conditioned aversion	Creating an association between livestock or crops and a negative taste (e.g. bitter) or outcome (e.g. nausea inducement, shock collars).	Using conditioned taste aversion, Gustavson *et al.* (1974) were able to suppress attacks by coyotes (*Canis latrans*) on lambs without affecting predation on other prey, while Baker *et al.* (2008) produced learned aversion towards maize in European badgers (*Meles meles*). Using a similar method, Andrewartha *et al.* (2023) have demonstrated the sustained aversion of wild red foxes to otherwise attractive baits. A beep and shock collar combination has also been trialled for deterring grey wolves from calves (Schultz *et al.*, 2005), which is similar to some of the virtual fences used for containing livestock (Campbell *et al.*, 2017).
Founder selection and targeted gene flow	Individuals with desirable traits/genes (e.g. shyness, preference for non-livestock prey) are selected for population establishment (as in the case of founder selection) or introduced to an existing population with the hope that those traits/genes will drive adaptation of the population in the desired direction (i.e. targeted gene flow).	To our knowledge, these tactics have not been applied to predators in an agricultural context, though there are some similarities with genetic biocontrol techniques (Teem *et al.*, 2020). An example from the conservation field demonstrates their potential applicability: Progeny of toad-savvy northern quolls (*Dasyurus hallucatus*) that had adapted to avoid eating the invasive and toxic cane toad (*Rhinella marina*) were less likely to eat cane toads than progeny from toad-naïve parents (Kelly and Phillips, 2019). Hybrid progeny (from one savvy and one naïve parent) were also less likely to eat cane toads, highlighting the potential to influence adaptation across populations (Kelly and Phillips, 2019).

(Continued)

Table 2 *(Continued)*

Tactic	Description	Examples
Habitat management	Altering habitat features (e.g. vegetation structure, floristics, availability of shelters) to reduce predation impact.	Fischer *et al.* (2018) suggested that their observation of reduced crop damage inflicted by rodents in organic (compared to conventional) wheat fields was due to the characteristically lower crop densities and greater crop heights of the organic fields they measured (perhaps mediated by the rodents' harvesting behaviour). Separately, the impacts of deer (Jarnemo *et al.*, 2014) and livestock predators (O'Farrell *et al.*, 2008) may be reduced by managing landscape heterogeneity to provide alternative food sources (see Section 4.2). See also 'trap cropping', a form of companion planting (Sarkar *et al.*, 2018).
Manipulating the landscape of fear	Changing the behaviour or distribution of predators by manipulating the environment they respond to with fear (e.g. avoidance, reduced predation rates).	Sullivan and Crump (1984) showed that the presence of the anal-gland compounds of mustelids that prey on snowshoe hares (*Lepus americanus*) could suppress the hare's feeding on coniferous tree seedlings. Widén *et al.* (2022) showed that playback of predator vocalisations (including human voices) could reduce crop damage by ungulates. See also 'hunting for fear' (Cromsigt *et al.*, 2013), livestock guardian dogs (van Bommel and Johnson, 2012), and the use of flags hung from ropes (i.e. fladry, Musiani *et al.*, 2003).
Misinformation tactics	Exposing predators to unrewarded livestock or crop odours.	Using odour camouflage, where the scent targeted by an animal to find a particular food is applied throughout the landscape, Parker *et al.* (2023) significantly reduced the predation of newly sown wheat seeds by wild house mice (*Mus musculus*).

Adapted from Evans *et al.* (2022).

areas in eastern Australia, a 5614 km-long dingo-proof fence was erected in the 1880s (with ongoing maintenance costs), and widespread culling of dingoes is still common (Prowse *et al.*, 2015).

Kangaroos also have a conflicting status as one of Australia's emblematic and culturally important species, and an economically valuable provider of meat and leather, but can also be deemed a nuisance that consumes crops, competes with livestock, and damages ecosystems when overabundant (Thomsen *et al.*, 2006; Boronyak *et al.*, 2013; Read *et al.*, 2021). Without a large apex predator, like the dingo, more human resources are required to manage kangaroo populations to protect agricultural and ecological assets and minimise the animal welfare impacts that arise when overpopulation leads to starvation (Read *et al.*, 2021; Wimpenny *et al.*, 2021). The resources needed to manage kangaroos, where dingoes are present, are not necessarily zero, but one can expect them to be less than in the alternative context, given the evidence that suggests dingoes can regulate kangaroo populations (Pople *et al.*, 2000). However, there are also costs required to manage dingoes, and the dynamics involved will likely vary in space and time.

For animal rewilding in agricultural landscapes to balance economic and environmental gains in Australia, the ecosystem services of dingoes and kangaroos need to be maximised, while minimising their disservices. For kangaroos, strategies are required to increase the appreciation and value of kangaroo meat and leather products, which could supplement, or replace, other income streams and reduce methane emissions (Gordon, 2019; Wilson and Edwards, 2019; Read *et al.*, 2021). Dingoes may aid the management of kangaroo populations through predation, thereby reducing competition with livestock for pasture (Letnic and Koch, 2010; Prowse *et al.*, 2015). In addition, dingoes may assist with the control of other predators and competitors in agricultural landscapes, such as foxes, European rabbits (*Oryctolagus cuniculus*), and wild pigs (Seward *et al.*, 2004; Fleming *et al.*, 2022), though there are still many knowledge gaps around their assumed biotic interactions (Allen *et al.*, 2013; Forsyth *et al.*, 2019; Dawson *et al.*, 2023), and dingoes do also consume threatened species (Doherty *et al.*, 2019). In contexts where the application of a dingo scent is intended to repel herbivores (Parsons and Blumstein, 2010), the presence of dingoes in the landscape may prevent habituation to that tactic, since the risk of predation associated with that scent is actually present. Underpinning these potential ecosystem services is the need to provide wildlife habitat across the landscape through land sharing strategies (Ampt and Baumber, 2014).

Another way dingoes may contribute to the regulation of grazing resources is by influencing the behaviour or abundance of invasive cats (*Felis catus*) and foxes, which consume a large number of the seed predators (e.g. native rodents) that modify vegetation composition (note that there are counter arguments and hypotheses on the complex relationship between these predators, and it is likely that their interactions vary depending on many environmental factors (Allen *et al.*, 2013; Morgan *et al.*, 2017; Fancourt *et al.*, 2019; Kreplins *et al.*, 2021; Menon *et al.*, 2024). Dingoes may alter the predation impacts of cats and foxes through direct attacks or by influencing the 'landscape of fear' (i.e. an individual's perception of variable predation risk across the landscape, which influences their distribution or behaviour, Manning *et al.*, 2009; Laundré *et al.*, 2010; Letnic and Koch, 2010; Letnic and Dworjanyn, 2011; Feit *et al.*, 2019). This could enable the resurgence or reintroduction of the small mammal prey that are more frequently targeted by cats and foxes, than by dingoes (Fleming *et al.*, 2022). The value of seed predation by these small mammals is evident in experiments undertaken by Mills *et al.* (2018) in arid Australia but may be easy to overlook. By consuming the seeds of woody shrubs, native rodents in arid ecosystems may reduce the human inputs needed to control undesirable shrub encroachment, and thereby increase the amount of pasture available for livestock (Mills *et al.*, 2018).

The innovations that may support the coexistence of dingoes with livestock are continuing to expand and improve (Smith and Appleby, 2018; Evans *et al.*, 2022; Boronyak and Jacobs, 2023). Where dingoes are presently absent or scarce in the landscape, one could choose to rewild with relatively shy individuals of that species (see 'Founder selection and targeted gene flow' in Table 2) that may be more responsive to a landscape of fear promoted by livestock guardian dogs (Rigg *et al.*, 2011; van Bommel and Johnson, 2012) or selective culling (i.e. targeted only at the individuals that attempt to interact with livestock, Smith and Appleby, 2018; Meuret *et al.*, 2020). On very large properties (thousands of square kilometres) where those patrolling measures are less feasible, simply ceasing non-targeted lethal control measures has been hypothesised to reduce dingo impacts on cattle by facilitating the social stabilisation among packs that reduces reproductive rates while keeping unlearned individuals out of the territory (Wallach *et al.*, 2017; Smith and Appleby, 2018). For individuals that persist in hunting livestock, are bold enough to do so, and are smart enough to avoid other defences, conditioned taste aversion could be an alternative tool that could potentially deter them from the livestock (see 'Conditioned aversion' in Table 2). The crucial factor here is that the predators that do not impose conflicts with agriculture are maintained in the landscape to pass on their genes and behaviours to drive positive adaptation and deliver a balance between economic and environmental gains (Evans *et al.*, 2021; Manning *et al.*, 2021).

8 Conclusion

Ecosystem services obtained by returning wild animals to agricultural landscapes can outweigh the disservices of those same animals. A resurgence of appreciation for wild animal products (e.g. food and leather) would provide opportunities for landholders to obtain new or supplementary income. Further, a greater balance between the different types of ecosystem services in agriculture could be achieved by coordinating broad spatial heterogeneity in land use, with various land sharing agricultural strategies (Fig. 1), forming a high-quality habitat matrix. Conventional agriculture and land sparing strategies will still dominate some landscapes, but where animal rewilding in agriculture is encouraged, those strategies may ultimately only occur as smaller patches within the aforementioned high-quality connective matrix. Finally, increased implementation of tactics that could drive or enable the adaptation of livestock, crops, their predators, as well as human perceptions and management strategies, could facilitate the long-term coexistence that will help to produce net positive outcomes from animal rewilding.

Alongside other advances in social, economic, ecological, and agricultural fields, we foresee that the coordination of wild provisioning services, broad spatial heterogeneity in land use, and coexistence conservation, together, will be key to facilitating sustainability through a balance of economic and environmental gains when undertaking animal rewilding in agricultural landscapes.

9 Where to look for further information

The following books provide more detail on some of the topics discussed in this chapter:

- Gordon IJ, Prins HHT and Squire GR (2017) Food Production and Nature Conservation: Conflicts and Solutions. Oxon: Routledge.
- Lindenmayer DB and Fischer J (2006) Habitat Fragmentation and Landscape Change: An Ecological and Conservation Synthesis. Washington, DC: Island Press.
- Norton D and Reid N (2013) Nature and Farming: Sustaining Native Biodiversity in Agricultural Landscapes. Victoria, Australia: CSIRO Publishing.

The Coexistence Conservation Laboratory at The Australian National University is open to possible collaborations in this field.

10 References

Allen, B. L., Fleming, P. J. S., Allen, L. R., *et al.* (2013) As clear as mud: a critical review of evidence for the ecological roles of Australian dingoes. *Biological Conservation* 159: 158–174.

Amador-Alcalá, S., Naranjo, E. J. and Jiménez-Ferrer, G. (2013) Wildlife predation on livestock and poultry: implications for predator conservation in the rainforest of south-east Mexico. *Oryx* 47(2): 243–250.

Ampt, P. and Baumber, A. (2014) Building connections between kangaroos, commerce and conservation in the rangelands. *Australian Zoologist* 33(3): 398–409.

Andrewartha, T., Evans, M., Blencowe, A., *et al.* (2023) Landscapes of nausea: successful conditioned taste aversion in a wild red fox population. *Conservation Science and Practice* 5(8): e12984.

Auster, R. E., Barr, S. W. and Brazier, R. E. (2020) Wildlife tourism in reintroduction projects: exploring social and economic benefits of beaver in local settings. *Journal for Nature Conservation* 58: 125920.

Baker, S.E., Ellwood, S. A., Slater, D., *et al.* (2008) Food aversion plus odor cue protects crop from wild mammals. *The Journal of Wildlife Management* 72(3): 785–791.

Beaufoy, G., Baldock, D. and Clark, J. (1994) *The nature of farming: low intensity farming systems in nine European countries.* London: Institute for European Environmental Policy.

Benton, T. G., Vickery, J. A. and Wilson, J. D. (2003) Farmland biodiversity: is habitat heterogeneity the key? *Trends in Ecology & Evolution* 18(4): 182–188.

Bignal, E. M. and McCracken, D. I. (1996) Low-intensity farming systems in the conservation of the countryside. *Journal of Applied Ecology* 33(3): 413–424.

Bommarco, R., Kleijn, D. and Potts, S. G. (2013) Ecological intensification: harnessing ecosystem services for food security. *Trends in Ecology & Evolution* 28(4): 230–238.

Boronyak, L., Ben-Ami, D. and Croft, D. D. R (2013) *Kanganomics: a socio–economic assessment of the commercial kangaroo industry.* Sydney, Australia: Report by the Centre for Compassionate Conservation at the University of Technology.

Boronyak, L. and Jacobs, B. (2023) Pathways to coexistence with dingoes across Australian farming landscapes. *Frontiers in Conservation Science* 4. https://doi.org/10.3389/fcosc.2023.1126140

Campbell, D. L. M., Lea, J. M., Farrer, W. J., *et al.* (2017) Tech-savvy beef cattle? How heifers respond to moving virtual fence lines. *Animals* 7. https://doi.org/10.3390/ani7090072

Campbell-Palmer, R., Gow, D., Schwab, G., *et al.* (2016) *The eurasian beaver handbook: ecology and management of Castor fiber.* Exeter, UK: Pelagic Publishing.

Carver S (2014) Making real space for nature: a continuumm approach to UK conservation. *ECOS 35* (3/4): 4–14.

Carver, S., Convery, I., Hawkins, S., *et al.* (2021) Guiding principles for rewilding. *Conservation Biology* 35(6): 1882–1893.

Cerqueira, Y., Navarro, L. M., Maes, J., *et al.* (2015) Chapter 3. Ecosystem services: the opportunities of rewilding in Europe. In Pereira, H. M. and Navarro, L. M. (eds) *Rewilding European landscapes.* Germany: SpringerOpen.

Charnley, S., Gosnell, H., Davee, R. and Abrams, J. (2020) Ranchers and beavers: understanding the human dimensions of beaver-related stream restoration on western rangelands. *Rangeland Ecology & Management* 73(5): 712–723.

Clough, Y., Barkmann, J., Juhrbandt, J., *et al.* (2011) Combining high biodiversity with high yields in tropical agroforests. *Proceedings of the National Academy of Sciences* 108(20): 8311–8316.

Corson, M. S., Mondière, A., Morel, L. and van der Werf, H. M. G. (2022) Beyond agroecology: agricultural rewilding, a prospect for livestock systems. *Agricultural Systems* 199: 103410.

Cromsigt, J. P. G. M., Kuijper, D. P. J., Adam, M., *et al.* (2013) Hunting for fear: innovating management of human–wildlife conflicts. *Journal of Applied Ecology* 50(3): 544–549.

Cromsigt, J. P. G. M., te Beest, M., Kerley, G. I. H., *et al.* (2018) Trophic rewilding as a climate change mitigation strategy? *Philosophical Transactions of the Royal Society B: Biological Sciences* 373(1761): 20170440.

Cunningham, S. A., Attwood, S. J., Bawa, K. S., *et al.* (2013) To close the yield-gap while saving biodiversity will require multiple locally relevant strategies. *Agriculture, Ecosystems & Environment* 173: 20–27.

Daily, G. C. (1997) *Nature's services: societal dependence on natural ecosystems.* Washington, DC: Island Press.

Daily, G. C. (1999) Developing a scientific basis for managing Earth's life support systems. *Conservation Ecology* 3(2): 14.

Dawson, S. J., Kreplins, T. L., Kennedy, M. S., *et al.* (2023) Land use and dingo baiting are correlated with the density of kangaroos in rangeland systems. *Integrative Zoology* 18(2): 299–315.

Doherty, T. S., Davis, N. E., Dickman, C. R., *et al.* (2019) Continental patterns in the diet of a top predator: Australia's dingo. *Mammal Review* 49(1): 31–44.

Edwards, D. P., Gilroy, J. J., Woodcock, P., *et al.* (2014) Land-sharing versus land-sparing logging: reconciling timber extraction with biodiversity conservation. *Global Change Biology* 20(1): 183–191.

Evans, M. J., Batson, W. G., Gordon, I. J., *et al.* (2021) The 'Goldilocks Zone' of predation: the level of fox control needed to select predator resistance in a reintroduced mammal in Australia. *Biodiversity and Conservation* 30(6): 1731–1752.

Evans, M. J., Weeks, A. R., Scheele, B. C., *et al.* (2022) Coexistence conservation: reconciling threatened species and invasive predators through adaptive ecological and evolutionary approaches. *Conservation Science and Practice* 4(7): e12742.

Fancourt, B. A., Cremasco, P., Wilson, C. and Gentle, M. N. (2019) Do introduced apex predators suppress introduced mesopredators? A multiscale spatiotemporal study of dingoes and feral cats in Australia suggests not. *Journal of Applied Ecology* 56(12): 2584–2595.

Feit, B., Feit, A. and Letnic, M. (2019) Apex predators decouple population dynamics between mesopredators and their prey. *Ecosystems* 22(7): 1606–1617.

Fischer, C., Gayer, C., Kurucz, K., *et al.* (2018) Ecosystem services and disservices provided by small rodents in arable fields: effects of local and landscape management. *Journal of Applied Ecology* 55(2): 548–558.

Fischer, J., Brosi, B., Daily, G. C., *et al.* (2008) Should agricultural policies encourage land sparing or wildlife-friendly farming? *Frontiers in Ecology and the Environment* 6(7): 380–385.

Fischer, J., Lindenmayer, D. B. and Manning, A. D. (2006) Biodiversity, ecosystem function, and resilience: ten guiding principles for commodity production landscapes. *Frontiers in Ecology and the Environment* 4(2): 80–86.

Fleming, P. A., Stobo-Wilson, A. M., Crawford, H. M., *et al.* (2022) Distinctive diets of eutherian predators in Australia. *Royal Society Open Science* 9(10): 220792.

Fleming, P. J. S., Ballard, G. and Cutter, N. (2021) There is no dingo dilemma: legislation facilitates culling, containment and conservation of dingoes in New South Wales. *Australian Zoologist* 41(3): 408–416.

Forsyth, D. M., Latham, A. D. M., Davis, N. E., *et al.* (2019) Interactions between dingoes and introduced wild ungulates: concepts, evidence and knowledge gaps. *Australian Mammalogy* 41(1): 12–26.

Gaba, S., Bretagnolle, F., Rigaud, T. and Philippot, L. (2014) Managing biotic interactions for ecological intensification of agroecosystems. *Frontiers in Ecology and Evolution* 2(29). https://doi.org/10.3389/fevo.2014.00029

Gabriel, D., Carver, S. J., Durham, H., *et al.* (2009) The spatial aggregation of organic farming in England and its underlying environmental correlates. *Journal of Applied Ecology* 46(2): 323–333.

Garrido, P., Elbakidze, M., Angelstam, P., *et al.* (2017) Stakeholder perspectives of wood-pasture ecosystem services: a case study from Iberian dehesas. *Land Use Policy* 60: 324–333.

Garvey, P. M., Banks, P. B., Suraci, J. P., *et al.* (2020) Leveraging motivations, personality, and sensory cues for vertebrate pest management. *Trends in Ecology & Evolution* 35(11): 990–1000.

Gordon, I. J. (2009) What is the future for wild, large herbivores in human-modified agricultural landscapes? *Wildlife Biology* 15(1): 1–9.

Gordon, I. J. (2019) Adopting a utilitarian approach to culling wild animals for conservation in National Parks. *Conservation Science and Practice* 1(10): e105.

Gordon, I. J., Pérez-Barbería, F. J. and Manning, A. D. (2021) Rewilding lite: using traditional domestic livestock to achieve rewilding outcomes. *Sustainability* 13. https://doi.org/10.3390/su13063347

Grass, I., Loos, J., Baensch, S., *et al.* (2019) Land-sharing/-sparing connectivity landscapes for ecosystem services and biodiversity conservation. *People and Nature* 1(2): 262–272.

Green, R. E., Cornell, S. J., Scharlemann, J. P. W. and Balmford, A. (2005) Farming and the fate of wild nature. *Science* 307(5709): 550–555.

Greenleaf, S. S. and Kremen, C. (2006) Wild bees enhance honey bees' pollination of hybrid sunflower. *Proceedings of the National Academy of Sciences* 103(37): 13890–13895.

Gustavson, C. R., Garcia, J., Hankins, W. G. & Rusiniak, K. W. (1974) Coyote predation control by aversive conditioning. *Science* 184(4136): 581–583.

Hansen, L. M. (2003) Insecticide-resistant pollen beetles (*Meligethes aeneus* F) found in Danish oilseed rape (*Brassica napus* L) fields. *Pest Management Science* 59(9): 1057–1059.

Hill, C. M. (1997) Crop-raiding by wild vertebrates: the farmer's perspective in an agricultural community in western Uganda. *International Journal of Pest Management* 43(1): 77–84.

Honda, T. (2021) Geographical personality gradient in herbivorous animals: implications for selective culling to reduce crop damage. *Ecological Research* 36(1): 145–151.

Iqbal, A., Khan, R. S., Khan, M. A., *et al.* (2021) Genetic engineering approaches for enhanced insect pest resistance in sugarcane. *Molecular Biotechnology* 63(7): 557–568.

IUCN/SSC (2013) *Guidelines for reintroductions and other conservation translocations. Version 1.0.* Gland, Switzerland: IUCN Species Survival Commission.

Jaffé, R., Pope, N., Carvalho, A. T., *et al.* (2015) Bees for development: Brazilian survey reveals how to optimize stingless beekeeping. *PLoS One* 10(3): e0121157.

Jarnemo, A., Minderman, J., Bunnefeld, N., *et al.* (2014) Managing landscapes for multiple objectives: alternative forage can reduce the conflict between deer and forestry. *Ecosphere* 5(8): art97.

Kelly, E. and Phillips, B. L. (2019) Targeted gene flow and rapid adaptation in an endangered marsupial. *Conservation Biology* 33(1): 112–121.

Kolkert, H., Smith, R., Rader, R. and Reid, N. (2020) Insectivorous bats foraging in cotton crop interiors is driven by moon illumination and insect abundance, but diversity benefits from woody vegetation cover. *Agriculture, Ecosystems & Environment* 302: 107068.

Kolkert, H., Smith, R., Rader, R. and Reid, N. (2021) Insectivorous bats provide significant economic value to the Australian cotton industry. *Ecosystem Services* 49: 101280.

Kremen, C. (2005) Managing ecosystem services: what do we need to know about their ecology? *Ecology Letters* 8(5): 468–479.

Kremen, C. (2020) Ecological intensification and diversification approaches to maintain biodiversity, ecosystem services and food production in a changing world. *Emerging Topics in Life Sciences* 4(2): 229–240.

Kreplins, T. L., Kennedy, M. S., O'Leary, R. A., *et al.* (2021) Fighting like cats and dogs? Dingoes do not constrain spatial and temporal movements of feral cats. *Food Webs* 27: e00173.

Laundré, J. W., Hernández, L., Ripple, W. J. (2010) The landscape of fear: ecological implications of being afraid. *The Open Ecology Journal* 3: 1–7.

Lemaire, G., Ryschawy, J., César de Faccio Carvalho, P. and Gastal, F. (2017) Chapter 6: agricultural intensification and diversity for reconciling production and environment: Role of integrated crop-livestock systems. In Gordon, I. J., Prins, H. H. T. and Squire, G. R. (eds) *Food production and nature conservation: conflicts and solutions*. Oxon: Routledge.

Lesslie, R. G. and Taylor, S. G. (1985) The wilderness continuum concept and its implications for Australian wilderness preservation policy. *Biological Conservation* 32(4): 309–333.

Letnic, M. and Dworjanyn, S. A. (2011) Does a top predator reduce the predatory impact of an invasive mesopredator on an endangered rodent? *Ecography* 34(5): 827–835.

Letnic, M. and Koch, F. (2010) Are dingoes a trophic regulator in arid Australia? A comparison of mammal communities on either side of the dingo fence. *Austral Ecology* 35(2): 167–175.

Leyequién, E., de Boer, W. F. and Toledo, V. M. (2010) Bird community composition in a shaded coffee agro-ecological matrix in Puebla, Mexico: the effects of landscape heterogeneity at multiple spatial scales. *Biotropica* 42(2): 236–245.

Lindenmayer, B. D. and Fischer, J. (2006) Chapter 14: the matrix and landscape heterogeneity. In Lindenmayer, B. D. and Fischer, J. (eds) *Habitat fragmentation and landscape change: an ecological and conservation synthesis*. Washington, DC: Island Press.

Lodberg-Holm, H. K., Garvik, E. S., Fountain, M. S., *et al.* (2022) Crop circles revealed spatio-temporal patterns of beaver foraging on cereal fields. *Agriculture, Ecosystems & Environment* 337: 108066.

López-Uribe, M. M., Morreale, S. J., Santiago, C. K. and Danforth, B. N. (2015) Nest suitability, fine-scale population structure and male-mediated dispersal of a solitary ground nesting bee in an urban landscape. *PLoS One* 10(5): e0125719.

Lund, J. F. and Jensen, F. S. (2017) Is recreational hunting important for landscape multi-functionality? Evidence from Denmark. *Land Use Policy* 61: 389–397.

Lyons, A., Ashton, P. A., Powell, I. and Oxbrough, A. (2017) Impacts of contrasting conservation grazing management on plants and carabid beetles in upland calcareous grasslands. *Agriculture, Ecosystems & Environment* 244: 22–31.

Manning, A. D., Andrewartha, T. A., Blencowe, A., *et al.* (2021) Bettering the devil you know: can we drive predator adaptation to restore native fauna? *Conservation Science and Practice* 3(8): e447.

Manning, A. D., Gordon, I. J. and Ripple, W. J. (2009) Restoring landscapes of fear with wolves in the Scottish Highlands. *Biological Conservation* 142(10): 2314–2321.

Manning, A. D. and Lindenmayer, D. B. (2009) Paddock trees, parrots and agricultural production: an urgent need for large-scale, long-term restoration in south-eastern Australia. *Ecological Management & Restoration* 10(2): 126–135.

Manning, A. D., Lindenmayer, D. B. and Nix, H. A. (2004) Continua and umwelt: novel perspectives on viewing landscapes. *Oikos* 104(3): 621–628.

Mattison, E. H. A. and Norris, K. (2005) Bridging the gaps between agricultural policy, land-use and biodiversity. *Trends in Ecology & Evolution* 20(11): 610–616.

McIntyre, S. and Hobbs, R. (1999) A framework for conceptualizing human effects on landscapes and its relevance to management and research models. *Conservation Biology* 13(6): 1282–1292.

McKee, S., Anderson, A., Carlisle, K. and Shwiff, S. A. (2020) Economic estimates of invasive wild pig damage to crops in 12 US states. *Crop Protection* 132: 105105.

Menichetti, L., Touzot, L., Elofsson, K., *et al.* (2019) Interactions between a population of fallow deer (*Dama dama*), humans and crops in a managed composite temperate landscape in southern Sweden: Conflict or opportunity? *PLoS One* 14(4): e0215594.

Menon, V., McGregor, H., Giljohann, K., *et al.* (2024) Ecological factors influencing invasive predator survival and movement: insights from a continental-scale study of feral cats in Australia. *Biological Invasions* 2024. https://doi.org/10.1007/s10530-024-03337-y

Meuret, M., Moulin, C. H., Bonnet, O., *et al.* (2020) Missing shots: has the possibility of shooting wolves been lacking for 20 years in France's livestock protection measures? *The Rangeland Journal* 42(6): 401–413.

Millennium Ecosystem Assessment (2003) *Ecosystems and human well-being: a framework for assessment.* Washington, DC: Island Press.

Mills, C. H., Gordon, C. E. and Letnic, M. (2018) Rewilded mammal assemblages reveal the missing ecological functions of granivores. *Functional Ecology* 32(2): 475–485.

Morgan, H. R., Hunter, J. T., Ballard, G., *et al.* (2017) Trophic cascades and dingoes in Australia: does the Yellowstone wolf–elk–willow model apply? *Food Webs* 12: 76–87.

Musiani, M., Mamo, C., Boitani, L., *et al.* (2003) Wolf depredation trends and the use of fladry barriers to protect livestock in western North America. *Conservation Biology* 17(6): 1538–1547.

Norton, D. A., Reid, N. and Young, L. (2013) Ultimate drivers of native biodiversity change in agricultural systems. *F1000Res* 2: 214.

O'Farrell, P. J., Donaldson, J. S., Hoffman, M. T. and Mader, A. D. (2008) Small mammal diversity and density on the Bokkeveld escarpment, South Africa – implications for conservation and livestock predation. *African Zoology* 43(1): 117–124.

Obermeier, C., Mason, A. S., Meiners, T., *et al.* (2022) Perspectives for integrated insect pest protection in oilseed rape breeding. *Theoretical and Applied Genetics* 135(11): 3917–3946.

Parker, F. C. G., Price, C. J., Bytheway, J. P. and Banks, P. B. (2023) Olfactory misinformation reduces wheat seed loss caused by rodent pests. *Nature Sustainability* 6(9): 1041–1044.

Parkes, J. P., Nugent, G. and Warburton, B. (1996) Commercial exploitation as a pest control tool for introduced mammals in New Zealand. *Wildlife Biology* 2(3): 171–177.

Parsons, M. H. and Blumstein, D. T. (2010) Familiarity breeds contempt: Kangaroos persistently avoid areas with experimentally deployed dingo scents. *PLoS One* 5(5): e10403.

Perfecto, I. and Vandermeer, J. (2017) Chapter 7: a landscape approch to integrating food production and nature conservation. In Gordon, I. J., Prins, H. H. T. and Squire, G. R. (eds) *Food production and nature conservation: conflicts and solutions*. Oxon: Routledge.

Phalan, B., Onial, M., Balmford, A. and Green, R. E. (2011) Reconciling food production and biodiversity conservation: land sharing and land sparing compared. *Science* 333(6047): 1289–1291.

Plieninger, T., Torralba, M., Hartel, T. and Fagerholm, N. (2019) Perceived ecosystem services synergies, trade-offs, and bundles in European high nature value farming landscapes. *Landscape Ecology* 34(7): 1565–1581.

Ponisio, L. C., de Valpine, P., M'Gonigle, L. K. and Kremen, C. (2019) Proximity of restored hedgerows interacts with local floral diversity and species' traits to shape long-term pollinator metacommunity dynamics. *Ecology Letters* 22(7): 1048–1060.

Pooley, S., Bhatia, S. and Vasava, A. (2021) Rethinking the study of human–wildlife coexistence. *Conservation Biology* 35(3): 784–793.

Pople, A. R., Grigg, G. C., Cairns, S. C., *et al.* (2000) Trends in the numbers of red kangaroos and emus on either side of the South Australian dingo fence: evidence for predator regulation? *Wildlife Research* 27(3): 269–276.

Power, A. G. (2010) Ecosystem services and agriculture: tradeoffs and synergies. *Philosophical Transactions of the Royal Society B: Biological Sciences* 365(1554): 2959–2971.

Prowse, T. A. A., Johnson, C. N., Cassey, P., *et al.* (2015) Ecological and economic benefits to cattle rangelands of restoring an apex predator. *Journal of Applied Ecology* 52(2): 455–466.

Puttock, A., Graham, H. A., Ashe, J., *et al.* (2021) Beaver dams attenuate flow: a multi-site study. *Hydrological Processes* 35(2): e14017.

Pywell, R. F., Heard, M. S., Woodcock, B. A., *et al.* (2015) Wildlife-friendly farming increases crop yield: evidence for ecological intensification. *Proceedings of the Royal Society B: Biological Sciences* 282(1816): 20151740.

Raudsepp-Hearne, C., Peterson, G. D. & Bennett, E. M. (2010) Ecosystem service bundles for analyzing tradeoffs in diverse landscapes. *Proceedings of the National Academy of Sciences* 107(11): 5242–5247.

Read, J. L., Wilson, G. R., Coulson, G., *et al.* (2021) Improving kangaroo management: a joint statement. *Ecological Management & Restoration* 22(S1): 186–192.

Rigg, R., Find'o, S., Wechselberger, M., *et al.* (2011) Mitigating carnivore–livestock conflict in Europe: lessons from Slovakia. *Oryx* 45(2): 272–280.

Robinson, R. A., Wilson, J. D. and Crick, H. Q. P. (2001) The importance of arable habitat for farmland birds in grassland landscapes. *Journal of Applied Ecology* 38(5): 1059–1069.

Salatino, A., Pereira, L. R. L. and Salatino, M. L. F. (2019) The emerging market of propolis of stingless bees in tropical countries. *MOJ Food Processing & Technology* 7(1): 27–29.

Sarkar, S. C., Wang, E., Wu, S. and Lei, Z. (2018) Application of trap cropping as companion plants for the management of agricultural pests: a review. *Insects* 9. https://doi.org/10.3390/insects9040128

Saunders, M. E., Peisley, R. K., Rader, R. and Luck, G. W. (2016) Pollinators, pests, and predators: recognizing ecological trade-offs in agroecosystems. *Ambio* 45(1): 4–14.

Schultz, R. N., Jonas, K. W., Skuldt, L. H. and Wydeven, A. P. (2005) Experimental use of dog-training shock collars to deter depredation by gray wolves. *Wildlife Society Bulletin* 33(1): 142–148.

Senapathi, D., Goddard, M. A., Kunin, W. E. and Baldock, K. C. R. (2017) Landscape impacts on pollinator communities in temperate systems: evidence and knowledge gaps. *Functional Ecology* 31(1): 26–37.

Seward, N. W., VerCauteren, K. C., Witmer, G. W. and Engeman, R. M. (2004) Feral swine impacts on agriculture and the environment. *Sheep & Goat Research Journal* 19: 34–40.

Shwiff, S. A., Kirkpatrick, K. N. and Godwin, K. (2011) Economic evaluation of beaver management to protect timber resources in Mississippi. *Human-Wildlife Interactions* 5(2): 306–314.

Smith, B. P. and Appleby, R. G. (2018) Promoting human–dingo co-existence in Australia: moving towards more innovative methods of protecting livestock rather than killing dingoes (*Canis dingo*). *Wildlife Research* 45(1): 1–15.

Smith, R. and Watson, A. (2018) Working with nature to improve the environment and profitability of irrigated cotton production at 'Kilmarnock', Namoi Valley, New South Wales. *Ecological Management & Restoration* 19(S1): 63–72.

Strohbach, M. W., Kohler, M. L., Dauber, J. and Klimek, S. (2015) High nature value farming: from indication to conservation. *Ecological Indicators* 57: 557–563.

Sullivan, T. P. and Crump, D. R. (1984) Influence of mustelid scent-gland compounds on suppression of feeding by snowshoe hares (*Lepus americanus*). *Journal of Chemical Ecology* 10(12): 1809–1821.

Sunderland, K. and Samu, F. (2000) Effects of agricultural diversification on the abundance, distribution, and pest control potential of spiders: a review. *Entomologia Experimentalis et Applicata* 95(1): 1–13.

Sutcliffe, L., Akeroyd, J., Page, N. and Popa, R. (2015) Combining approaches to support high nature value farmland in southern Transylvania, Romania. *Hacquetia* 14(1): 53–63.

Tahmasebi, T., Karami, E. and Keshavarz, M. (2020) Agricultural land use change under climate variability and change: drivers and impacts. *Journal of Arid Environments* 180: 104202.

Teem, J. L., Alphey, L., Descamps, S., *et al.* (2020) Genetic biocontrol for invasive species. *Frontiers in Bioengineering and Biotechnology* 8. https://doi.org/10.3389/fbioe.2020.00452

Thies, C., Steffan-Dewenter, I. and Tscharntke, T. (2003) Effects of landscape context on herbivory and parasitism at different spatial scales. *Oikos* 101(1): 18–25.

Thomas, V., Mondière, A., Corson, M. S. and van der Werf, H. M. G. (2022) 24. Domesticating rewilding: combining rewilding and agriculture offers environmental and human benefits. In Bruce, D. and Bruce, A. (eds) *Transforming food systems: ethics, innovation and responsibility*. The Netherlands: Wageningen Academic Publishers, pp. 165–170.

Thompson, S., Vehkaoja, M., Pellikka, J. and Nummi, P. (2021) Ecosystem services provided by beavers *Castor* spp. *Mammal Review* 51(1): 25–39.

Thomsen, D. A., Muir, K. and Davies, J. (2006) Aboriginal perspectives on kangaroo management in South Australia. *The Rangeland Journal* 28(2): 127–137.

Tisdell, C. A. (1982) *Wild pigs: environmental pest or economic resource?* NSW, Australia: Pergamon Press.

Tschumi, M., Ekroos, J., Hjort, C., *et al.* (2018) Predation-mediated ecosystem services and disservices in agricultural landscapes. *Ecological Applications* 28(8): 2109–2118.

Urlich, S. C. (2015) What's the end-game for biodiversity: is it time for conservation evolution? *New Zealand Journal of Ecology* 39(1): 133–142.

van Bommel, L. and Johnson, C. N. (2012) Good dog! Using livestock guardian dogs to protect livestock from predators in Australia's extensive grazing systems. *Wildlife Research* 39(3): 220–229.

van Eeden, L. M., Dickman, C., Crowther, M. and Newsome, T. (2021a) A theory of change for promoting coexistence between dingoes and livestock production. *Conservation Science and Practice* 3(3): e304.

Van Eeden, L. M., Crowther, M. S., Dickman, C. R. and Newsome, T. M. (2021b) Wicked "wild dogs": Australian public awareness of and attitudes towards dingoes and dingo management. *Australian Zoologist* 41(3): 467–479.

van Eeden, L. M., Eklund, A., Miller, J. R. B., *et al.* (2018) Carnivore conservation needs evidence-based livestock protection. *PLoS Biology* 16(9): e2005577.

Viana, B. F., Coutinho, J. G. d. E., Garibaldi, L. A., *et al.* (2014) Stingless bees further improve apple pollination and production. *Journal of Pollination Ecology* 14(0): 261–269.

Wallach, A. D., Ramp, D. and O'Neill, A. J. (2017) Cattle mortality on a predator-friendly station in central Australia. *Journal of Mammalogy* 98(1): 45–52.

Weibull, A.-C, Östman, Ö. and Granqvist, Å. (2003) Species richness in agroecosystems: the effect of landscape, habitat and farm management. *Biodiversity & Conservation* 12(7): 1335–1355.

Widén, A., Clinchy, M., Felton, A. M., *et al.* (2022) Playbacks of predator vocalizations reduce crop damage by ungulates. *Agriculture, Ecosystems & Environment* 328: 107853.

Wilson, G. R. and Edwards, M. (2019) Professional kangaroo population control leads to better animal welfare, conservation outcomes and avoids waste. *Australian Zoologist* 40(1): 181–202.

Wimpenny, C., Hinds, L. A., Herbert, C. A., *et al.* (2021) Fertility control for managing macropods – current approaches and future prospects. *Ecological Management & Restoration* 22(S1): 147–156.

Zhang, W., Ricketts, T. H., Kremen, C., *et al.* (2007) Ecosystem services and dis-services to agriculture. *Ecological Economics* 64(2): 253–260.

Zheng, H., Wang, L. and Wu, T. (2019) Coordinating ecosystem service trade-offs to achieve win–win outcomes: a review of the approaches. *Journal of Environmental Sciences* 82: 103–112.

Chapter 15

Animal rewilding in theory and practice: Australia and New Zealand

Christopher R. Dickman, Aaron C. Greenville and Glenda M. Wardle, The University of Sydney, Australia

1 Introduction

The term 'rewilding' is relatively new, having been used in an ecological context for the first time just over 30 years ago (Foreman et al., 1992; Noss, 1992). Yet, despite the novelty of the term, the concept of rewilding is not new, and in recent practice, rewilding has taken on a wide variety of meanings. The term itself harkens both to the early German word 'will', which described places or creatures not under human domination, and wild-doer-ness (or naess in Old English) that connoted a place of self-willed creatures (Johns, 2019). Wildlands, then, can be considered self-willed lands or wilderness, and rewilding the process by which this state is recovered (Johns, 2019).

1.1 What is rewilding?

In first using the term 'rewilding', Foreman et al. (1992) argued that it was time to rewild North America and in their definition they foreshadowed many of the concepts that rewilding is now considered to include: 'To rewild is to ensure designated areas can support in perpetuity all native species – including top

http://dx.doi.org/10.19103/AS.2024.0134.25

predators, other keystone species, and wide-ranging species – all ecosystem types and processes, disturbance regimes and resilience. ... Whatever term is used – restore, restoration, recovery, or rewilding – the centrality of the very large-scale, the inclusion of all native species including highly interactive ones (keystone), and the importance of natural process, are part of a consistent thread' (Johns, 2019, 18–19). Many subsequent definitions of rewilding have emphasised different aspects of the early concept, suggesting that the aim of rewilding should be to return non-wild areas back to a wild state (Corlett, 2016), to introduce or reintroduce wild organisms to places where they are absent (Prior and Brady, 2017), especially if they are functionally important (Naundrup and Svenning, 2015), or to ensure that active measures are taken to ensure renewed ecosystem function (Law et al., 2017). Other workers have expanded the concept of rewilding (Foreman et al. 1992) by proposing that rewilding can take place at relatively small scales, such as in urban or agricultural environments where very large-scale interventions are difficult to achieve (Owens and Wolch, 2019), or that rewilding can be a passive process of recovery that occurs when human disturbance in an area ceases (e.g. Regos et al., 2016; Clancy and Ward, 2020). Although there are sometimes calls to restore places to pre-existing states, such as to a time before humans arrived or before human activities became highly destructive, it is generally recognised that this is an unrealistic and elusive goal due to the extinction of key species and uncertainty about how earlier environments functioned (Law et al., 2017; du Toit, 2019). Differing perspectives on what rewilding is, and should be, are summarised in useful reviews by Jørgensen (2015) and Pettorelli et al. (2018, 2019).

A key factor driving the rise of rewilding is that it can provide, at least potentially, a partial antidote to the demise of biodiversity and ecological processes at local and regional scales (Fernández et al., 2017). Because rewilding often involves large and charismatic species, it is usually also seen as positive conservation activity that fosters public support and engagement (Monbiot, 2013; Dunn-Capper et al., 2024). A useful framework for capturing the different perceptions about rewilding, and what it can achieve, was presented by Schulte to Bühne et al. (2022) (Fig. 1).

1.2 What this chapter covers

In this chapter, we focus on rewilding in Australia and Aotearoa New Zealand. We first discuss the background rationale for why rewilding is necessary and consider some of the theoretical considerations that have informed practical approaches to rewilding in this part of the world. We then describe three case studies of rewilding that show its utility in recovering native species and ecological processes before advocating a novel and digitally enabled approach that could be used to inform the rewilding of agricultural landscapes in future.

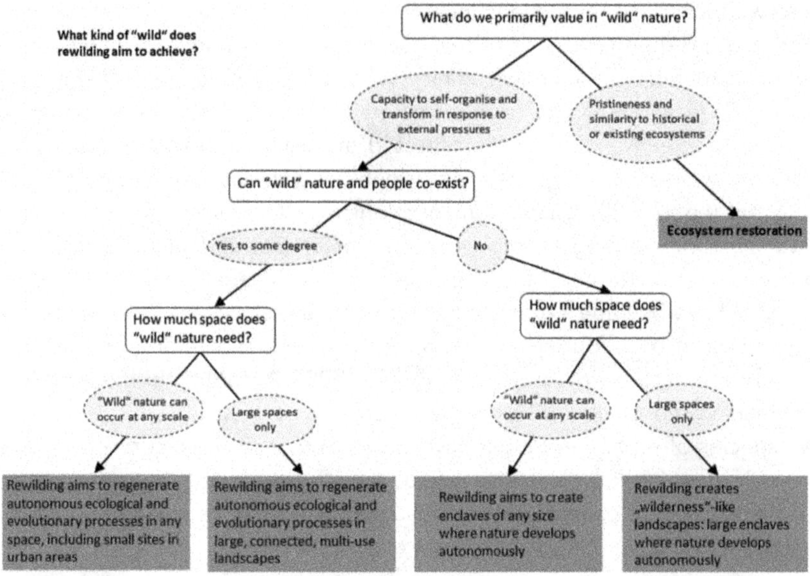

Figure 1 Flowchart showing different perceptions about rewilding and the decisions that need to be taken to achieve different outcomes. Because of the need to involve different stakeholder groups in rewilding, the branches leading to the left-hand side of the flowchart are most relevant for the discussions in this chapter. Source: From Schulte to Bühne et al. (2022), unchanged; CC Attribution 4.0, see: https://creativecommons.org/licenses/by/4.0/.

Following the multiplicity of perceptions of what rewilding is and what it should do, as summarised above, here we use a simple operational definition that is relevant to the Australian and New Zealand context: *rewilding is the deliberate (re)introduction of animal species into local or regional areas with the joint aims to assist the conservation of those species and promote the re-establishment of lost or degraded ecological processes*. As people must be involved in setting up, managing and monitoring the outcomes of rewilding under this definition, the pathways involved lie principally to the left of the decision tree depicted in Fig. 1. Our focus is on animals for reasons that we describe in the next section.

2 The benefits of rewilding

2.1 Biodiversity in Australia and New Zealand: impacts of humans

The arrival of humans in Australia ~60 000–65 000 years ago and in New Zealand just ~700 years ago coincided with sudden and dramatic losses of large animals. In Australia, these 'megafauna' included giant herbivorous and carnivorous marsupials, large flightless birds, eagles and reptiles, whereas in New Zealand the extinctions included giant birds such as moas and their predators, including Haast's eagle *Hieraaetus moorei*. Apart from an enigmatic Mesozoic mammal of unknown affinity (Worthy et al., 2006), New Zealand has historically lacked native terrestrial mammals. Although the role of humans in causing these early extinctions is contested, at least in Australia (Hocknull et al., 2020; Johnson et al., 2021), there is no doubt that the direct and indirect effects of recent (i.e. last 200 years) human colonists in Australia and New Zealand have been devastating to large numbers of animal and plant species and the communities to which they belonged. In Australia, at least 43 species of mammal have become extinct since European arrival, as have 22 species and subspecies of bird, four species of amphibian, one reptile and an unknown but probably large number of invertebrates (EPBC Act, 2023; Burbidge, 2024; Newman-Martin et al., 2023). In New Zealand, at least 20 species of bird, two species of reptile and a bat have been extirpated since European arrival (Miskelly et al., 2008; Nelson et al., 2015). In both countries, many plant species have also become extinct, and disruptions and losses of ecological functions and processes have been severe (Stow et al., 2015).

2.2 Biodiversity in Australia and New Zealand: impacts of biodiversity loss

Losses of biological diversity in Australia and New Zealand are of particularly grave concern. This is because there is a high prevalence of endemic species (and often genera) in most major taxon groups, so extinctions, if they occur, are irrecoverable. The nations of Australia and New Zealand therefore bear sole responsibility for the conservation of these species. However, the widespread erosion of biological diversity has deleterious effects more broadly on the aesthetic, functional, economic, social and cultural values of human society and can also have damaging effects on human health (Robin, 2012; Maller et al., 2019). Among animals, for example, several species within the families Peramelidae and Potoroidae in Australia regularly turn over large amounts of topsoil as part of their foraging and burrowing activities, burying seeds, leaf litter and other organic material that helps to maintain soil fertility and structure and provide improved conditions for plant growth (Dickman, 2015; Eldridge and Frankham, 2015). Losses of these 'engineer' species – regionally for some, wholly for others – is likely to have led to long-term deterioration of natural and agricultural soils in many areas. Other ecological functions can also be diminished when key species are lost. For example, many vertebrate species act as vectors for pollen, seeds or the spores of mycorrhizal fungi, while many invertebrates play critical roles in nutrient recycling by consuming dead plant and animal biomass (Augee and Fox, 2000); top predators may keep herbivores in check, increasing both carbon sequestration and the diversity of species that exploit structurally complex habitats (Wilmers et al., 2012).

Biodiversity losses can also be economically costly. Consider one exemplar species, the koala *Phascolarctos cinereus*. In the late 1990s, surveys by Tor Hundloe and Clive Hamilton of foreign tourists departing Australia showed that 72% of visitors had arrived with a particular wish to see koalas, while 66% nominated kangaroos, and 11% of visitors would not have visited Australia if there was no unique wildlife to be seen. Extrapolating from these percentages and overall tourism revenue in 1996 of $16.1 billion, Hundloe and Hamilton (1997) estimated that Australia's unique wildlife returned ~$1.8 billion to the national economy that year, with ~$1.1 billion attributable to the koala alone. If the koala were to decline or disappear, it is likely that opportunities to see other charismatic species and visit functional natural systems would keep nature-based tourism alive, but tourist revenue would be clearly diminished. Losses of other natural icons such as the Great Barrier Reef in north-eastern Australia or sanctuary sites in New Zealand such as Tiritiri Matangi near Auckland or Zealandia in Wellington would also have major impacts on the national economies of both countries, as well as, of course, on aesthetic, cultural and social values. It is not by accident that a stylised emu and kangaroo feature on the Australian coat of arms or that native ferns appear on the coat of arms of New Zealand.

Realisation of the intrinsic and multiple human-centered values of biological diversity has led to wide-ranging efforts to protect and conserve what is left and, where possible, to restore species and ecological processes at local to regional scales. Rewilding forms a key part of this collective enterprise. In Australia, mammals are usually the main focus of rewilding (Sweeney et al., 2019). This reflects the dire conservation status of many surviving species, the importance of the ecological functions they carry out, their status as icons or symbols of regional or national pride, and the cultural importance of species, such as the greater bilby *Macrotis lagotis*, for the Traditional Owners of many rewilding sites. In New Zealand, where native terrestrial mammals are absent, birds, reptiles and sometimes invertebrates are foci for rewilding for the same panoply of reasons as for their counterparts in Australia.

3 The challenges of rewilding

Despite the potential benefits that flow from rewilding, it can be difficult to achieve successful results. At the minimum, rewilding needs clear goals and detailed planning with provision for contingencies and adaptive management, adequate long-term resourcing, and engagement and agreement among many stakeholders.

3.1 Setting the objectives: the role of stakeholders

In the first instance, the need for or desirability of a rewilding programme must be considered and should proceed if most or all participants in the programme agree that the benefits are likely to outweigh the costs. This step should engage people who will be directly and indirectly involved in the programme such as site owners or managers, neighbours, Traditional Knowledge custodians, Western scientists and practitioners. Additional stakeholders will often include representatives from academic institutions, government and non-government organisations that provide funding, material infrastructure, legal advice, labour or other resources needed for a potential rewilding programme to be successful (Dalrymple and Bellis, 2023). Other stakeholders are perhaps less obvious but include the broader public and future generations of the community who will inherit the rewilding programme. For contentious rewilding initiatives, such as the reintroduction of grey wolves *Canis lupus*, the need to engage with all stakeholders is particularly obvious. In the USA, for example, wolves were reintroduced to Yellowstone National Park after 21 years of debate and occurred when the interests of ranchers and concerned citizens outside Yellowstone had been accommodated (Bangs and Fritts, 1996; Smith et al., 2006). In Japan, by contrast, debate about reintroducing wolves to Shiretoko National Park, on Hokkaido, has remained unresolved for more than two decades, with public engagement and consultation having been minimal (Kameyama, 2006).

When a stakeholder group has been assembled, it needs to meet to specify what the programme aims to achieve, to set clear objectives and timelines, ensure adherence to relevant state and national legislation concerning land use and wildlife translocations, and confirm that resources are adequate to carry the programme to completion. In some programmes, just one or two focal species may be targeted for reintroduction because of their perilous conservation status or cultural importance, the primary aim being to secure their populations (e.g. Eason et al., 2006; Wise et al., 2019). In other programmes the aim may be to reintroduce a suite of species that should help to recreate lost ecological processes. Whatever the objective(s) of the rewilding programme, these should be clearly stated and quantitative milestones specified to allow ongoing assessment of programme success. In species reintroductions, for example, milestones may be set that a founder population increases in size by 10% each year for 5 years before levelling off; a similar threshold may be set to ensure the recovery of an ecological process (Burbidge, 1996; Saunders, 1996). Monitoring is essential to track whether milestones are being met, and if they are not, a review of programme protocols needs to be made and management changed to accommodate and respond to the findings of the review. Monitoring results with interpretive commentary should also be published, ideally in widely accessible fora, so that the 'silent' stakeholders in the broader public are kept informed and engaged with the rewilding programme. Examples of this are the newsletters disseminated by the National Environmental Science Program to keep the community informed about reintroductions of mammals to Booderee National Park on the south coast of New South Wales (Robinson, 2018), and the FrogID programme run by the Australian Museum that provides regular updates on frog populations – including declining, recovering and common species – across Australia, using records submitted to the programme by 'citizen scientists' in the broader public (Australian Museum, 2024).

3.2 Selection of sites for rewilding

At the same time that rewilding objectives and protocols are being established, stakeholders should also identify a site or sites where the programme will take place. A potentially suitable site is one that would have security of land tenure and access and be large enough to support viable populations of the species that are targeted for reintroduction. For populations to be viable and self-sustaining, the site would need to contain sufficient habitat, food, shelter and other material resources that the different target species require, and it would also need to be cleared of the causes for the species' extinction there in the past. Evaluating whether a site is likely to support viable populations of target species can be challenging and ideally requires information on the ecological requirements, behaviour and life history parameters of these species. An approach that can be used to make such evaluations is population viability analysis, as described in the next section. In addition, it is clearly desirable that native species that occur already in rewilding sites are not likely to be affected detrimentally by the arrival of new species, such as a predator (Stepkovitch et al., 2023). Even if the species to be reintroduced were known to be present at some time in the past, it is possible that the environment changed during the period when they were absent such that other species moved in or gained ascendancy. In many programmes, for example, ground-dwelling arthropods are often overlooked but may be at risk from the foraging or digging activities of small- or medium-sized mammals when these are reintroduced (Taylor et al., 2017; Coulter et al., 2023).

Even when rewilding sites have been thoroughly assessed as suitable, successful reintroduction is not guaranteed. For example, eastern quolls *Dasyurus viverrinus* became extinct on Australia's east coast in the 1960s, ostensibly due to habitat loss and predation from the European red fox *Vulpes vulpes* and feral domestic cat *Felis catus* (Caughley, 1980), and were reintroduced to Booderee National Park on the south coast of New South Wales in 2018. The site had been selected after extensive consultation with the broader public and the local community, and had been exhaustively evaluated to ensure that it provided the necessary resources for quolls to persist. The entire area had also been subject to long-term fox control, and populations of the southern brown bandicoot *Isoodon obesulus* and long-nosed potoroo *Potorous tridactylus* had already been reintroduced successfully. Unfortunately, despite the reintroduction of 20 quolls in mid-2018 and another 40 quolls a year later, and despite evidence from the monitoring programme that some quolls bred successfully, it appears that the population had died out by 2021 (Glenday and Kennedy, 2021). Several quolls were killed on roads, others by foxes, and yet more by domestic dogs and native predators. Although this rewilding attempt failed, it did identify the key threats that quolls face in Booderee and will allow management to be modified to increase the odds of future reintroductions succeeding (Glenday and Kennedy, 2021).

In practice, most sites for rewilding to date in Australia and New Zealand have been designated on protected land such as nature reserves, national parks or Indigenous Protected Areas (IPAs) or land that is privately owned by conservation organisations such as the Australian Wildlife Conservancy (Dickman, 2012; Legge et al., 2018a). Despite the potential benefits accruing from rewilding, relatively few sites have been established in agricultural or suburban/peri-urban areas, although notable exceptions exist such as Mulligans Flat Woodland Sanctuary on the outskirts of Canberra and Zealandia in Wellington.

3.3 Selection of species for rewilding

While sites are being considered for rewilding, advocates will usually have compiled a list of species that are candidates to be reintroduced there. There are several challenges associated with identifying such candidates. The first is to be clear about whether the rewilding programme aims to focus on translocating species that are primarily of conservation concern, have important functional roles in the system, fulfill dual roles or whether a range of species is ultimately to be rewilded. The selected species sometimes have complementary roles. At Booderee National Park, for example, rewilded bandicoots improve soil health via their digging activities, potoroos are likely to be vectors of the spores of mycorrhizal fungal that assist the growth of vascular plants, and eastern quolls are carnivores. All three species are nationally threatened. If the programme proposes to reintroduce species that were suspected to occur formerly at the rewilding site, it is important to verify that these species were once present using literature records, historical accounts, museum specimens or other evidence such as species distribution modelling, and also to establish that these species are now absent. If small, remnant populations of any target species are present in a rewilding site, release of new individuals from elsewhere could disrupt the social structure of the remnant population and lead to intraspecific aggression, or have unwanted genetic consequences such as outbreeding depression (e.g. Marshall and Spalton, 2000). It is common practice in reinforcement programmes to augment small populations by the addition of new individuals, but in these situations, the demographic and genetic statuses of both the donor and recipient populations are well known and the aim is to bolster population size and growth (IUCN/SSC, 2013).

When species have been identified as once occurring in the rewilding site, another step is to ensure that known or suspected causes of their demise have been removed or can be adequately managed. In Australia and New Zealand, common causes of local (and regional) extinctions of vertebrates are the loss of habitat and the introduction of invasive predators, competitors or pathogens (Dickman, 2018; Woinarski et al., 2019; Glen et al., 2023). If these disturbances are ongoing and have not been abated, it is unlikely that any reintroduction attempts will succeed. In Australia, for example, Short et al. (1992) showed that the reintroduction success of macropodid marsupials was just 8% in sites where invasive predators were present and 82% in sites where these predators were absent. Similar results were reported by Finlayson et al. (2010) for potoroid marsupials and by Moseby et al. (2011) for several additional species of native mammals, implicating the primacy of predation in determining reintroduction success.

If such threats have been nullified in a rewilding site, a further requirement at this stage is to carry out surveys to ensure that the key resource requirements of the target species can be met. This is important as changes may have occurred in the habitat condition of the rewilding site since the target species last occurred there; food and shelter sources may have altered, and new potential threats may have appeared. If key resources have been eroded, it may be necessary to supplement them during the rewilding process or to ensure that the rewilding site is large enough to hold lower density but still viable populations of the target species.

Gauging whether resources are adequate to support populations within prescribed rewilding sites can be difficult, especially if the sites are characterised by high levels of spatial and temporal variation in resource quantity or quality. One simple method to assess potential site-suitability is to use published density estimates for the target species in similar habitats (e.g. Dickman, 2014; van Eeden et al., 2020) and estimate how many individuals could be accommodated in the rewilding site. Although minimum numbers for population viability have been much debated (Ballou et al., 1995), empirical estimates suggest that ~3700 – 5400 individuals are needed for long-term persistence of vertebrates and vascular plants (Traill et al., 2007) although numbers in the 100s often allow at least short-term (e.g. 20–50 years) persistence.

Another commonly used method to overcome these difficulties is population viability analysis (PVA). This tool allows the user to input genetic and demographic parameters such as a species' reproductive rate, survival rate, age at first reproduction, breeding system, expected population size under different resource conditions, as well as the effects of threats or disturbance events, and to then predict population size (or the probability of population persistence) when these parameters are varied. In one example, at a rewilding site in Sydney, Banks (2004) used PVA to show that increasing levels of adult mortality and environmental variation markedly reduced the probability of long-term (50-year) persistence of an existing population of the southern long-nosed bandicoot *Perameles nasuta* (Fig. 2). The results suggested that, to maximise the likelihood of bandicoot persistence, practical management options could focus most effectively on reducing adult mortality due to predation from the invasive European red fox and from roadkill.

Rewilding sites in Australia and New Zealand are often limited in area, and this can place constraints on the number and composition of species that are reintroduced. Large species generally need more space than small species, and large areas with a high diversity of habitats will usually provide a greater diversity of niches for many species than will small areas or sites with structurally simple habitats (MacArthur and Wilson, 1967). Consideration also needs to be given to the interactions that may occur between rewilded species. For example, overlap in habitat components used by the brush-tailed bettong, or woylie *Bettongia penicillata*, and the burrowing bettong *Bettongia lesueur* and greater bilby, suggested the possibility of competition for space, with numbers

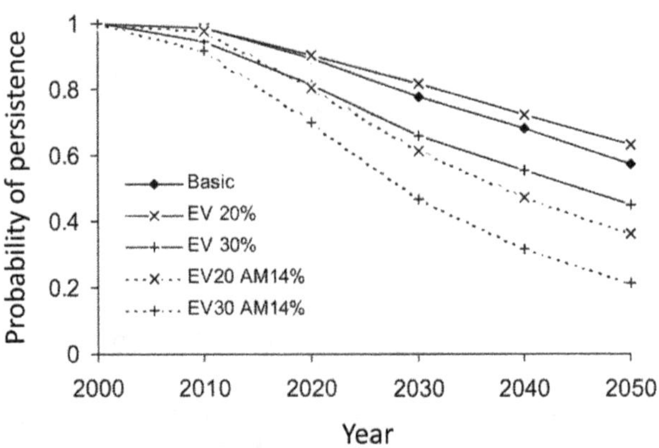

Figure 2 The probability of persistence of the southern long-nosed bandicoot *Perameles nasuta* in a confined population at North Head, Sydney, over a period of 50 years, as forecast by population viability analysis. The forecasts have been made using different levels of environmental variation in carrying capacity (EV) and adult mortality (AM); the basic model assumes a value of 25% EV. Increasing EV with additional adult mortality, such as from predation by the red fox and roadkill in this urban population, together reduce the probability of persistence of this population. Source: Redrawn from: Banks (2004), with permission from the Royal Zoological Society of New South Wales.

of the first species declining in a rewilding programme as those of the last two increased (Finlayson et al., 2008). Similarly, a reintroduction of the endangered woma python *Aspidites ramsayi* was thwarted by predation from another snake, the mulga snake *Pseudechis australis* (Read et al., 2011), and a reintroduction of the western quoll *Dasyurus geoffroii* led to population declines of both the burrowing bettong and the Shark Bay bandicoot *Perameles bougainville* that had been reintroduced to the same rewilding area (Stepkovitch et al., 2023). Such results emphasise the need to understand ecological interactions that are likely to occur between taxa that are targeted for rewilding (Moseby et al., 2023), and to manage these – such as with the provision of extra shelter or other resources, or even with a reduction in the population of 'problem' or overabundant native species – if they are likely to compromise the success of the programme.

Although rewilding most commonly seeks to move species into sites where they formerly occurred, species may sometimes be introduced to sites where there is no evidence of their former occurrence. There are two reasons for this. Firstly, ecological analogues of extinct species may be moved to new sites if they perform important ecological functions that could otherwise not be restored. This is sometimes known as an ecological replacement (Gaywood and Stanley-Price, 2023). Secondly, species distribution modelling may show that the favoured climate of a target species will shift in the future to areas where it never naturally occurred; rewilding sites located in these areas may then be more appropriate than sites within the species' current range that will become less favourable in the future. Translocations of species into geographically new areas based on future projections of climate suitability – often termed assisted colonisation – are becoming more common, especially if natural range shifts are precluded by barriers or fragmentation of species' habitats (Hällfors and Dalrymple, 2023). Whatever the reason for contemplating introductions of species to rewilding sites outside their former known ranges, assessments of the suitability of the new sites need to be made in a similar manner to assessments that are made for species being reintroduced (IUCN/SSC, 2013).

A good example of assisted colonisation is that of Gilbert's potoroo *Potorous gilbertii*. Initially thought to have become extinct in the nineteenth century, a small population of this potoroo (30–40 individuals) was discovered in 1994 in dense coastal heath and closed woodland on the south coast of Western Australia. Occurring in very small numbers in a single location that was at risk of disturbances such as wildfire automatically qualified this species as Critically Endangered. To boost its numbers and reduce the risk of a catastrophic event extirpating the population, small numbers of animals were moved in 2005 to a new site – Bald Island, 1.5 km offshore in the Great Australian Bight – where the species had not previously been known to occur (Friend, 2023). A wildfire in 2015 reduced the mainland population of potoroos to just seven individuals and rendered 90% of the habitat unsuitable in the near term (Friend, 2023). Fortunately, the population on Bald Island has thrived since 2005, with this assisted colonisation programme almost certainly saving Gilbert's potoroo from extinction (Bolam et al., 2021). Experimental releases of Gilbert's potoroo on Middle Island, near Esperance, in 2017 and 2018, have confirmed that suitable habitat is present there (Friend, 2023), thus paving the way for a further assisted colonisation programme for this species in the future.

3.4 Selection of individuals for rewilding

A further consideration in rewilding concerns the origin and numbers of animals to be moved. If donor individuals are sourced from wild populations care must be taken not to deplete those populations, especially if the species is subject to rewilding because it is of conservation concern. Population viability modelling can be used profitably to forecast the numbers, sexes and age classes of individuals that can be removed from source populations without deleterious effects. By contrast, if donor individuals are sourced from captive breeding facilities care will be needed to ensure that they are well prepared to live unassisted when translocated to the wild. These individuals should have been exposed in captivity to ambient conditions, to food types and habitats similar to those in the rewilding site, and be minimally habituated to human presence. For example, to maintain wild behaviours in Tasmanian devils *Sarcophilus harrisii* destined for later release, Izzard et al. (2019) recommended the use of free-range enclosures (9–22 ha) where devils are able to hunt their usual prey (small to large vertebrates), have access to fresh water and natural denning materials such as hollow logs, and experience minimal interference from people.

Whether animals are sourced from the wild or captivity, sufficient numbers of each sex and of different age classes (juveniles, sub-adults or adults) will need to be translocated to the rewilding site to maximise the chances that the move will be successful. Again, PVA, or extensions such as population and habitat viability analysis (PHVA), is often used to project the results of adding different numbers and combinations of sex/age groups, and whether one or more translocation efforts will be needed to ensure establishment success (Lacy, 2019; Lees et al., 2019). A recent review of published field studies found that release success increased up to about 20–50 individuals but – paradoxically – declined slightly above several hundred individuals (Morris et al., 2021). In practice, in many reintroductions of vertebrates, adult animals – especially breeding females – are preferred because this demographic component is most likely to survive and contribute to population growth (Fischer and Lindenmayer, 2000; Seddon et al., 2007). Further considerations when selecting animals to translocate to a rewilding site include ensuring, as far as possible, that individuals are not closely related and collectively have at least moderate levels of genetic diversity and are healthy and free of parasites or pathogens that could otherwise compromise individual welfare and population growth (Dunlop and Watson, 2022). For example, several failed attempts to reintroduce the endangered green and golden bell frog *Litoria aurea* in eastern Australia implicated the cause of failure as chytridiomycosis, a usually fatal disease caused by the chytrid fungus *Batrachochytrium dendrobatidis* (Daly et al., 2008; Stockwell et al., 2008; White and Pyke, 2008). These researchers emphasised the need to check for this fungus in translocated frogs, at the recipient sites, and moreover to ensure that the researchers themselves do not act as inadvertent vectors of the pathogen.

Founding rewilding sites with related animals is likely to lead to inbreeding and genetic drift, and potentially to problems associated with individual development, growth, reproduction and susceptibility to disease (Keller et al., 2012). In contrast, genetically diverse founders can be expected to be individually less prone to such problems and to be more resilient at the population level to environmental changes and other challenges (Groombridge et al., 2012). Animals that are selected for translocation should, finally, be checked by veterinary personnel to ensure that they are in good health before being moved, with a period of quarantine indicated if there is uncertainty about any aspects of health or condition (Beckmann et al., 2022). Tarszisz et al. (2014) argued that this last step should include assessment of physiological well-being, with animals being screened to check levels of stress, nutritional and metabolic status. In most rewilding translocations, animals are moved to the new site within days of passing veterinary and health checks and then liberated in either 'soft' or 'hard' release protocols. In soft protocols, animals may be kept temporarily within pens on site to adjust to local conditions or are provided with supplementary resources for a period after release. In hard releases, animals are liberated directly into the rewilding site.

When animals have been selected for rewilding, a final consideration is when to release them. There are some common sense guidelines that should be followed, such as not releasing amphibians or other animals with requirements for free water during times of drought, or reptiles during winter when they are usually inactive. However, optimal timing for release is likely to vary, even between related taxa, and here the attributes of individual species may need to be considered. For example, chytrid infections in wild species of frog are generally low during summer, so chytrid-susceptible species such as the green and golden bell frog should have the best odds of successful establishment if released then (Daly et al., 2008). If dasyurid marsupials are being rewilded and females with pouch young have been selected for release, the release time will be constrained to spring or summer owing to the strict seasonality in time of breeding in most species (Baker and Dickman, 2018). For these reasons – among many others – it is important to have a sound knowledge of the ecology of species to be rewilded before release dates are set.

3.5 Monitoring

To understand whether a rewilding programme is achieving its stated goals, or whether management intervention is needed, released individuals need to be monitored after their release into the rewilding site. Monitoring allows managers to gauge whether individuals are finding sufficient food, maintaining body condition, avoiding negative interactions with other animals, growing and reproducing successfully, and whether milestones that have been set for rewilding success at the population level across multiple species are being met. This monitoring also needs to include the state of the rewilding site and external drivers of rewilding success such as resource availability and climatic influences. In adaptive management, monitoring also provides an important feedback loop that permits managers to learn about the system and to change their actions as needed to better achieve programme success (Lyons et al., 2008). The methods used to monitor fauna and environmental conditions will differ depending on the taxa and the available resources but should always be deployed for long enough to ensure that the goals of the programme have been achieved. Monitoring results should also be published to allow external critique and to keep the broader public aware and supportive of the outcomes of the rewilding programme. Publication of results also informs managers of what works, and what doesn't, in rewilding programmes. For example, up until 2004, monitoring implicated feral cats, European red foxes and dogs *Canis familiaris* in the failure of 11 of 13 translocations (85%) of potoroid marsupials where a cause of failure could be identified (Finlayson et al., 2010). After 2004, monitoring and intensive control of introduced predators has seen failure rates drop to below 50% (Finlayson et al., 2010). There has been much research on design principles for efficient monitoring, a selection of which includes Lindenmayer et al. (2012, 2014), Armstrong et al. (2015), Legge et al. (2018b) and Moseby et al. (2023).

4 Case studies

In many parts of the world rewilding takes place in open landscapes and uses large herbivores and carnivores to restore ecosystem processes, but in Australia and New Zealand rewilding occurs most commonly in confined areas and focuses on small- and medium-sized vertebrates (Innes et al., 2012, 2019; Sweeney et al., 2019). The necessity to use confined areas arises from the pervasive impacts of invasive species such as feral cats on native biota in open landscapes, such that successful rewilding is possible only where invasive species are continuously controlled or have been eradicated (Woinarski et al.,

2019; Glen et al., 2023). In practice, this means that rewilding initiatives mostly occur within predator-proof fences or on offshore islands from which invasive predators have been removed. Despite the high capital costs of building predator-proof fences and maintaining them in perpetuity, the problems of small scale, limitations of numbers of species that can be reintroduced and the risks of stochastic events and erosion of genetic diversity (e.g. Dickman, 2012; Stiglingh et al., 2023), there has been explosive growth in the construction of predator-proof fences and of eradications of invasive species from mainland enclosures and offshore islands in recent years (Fig. 3). In Australia, for example, a stocktake by Legge et al. (2018a) showed that 17 fenced areas covering a total area of 346 km^2 and 101 offshore islands covering 2152 km^2 protected 188 populations of 38 threatened mammal species that are otherwise at risk from invasive predators. In 2018, cats were successfully eradicated from the largest island in Western Australia, Dirk Hartog Island (633 km^2), paving the way for an ambitious rewilding programme to return the island environment to a more natural state and restore long-lost ecological processes (Algar et al., 2020).

In the case studies discussed in this chapter, we firstly describe rewilding in confined predator-free sites in Australia and New Zealand before discussing a controversial proposal to rewild regional areas in Australia, including agricultural

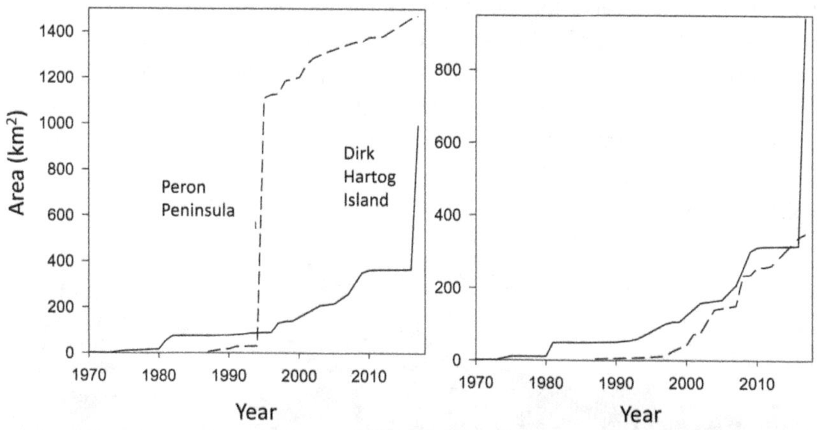

Figure 3 The cumulative expansion of fenced areas and offshore islands in the Australian region whence invasive species such as the European red fox *Vulpes vulpes* and feral cat *Felis catus* have been eradicated to improve the success of subsequent rewilding initiatives. Solid lines indicate islands, dashed lines indicate fenced areas on the mainland. The left panel shows the areal expansion of all projects where eradication of invasive species was attempted; the right panel shows the results of those projects that were successful. The failure of a predator-proof fence at Peron Peninsula markedly reduced the area of successfully fenced mainland areas, whereas the successful eradication of invasive species from Dirk Hartog Island (633 km^2) in Western Australia dramatically increased the overall area for rewilding of offshore islands. Source: Redrawn from: Legge et al. (2018a), with permission from CSIRO Publishing.

and pastoral landscapes, using the dingo *Canis dingo*. We then present a fourth case study from a mixed agricultural landscape that describes initial steps in documenting and monitoring existing fauna as a prelude to possible rewilding.

4.1 Case study 1: Tiritiri Matangi Island

Situated in Hauraki Gulf north of Auckland, Tiritiri Matangi Island has a long history of exploitation and degradation from human activity (Fig. 4). The island environment, covering just 220 ha, has been subject to intensive farming, overgrazing, soil compaction from livestock, regular burning and has endured the destruction of the lower strata of the native forest (Mitchell, 2013). Following public interest in the concept of restoring the island's ecosystems and returning some of the animal species that had once occurred there, the release of kakariki *Cyanoramphus novaezelandiae* onto Titritiri Matangi in 1973–74 heralded the beginning of an ambitious rewilding programme that has continued until the present (Craig et al., 1995; Towns et al., 2018). Legal steps were taken between 1975 and 1978 to incorporate the island into the Hauraki Gulf Maritime Park, with gazettal as a recreational and then scientific reserve following by 1980 (Galbraith and Cooper, 2013). The community-led group, Supporters of Tiritiri Matangi, formed in 1988 and, in collaboration with the New Zealand Department of Conservation, developed a long-term restoration plan involving both the planting of 280 000 trees and regeneration of vegetation via natural successional processes (Towns et al., 2018). Importantly, invasive weeds have been controlled since 1984 to help speed the recovery of native vegetation, the kiore or Pacific rat *Rattus exulans* was eradicated in 1993 and surveillance

Figure 4 Aerial view of Tiritiri Matangi Island in the Hauraki Gulf, New Zealand, showing extensive areas of forest following an extensive planting and revegetation campaign to restore the formerly degraded agricultural environment. CC licence.

monitoring for other invasive species, such as the Argentine ant *Linepithema humile*, has continued (Galbraith and Cooper, 2013). With the cessation of livestock grazing and other farming activities in the early 1970s, these actions helped to remove some of the threats that had contributed to the erosion of biodiversity on Tiritiri Matangi and paved the way for the recovery and reintroduction of a wide array of animal species.

Two native animal species that had not been recorded previously on Tiritiri Matangi, the common gecko *Woodworthia maculata* and a giant centipede *Cormocephalus rubriceps*, were detected in the mid-2000s when their increased numbers made them more conspicuous, and a further 15 species have been introduced or reintroduced from the mid-1980s following the earlier release of kakariki (Baling et al., 2013; Galbraith and Cooper, 2013). Eleven of these species were birds, three species were reptiles – including the northern tuatara *Sphenodon punctatus* – and one was an invertebrate, the wetapunga or Little Barrier giant weta *Deinacrida heteracantha*, introduced in 2011. Some of these species, such as the vulnerable stitchbird or hihi *Notiomystis cincta* and endangered South Island takahē *Porphyrio hochstetteri*, are of conservation concern and their translocation to Tiritiri Matangi has assisted in boosting their numbers. Others contribute to the recovery of ecological functions such as seed dispersal, pollination and nutrient recycling. It is unlikely that populations of any of these species could have recruited naturally onto Tiritiri Matangi owing to their limited dispersal abilities or lack of nearby source populations on the adjacent mainland.

The case of Tiritiri Matangi is particularly instructive for several reasons. In the first instance, the island was subject to rewilding as part of a broader restoration initiative well before the concept of rewilding had become established. It was, and continues to be, therefore, a visionary and pioneering initiative. Secondly, the rewilding of Tiritiri Matangi was fundamentally a community-led initiative that has continued to involve, and rely upon, voluntary public participation. Community involvement has occurred at all levels from planning to assistance with animal releases through to monitoring and management, with over 25 000 people a year now visiting Tiritiri Matangi to observe the successful outcomes of the restoration and rewilding programmes (Galbraith, 2013). In addition to providing an important source of revenue that assists continued restoration efforts and the local economy, tourist visits to the island increase awareness of the importance of rewilding in the broader community. Thirdly, perhaps because of its ease of access and proximity to a major city, Tiritiri Matangi has served as a major natural laboratory for ecological research by tertiary students, professional scientists and citizen scientists. The large body of research covering almost five decades is too voluminous to present in detail here, but the research findings have identified *inter alia* best practice protocols for translocating diverse animal species, illuminated the roles of genetics, demography, extrinsic factors, uncertainty and even luck in achieving successful outcomes, and uncovered the conditions needed for the re-establishment of key ecological processes. Long-term monitoring and adaptive management to improve the appropriateness of interventions have also been hallmarks of the rewilding programme on Tiritiri Matangi. The success of these endeavours can be gauged by the many animal species that have successfully established on the island, and by several species – including the kakariki and at least four other bird species – whose populations have increased to the point that they can now be 'harvested' to contribute to rewilding programmes elsewhere (Galbraith and Cooper, 2013; Graham et al., 2013). Armstrong et al. (2023) presented the results of long-term (>20-year) monitoring of hihi and toutouwai or New Zealand robin *Petroica longipes* on Tiritiri Matangi, elaborating several benefits that can accrue for improved management, of understanding of the dynamics or reintroduced populations, and for informing best practice for future reintroductions elsewhere.

4.2 Case study 2: Scotia Sanctuary

Western New South Wales has a long history of pastoralism, with leases for livestock – principally sheep – being taken up from the 1840s and now covering almost all of the 325 000 km^2 Western Division (Lunney, 1994). A combination of many factors including overstocking, overgrazing by the European rabbit *Oryctolagus cuniculus* and, prior to the turn of the twentieth century, severe droughts, led to considerable degradation of the land and to a Royal Commission in 1900 to 'Inquire into the condition of the Crown Tenants' in the Western Division (Royal Commission, 1901). Uptake of many of the recommendations made by the Royal Commission, especially the proclamation of the *Western Lands Act 1901*, has led to significant improvements in land management (Lunney, 1994). Much of the shrubby and taller woody vegetation that had been removed has regenerated, although overgrazing by feral herbivores such as goats *Capra hircus*, European rabbits and native herbivores such as red kangaroos *Osphranter rufus* and grey kangaroos *Macropus* spp. reduces recruitment of palatable shrubs as well as vegetative biomass (Auld, 1993, 1995; Barnes and Wise, 2010).

The pressures associated with the pastoral industry and predation from the European red fox and feral cat have taken a substantial toll on small and medium-sized native vertebrates in the Western Division of New South Wales (Lunney, 2001). Thus, 27 of the original 71 native mammals are regionally or wholly extinct, with most of these species – solely comprising marsupials and native rodents – falling within a 'critical weight range' of 35 – 5500 g (Dickman et al., 1993). Importantly, some of the now-extinct species were engineers that would have contributed to soil health via their digging and burrowing activities and acted as vectors for plant seeds and mycorrhizal fungi. In addition to the parlous state of native mammals, many of the birds, reptiles and amphibians of western New South Wales are also in peril (Sadlier and Pressey, 1994; Smith et al., 1994), marking the Western Division of New South Wales as being in urgent need of improved management and remediation.

Although rewilding across most or all of the Western Division would be challenging (but see case study 3, below), rewilding at smaller regional scales has commenced, such as at Scotia Sanctuary. Lying on the border with South Australia, 150 km south of Broken Hill in southwestern New South Wales, Scotia Sanctuary – covering almost 65 000 ha – was purchased and destocked by Earth Sanctuaries Ltd in the mid-1990s before being acquired by the Australian Wildlife Conservancy in 2002. It is managed with Traditional Owners to protect extensive areas of mallee woodland and spinifex grassland on predominantly sand dune and loamy plain landforms. For rewilding to be successful, a predator-proof fence enclosing 4000 ha was erected after Scotia became a sanctuary, and red foxes, feral cats, goats and European rabbits were eradicated (Finlayson et al., 2008). With the removal of these threats, the numbat *Myrmecobius fasciatus* was released in 1999 and then, from November 2004, a series of further reintroductions was carried out of brush-tailed bettongs, burrowing bettongs, greater bilbies, the bridled nailtail wallaby *Onychogalea fraenata* and, in 2006, the greater stick-nest rat *Leporillus conditor* (Finlayson et al., 2008). Intensive live-trapping and radio-tracking of these species confirmed the short-term success of the reintroductions, although numbers of brush-tailed bettongs declined a couple of years after the initial releases, perhaps due to drought, competition with burrowing bettongs or the increasing age of the founder population (Finlayson, 2010). A series of studies documented further aspects of the density, habitat use and diet of the reintroduced species (Pizzuto et al., 2007; Vieira et al., 2007; Navnith et al., 2009), indicating that each species was re-establishing successfully, at least in the early stages post-reintroduction. Construction of a second 4000 ha predator-proof fence at Scotia has allowed further expansion of the populations of these mammals.

The re-establishment in Scotia of mammals that once occurred in the broader region improves their overall conservation status and thus fulfils one of the key goals of rewilding. The second goal, of re-establishing lost or degraded ecological processes, is likely also to have been attained in the Scotia system owing to the engineering habits of several of the reintroduced species. Thus, the digging activities of greater bilby, bettongs and others create foraging pits that increase the infiltration of water and the presence of leaf litter in the soil, increasing soil organic content and reducing soil temperature and evaporation (Eldridge and Mensinga, 2007; Eldridge and James, 2009). The foraging pits serve as traps for seeds, and the increased organic matter and associated microbial activity in the adjacent soil provide fertile conditions that favour plant growth (Eldridge et al., 2017; Radnan and Eldridge, 2018). Further research at Scotia has shown that the reintroduction of native mammals has had cascading effects on invertebrates, increasing populations of some functional groups and reducing populations of others, but increasing such functions as dung removal, litter breakdown and pollination (Gibb, 2012; Silvey et al., 2015; Coggan et al., 2016). The overall effect of the return of native mammals to Scotia is still an active area of research (Michael et al., 2022; Bauer et al., 2023). However, improvements in soil health, regeneration of vegetation and creation of new microhabitats such as foraging pits and burrows suggest that the rewilding programme has successfully met the twin goals of species conservation and restoration of ecological function to a formerly degraded landscape. These improvements have been achieved in the absence of current agricultural activity but could be conceivably extended to surrounding production landscapes if medium-sized mammals were able to persist outside the predator-proof fence.

Scotia could, in addition, be a valuable refuge for several threatened bird species in the Mallee Bird Community of the Murray Darling Depression Bioregion that was listed recently as endangered under the *Environment Protection and Biodiversity Conservation Act 1999* (DAWE, 2021). The approved conservation advice for this community listed 20 mallee-specialist or mallee-dependent bird species that would be likely to occur in the extensive mallee woodland on Scotia, including the endangered black-eared miner *Manorina melanotis*, mallee emu-wren *Stipiturus mallee* and vulnerable malleefowl *Leipoa ocellata* (DAWE, 2021). Scotia also provides an important refuge for mallee-dependent reptiles such as the western blue-tongue lizard *Tiliqua occipitalis* and marble-faced delma *Delma australis* that are threatened in New South Wales (Bauer et al., 2023).

4.3 Case study 3: Dingoes in the rangelands

As shown by the Tiritiri Matangi Island and Scotia Sanctuary case studies, as well as by many other examples (e.g. Short et al., 1992; Dickman, 2012; Legge et al., 2018a; Innes et al., 2019), successful rewilding is possible if feral herbivores and introduced predators are removed from areas where reintroductions of native animals are to take place. However, disadvantages of rewilding within fenced areas include high maintenance and other costs and the relatively small scales at which they can be operated. Offshore islands can provide larger areas for rewilding once they have been cleared of threats to native fauna, but are unlikely to afford protection for large numbers of species (Legge et al., 2018a). A potential, albeit hotly debated solution to landscape-scale rewilding in Australia, is to use the dingo as a 'threat-controller' (Fig. 5). Dingoes occur throughout mainland Australia but are rare or absent over large areas in the southeast and southwest of the continent due to active control by humans.

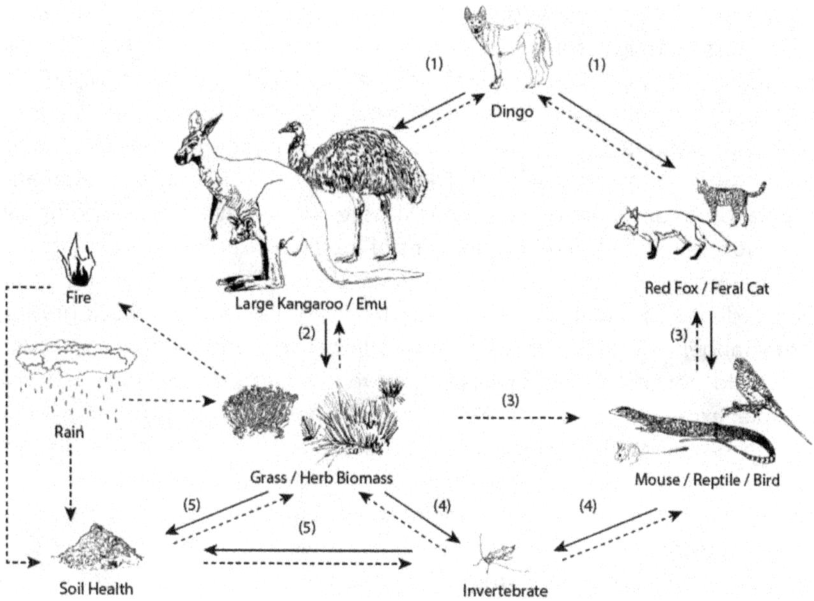

Figure 5 Potential interactions between dingoes, other species and the abiotic environment if dingoes were to be allowed to move back (or were reintroduced) into Sturt National Park, New South Wales, Australia. The solid lines represent effects associated with dingo recovery, the dashed lines represent reciprocal interactions between different components in the ecosystem, and numbers in parentheses represent the expected sequence of events. Two pathways are depicted. To the left of the figure, dingo suppression of large herbivores such as kangaroos and emus would be expected to increase plant biomass and cover, with flow-on effects for animals that use the vegetation, as well as soil quality. To the right of the figure, dingo suppression of red foxes and feral cats would be expected to reduce predatory pressure on small vertebrates and allow their populations to increase, with flow-on effects for invertebrates. The timing and strength of all interactions is likely to be influenced by large rainfall events and fires. Source: From Newsome et al. (2015), unchanged, with permission (License no. 5742400708780).

Much theory and some empirical evidence indicates that dingoes might be able to suppress the activity and abundance of smaller predators such as the red fox and feral cat, reducing the impacts of these species on smaller native prey (Glen and Dickman, 2005; Letnic et al., 2009; but see also Menon et al., 2024). Further evidence demonstrates that dingoes can reduce populations of introduced herbivores such as feral goats and native herbivores such as large kangaroos and emus *Dromaius novaehollandiae*, especially when populations of these species have already been reduced by factors such as drought (Newsome et al., 2001; Letnic et al., 2012).

There is debate about the quality of the evidence for these suppressive effects to reduce the prevalence of documented biotic threats to small- and medium-sized native vertebrates (e.g. Allen and Fleming, 2012; Hayward and Marlow, 2014), but the potential beneficial effect of dingoes has led to calls to restore dingoes in areas where they are currently controlled to assist with regional-scale rewilding (Dickman et al., 2009; Ritchie et al., 2012). Indeed, at the continental scale, Letnic et al. (2009) estimated that restoring dingo populations would produce a net benefit for the conservation of native mammals across 2.42×10^6 km^2, primarily in the arid and semi-arid rangelands. In addition to these benefits, the suppressive effects of dingoes on red foxes could allow reintroductions of medium-sized mammals that have become regionally extinct in arid regions, thus facilitating rewilding at an unprecedented scale. It is possible that the long history of livestock grazing in the rangelands has created a new, alternative state for the ecosystem that would not necessarily be reversed even if dingoes were rewilded. For example, experimental work in Yellowstone National Park has shown recently that the return of large carnivores decades after their extirpation did not lead to restoration of the original riparian plant communities; changes in the composition of the herbivore community following the initial loss of predators drove these communities into a stable, alternative, state (Hobbs et al., 2024). Replicated, small-scale reintroduction experiments would therefore be valuable to quantify the putative benefits of dingo restoration before rewilding attempts are made at regional scales.

Despite its appeal, any rewilding initiative using dingoes is not currently possible because of active dingo control. Throughout the arid and semi-arid rangelands dingoes are shot, trapped or poisoned to reduce their attacks on livestock, and long fences exclude dingoes from sheep flocks in the southwestern and southeastern corners of the country (Newsome et al., 2015). Although the land in these areas is marginal for sheep and calls have been made to remove the fence in the southeast to allow dingoes to return (Philip, 2022), it is unlikely that sympathetic management of the dingo could occur without appropriate engagement of the pastoral industry. Several studies have shown that alternative methods of livestock protection are available (e.g. van Eeden et al., 2018; Smith et al., 2021), with some methods, such as the use of guardian dogs that are bonded to sheep in open range paddocks showing potential for reliable and cost-effective protection from dingo attacks (van Bommel and Johnson, 2023). With appropriate industry or government assistance to help demonstrate the effectiveness of such alternative methods at the property level, dingo rewilding may be possible at scale in rangeland areas where sheep grazing currently predominates. Proposals have been put forward to shift the dingo fence to allow dingoes to move into a currently dingo-free zone in Sturt National Park in northern New South Wales to quantify the biodiversity and ecosystem benefits of dingoes (Newsome et al., 2015), but these ideas are yet to be followed through.

In lower rainfall regions where beef cattle are grazed, arguments to restore dingoes are particularly compelling. Here, evidence is accumulating that dingoes are a minor source of cattle mortality compared with poor environmental conditions (Emmott, 2021; Pollock, 2021), and that shifting to 'predator-friendly' management that precludes lethal control of dingoes, coupled with improved husbandry standards, can improve both cattle survival and welfare (Wallach et al., 2017). Reports from working cattle stations that eschew lethal dingo management indicate that dingo predation reduces populations of large kangaroos, feral goats and feral pigs *Sus scrofa*, in turn reducing total grazing pressure and soil erosion, while also facilitating increased plant regeneration and cover (Emmott, 2021; Pollock, 2021). These conditions improve cattle health and welfare, which benefits producers, and also improve environmental health, which benefits biodiversity. Given the economic, social, cultural and environmental benefits to stakeholders that appear to accrue from dingo-friendly management practices on large, open-range properties (Prowse et al., 2015), it is unclear why dingo rewilding is not practised more widely. As noted by Emmott (2021, 462), a third generation grazier working in western Queensland: 'We would all be much better served by a more thoughtful and nuanced approach to Dingoes. Indiscriminate killing across large areas is not serving the best interests of our agri-businesses, nor is it keeping our landscapes healthy. We really need to rethink the sustainable long term use of our ecosystems, and our role in them.'

4.4 Case study 4: DigiFarm Narrabri

The abovementioned three case studies demonstrate how rewilding can improve the conservation status of threatened species and help to restore lost or degraded ecological processes. The first two case studies reflect the common reality in Australia and New Zealand that rewilding is most likely to succeed if land is protected from invasive species using barriers, whereas the third describes a landscape-scale concept mired in controversy. In mixed-agricultural areas where cropping and livestock grazing are to continue, how might a rewilding programme be initiated? Here, we describe a practical example of the initial steps that can be taken, using work carried out by the authors as part of the DigiFarm – Biodiversity project on Llara, a working farm owned and run as a profitable agribusiness by the University of Sydney at Narrabri in northwestern New South Wales. DigiFarm, funded by the National Landcare Programme between 2019 and 2023, sought to develop an integrated digital network of sensors to monitor soil health, farm production, biodiversity and ecosystem health, with the aim of ensuring a rich and sustainable environment, i.e. one that would be resilient to future climate change and other challenges (https://www.sydney.edu.au/agriculture/our-research/Digifarm.html, accessed 15 November 2023).

Lying on Gomilaroi country, the soils of Llara Farm (2600 ha) are predominantly vertosols that are used to grow a range of dryland crops as well as to support pastures for beef cattle. There are also patches of remnant open forest dominated by *Callitris* and *Eucalyptus* spp. (Fig. 6). For the biodiversity component of DigiFarm, a grid of 20 remote camera traps, 20 audio devices and 12 ultrasonic bat detectors with approximately 1 × 1 km spacing was set up to monitor vertebrate diversity across the different habitats. Over the 3½-year period of study these remote sensors captured 1 million images of 55 native and introduced vertebrate species, almost 441 000 call files of 14 species of bat (of which five species are threatened), and allowed the identification of 140 bird species from their songs (https://www.sydney.edu.au/agriculture/our-research/Digifarm.html, accessed 15 November 2023). Because of the very large numbers of photographs and acoustic recordings that were obtained, artificial intelligence algorithms were trained to identify the species and automate the collection of biodiversity data, with on-ground surveys by people helping to validate the results. Further surveys on Llara focused on vegetation and invertebrates, including targeted searches for noxious weeds and pests such as cotton mirids.

Although results are still being synthesised from the DigiFarm project, the research has yielded deep insight into patterns in the spatial and temporal distributions and relative activity of all main fauna groups on Llara. The data do not at this stage confirm that the different animal species are resident on the farm or have viable populations there, or elucidate any role of neighbouring properties in supporting faunal activity on Llara, but they do furnish a list of candidate species to target for more focused research on these topics. If rewilding were to go ahead, this information should allow farm managers to select species to reintroduce in the knowledge that they do not currently occur on the property, identify areas where threats from species such as feral pigs and red foxes should be mitigated, and highlight parts of the property where targeted protection measures could be taken. For example, the activity of all or most bat species was concentrated in remnant woodland, as was the activity of some native birds and feral pigs. Protection of remnant woodland on Llara, and perhaps on neighbouring properties, is therefore important to maintain the diversity of native species and to ensure that they remain as integral parts of the farm's biodiversity. Microbats and insectivorous birds can also be important consumers of crop pest insects (Kolkert et al., 2020, 2021). At the same time, the woodland and woodland-crop or pasture ecotones provide foci for managing feral mammals such as pigs.

An important outcome of DigiFarm is not just that it has produced a valuable database on the production and biodiversity assets of Llara Farm, but also that it has pioneered a system that allows land owners to install sensors on their properties and to then feed the data into pipelines that provide

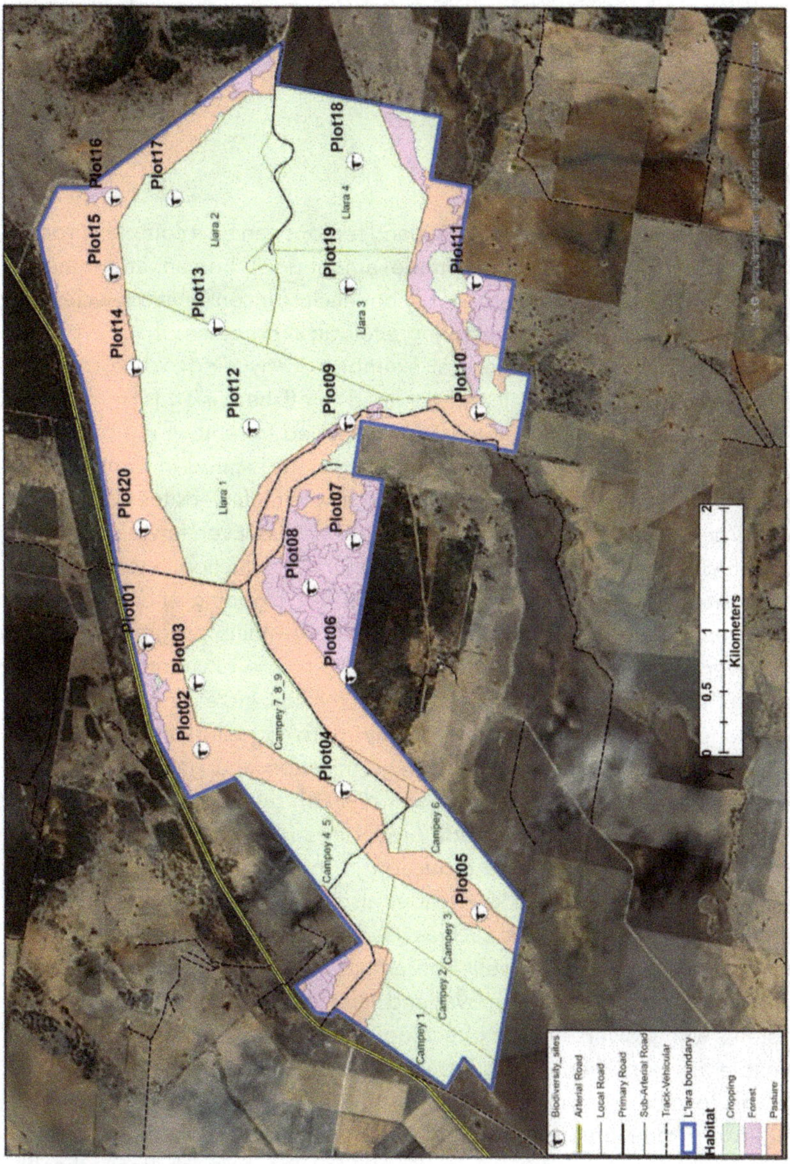

Figure 6 Major habitat types and monitoring plots for biodiversity surveys on Llara Farm, Narrabri, New South Wales.

fast updates on which species the sensors have detected. If this system was expanded onto other properties, it has the potential to document at scale across mixed agricultural landscapes the species that are present, their distribution and activity, and also identify which former species are absent and hence are potential candidates for future rewilding. Such information should comprise key first steps in attempts to achieve the triple bottom line of environmental, social and financial accounting.

5 Conclusion

Rewilding in the Australian and New Zealand regions can be a complex process involving many steps and considerable amounts of human and material resourcing, but if successful it can be very beneficial for conserving species that are at risk of extinction and for restoring ecological processes that have been eroded or lost from the landscape. In all examples of rewilding to date, rewilded areas have been relatively small and confined to offshore islands or mainland sites that have been fenced or intensively managed to remove the damaging effects of invasive species. Many rewilding sites are located in remote areas and focus largely on conservation and research, with far fewer sites occurring in suburban/peri-urban areas and allowing public access. Even fewer sites occur on land that is used concurrently for agricultural production, although most rewilding sites lie on land that was once used for cropping and/or livestock grazing. Depending on the mix of species used, rewilding programmes often produce very clear and demonstrable benefits to the environment such as increasing the retention of organic material in the soil, increasing infiltration of rainwater and cycling of nutrients, thus greatly improving soil health and stability. These benefits flow to increasing plant germination and growth, especially if rewilded species include vectors of soil mycorrhizal fungi, plant seeds and pollen. If rewilded landscapes include predators such as quolls *Dasyurus* spp. or the dingo, there should be a reduction in total grazing pressure, further benefitting plant growth. If the landscapes also include invertebrates, many of the above ecological processes such as seed dispersal and pollination will be replicated and expanded to benefit a potentially greater range of plant species, and functions such as dung removal and litter breakdown will be included. Ecosystems with a diverse range of ecological processes are likely to be more sustainable, more resilient to disturbances such as climate change and also better able to capture and sequester carbon.

Apart from the example in case study 4, it is clear that rewilding is uncommon in agricultural landscapes. In view of the improvements in landscape health, productivity and sustainability that can accrue from rewilding, it would therefore seem prudent to explore pathways that would allow rewilding concepts to be taken up more broadly to enhance crop and livestock production. Although we do not explore the reasons for the lack of integration between agricultural and conservation concerns here, we emphasise that rewilding offers potential opportunities for both enterprises to more rapidly achieve their mutual goals of sustainability and resilience than if they continue to work on their own and in parallel with each other.

6 Future trends in research

For rewilding, as for agriculture, a key question for the future is how climate change will influence where species can live and thrive. New models are now available that can project future climates at very high resolution (10 × 10 km) under different scenarios (e.g. Chapman et al., 2023). Used together with existing species distribution models, it should be possible to identify suites of species that are most likely to persist in areas designated for future rewilding programmes and which crops and livestock are likely to be most productive in future regional and continental landscapes. As noted earlier, the buffering effects that rewilding may have on agricultural enterprises should also be factored into climate modelling. To provide robust biological data to incorporate into climate models, we suggest that joint rewilding and agricultural enterprises be set up at different scales (e.g. 10–1000 km^2) in different regions, and changes in crop or livestock production measured where suites of different native species have been rewilded. For control comparisons, parallel sites should also be established where no rewilding takes place. Sensor arrays to measure soil moisture, nutrients organic activity and overall health, farm productivity and biodiversity should be established at these sites, using previous examples of digitally enabled networks such as DigiFarm (https://www.sydney.edu.au/agriculture/our-research/Digifarm.html, accessed 16 November 2023) to gather information on how rewilding affects key measures of environmental health. If, as anticipated, rewilding improves sustainability and resilience of production and natural capital, these buffering effects could be factored into climate models to identify additional areas where agricultural and conservation enterprises could be run in future.

7 Where to look for further information

The following publications provide a helpful overview of rewilding:

- Gaywood, M. J., Ewen, J. G., Hollingsworth, P. M. and Moehrenschlager, A. (Eds) (2023). *Conservation Translocations*. Cambridge University Press, Cambridge.
- IUCN/SSC (2013). *Guidelines for Reintroductions and Other Conservation Translocations*, version 1.0. IUCN Species Survival Commission, Gland.

- Pettorelli, N., Durant, S. M. and du Toit, J. T. (Eds) (2019). *Rewilding*. Cambridge University Press, Cambridge.
- Sweeney, O. F., Turnbull, J., Jones, M., Letnic, M., Newsome, T. M. and Sharp, A. (2019). An Australian perspective on rewilding. *Cons. Biol.* 33: 812–820.

Key research on rewilding can be found at the following organisations:

- International Union for Conservation of Nature (IUCN) (https://www.iucn.org).
- World Wide Fund for Nature-Australia (WWF-A) (https://wwf.org.au/what-we-do/rewilding-australia).
- Rewild Aotearoa (https://rewildaotearoa.org.nz).
- Predator Free 2050 (https://www.doc.govt.nz/nature/pests-and-threats/predator-free-2050).
- Australian Wildlife Conservancy (https://www.australianwildlife.org).
- The University of Sydney Institute of Agriculture (https://www.sydney.edu.au/agriculture/our-research/Digifarm.html).
- National Parks Association of NSW (https://npansw.org.au/2021/08/31/rewilding-a-review-of-risks-and-benefits).
- Land for Wildlife (https://www.rewildingmag.com/land-for-wildlife-australia).
- Rewilding Earth (https://rewilding.org/what-is-rewilding).

8 Acknowledgements

We thank Guy Roth for assistance with the DigiFarm project, and Guy Ballard, David Paton and Nick Reid for their helpful comments on the initial draft of this chapter.

9 References

Algar, D., Morris, K., Asher, J. and Cowen, S. (2020). Dirk Hartog Island 'Return to 1616' Project – the first six years (2014 to 2019). *Ecol. Manage. Restor.* 21(3): 173–183.

Allen, B. L. and Fleming, P. J. S. (2012). Reintroducing the dingo: the risk of dingo predation to threatened vertebrates of western New South Wales. *Wildl. Res.* 39(1): 35–50.

Armstrong, D. P., Hayward, M. W., Moro, D. and Seddon, P. J. (Eds) (2015). *Advances in Reintroduction Biology of Australian and New Zealand Fauna*. CSIRO Publishing, Melbourne.

Armstrong, D. P., Parlato, E. H. and Ewen, J. G. (2023). Five reasons to consider long-term monitoring – case studies from bird reintroductions on Tiritiri Matangi Island. In: Gaywood, M. J., Ewen, J. G., Hollingsworth, P. M. and Moehrenschlager, A. (Eds). *Conservation Translocations*. Cambridge University Press, Cambridge, pp. 429–435.

Augee, M. and Fox, M. (2000). *Biology of Australia and New Zealand*. Benjamin Cummings, Sydney.

Auld, T. D. (1993). The impact of grazing on regeneration of the shrub *Acacia carnei* in arid Australia. *Biol. Cons.* 65(2): 165–176.

Auld, T. D. (1995). Seedling survival under grazing in the arid perennial *Acacia oswaldii*. *Biol. Cons.* 72(1): 27–32.

Australian Museum (2024). FrogID. https://australian.museum/publications/explore_summer2024/frogid/, (accessed 18 February 2024).

Baker, A. and Dickman, C. (2018). *Secret Lives of Carnivorous Marsupials*. CSIRO Publishing, Melbourne.

Baling, M., van Winkel, D., Rixon, M., Ruffell, J., Ji, W. and Ussher, G. (2013). A review of reptile research and conservation management on Tiritiri Matangi Island, New Zealand. *N. Z. J. Ecol.* 37: 272–281.

Ballou, J. D., Gilpin, M. and Foose, T. J. (Eds) (1995). *Population and Management for Survival and Recovery: Analytical Methods and Strategies in Small Population Conservation*. Columbia University Press, New York.

Bangs, E. E. and Fritts, S. H. (1996). Reintroducing the gray wolf to central Idaho and Yellowstone National Park. *Wildl. Soc. Bull.* 24: 402–413.

Banks, P. B. (2004). Population viability analysis in urban wildlife management: modelling management options for Sydney's quarantined bandicoots. In: Lunney, D. and Burgin, S. (Eds). *Urban Wildlife: More Than Meets the Eye*. Royal Zoological Society of New South Wales, Sydney, pp. 70–77.

Barnes, M. and Wise, G. (2010). *Celebrating 100 Years of Natural Resource Progress in the Western Division of NSW*, updated reprint of 2002 original. Western Catchment Management Authority, Cobar.

Bauer, T., Weerasena, C., Ladd, R., Holland, G., Wauchope, M., Joseph, L. and Kanowski, J. (2023). *Scotia Wildlife Sanctuary Ecohealth Report for 2022*. Australian Wildlife Conservancy, Perth.

Beckmann, K. M., Cromie, R. L., Sainsbury, A. W., Hilton, G. M., Ewen, J. G., Soorae, P. S. and Kock, R. A. (2022). Wildlife health outcomes and opportunities in conservation translocations. *Ecol. Solut. Evid.* 3(3): e12164.

Bolam, F. C., Mair, L., Angelico, M., Brooks, T. M., Burgman, M., Hermes, C., Hoffmann, M., Martin, R. W., McGowan, P. J. K., Rodrigues, A. S. L., Rondinini, C., Westrip, J. R. S., Wheatley, H., Bedolla-Guzmán, Y., Calzada, J., Child, M. F., Cranswick, P. A., Dickman, C. R., Fessl, B., Fisher, D. O., Garnett, S. T., Groombridge, J. J., Johnson, C. N., Kennerley, R. J., King, S. R. B., Lamoreux, J. F., Lees, A. C., Lens, L., Mahood, S. P., Mallon, D. P., Meijaard, E., Méndez-Sánchez, F., Percequillo, A. R., Regan, T. J., Renjifo, L. M., Rivers, M. C., Roach, N. S., Roxburgh, L., Safford, R. J., Salaman, P., Squires, T., Vázquez-Domínguez, E., Visconti, P., Woinarski, J. C. Z., Young, R. P. and Butchart, S. H. M. (2021). How many bird and mammal extinctions has recent conservation action prevented? *Cons. Lett.* 14(1): e12762.

Burbidge, A. A. (1996). Essentials of a good recovery plan. In: Stephens, S. and Maxwell, S. (Eds). *Back from the Brink: Refining the Threatened Species Recovery Process*. Surrey Beatty and Sons, Sydney, pp. 55–62.

Burbidge, A. A. (2024). Australian terrestrial mammals: how many modern extinctions? *Aust. Mammal.* in press.

Caughley, J. (1980). Native quolls and tiger quolls. In: Haigh, C. (Ed.). *Endangered Animals of New South Wales*, NSW National Parks and Wildlife Service, Sydney, pp. 45–48.

Chapman, S., Syktus, J., Trancoso, R., Thatcher, M., Toombs, N., Wong, K. K.-H. and Takbask, A. (2023). Evaluation of dynamically downscaled CMIP6-CCAM models over Australia. *Earth's Future* 11(11): e2023EF003548.

Clancy, C. and Ward, K. (2020). Auto-rewilding in post-industrial cities: the case of inland cormorants in urban Britain. *Cons. Soc.* 18(2): 126–136.

Coggan, N. V., Hayward, M. W. and Gibb, H. (2016). Termite activity and decomposition are influenced by digging mammal reintroductions along an aridity gradient. *J. Arid Environ.* 133: 85–93.

Corlett, R. T. (2016). The role of rewilding in landscape design for conservation. *Curr. Landsc. Ecol. Rep.* 1(3): 127–133.

Coulter, B., West, R. S., Kingsford, R. T. and Moseby, K. E. (2023). The influence of drought, precipitation and fossorial mammal reintroduction on the density of fossorial arthropods and their burrows in arid Australia. *Aust. Ecol.* 48(7): 1252–1273.

Craig, J., Mitchell, N., Walter, W., Galbraith, M. and Chalmers, G. (1995). Involving people in the restoration of a degraded island: Tiritiri Matangi Island. In: Saunders, D. A., Craig, J. L. and Mattiske, E. M. (Eds). *Nature Conservation 4: The Role of Networks*. Surrey Beatty and Sons, Sydney, pp. 534–541.

Dalrymple, S. E. and Bellis, J. M. (2023). Conservation translocations: planning and the initial appraisal. In: Gaywood, M. J., Ewen, J. G., Hollingsworth, P. M. and Moehrenschlager, A. (Eds). *Conservation Translocations*. Cambridge University Press, Cambridge, pp. 43–74.

Daly, G., Johnson, P., Malolakis, G., Hyatt, A. and Pietsch, R. (2008). Reintroduction of the green and golden bell frog *Litoria aurea* to Pambula on the south coast of New South Wales. *Austral. Zool.* 34(3): 261–270.

DAWE (2021). *Approved Conservation Advice for the Mallee Bird Community of the Murray Darling Depression Bioregion*. Australian Government Department of Agriculture, Water and the Environment, Canberra.

Dickman, C. R. (2012). Fences or ferals? Benefits and costs of conservation fencing in Australia. In: Somers, M. J. and Hayward, M. W. (Eds). *Fencing for Conservation: Restriction of Evolutionary Potential or a Riposte to Threatening Processes?* Springer, New York, pp. 43–63.

Dickman, C. R. (2014). Invasion ecology of Australasian marsupials. In: Prins, H. H. T. and Gordon, I. J. (Eds). *Invasion Biology and Ecological Theory: Insights from a Continent in Transformation*. Cambridge University Press, Cambridge, pp. 159–195.

Dickman, C. R. (2015). Family Peramelidae (bandicoots and echymiperas). In: Wilson, D. E. and Mittermeier, R. A. (Eds). *Handbook of the Mammals of the World. Volume 5. Monotremes and Marsupials*. Lynx Edicions, Barcelona, pp. 362–398.

Dickman, C. R. (2018). Biodiversity in Australia: an overview. In: Pullaiah, T. (Ed.). *Global Biodiversity. Volume 4. Selected Countries in the Americas and Australia*. Apple Academic Press, Oakville, pp. 515–558.

Dickman, C. R., Glen, A. S. and Letnic, M. (2009). Reintroducing the dingo: can Australia's conservation wastelands be restored? In: Hayward, M. W. and Somers, M. J. (Eds). *Reintroduction of Top-Order Predators*. Blackwell Publishing, Oxford, pp. 238–269.

Dickman, C. R., Pressey, R. L., Lim, L. and Parnaby, H. E. (1993). Mammals of particular conservation concern in the Western Division of New South Wales. *Biol. Cons.* 65(3): 219–248.

Dunlop, J. A. and Watson, M. J. (2022). The hitchhiker's guide to Australian conservation: a parasitological perspective on fauna translocations. *Aust. Ecol.* 47(4): 748–764.

Dunn-Capper, R., Giergiczny, M., Fernández, N., Marder, F. and Pereira, H. M. (2024). Public preference for the rewilding framework: a choice experiment in the Oder Delta. *People Nat.* 6: in press.

du Toit, J. T. (2019). Pleistocene rewilding: an enlightening thought experiment. In: Pettorelli, N., Durant, S. M. and du Toit, J. T. (Eds). *Rewilding*. Cambridge University Press, Cambridge, pp. 55–72.

Eason, D. K., Elliott, G. P., Merton, D. V., Jansen, P. W., Harper, G. A. and Moorhouse, R. J. (2006). Breeding biology of kakapo (*Strigops habroptilus*) on offshore island sanctuaries, 1990–2002. *Notornis* 53: 27–36.

Eldridge, D. J., Delgado-Baquerizo, M., Woodhouse, J. N. and Neilan, B. A. (2017). Contrasting effects of two mammalian soil engineers on microbial communities. *Aust. Ecol.* 42(4): 380–384.

Eldridge, D. J. and James, A. I. (2009). Soil disturbance by native animals plays a critical role in maintaining healthy Australian landscapes. *Ecol. Manage. Restor.* 10(s1): S27–S34.

Eldridge, D. J. and Mensinga, A. (2007). Foraging pits of the short-beaked echidna *Tachyglossus aculeatus* as small-scale patches in a semi-arid Australian box woodland. *Soil Biol. Biochem.* 39(5): 1055–1065.

Eldridge, M. D. B. and Frankham, G. J. (2015). Family Potoroidae (bettongs and potoroos). In: Wilson, D. E. and Mittermeier, R. A. (Eds). *Handbook of the Mammals of the World. Volume 5. Monotremes and Marsupials*. Lynx Edicions, Barcelona, pp. 600–628.

Emmott, A. (2021). The dingo as a management tool on a beef cattle enterprise in western Queensland. *Aust. Zool.* 41(3): 459–466.

EPBC Act (2023). *Environment Protection and Biodiversity Conservation Act 1999 List of Threatened Fauna*. Available at: https://www.environment.gov.au/cgi-bin/sprat/public/publicthreatenedlist.pl?wanted=fauna, accessed 15 October 2023.

Fernández, N., Navarro, L. M. and Pereira, H. M. (2017). Rewilding: a call for boosting ecological complexity in conservation. *Cons. Lett.* 10: 301–306.

Finlayson, G. R. (2010). The reintroduction of medium-sized marsupials in western New South Wales. PhD thesis, University of Sydney, Sydney.

Finlayson, G. R., Finlayson, S. T. and Dickman, C. R. (2010). Returning the rat-kangaroos: translocation attempts in the Family Potoroidae (superfamily Macropodoidea) and recommendations for conservation. In: Coulson, G. and Eldridge, M. (Eds). *Macropods: the Biology of Kangaroos, Wallabies and Rat-Kangaroos.* CSIRO Publishing, Melbourne, pp. 245–262.

Finlayson, G. R., Vieira, E. M., Priddel, D., Wheeler, R.. and Dickman, C. R. (2008). Multi-scale patterns of habitat use by re-introduced mammals: a case study using medium-sized marsupials. *Biol. Cons.* 141: 320–331.

Fischer, J. and Lindenmayer, D. B. (2000). An assessment of the published results of animal relocations. *Biol. Cons* 96(1): 1–11.

Foreman, D., Davis, J., Johns, D., Noss, R. and Soulé, M. (1992). The Wildlands Project mission statement. *Wild Earth Special Issue*, 3–4.

Friend, J. A. (2023). Gilbert's potoroo, *Potorous gilbertii* (Gould, 1841). In: Baker, A. M. and Gynther, I. C. (Eds). *Strahan's Mammals of Australia* (4th edn.). Reed New Holland Publishers, Sydney, pp. 300–301.

Galbraith, M. (2013). Public and ecology – the role of volunteers on Tiritiri Matangi Island. *N. Z. J. Ecol.* 37: 266–271.

Galbraith, M. and Cooper, H. (2013). Tiritiri Matangi – an overview of 25 years of ecological restoration. *N. Z. J. Ecol.* 37: 258–260.

Gaywood, M. J. and Stanley-Price, M. (2023). Moving species: reintroductions and other conservation translocations. In: Gaywood, M. J., Ewen, J. G., Hollingsworth, P. M. and Moehrenschlager, A. (Eds). *Conservation Translocations*. Cambridge University Press, Cambridge, pp. 3–42.

Gibb, H. (2012). How might terrestrial arthropod assemblages have changed after the dramatic decline of critical weight range (CWR) mammals in Australia? Using reintroductions at Scotia Sanctuary as a model for pre-European ecosystems. *Proc. Linn. Soc. New S. W.* 134: A19–A26.

Glen, A. S. and Dickman, C. R. (2005). Complex interactions among mammalian carnivores in Australia, and their implications for wildlife management. *Biol. Rev.* 80(3): 387–401.

Glen, A. S., Edwards, S., Finlay-Smits, S., Jones, C., Niebuhr, C. N., Norbury, G. L. and Samaniego, A. (2023). Management of cats in Aotearoa New Zealand: a review of current knowledge and research needs. *N. Z. J. Ecol.* 47: 3550.

Glenday, J. and Kennedy, A. (2021). Can extinct eastern quolls really return to the wild? Available at: https://www.abc.net.au/news/2021-05-22/can-extinct-australian-eastern-quolls-return-to-wild/100140048, accessed 29 January 2024.

Graham, M., Veitch, D., Aguilar, G. and Galbraith, M. (2013). Monitoring terrestrial bird populations on Tiritiri Matangi Island, Hauraki Gulf, New Zealand, 1987–2010. *N. Z. J. Ecol.* 37: 359–369.

Groombridge, J. J., Raisin, C., Bristol, R. and Richardson, D. S. (2012). Genetic consequences of reintroductions and insights from population history. In: Ewen, J. G., Armstrong, D. P., Parker, K. A. and Seddon, P. J. (Eds). *Reintroduction Biology: Integrating Science and Management*. Wiley, New Jersey, pp. 395–440.

Hällfors, M. and Dalrymple, S. E. (2023). Assisted colonisation and ecological replacement. In: Gaywood, M. J., Ewen, J. G., Hollingsworth, P. M. and Moehrenschlager, A. (Eds). *Conservation Translocations*. Cambridge University Press, Cambridge, pp. 331–353.

Hayward, M. W. and Marlow, N. (2014). Will dingoes really conserve wildlife and can our methods tell? *J. Appl. Ecol.* 51(4): 835–838.

Hobbs, N. T., Johnston, D. B., Marshall, K. N., Wolf, E. C. and Cooper, D. J. (2024). Does restoring apex predators to food webs restore ecosystems? Large carnivores in Yellowstone as a model system. *Ecol. Monogr.* e1598.

Hocknull, S. A., Lewis, R., Arnold, L. J., Pietsch, T., Joannes-Boyau, R., Price, G. J., Moss, P., Wood, R., Dosseto, A., Louys, J., Olley, J. and Lawrence, R. A. (2020). Extinction of eastern Sahul megafauna coincides with sustained environmental deterioration. *Nat. Commun.* 11(1): 2250.

Hundloe, T. and Hamilton, C. (1997). Koalas and tourism: an economic evaluation. Discussion paper 13, The Australia Institute, Canberra.

Innes, J., Fitzgerald, N., Binny, R., Byrom, A., Pech, R., Watts, C., Gillies, C., Maitland, M., Campbell-Hunt, C. and Burns, B. (2019). New Zealand ecosanctuaries: types, attributes and outcomes. *J. R. Soc. N. Z.* 49(3): 370–393.

Innes, J., Lee, W. G., Burns, B., Campbell-Hunt, C., Watts, C., Phipps, H. and Stephens, T. (2012). Role of predator-proof fences in restoring New Zealand's biodiversity: a response to Scofield et al. (2011). *N. Z. J. Ecol.* 36: 232–238.

IUCN/SSC (2013). *Guidelines for Reintroductions and Other Conservation Translocations*, version 1.0. IUCN Species Survival Commission, Gland.

Izzard, S., Barnard, O. and Schaap, D. (2019). Managing and maintaining wild temperament and behaviours in captivity. In: Hogg, C. J., Fox, S., Pemberton, D. and Belov, K. (Eds). *Saving the Tasmanian Devil: Recovery Through Science-Based Management*. CSIRO Publishing, Melbourne, pp. 207–221.

Johns, D. (2019). History of rewilding: ideas and practice. In: Pettorelli, N., Durant, S. M. and du Toit, J. T. (Eds). *Rewilding*. Cambridge University Press, Cambridge, pp. 12–33.

Johnson, C. N., Dortch, J. and Worthy, T. H. (2021). Interactions with megafauna. In: McNiven, I. J. and David, B. (Eds). *The Oxford Handbook of the Archaeology of Indigenous Australia and New Guinea*. Oxford University Press, Oxford, pp. 1–30.

Jørgensen, D. (2015). Rethinking rewilding. *Geoforum* 65: 482–488.

Kameyama, A. (2006). Extirpation of wolves in Hokkaido and possibility of wolf reintroduction in Shiretoko National Park. In: McCullough, D. R., Kaji, K. and Yamanaka, M. (Eds). *Wildlife in Shiretoko and Yellowstone National Parks: Lessons in Wildlife Conservation from Two World Heritage Sites*. Shiretoko Nature Foundation, Hokkaido, pp. 254–256.

Keller, L. F., Biebach, I., Ewing, S. R. and Hoeck, P. E. A. (2012). The genetics of reintroductions: inbreeding and genetic drift. In: Ewen, J. G., Armstrong, D. P., Parker, K. A. and Seddon, P. J. (Eds). *Reintroduction Biology: Integrating Science and Management*, Wiley, New Jersey, pp. 360–394.

Kolkert, H., Andrew, R., Smith, R., Rader, R. and Reid, N. (2020). Insectivorous bats selectively source moths and eat mostly pest insects on dryland and irrigated cotton farms. *Evol. Ecol.* 10(1): 371–388.

Kolkert, H. L., Smith, R., Rader, R. and Reid, N. (2021). Prey removal in cotton crops next to woodland reveals periodic diurnal and nocturnal invertebrate predation gradients from the crop edge by birds and bats. *Sci. Rep.* 11(1): 5256.

Lacy, R. C. (2019). Lessons from 30 years of population viability analysis of wildlife populations. *Zoo Biol.* 38(1): 67–77.

Law, A., Gaywood, M. J., Jones, K. C., Ramsay, P. and Willby, N. J. (2017). Using ecosystem engineers as tools in habitat restoration and rewilding: beaver and wetlands. *Sci. Tot. Environ.* 605: 1021–1030.

Lees, C. M., Andrew, P., Spindler, R., Jakob-Hoff, R., Traylor-Holzer, K. and Byers, O. (2019). A One Plan approach to saving the devil: population and habitat viability assessment. In: Hogg, C. J., Fox, S., Pemberton, D. and Belov, K. (Eds). *Saving the Tasmanian Devil: Recovery Through Science-Based Management*. CSIRO Publishing, Melbourne, pp. 175–184.

Legge, S., Woinarski, J. C. Z., Burbidge, A. A., Palmer, R., Ringma, J., Radford, J. Q., Mitchell, N., Bode, M., Wintle, B., Baseler, M., Bentley, J., Copley, P., Dexter, N., Dickman, C. R., Gillespie, G. R., Hill, B., Johnson, C. N., Latch, P., Letnic, M., Manning, A., McCreless, E. E., Menkhorst, P., Morris, K., Moseby, K., Page, M., Pannell, D. and Tuft, K. (2018a). Havens for threatened Australian mammals: the contributions of fenced areas and offshore islands to the protection of mammal species susceptible to introduced predators. *Wildl. Res.* 45(7): 627–644.

Legge, S., Lindenmayer, D. B., Robison, N. M., Scheele, B. C., Southwell, D. M. and Wintle, B. A. (Eds) (2018b). *Monitoring Threatened Species and Ecological Communities*. CSIRO Publishing, Melbourne.

Letnic, M., Koch, F., Gordon, C., Crowther, M. S. and Dickman, C. R. (2009). Keystone effects of an alien top-predator stem extinctions of native mammals. *Proc. R. Soc. B* 276(1671): 3249–3256.

Letnic, M., Ritchie, E. G. and Dickman, C. R. (2012). Top predators as biodiversity regulators: the dingo *Canis lupus dingo* as a case study. *Biol. Rev.* 87(2): 390–413.

Lindenmayer, D. B., Burns, E., Thurgate, N. and Lowe, A. (Eds) (2014). *Biodiversity and Environmental Change: Monitoring, Challenges and Direction*. CSIRO Publishing, Melbourne.

Lindenmayer, D. B., Gibbons, P., Bourke, M., Burgman, M., Dickman, C. R., Ferrier, S., Fitzsimons, J., Freudenberger, D., Garnett, S. T., Groves, C., Hobbs, R. J., Kingsford, R. T., Krebs, C., Legge, S., Lowe, A. J., Mclean, R., Montambault, J., Possingham, H., Radford, J., Robinson, D., Smallbone, L., Thomas, D., Varcoe, T., Vardon, M., Wardle, G., Woinarski, J. and Zerger, A. (2012). Improving biodiversity monitoring. *Aust. Ecol.* 37(3): 285–294.

Lunney, D. (1994). Review of official attitudes to western New South Wales 1901–93 with particular reference to the fauna. In: Lunney, D., Hand, S., Reed, P. and Butcher, D. (Eds). *Future of the Fauna of Western New South Wales*. Royal Zoological Society of New South Wales, Sydney, pp. 1–26.

Lunney, D. (2001). Causes of the extinction of native mammals of the Western Division of New South Wales: an ecological interpretation of the nineteenth century historical record. *Rangel. J.* 23(1): 44–70.

Lyons, J. E., Runge, M. C., Laskowski, H. P. and Kendall, W. L. (2008). Monitoring in the context of structured decision-making and adaptive management. *J. Wildl. Manag.* 72(8): 1683–1692.

MacArthur, R. H. and Wilson, E. O. (1967). *The Theory of Island Biogeography*. Princeton University Press, Princeton, New Jersey.

Maller, C., Mumaw, L. and Cooke, B. (2019). Health and social benefits of living with 'wild' nature. In: Pettorelli, N., Durant, S. M. and du Toit, J. T. (Eds). *Rewilding*. Cambridge University Press, Cambridge, pp. 165–181.

Marshall, T. C. and Spalton, J. A. (2000). Simultaneous inbreeding and outbreeding depression in reintroduced Arabian oryx. *Anim. Cons.* 3(3): 241–248.

Menon, V., McGregor, H., Giljohann, K., Wintle, B., Pascoe, J., Robley, A., Johnston, M., Fancourt, B., Bengsen, A., Buckmaster, T., Comer, S., Hamer, R., Friend, T., Jansen, J., Zewe, F., Fleming, P., Ballard, G., Moseby, K., Gentle, M., Scomparin, C., Speed, J., Clausen, L., Le Pla, M. and Hradsky, B. (2024). Ecological factors influencing invasive predator survival and movement: Insights from a continental-scale study of feral cats in Australia. *Biol. Inv.* in press.

Michael, K. H., Leonard, S. W. J., Decker, O. and Verdon, S. J. and Gibb, H. (2022). Testing the effects of ecologically extinct mammals on vegetation in arid Australia: a long-term experimental approach. *Aust. Ecol.* 47: 226–238.

Miskelly, C. M., Dowding, J. E., Elliott, G. P., Hitchmough, R. A., Powlesland, R. G., Robertson, H. A., Sagar, P. M., Scofield, R. P. and Taylor, G. A. (2008). Conservation status of New Zealand birds, 2008. *Notornis* 55: 117–135.

Mitchell, N. D. (2013). Tiritiri Matangi Island: what if nothing had been done? *N. Z. J. Ecol.* 37: 261–265.

Monbiot, G. (2013). *Feral: Searching for Enchantment on the Frontiers of Rewilding*. Penguin, London.

Morris, S. D., Brook, B. W., Moseby, K. E. and Johnson, C. N. (2021). Factors affecting success of conservation translocations of terrestrial vertebrates: a global systematic review. *Global. Ecol. Cons.* 28: e01630.

Moseby, K. E., Read, J. L., Paton, D. C., Copley, P., Hill, B. M. and Crisp, H. A. (2011). Predation determines the outcome of 10 reintroduction attempts in arid South Australia. *Biol. Cons.* 144(12): 2863–2872.

Moseby, K. E., Read, J. L., Tuft, K. and Van der Weyde, L. K. (2023). Influence of interactive effects on long-term population trajectories in multispecies reintroductions. *Cons. Biol.* in press.

Naundrup, P. J. and Svenning, J.-C. (2015). A geographic assessment of the global scope for rewilding with wild-living horses (*Equus ferus*). *PLoS ONE* 10(7): e0132359.

Navnith, M., Finlayson, G. R., Crowther, M. S. and Dickman, C. R. (2009). The diet of the re-introduced greater bilby *Macrotis lagotis* in the mallee woodlands of western New South Wales. *Austral. Zool.* 35(1): 90–95.

Nelson, N. J., Hitchmough, R. and Monks, J. M. (2015). New Zealand reptiles and their conservation. In: Stow, A., Maclean, N. and Holwell, G. I. (Eds). *Austral Ark: The State of Wildlife in Australia and New Zealand*. Cambridge University Press, Cambridge, pp. 382–404.

Newman-Martin, J., Travouillon, K. J., Warburton, N., Barham, M. and Blyth, A. J. (2023). Taxonomic review of the genus *Dasycercus* (Dasyuromorphia: Dasyuridae) using modern and subfossil material; and the description of three new species. *Alcheringa* 47: 624-661.

Newsome, A. E., Catling, P. C., Cooke, B. D. and Smyth, R. (2001). Two ecological universes separated by the dingo barrier fence in semi-arid Australia: interactions between landscapes, herbivory and carnivory, with and without dingoes. *Rangel. J.* 23(1): 71–98.

Newsome, T. M., Ballard, G.-A., Crowther, M. S., Dellinger, J. A., Fleming, P. J. S., Glen, A. S., Greenville, A. C., Johnson, C. N., Letnic, M., Moseby, K. E., Nimmo, D. G., Nelson, M. P., Read, J. L., Ripple, W. J., Ritchie, E. G., Shores, C. R., Wallach, A. D., Wirsing, A. J. and Dickman, C. R. (2015). Resolving the value of the dingo in ecological restoration. *Restor. Ecol.* 23(3): 201–208.

Noss, R. F. (1992). The Wildlands Project: land conservation strategy. *Wild Earth Special Issue*, 10–25.

Owens, M. and Wolch, J. (2019). Rewilding cities. In: Pettorelli, N., Durant, S. M. and du Toit, J. T. (Eds). *Rewilding*. Cambridge University Press, Cambridge, pp. 280–302.

Pettorelli, N., Barlow, J., Stephens, P. A., Durant, S. M., Connor, B., Schulte to Bühne, H., Sandom, C. J., Wentworth, J. and du Toit, J. T. (2018). Making rewilding fit for policy. *J. Appl. Ecol.* 55(3): 1114–1125.

Pettorelli, N., Durant, S. M. and du Toit, J. T. (2019). Rewilding: a captivating, controversial, twenty-first-century concept to address ecological degradation in a changing world. In: Pettorelli, N., Durant, S. M. and du Toit, J. T. (Eds). *Rewilding*. Cambridge University Press, Cambridge, pp. 1–11.

Philip, J. (2022). The dog fence. *Aust. Geog.* May–June 2022: 72–79.

Pizzuto, T. A., Finlayson, G. R., Crowther, M. S. and Dickman, C. R. (2007). Microhabitat use of the brush-tailed bettong (*Bettongia penicillata*) and burrowing bettong (*B. lesueur*) in semi-arid New South Wales: implications for reintroduction programs. *Wildl. Res.* 34(4): 271–279.

Pollock, D. (2021). Rediscovering a role for dingoes on Wooleen Station. *Aust. Zool.* 41(3): 480–486.

Prior, J. and Brady, E. (2017). Environmental aesthetics and rewilding. *Environ. Values* 26(1): 31–51.

Prowse, T. A. A., Johnson, C. N., Cassey, P., Bradshaw, C. J. A. and Brook, B. W. (2015). Ecological and economic benefits to cattle rangelands of restoring an apex predator. *J. Appl. Ecol.* 52(2): 455–466.

Radnan, G. N. and Eldridge, D. J. (2018). Does the morphology of animal foraging pits influence secondary seed dispersal by ants? *Aust. Ecol* 42: 920–928.

Read, J. L., Johnston, G. R. and Morley, T. P. (2011). Predation by snakes thwarts trial reintroduction of the endangered woma python *Aspidites ramsayi*. *Oryx* 45(4): 505–512.

Regos, A., Domínguez, J., Gil-Tena, A., Brotons, L., Ninyerola, M. and Pons, X. (2016). Rural abandoned landscapes and bird assemblages: winners and losers in the rewilding of a marginal mountain area (NW Spain). *Reg. Environ. Change* 16(1): 199–211.

Ritchie, E. G., Elmhagen, B., Glen, A. S., Letnic, M., Lugwig, G. and McDonald, R. A. (2012). Ecosystem restoration with teeth: what role for predators? *Trends Ecol. Evol.* 27(5): 265–271.

Robin, L. (2012). Global ideas in local places: the humanities in environmental management. *Environ. Humanit.* 1(1): 69–84.

Robinson, N. (2018). Using reintroductions to understand causes of mammal declines and extinctions at Booderee National Park. National Environmental Science Program Threatened Species Recovery Hub. https://www.nespthreatenedspecies.edu.au/media/niujnemc/3-2-2-1-booderee-facthseet_v8.pdf, (accessed 18 February 2024).

Royal Commission (1901). Western Division of New South Wales. Royal Commission to inquire into the condition of the Crown Tenants: Report and summary of evidence, Parts I and II. William Applegate Gullick, Government Printer, Sydney.

Sadlier, R. A. and Pressey, R. L. (1994). Reptiles and amphibians of particular conservation concern in the Western Division of New South Wales. *Biol. Cons* 69(1): 41–54.

Saunders, A. (1996). Species recovery planning: some New Zealand case studies. In: Stephens, S. and Maxwell, S. (Eds). *Back from the Brink: Refining the Threatened Species Recovery Process*. Surrey Beatty and Sons, Sydney, pp. 83–86.

Schulte to Bühne, H., Pettorelli, N. and Hoffmann, M. (2022). The policy consequences of defining rewilding. *Ambio* 51(1): 93–102.

Seddon, P. J., Armstrong, D. P. and Maloney, R. F. (2007). Developing the science of reintroduction biology. *Cons. Biol.* 21(2): 303–312.

Short, J., Bradshaw, S. D., Giles, J., Prince, R. I. T. and Wilson, G. R. (1992). Reintroduction of macropods (Marsupialia: Macropodoidea) in Australia–a review. *Biol. Cons.* 62(3): 189–204.

Silvey, C. J., Hayward, M. W. and Gibb, H. (2015). Effects of reconstruction of a pre-European vertebrate assemblage on ground-dwelling arachnids in arid Australia. *Oecologia* 178(2): 497–509.

Smith, B. P., Appleby, R. G. and Jordan, N. R. (2021). Co-existing with dingoes: challenges and solutions to implementing non-lethal management. *Austral. Zool.* 41(3): 491–510.

Smith, D., Stahler, D., Guernsey, D. and Bangs, E. (2006). Wolf restoration in Yellowstone National Park. In: McCullough, D. R., Kaji, K. and Yamanaka, M. (Eds). *Wildlife in Shiretoko and Yellowstone National Parks: Lessons in Wildlife Conservation from Two World Heritage Sites*. Shiretoko Nature Foundation, Hokkaido, pp. 242–254.

Smith, P. J., Pressey, R. L. and Smith, J. E. (1994). Birds of particular conservation concern in the Western Division of New South Wales. *Biol. Cons.* 69(3): 315–338.

Stepkovitch, B., Neave, G., Jensen, M. A., Tuft, K. and Moseby, K. E. (2023). From threatened to threatening: impacts of a reintroduced predator on reintroduced prey. *Anim. Cons.* in press.

Stiglingh, A. D., Mosley, L. M., Smernik, R. J. and Fitzpatrick, R. W. (2023). Mapping soil corrosivity in an Australian arid environment and comparing corrosion rates of exclusion fence netting with different zinc coatings. *Geoderma Reg.* 34: e00670.

Stockwell, M. P., Clulow, S., Clulow, J. and Mahony, M. (2008). The impact of the amphibian chytrid fungus *Batrachochytrium dendrobatidis* on a green and golden bell frog *Litoria aurea* reintroduction program at the Hunter Wetlands Centre Australia in the Hunter region of NSW. *Austral. Zool.* 34(3): 379–386.

Stow, A., Maclean, N. and Holwell, G. I. (Eds.) (2015). *Austral Ark: The State of Wildlife in Australia and New Zealand*. Cambridge University Press, Cambridge.

Sweeney, O. F., Turnbull, J., Jones, M., Letnic, M., Newsome, T. M. and Sharp, A. (2019). An Australian perspective on rewilding. *Cons. Biol.* 33(4): 812–820.

Tarszisz, E., Dickman, C. R. and Munn, A. J. (2014). Physiology in conservation translocations. *Cons. Physiol.* 2(1): cou054.

Taylor, G., Canessa, S., Clarke, R. H., Ingwersen, D., Armstrong, D. P., Seddon, P. J. and Ewen, J. G. (2017). Is reintroduction biology and effective applied science? *Trends Ecol. Evol.* 32(11): 873–880.

Towns, D., Broome, K. and Saunders, A. (2018). Ecological restoration on New Zealand islands: a history of shifting scales and paradigms. In: Moro, D., Ball, D. and Bryant, S. (Eds). *Australian Island Arks: Conservation, Management and Opportunities*. CSIRO Publishing, Melbourne, pp. 205–220.

Traill, L. W., Bradshaw, C. J. A. and Brook, B. W. (2007). Minimum viable population size: a meta-analysis of 30 years of published estimates. *Biol. Cons.* 139(1–2): 159–166.

van Bommel, L. and Johnson, C. N. (2023). Still a good dog! Long-term use and effectiveness of livestock guardian dogs to protect livestock from predators in Australia's extensive grazing systems. *Wildl. Res.*51: 23008.

van Eeden, L. M., Crowther, M. S., Dickman, C. R., Macdonald, D. W., Ripple, W. J., Ritchie, E. G. and Newsome, T. M. (2018). Managing conflict between large carnivores and livestock. *Cons. Biol.* 32(1): 26–34.

van Eeden, L. M., Nimmo, D., Mahony, M., Herman, K., Ehmke, G., Driessen, J., O'Connor, J., Bino, G., Taylor, M. and Dickman, C. R. (2020). *Impacts of the Unprecedented 2019-2020 Bushfires on Australian Animals*. WWF-Australia, Sydney.

Vieira, E. M., Finlayson, G. R. and Dickman, C. R. (2007). Habitat use and density of numbats (*Myrmecobius fasciatus*) reintroduced in an area of mallee vegetation, New South Wales. *Austral. Mammal.* 29(1): 17–24.

Wallach, A. D., Ramp, D. and O'Neill, A. J. (2017). Cattle mortality on a predator-friendly station in central Australia. *J. Mammal.* 98(1): 45–52.

White, A. W. and Pyke, G. H. (2008). Frogs on the hop: translocations of green and golden bell frogs *Litoria aurea* in Greater Sydney. *Austral. Zool.* 34(3): 249–260.

Wilmers, C. C., Estes, J. A., Edwards, M., Laidre, K. L. and Konar, B. (2012). Do trophic cascades affect the storage and flux of atmospheric carbon? An analysis of sea otters and kelp forests. *Front. Ecol. Environ.* 10(8): 409–415.

Wise, P., Peck, S., Clarke, J. and Hogg, C. J. (2019). Conservation introduction of Tasmanian devils to Maria Island: a managed response to DFTD. In: Hogg, C. J., Fox, S., Pemberton, D. and Belov, K. (Eds). *Saving the Tasmanian Devil: Recovery Through Science-Based Management*. CSIRO Publishing, Melbourne, pp. 223–236.

Woinarski, J. C. Z., Legge, S. M. and Dickman, C. R. (2019). *Cats in Australia: Companion and Killer*. CSIRO Publishing, Melbourne.

Worthy, T. H., Tennyson, A. J. D., Archer, M., Musser, A. M., Hand, S. J., Jones, C., Douglas, B. J., McNamara, J. A. and Beck, R. M. D. (2006). Miocene mammal reveals a Mesozoic ghost lineage on insular New Zealand, southwest Pacific. *Proc. Natl Acad. Sci. USA* 103(51): 19419–19423.

Part 5

Looking ahead

Chapter 16

Biodiversity and agricultural landscapes: where are the gaps?

Nick C. H. Reid, University of New England, Australia; and David C. Paton, University of Adelaide, Australia

1 Introduction

One of the many challenges in agricultural regions worldwide is how to conserve the native biodiversity of each region, given that most of the land and associated water resources in these regions are managed for food and fibre production. In some parts of the world, the opportunity has long since passed: for instance, Ireland had lost virtually all its primary tree cover by 1600 (Montgomery and Reid, 2024). In other regions where agricultural expansion is still ongoing, such as Amazonia, the challenge is to cease illegal deforestation and improve the well-being of the indigenous human population that has long co-existed with their rainforest home. This will require law enforcement and conservation financing to plan a future where the human and the natural world can coexist indefinitely (Hanusch, 2023).

http://dx.doi.org/10.19103/AS.2024.0134.26

Most agricultural regions of the world exist somewhere between these two extremes, where transformation has led to the loss of a large part of the natural landscape in favour of agriculture but where varying amounts of wild nature persist in varying degrees of modification. The natural areas in such agricultural landscapes are usually much reduced in extent, fragmented and their habitat condition degraded. Such regions are in the grip of an extinction spasm (as defined by Myers, 1987), with biotic relaxation (loss of species) occurring steadily through time (Tilman et al., 1994; Blanchard and Munoz, 2023). This extinction debt is caused by earlier habitat loss, fragmentation and modification due to agricultural transformation and is exacerbated by further reduction and alteration of the native habitats that remain.

The challenge is therefore how to safeguard what remains of the native biodiversity associated with pre-agricultural landscapes in the face of extensive and sometimes almost complete transformation of land and water resources for agriculture. The 15th Conference of the Parties to the UN Convention on Biological Diversity (CBD) in Montreal in December 2022 responded with what is known as the Kunming-Montreal Global Biodiversity Framework. The framework supported by over 190 countries committed the parties to (CBD, 2022):

- cease transforming natural areas ('Reduce to near zero the loss of areas of high biodiversity importance and high ecological integrity' (Target 1 of the framework);
- restore at least 30% of degraded terrestrial, inland water and marine and coastal ecosystems by 2030 to enhance biodiversity and ecosystem functions and services, ecological integrity and habitat connectivity (Target 2); and
- conserve and manage for biodiversity conservation at least 30% of the world's lands, coastal areas and oceans by 2030 (Target 3), noting that only 17% of land and 8% of marine areas were adequately protected and managed for biodiversity globally in 2022.

Target 3 had its origins at the 74th United Nations General Assembly in September 2019, where a handful of countries announced their intention to form a coalition for nature. This group officially launched as the High Ambition Coalition for Nature and People (2024) in January 2021.

While there is some evidence that 30% represents a tipping point for the rate of extinctions once the area of remaining habitat drops below this figure (Andrén, 1994), species–area relationships are in fact a continuum. The less habitat that remains, the greater the eventual loss of species. The contribution of a particular percentage of remaining habitat managed for biodiversity outcomes is contingent on the biota and the life-history characteristics of the component species, the size and configuration of the remnant habitats, their management, threats and the ecological dynamics of the broader landscape matrix (Carroll and Noss, 2022). Targets such as 30% by 2030 are thus empirical generalisations and do not pretend to guarantee the conservation of all biodiversity into the future. Indeed, Target 3 is simply an interim goal given the 2050 Vision of the CBD (Watson et al., 2023).

However, Carroll and Noss (2022) have argued that percentage-protected targets such as 30% by 2030 can anchor comprehensive national and regional biodiversity strategies and communicate the level of ambition necessary to reverse or at least slow the current trends of biodiversity loss. They accepted that what is possible to achieve for conservation in an ecoregion with abundant remaining wild area is quite different from what can be achieved in an ecoregion dominated by intensive agriculture. Regions where restoration of degraded and denuded areas is needed will require substantial effort, funds and time to meet the 30% target. In such cases, the 2030 timeline for delivering these targets is unrealistic.

Following the lead of the High Ambition Coalition for Nature and People and the Global Biodiversity Framework, several countries have developed plans, strategies or regulations broadly embracing the 30% target by 2030. In May 2021, the US announced a locally led, voluntary, nationwide conservation goal to conserve 30% of US lands and waters by 2030 (Anon, 2021). In June 2021, the Parliament of the European Union (EU) adopted the *EU Biodiversity Strategy for 2030*, which proposed legal protection of a minimum of 30% of Europe's land (including inland waters) and sea, with at least one-third under strict protection, including all remaining primary and old-growth forests (EC, 2021; EU, 2021). In September 2022, Australia's *Threatened Species Strategy Action Plan 2022–2032* (DCCEEW, 2022b) was launched by the federal government, committing to conserve at least 30% of Australia's land mass by 2030, a commitment agreed to by the state and territory ministers in the following year (DCCEEW, 2024).

2 How to reach the 30% target for biodiversity conservation of the world's lands by 2030

In most agricultural regions across the globe, the current UN-sanctioned challenge is to achieve effective biodiversity conservation of 30% of each of the major land and habitat types, including wetlands, on a regional basis. The precise wording of Target 3 ('Conserve 30% of Land, Waters and Seas') in the Global Biodiversity Framework with respect to terrestrial ecosystems is to (CBD, 2022)

> *Ensure and enable that by 2030 at least 30 per cent of terrestrial and inland water areas ... and coastal areas, especially areas of particular importance for biodiversity*

> *and ecosystem functions and services, are effectively conserved and managed through ecologically representative, well-connected and equitably governed systems of protected areas and other effective area-based conservation measures, recognizing indigenous and traditional territories, where applicable, and integrated into wider landscapes … while ensuring that any sustainable use, where appropriate in such areas, is fully consistent with conservation outcomes, recognizing and respecting the rights of indigenous peoples and local communities, including over their traditional territories.*

The question posed in this chapter thus becomes 'How best to restore habitats in extensively transformed agricultural regions such that at least 30% of the land within these regions is managed for biodiversity conservation?' Two approaches can improve biodiversity outcomes:

- bottom-up approaches and
- top-down approaches.

Bottom-up approaches are the management actions undertaken by farmers on their properties. Many of the chapters in this book describe actions and activities that farmers can take to maintain or restore native biodiversity on their farms not only often to their material benefit but also simply because they think it is the 'right' thing to do (Klebl et al., 2024). Examples include:

- regenerative or eco-farming (Bennett et al., 2024; Squire and Hawes, 2024);
- coordinated activities of networks of farmers to address high-priority issues, such as weed control in Victoria (Australia) and conservation farming initiatives in the Chesapeake watershed (USA) (Alter et al., 2024);
- prioritising soil health through a range of farming practices (Day et al., 2024);
- taking a spatially hierarchical approach to restoring biodiversity in farmland (Lindenmayer et al., 2024; Bennett et al., 2024), given the importance of habitat heterogeneity at multiple scales, from:
    - individual structures such as peatlands (van Diggelen, 2024), farm dams, buildings and large old trees (Lindenmayer et al., 2024; Bennett et al., 2024), to
    - in-field or paddock scale, by creating semi-natural field margins (Morrison, 2024), or establishing and managing hedgerows (Montgomery and Reid, 2024) and native plantings for native biodiversity (Lindenmayer et al., 2024); and
    - whole farm and landscape (cross-tenure) scale, where different agricultural enterprises, the conservation of remnant native vegetation for wildlife habitat and habitat connectivity and coordinated threat

management are critical considerations (Bennett et al., 2024; Lindenmayer et al., 2024).

These approaches can range from activities on individual farms to large-scale collaborations by groups of farmers and others across regions or watersheds (Alter et al., 2024).

Top-down approaches are governmental statutory and policy interventions that prohibit, encourage or require farmers to undertake specified management actions on their farms. Examples of top-down approaches to maintaining or restoring native biodiversity on farms are outlined in multiple chapters in this book and include:

- government-funded extension and training programs for farmers to establish and manage hedgerows on-farm (Montgomery and Reid, 2024);
- government payments to farmers for biodiversity-positive environmental management, and the preservation or restoration of natural ecosystems on their farms (Dodsworth et al., 2024; Rausser and Zilberman, 2024; Lewis-Reddy, 2024; Jones et al., 2024);
- establishment of nature-positive markets by governments to encourage philanthropy and the private sector to incentivise farmers to conserve biodiversity on-farm (Ansell et al., 2024);
- government reversal of pest animal policies and regulations that permit the return of wild animals to agricultural regions (Smith et al., 2024; Dickman et al., 2024); and
- government purchase of farms with high potential for management and restoration of high-conservation-value species and ecosystems (Dickman et al., 2024).

Other top-down approaches include legislation that outlaws native vegetation clearance, preventing natural habitats from being converted into farmland. While cessation of habitat clearance is essential to prevent even more losses of native biodiversity in the future, command-and-control regulations are often found wanting and can provoke even greater loss of habitat than might have otherwise been the case, taking several iterations to achieve the level of control and habitat preservation desired (e.g. Bartel, 2003, 2004; Fowler, 1986; Harris, 2024). The design of new natural capital markets can also be imperfect, resulting in suboptimal outcomes and requiring adaptive finetuning over time (Chubb et al., 2022; Macintosh and Butler, 2023; Rausser and Zilberman, 2024; Ansell et al., 2024).

In addition to bottom-up and top-down approaches, third parties such as philanthropists, non-profit organisations, researchers and research institutions can play a role in helping farmers to conserve native biodiversity on their farms.

They may provide biodiversity data that enable farmers to apply for payment for biodiversity-positive management or trade in biodiversity markets (Dickman et al., 2024). Philanthropic individuals and consortia, non-profits and research institutions may also purchase and manage farms exclusively for biodiversity (e.g. Dickman et al., 2024).

However, the problem with all these approaches is that, in practice, they are unlikely to make much progress towards 30% by 2030 (Cozim-Melges et al., 2024). Either biodiversity-positive management is confined to small areas of farms (e.g. field margins or boundaries), or only a few farmers are convinced by the available incentives to adopt biodiversity-positive management or conserve and manage land for conservation, or few farms are purchased to restore native biodiversity. For instance, the private land listed under heritage agreements dating back to 2010 in New South Wales to conserve biodiversity in perpetuity is less than 0.5% of the area of the state (NSW DCCEEW, 2024), and non-profits manage only 1% of Australia for biodiversity conservation (DEECCW, 2022a).

The lack of adoption of biodiversity-positive management is due to many factors (FAO, 2019), including:

- the complexity of many biodiversity-focused practices;
- their context specificity and the need for a good understanding of the local ecosystem;
- often high short-term costs compared to benefits that only accrue in the medium to long term; and
- the joint need to both build landowner capacity to manage biodiversity assets and develop new policy frameworks.

In addition, many species cannot thrive in human-modified landscapes and need undisturbed natural areas, and sometimes large areas of native habitat, to persist in predominantly agricultural regions (Pauli et al., 2017; Dietz et al., 2020; Williams et al., 2022).

In many agricultural regions worldwide, achieving 30% of land set aside for biodiversity conservation by 2030 will therefore be difficult, if not impossible, because the extent of remaining natural areas in the most intensively farmed regions is often well under 30% and sometimes closer to 1%. South Australia is a global example of the over-zealous transformation of nature for farming. The amount of remnant native vegetation across some parts of the state (e.g. the Mount Lofty Ranges and Adelaide Plains, the lower South-East and Yorke Peninsula) in 1976 was estimated at less than 10% of the original vegetation cover, prior to the introduction of vegetation clearance controls (Fowler, 1986). Remnant native vegetation is in a similarly parlous state in many other agricultural regions across southern and eastern Australia and other parts of the world.

For regions like these, the types of initiatives described in many of the chapters in this book will be unable to bridge the gap in land managed for biodiversity conservation. While generally positive for biodiversity, the initiatives described in earlier chapters do not advocate large increases in the area of biodiverse habitat on farms or across farming regions. Rather, initiatives such as the establishment of hedgerows, biodiverse field margins, wildlife corridors and farm dams and maintenance of remnant native vegetation are likely to lead to only minor increases in biodiverse habitats on farms and across agricultural districts, and management in most cases will continue to be agriculturally focused. This is essentially a 'business as usual' approach and will not arrest much of the imminent loss of biodiversity predicted for regions that have been recently heavily cleared.

This leaves only one option in regions that have fallen well below the 30% threshold: restoration of large tracts of existing farmland to habitats managed for native biodiversity. This can be achieved by paying existing farmers to revegetate and manage land and wetlands for biodiversity. Alternatively, government or philanthropic institutions can purchase farmland, including whole farms and clusters of farms, to restore native biodiversity. The need for ecosystem restoration on a massive global scale was anticipated by the United Nations in 2019, with its declaration of the Decade on Ecosystem Restoration, 2021–2030 (https://www.decadeonrestoration.org/what-decade). The aim of this decadal program is to restore ecosystems on an immense scale of hundreds of millions of hectares globally (UNEP and FAO, 2020). The justification for the Decade on Ecosystem Restoration was to mobilise political and popular support, and technical and scientific capacity in the public and private sectors, to invest in the many hundreds of thousands of ecosystem restoration initiatives worldwide needed to address the biodiversity and climate crises and achieve the UN Sustainable Development Goals (Meli et al., 2023).

The following case study of ecosystem restoration at scale originated when the South Australian Government purchased a large area of farmland decades ago in the Adelaide – Mount (Mt) Lofty region for a new city that was not finally built. A biodiversity conservation non-profit, BioR, subsequently commenced broad-scale revegetation of woodland and mallee habitats in the region, where retention of native vegetation is currently well below the 30% threshold.

3 Case study: conserving biodiversity in agricultural landscapes – the need for a regional approach in southern Australia

The Mt Lofty region in South Australia was settled by Europeans in the 1830s and subsequently transformed into a productive agricultural region, with areas of viticulture, horticulture and extensive cropping and grazing. At the time of settlement, the region was covered in a diversity of woody habitats from coastal scrubs to heathland and a range of different eucalypt woodlands and forests (e.g. Paton et al., 2000). In the following 150 years, much of this original vegetation was cleared for agriculture and, to a much lesser extent, housing. Adelaide, the capital of South Australia, with its expanding suburbs, lies within the region.

By the early 1980s, less than 10% of the original vegetation remained. The vegetation across the lower elevations, which were more amenable for agricultural development, was disproportionately cleared. In these areas just 2% of the original vegetation remained by 1980 while, at higher elevations with steeper slopes and ridges and more skeletal soils, around 16% of the native vegetation remained (Paton et al., 2000). The loss of native vegetation has had profound impacts on the biota, with many species experiencing substantial declines in distribution and abundance and important ecological processes like pollination being disrupted (e.g. Paton, 2000; Paton et al., 2004). Although introduced honeybees can offset losses of pollination services for some of the native plants, honeybees can equally exacerbate the damage by competing with native fauna for floral resources (Paton, 1996).

The extent and pattern of vegetation clearance in the Mt Lofty region is replicated in other parts of temperate South Australia, as well as southern Australia in general. In the early and mid-1980s, the South Australian Government introduced legislation to initially reduce and ultimately prevent further large-scale clearance of native vegetation. This largely prevented further reductions in the cover of remnant native vegetation across the arable regions of South Australia, including the Mt Lofty region, but small areas continue to be cleared under permit.

Despite the extent of vegetation clearance, the Mt Lofty region along with Kangaroo Island was listed as one of 15 biodiversity hotspots across the continent by the Australian Government in 2003 (e.g. see Taylor et al., 2018). This reflects both the isolation of the woodlands in this region from other similar woodlands in south-eastern Australia, and the impending loss of biodiversity from the region. Ford and Howe (1980) were the first to point out that with just 10% of the native vegetation remaining, 35–50 of the 115 woodland bird species that originally occurred in the Mt Lofty region were likely to disappear, irrespective of how well the remnant vegetation was managed, because the amount of habitat remaining was insufficient to support all the species. The predicted losses of woodland bird species might be higher than this, given that the remaining vegetation is no longer pristine and is threatened by a range of perturbations, including incursions of weeds, excessive browsing of the understorey (e.g. Prowse et al., 2019, particularly by over-abundant Western Grey Kangaroos *Macropus fuliginosus*) and climate change.

As predicted by Ford and Howe (1980), most of the Mt Lofty region's woodland bird species have continued to decline since the early 1980s, despite there being no further vegetation clearance (Paton et al., 1994, 2004; Szabo et al., 2011; Prowse et al., 2021). Ten species are now considered regionally extinct and, for many others, regional extinction is imminent unless substantial amounts of additional woodland habitat are re-instated across the region in a timely fashion. The reason why many of the woodland bird species have not yet disappeared is because there is a time lag between the initial clearance of native vegetation and the loss of species: the disruption to demographic processes (reproduction, survival and dispersal) caused by vegetation clearance, which then leads to the disappearance of species, takes time to unfold. This lag provides a unique opportunity to reverse imminent species loss by reinstating sufficient additional habitat before it is too late. Delaying the re-instatement of habitat reduces the numbers of species that are ultimately retained.

The critical requirement is to put back substantial amounts of appropriate habitat. However, revegetation efforts within the region, although promoted as benefitting woodland birds, have been grossly inadequate. Most revegetation efforts to date have produced patches of vegetation that were too small (often less than 1 ha), linear in shape (e.g. along fence-lines and creek-lines), on poorer soils, consisting of a few plant species often planted in rows and at very high densities that were largely disconnected from other patches of vegetation (Paton et al., 2004). These revegetation efforts supported small numbers of widespread bird species and were rarely used consistently by declining species (Paton et al., 2004).

This lack of use is not surprising given that most of the declining woodland bird species have home ranges that are 3–20 ha and as much as 200 ha (Paton et al., 2010; Allan, 2016), much larger than the individual patches of revegetation. Furthermore, woodland birds require structurally and floristically diverse habitats not just to meet differing seasonal requirements for food and nesting but to meet the differing ecological requirements of diverse species (Paton et al., 2004). Future restoration efforts therefore need to reconstruct more diverse habitats at patch scales of 20–100 ha with priority given to revegetation on better quality soils. This is unlikely to be delivered by habitat reconstruction within individual farms and will likely require a proportion of existing farms, if not whole farms or districts, to be retired from agriculture and set aside for biodiversity conservation.

The critical question that arises from this is: how much additional habitat needs to be reinstated to prevent ongoing losses of woodland bird species from the Mt Lofty region? Paton (2010a) used Ford and Howe's (1980) species–area relationship model to assess woodland birds in the Mt Lofty region and estimated that increasing the area of 'native' habitat from a cover of <10% to a cover of 30% would prevent the loss of 35–40 woodland bird species. This

Published by Burleigh Dodds Science Publishing Limited, 2024.

equates to re-establishing woodland habitats across 150 000 ha of cleared land across the region. Delivering this new habitat would transform the region, require substantial changes to farming and require broad and on-going community support to deliver. This might seem like an ambitious target, but it is no more than called for in the UN Global Biodiversity Framework, that is, to effectively manage at least 30% of all land and water for biodiversity conservation (CBD, 2022). Australia was one of over 190 countries that supported this decision and, in November 2023, Australia's environment ministers agreed to protect and conserve 30% of Australia's land and oceans by 2030 (DCCEEW, 2024). This will not happen overnight, as there are many hurdles to overcome, ranging from securing funding, acquiring the land and learning how to build structurally and floristically diverse habitats for the future at ever-increasing scales.

4 Habitat restoration at scale: the role of BioR

BioR (https://www.bior.org.au/) is a not-for-profit established to reconstruct self-sustaining and resilient habitats for biodiversity outcomes. BioR is currently restoring habitat to a 550-ha farm near Monarto, on the eastern scarp of the Mt Lofty region, about 60 km east of Adelaide. The property, Frahns Farm, is crown land and was grazed by sheep until 2016, when it was set aside for habitat re-establishment. The area is in a rain shadow with low annual rainfall and is burdened with a range of environmental weeds and high densities of Western Grey and Red Kangaroos (*Osphranter rufus*).

Initial attempts to re-establish habitat on this property in 2016–2018 consisted of planting seedlings of a range of large shrubs and trees. This approach involved spot spraying to control weeds at the places where seedlings were to be planted and protecting the planted seedlings from grazing using corflute or wire mesh guards. The outcomes from these standard revegetation attempts were poor (Fig. 1h), with fewer than 10% of the seedlings establishing. A different approach was required and, in 2019, BioR changed its approach and methods.

First, an extensive area of the farm (~170 ha) was enclosed in a kangaroo-proof fence and over 100 kangaroos were removed from inside the enclosure. This approach avoids the need to individually guard each seedling and was cost-effective at this scale. The exclusion of kangaroos had other benefits in that there were immediate improvements in the condition of the small amounts of degraded remnant vegetation that remained within the fenced area, with some plants being able to recruit (reproduce and expand coverage) for the first time in decades (Fig. 1a). However, there was no recolonisation of the more extensive areas that had been cleared and managed as improved pasture for sheep.

Figure 1 (a) Fencing to exclude kangaroos enables the native vegetation remnants at Frahns Farm to recover from overgrazing (left side of image) while areas where kangaroos still have access remain heavily grazed (right) (photo credits: David Paton, unless otherwise indicated). (b) Grading the top 5–10 cm of topsoil reduces the high seed densities of environmental weeds, minimising subsequent re-establishment of weeds and improving revegetation success. (c) Auguring holes in preparation for planting seedlings. (d) Native grasses have germinated from direct seeding and are establishing in the area beyond the planting bowl into which a eucalypt seedling has been planted. (e) Planting bowls capture runoff for planted seedlings (credit: Tristan Avella-O'Brien). (f) Successful establishment of various native Australian plants 28 months after they were planted as seedlings at Frahns Farm. A standard corflute guard is shown against one of the plants. Corflute guards are frequently used to protect seedlings from grazing in Australia, but these guards restrict the lateral growth of most woody species at Frahns Farm, and BioR has ceased using them. One of the many advantages of exclusion fencing and removing overabundant herbivores is that individually guarding seedlings is no longer required. (g) Complementary planting of species in a planting bowl: *Allocasuarina verticillata* (tree to 5–10 m) with *Acacia argyrophylla* (shrub to 2 m), and *Atriplex semibaccata* (low spreading shrub to 20 cm); note the galenia infestation on the mound in the middle distance. (h) A planted seedling unable to grow beyond its corflute guard due to excessive kangaroo browsing in an unfenced area of Frahns Farm prior to 2019 (credit: Fiona Paton).

Second, for these pastured areas, approximately 5 cm of topsoil was removed with a grader to reduce the extensive seed banks that had accumulated over many years by various pasture grasses and weeds (e.g. *Avena fatua*, *Galenia pubescens*; Fig. 1b). Grading also reduced any allelopathic chemicals left in the surface soil by the weeds and reduced any added nutrients left over from past farming practices. A series of holes for planting seedlings was dug across the graded area with varied density and spacing for structural diversity (i.e. to produce areas with different densities of larger plants; Fig. 1c). The graded area was then direct-seeded with a mix of native grass and herb seeds prior to planting seedlings of larger shrubs and trees in the holes (Fig. 1d). The allocation of seedlings, involving more than 100 species of plants, considered topographical and edaphic features that might influence their performance. In addition, wherever possible at least five individuals of a species were planted within 20 m of each other to improve prospects of cross-pollination (McCallum et al., 2019).

The seedlings were planted by teams of volunteers under guidance, helping to reconnect people to nature as well as hopefully building lasting ownership. An important aspect was to form a shallow planting bowl to capture local runoff from rainfall, into which each seedling was planted (Fig. 1e). This helped offset low rainfall. Once planted, the seedlings were watered in and then watered again once or twice during the drier months in their first year in the absence of rainfall. This planting technique resulted in a dramatic improvement in planting success, with more than 80% and sometimes over 90% seedling survival and establishment in the first year (Fig. 1f).

In 2019, only 6 ha and 6000 seedlings were planted. However, by 2024, BioR's capacity had grown to re-establishing plants on over 20 ha of cleared land, direct seeding these areas with a diverse mix of seeds from native grasses, herbs and small shrubs, as well as planting more than 45 000 seedlings of shrubs and trees. Over 180 native plant species are now incorporated into the planting programs. Most of these are local species with seeds generally sourced from multiple areas within a 10-km radius of the farm but also including species from more xeric (dry) locations in the region. To further improve the efficiency of the planting programs, BioR increased the diameter of the planting bowls from 10 cm in 2019 to 45 cm in 2022 and now plants seedlings of two or three complementary plants in each planting bowl. Complementary plants might include a grass, a small herb or shrub and a large shrub or tree (Fig. 1g). BioR intends to grow the areas being planted by 20% per annum for the next 25 years so that, by 2050, more than 1000 ha of cleared land will be planted per annum. Even so, more than 1000 ha of land would need to be planted annually for another 50 years before the 150 000-ha target is reached.

Given the scale of the plantings needed to prevent the losses of species across the Mt Lofty region, any plantings that are undertaken need to be self-sustaining and resilient. To be self-sustaining, the plants need to flower and set seeds, with the seeds able to germinate and establish the next generation of plants, with recruitment matching losses of plants. Resilience refers to the capacity of the new vegetation to cope with and recover from perturbations like fire and drought. Although many of the shorter-lived plants in the revegetation at Frahns Farm have already begun flowering and setting seeds, assessing whether the new plantings are self-sustaining and resilient will need to wait until the vegetation matures: woodland habitats take decades to mature, with some habitat features like tree-hollows not appearing for more than 100 years. These planting programs therefore require intergenerational ownership to be delivered.

Equally, it is premature to assess whether declining woodland birds will colonise the new habitats, at Frahns Farm. However, similar plantings at Cygnet Park on Kangaroo Island, South Australia, commenced a decade earlier, are already being used by a wide range of woodland bird species, including species that have experienced significant declines in distribution and abundance (Paton et al., 2018; Figs 2 and 3).

Figure 2 (a) Cygnet Park consisted of about 100 ha of riparian woodland with the surrounding 200 ha cleared and grazed by sheep and cattle until 2004. From 2008 to 2012, (b) volunteers assisted with planting seedlings of a wide diversity of native plant species; (c) 2-year-old plantings (credit: Tom Hunt); culminating in (d) the re-establishment of 170 ha of diverse native vegetation on the cleared areas of this farm (credit: Blair Wickham).

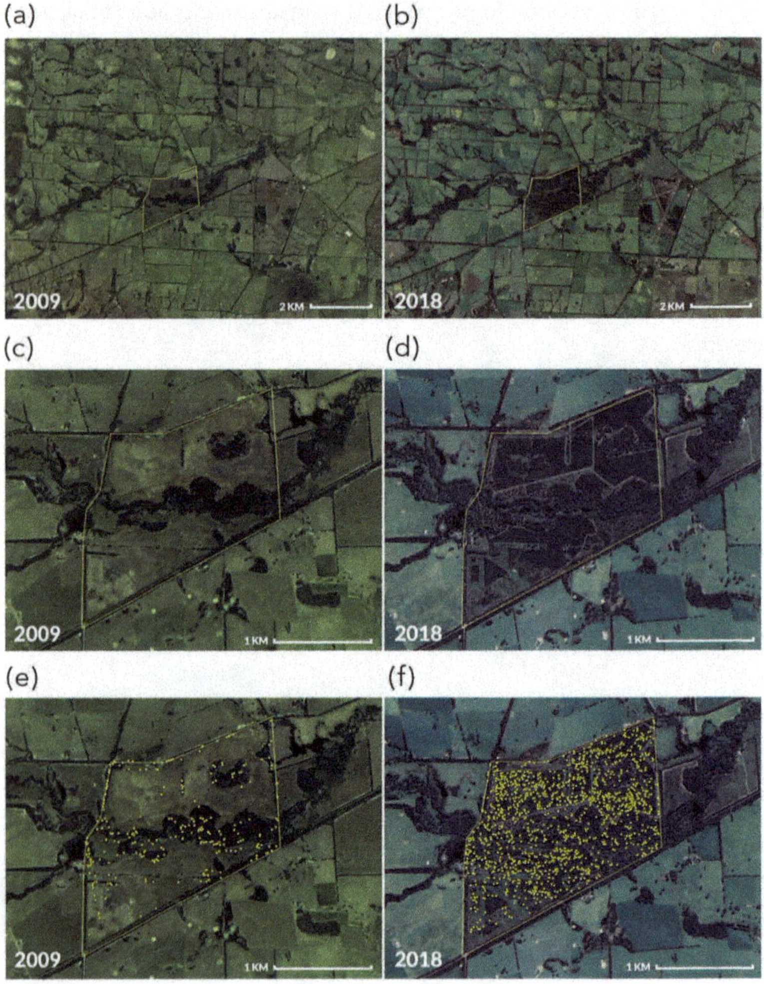

Figure 3 Aerial images of Cygnet Park (delimited by the yellow border) in (a, c, e) 2009 and (b, d, f) 2018, showing the cover of native vegetation (dark areas) pre- and post-planting at different scales and the distribution of woodland birds (shown as yellow dots) across the property in (e) spring 2009 and (f) spring 2018. Woodland birds were assessed using systematic area searches where the locations of all birds that were detected were mapped using a GPS (Paton et al., 2018).

5 Other broad-scale impacts of farming on native biodiversity: water extraction for irrigated agriculture

In addition to the almost complete transformation of terrestrial natural areas for agriculture, water extraction for irrigated agriculture can also have major impacts on the native biodiversity of rivers, their catchments, lakes and inland seas where extraction has exceeded sustainable limits. Well-known examples include:

- the de-watering of the Aral Sea for irrigation and the subsequent deterioration of a large portion of the former sea into the Aralkum Desert over the past 60 years (Micklin, 2007; Tussopova et al., 2020);
- the Colorado River (Schmidt et al., 2023);
- the Chesapeake Bay watershed (Alter et al., 2024); and
- the Murray–Darling Basin in south-eastern Australia (MDBA, 2010).

The obvious solution to over-allocation of water resources is to re-allocate sufficient quantities of water for the environment in order to sustain the native biodiversity that persists. This has been the approach in the 1 600 000-km^2 Murray–Darling Basin, where over-allocation of water principally for irrigated agriculture has seen a decline in biodiversity across the basin. Of the 13 000 GL of (long-term average) annual divertible flow in the basin, 11 500 GL (88%) is removed for irrigation, industrial use and domestic supply, with agricultural irrigation accounting for 95% of extractions.

The Murray–Darling Basin Plan, legislated in 2012, proposed to return 2750 GL of water to the environment. This was well below the 3000–7600 GL/year originally advised by the Murray–Darling Basin Authority (2010) prior to public consultation. After a pause in water buy-backs under the previous federal government, a reduced interim target of 2075 GL by 30 June 2024 was announced by the current government (Wheeler, 2023). In May 2023, there was still a shortfall of 49.2 GL to reach this target (MDBA, 2023). The following case study of the Murray–Darling Basin, with a focus on the ecological health of the terminal system of lakes and wetlands, explains how these current measures to return water in the basin to the environment may be insufficient to save the lakes and a Ramsar-listed wetland, the Coorong, from irreversible environmental decline (Ramsar relates to an international treaty for wetland conservation signed in the city of Ramsar in Iran in 1971).

6 Case study: over-extraction of water from the Murray-Darling Basin and consequences for biodiversity

The Murray–Darling Basin has been described as Australia's food bowl (Davidson, 2023). The Basin covers a seventh of the Australian continent and is spread across four Australian states: Queensland, New South Wales, Victoria and South Australia. The name is taken from the two major rivers that run through the basin: the Murray and Darling Rivers. The Darling River runs across New South Wales but has tributaries that stretch into southern Queensland. It joins the Murray River at Wentworth in New South Wales. The 2500-km-long Murray River runs from the Australian Alps westwards where it forms the boundary between the states of Victoria and New South Wales, then through eastern South Australia, before discharging into Lake Alexandrina immediately upstream of the Murray Mouth and ultimately into the Southern Ocean. A series of major tributaries join the Murray before reaching South Australia. Some of these tributaries include the Murrumbidgee, Lachlan and Darling Rivers on the New South Wales side of the river, and the Ovens, Goulburn and Loddon Rivers on the Victorian side.

It is hard to understate the environmental or socio-economic significance of Australia's Murray–Darling Basin (MDBA, 2010). Some 30 000 wetlands occur across the basin, of which 16 are Ramsar-listed, providing critical habitat for 95 state and Commonwealth-listed threatened inundation-dependent fauna species. The basin contributes 39% of Australia's agricultural production and provides critical water supplies for more than three million people.

Prior to any water being extracted from the basin, around 12 500 GL flowed annually to the sea, although the volumes varied dramatically from year to year. Extraction of water for irrigation from the Murray River commenced in the late nineteenth century and, within a few decades, there were concerns about increasing marine incursions into Lake Alexandrina, particularly during autumn when river flows were lowest. This ultimately led to a series of barrages being built along the southern edge of Lake Alexandrina in the late 1930s to prevent incursions of marine water upstream. The volume of water extracted from the basin grew steadily through the twentieth century, aided by the construction of a series of locks along the Murray and by large storage dams (e.g. Hume Dam). By the turn of the century, more than 13 000 GL per annum of surface flows was being extracted on average from the basin. Water harvested and stored in dams during years of higher flows was used to supplement the volumes being taken in low-flow years. Most of the water extracted was used to irrigate pastures and various perennial (e.g. almonds, citrus and grapes) and annual (e.g. rice and cotton) crops.

This extraction of water had a dramatic influence on the volumes of water that flowed down the river and eventually reached the Murray Mouth. During the 1980s and 1990s, only 4000–5000 GL per annum, on average, flowed out to sea (e.g. Paton, 2010b). The reduced volumes and duration of flows along the river in turn reduced the frequency, duration and extent that floodplain wetlands along the Murray were inundated, leading to profound negative effects on terrestrial and aquatic vegetation, aquatic invertebrates, fish and birds (e.g. Kingsford, 2000, Kingsford et al., 2011; Paton, 2010b). Flows to the Murray Mouth reduced even further during the first decade of the twenty-first century when south-eastern Australia experienced an extended drought from 1997 to 2009 (Bureau of Meteorology, 2024). During this Millennium Drought, flows to the Murray Mouth were inadequate to keep the mouth open, requiring continuous dredging by 2002. Keeping the mouth open was important for a range of diadromous fish that spend parts of their life cycle in freshwater and other parts in marine waters (Bice et al., 2018).

An open Murray Mouth was also important to maintain a regular tidal prism in the upper reaches of the Coorong. The Coorong is a long narrow lagoon wedged between the mouth and Lake Alexandrina and stretching 100 km to the southeast behind the coastal dunes of Younghusband Peninsula that separates the Coorong from the Southern Ocean. The Coorong is an inverse estuary (where the salinity increases with distance from the mouth) and differs from the water regimes of other wetlands in the Murray–Darling Basin. Salinities increase along the length of the Coorong according to distance from the Murray Mouth: the upper reaches closest to the mouth are estuarine while salinities at the southern end are two to three times saltier than sea water. During summer, the Coorong supports large numbers of migratory Palaearctic shorebirds (Figs 4a and b). Without a tidal prism – a volume of water that enters and leaves the estuary with the tides – these shorebirds would not be able to access the productive intertidal mudflats in the estuarine areas of the Coorong.

Reductions in flows in the lower Murray and, particularly, the timing of flows have had a profound impact on the ecology and health of the Coorong, particularly the hypersaline southern end. The southern Coorong with its higher salinities supports a lower diversity of aquatic organisms than the less salty areas but is still highly productive (e.g. Paton, 2010b). A key biotic component of these hypersaline waters is an annual plant, *Ruppia tuberosa*, which grows in shallow water around the margins of the southern Coorong (Fig. 4c). These margins are ephemeral in that they are usually covered with shallow water from late autumn until late summer, allowing the plant to establish, grow and reproduce before the water recedes. Critical to preventing the water from receding prior to autumn are flows over the barrages. Prior to any extraction of water, flows to the Murray Mouth increased during winter, peaked in late spring and then gradually declined over summer. Although more than 99% of this water flows out to sea, the flows elevate water levels in the northern Coorong, such that water levels are maintained in the southern Coorong, enabling the plant to complete its life cycle.

With extraction for irrigation, however, the flows are truncated and, although still increasing over winter and into spring, they often cease in late spring before the plants have finished flowering and setting seeds. Under these changed conditions, *R. tuberosa* struggles to maintain healthy vigorous populations. This was exacerbated during the Millennium Drought when little water reached the Murray Mouth over multiple years and *R. tuberosa*, a key component in food chains, was all but eliminated from the southern Coorong (Paton, 2010b).

Figure 4 Images of the Coorong and Lake Alexandrina. The abundance of Palaearctic shorebirds including (a) Red-necked Stints (*Calidris ruficollis*) and (b) Black-tailed Godwits (*Limosa limosa*) have declined in the Coorong and Murray lakes in recent years (credits: Darcy Whittaker). (c) Ecologists sampling for *Ruppia tuberosa* in the southern Coorong in winter (credit: Garry Trethewey). (d) Fairy Terns are listed as endangered in South Australia. They breed in the southern Coorong but only when the salinities are not so high to exclude their fish prey (credit: Tom Hunt). (e) So little water has flowed into Lake Alexandrina in recent droughts that the water level of the Lake has fallen below sea level for extended periods, exposing large areas of bare lake bed and separating the fringing reeds from the water. This image of the Mud Islands near the Murray Mouth shows the line of barrages that separate Lake Alexandrina from the Coorong stretching southwards in the top righthand corner of the image (credit: Russell Seawright).

The lack of adequate flows during the Millennium Drought also resulted in excessively high salinities in the southern Coorong. Instead of maximum salinities two to three times those of seawater, salinities in the southern Coorong were more than five times those of seawater. These levels exceeded the salinity tolerances of two highly salt-tolerant and formerly highly abundant aquatic species in the southern Coorong: the hardyhead fish, *Atherinosoma microstoma*, and the chironomid or midge, *Tanytarsus barbitarsis*. Both disappeared from the southern Coorong but maintained populations further north in the Coorong where salinities, although also higher, remained within tolerable ranges (Paton, 2010b; Wedderburn et al., 2016). With the loss of fish from the southern Coorong, many piscivorous birds like the threatened Fairy Tern (*Sternula nereis*) vacated the southern Coorong (Paton, 2010b; Fig. 4d).

The reason for the heightened salinities relates to the saltiness of water that offsets the water that evaporates off the surface of the Coorong over summer. When there is a flow of river water over the barrages into the northern Coorong, relatively fresh water with little salt is drawn into the Coorong to offset the evaporative losses. However, when there is no flow, seawater is drawn into the Coorong to offset evaporative losses. If there is an extended period with limited river flows, as there was during the Millennium Drought, salt accumulates in the southern Coorong and salinities rise.

The other conspicuous change during the Millennium Drought was the drop in water levels in Lake Alexandrina to more than a metre below sea level, stranding the extensive reed beds around the margins of the lakes, potentially exposing acid sulphate soils, and affecting a range of freshwater biota (e.g. Kingsford et al., 2011). Clearly the available water had been overallocated and there was insufficient reduction in the volume of water extracted upstream to prevent this (Fig. 4e).

Assessments and modelling indicated that between 3000 GL and 7600 GL needed to be returned to the environment to secure the health of the basin ecosystems, the lower value giving a low probability of success and the higher value a high probability of success (MDBA, 2010). Returning water to the environment is contentious, particularly given the large volumes proposed and the potential impact on irrigators and rural communities. The Australian Government introduced new legislation, the *Water Act 2007*, to help drive water reform, and the Murray–Darling Basin Plan was legislated in 2012. The Basin Plan aimed to return 2750 GL of water to the environment by 2024, by buying back water from irrigators and introducing water efficiency measures and other engineering solutions. The Water Act stipulated that the best available science should be used to determine the environmentally sustainable level of water that could be taken, but socioeconomic and political pressures resulted in the volumes to be returned to the environment being much less than the minimum needed. The intensity of public debate led to a Royal Commission (Walker, 2019) and water reform in the Murray–Darling Basin remains challenging.

More than 10 years after the Murray–Darling Basin Plan commenced, the full 2750 GL has still not been delivered. One of the enduring challenges is convincing upstream users that their water extraction contributes to the demise of unique wetland systems a thousand or more kilometres downstream. To date the reforms have been inadequate, as the Murray Mouth still needs dredges to keep it open, and the Coorong continues to be in poor condition (e.g. Paton et al., 2022; Paton and Paton, 2022).

7 Conclusion

According to the FAO (2019), a wide range of farm management practices and approaches favourable to the sustainable use and conservation of biodiversity for food and agriculture are being increasingly applied in many countries. A much higher proportion of OECD countries report the use of biodiversity-positive practices than non-OECD countries in the FAO's (2019) survey. The chapters in this book demonstrate that many farm management practices are positive both for farmers and native biodiversity. New regulations and policy frameworks have been developed or are being introduced in many jurisdictions to stop further destruction of native habitats in agricultural regions and to change environmental decision-making by farmers to align self-interest with biodiversity conservation. While there will inevitably be missteps in the development of new laws and agri-environmental schemes, these top-down approaches will evolve and improve, and successive generations of farmers will likely be increasingly accepting.

However, in the most recently and extensively transformed agricultural regions of the world where much wild biodiversity still persists (albeit for a short time before the extinction debts fully register), broad-acre restoration of natural environments and habitats is likely to be the only option to achieve the target of 30% of land set aside and managed for biodiversity by 2030 and even more ambitious native biodiversity conservation targets for 2050 and beyond. This will have to occur either through purchase of large areas of farmland by governments and the non-profit and private sectors and their ecological restoration, or the introduction of schemes that see farmers paid to undertake the restoration. Similarly, water entitlements will have to be returned to the environment in order to reduce and hopefully reverse biodiversity declines in the most over-allocated catchments around the world.

It remains to be seen whether setting targets such as 30% by 2030 in heavily transformed regions will spur governments, industry and the non-profit and philanthropic sectors to action or will be ignored as unrealistic. New regulations and policies will be required to facilitate land acquisition and broad-acre set-asides as well as the massive amounts of ecological, habitat and biodiversity restoration required to achieve 30% of land being able to support biodiversity in the world's most intensively farmed regions. Meeting the 30% target by 2030 is highly unlikely but the legislative and policy frameworks needed to allow this to happen should be advanced quickly, as time is of the essence.

New markets in natural capital (biodiversity, carbon, ecosystem services, etc.) could help generate the funds to rapidly upscale ecosystem restoration efforts. Large investments in education and training to develop a workforce to deliver and manage 30% of lands for biodiversity will be required, both for hands-on land acquisition and restoration work, as well as for biodiversity monitoring, planning, design and procurement. We have limited long-term evidence of how biodiversity is tracking across continents, countries, jurisdictions or regions, just at the time when such information is most critical to identify the species and populations spiralling towards local, regional or global extinction and thus most in need of prioritisation. Due to the growing importance of environmental, social and governance issues in board rooms and demanded by shareholders at corporate annual general meetings, agricultural industries should be encouraged to assume proactive responsibility for the native biodiversity impacted by their activities.

It is important to highlight that agricultural regions that have undergone the most complete transformation for food and fibre production away from wild nature, or have overallocated all or most of the available fresh water, are those where native terrestrial and aquatic biodiversity has and will suffer most. The collapse of regional biotas has already occurred in parts of the world with the longest history of agriculture and is underway in the most recently developed regions. Science has identified many biodiversity-friendly strategies and practices to accommodate many terrestrial native plants and animals in farmland, as the other chapters in this book testify, but the adoption of such approaches is often less than expected or desirable from a biodiversity conservation perspective, for a plethora of reasons.

Most worrying, however, is the biodiversity conservation challenge posed by the farming regions that have been most extensively cleared in the recent past. Even interim targets adopted by the global community, such as 30% of regional lands and waters managed for biodiversity conservation by 2030, are unlikely to be achievable in such regions, given the extent of farmland that needs to be repurposed, without increased intervention by the public, private or philanthropic sectors by at least an order of magnitude. The significant opposition of farming and regional communities to even the relatively modest and inadequate amount of water buy-backs for the environment in the Murray–Darling Basin to date suggests that buying up farmland to the 30% threshold could be politically impossible in all but the most unproductive agricultural regions, let alone more ambitious biodiversity targets for 2050 and beyond.

8 Future trends in research

Typically, ecologists undertake research on biodiversity in agricultural systems whereas agricultural scientists carry out the production-focused research, and economists, political scientists and lawyers study land clearance regulations and agri-environmental schemes. The FAO (2019) argued that research on food and agricultural systems should become more multi-disciplinary and focus more on interactions between the different components of such systems, including biodiversity. Improvements for the sustainable use and conservation of biodiversity are often constrained by a lack of understanding of interactions between sectors (e.g. between the production of crops and livestock and the new environmental commodities such as carbon and biodiversity), between wild and domesticated biodiversity and between the biophysical and socio-economic components of production systems and the policy settings that influence them. Cooperation across disciplines and greater involvement of farmers and their advisers in biodiversity conservation and restoration research in agricultural systems are required to overcome these knowledge gaps.

Nisbet (2007) dubbed long-term environmental monitoring 'Cinderella' science, being unloved and poorly funded. Reasons for the neglect of environmental monitoring by scientists and governments alike include a lack of funding, lack of appreciation of the value of long-term ecological monitoring in identifying the causes of environmental change and societal fascination with new and innovative science (Lindenmayer et al., 2014a). The sad fact remains that most of the world's biodiversity remains unmeasured. Despite clarion calls over the past two decades to invest more in long-term biodiversity monitoring (e.g. Legge et al., 2018; Lindenmayer et al. 2014b, 2022; Lovett et al., 2007; Takeuchi et al., 2021), we remain ignorant of short- and long-term trends in the abundance and distribution of most species and populations of wildlife the world over. We are thus almost certainly losing populations and species without knowing it, indeed before many taxa are known to science, which means the biodiversity crisis could be unfolding more quickly than most dare to believe. This must change and the development and implementation of new technologies to identify and continuously monitor biodiversity deserves the highest priority in environmental science.

Finally, research is required about the best ways to create new markets in biodiversity and related natural capital assets, such as carbon, in order to encourage private capital investment in the ecosystem restoration of wild nature and the protection and enhancement of native biodiversity at scale (Karolyi and Tobin-de la Puente, 2023). Public finance is insufficient. The difference between what is spent on nature conservation and what is needed, if most biodiversity is to be conserved and sustainably managed for the long term (the so-called 'biodiversity financing gap'), is significant (Deutz et al., 2020). Closing the gap and mobilising private investment to maintain ecosystem integrity and biodiversity, and the services they provide, requires hundreds of billions of dollars per year. UNEP (2021) estimated that the US$133 billion spent annually on nature-based solutions will need to triple by 2030 to meet global climate change, biodiversity and land degradation targets. Yet, the development of new markets in biodiversity to attract and funnel private capital to stem and reverse the global biodiversity crisis is in its infancy. Unfortunately, there are important barriers to the development of biodiversity credit markets, given the complexity of natural ecosystems, including issues around policy objectives, biodiversity metrics and equivalence, and market transaction data and transparency (Karolyi and Tobin-de la Puente, 2023). Research is required to guide the way markets value nature, nature-based assets and natural capital, and identify innovative financing and policy mechanisms to mobilise funding to conserve nature and reform the harmful agricultural policies that result in biodiversity loss.

9 Where to look for further information

Various books provide a comprehensive introduction to biodiversity and agriculture in different parts of the world. Norton and Reid (2013) introduce the reader to the native biodiversity in Australian and New Zealand farmland and provide many case studies of farmers who have incorporated different elements of wild nature on their farms. Macdonald and Feber's (2015) multi-authored collection describes the management of lowland farms for wildlife in the UK. McNeely and Scherr (2012) provide a global perspective of 'eco-agriculture', first describing the challenge to wild nature presented by agricultural intensification and rural poverty, then describing the opportunities for integrating biodiversity in agriculture and policies for encouraging eco-agriculture.

Seminal papers about the global biodiversity crisis and its causes include Dudley and Alexander's (2017) short review of agriculture and biodiversity. They find that agriculture is the largest contributor to biodiversity loss globally and that the impacts are expanding due to changing consumption patterns and growing populations. Vanbergen et al. (2020) document how land use intensification is undermining the natural foundations on which agriculture depends. They argue that agriculture will have to change to maximise productivity, sustainability and resilience, but this will ultimately depend on people's capacity to accept new ways of operating in response to the current environmental crisis.

Readers interested in the practice of ecosystem restoration should start with the joint publication by FAO, the Society of Ecological Restoration (SER) and the IUCN, setting out the *Standards of Practice to Guide Ecosystem*

Restoration (Nelson et al., 2024), and SER's 2nd edition of their *International Principles and Standards for the Practice of Ecological Restoration* (Gann et al., 2019). The specialist journals, *Ecological Restoration*, *Restoration Ecology* and *Ecological Management and Restoration*, chart the development of the applied ecological discipline of ecosystem restoration and its continuing evolution.

10 References

Allan, J. (2016), *Habitat Reconstruction Guidelines for Woodland Birds: A Bird-Orientated Approach*, PhD thesis, University of Adelaide, Adelaide, SA., Aust.

Alter, T., Hull, L., Laddin, K. F., Reid, M. and Kleinert, H. (2024), 'Engaging local voices: farmers, rural communities, and ecological restoration'. In N. Reid and R. Smith (Eds), *Managing Biodiversity in Agricultural Landscapes: Conservation, Restoration and Rewilding Practices in Farming*, Burleigh Dodds Science Publishing Limited, Cambridge, UK, pp. 97–136.

Andrén, M. (1994), 'Effects of habitat fragmentation on birds and mammals in landscapes with different proportions of suitable habitat: a review', *Oikos*, 71, 355–366.

Anon. (2021), *Biden-Harris Administration Outlines 'America the Beautiful' Initiative*, The White House, Washington, DC, https://www.whitehouse.gov/ceq/news-updates/2021/05/06/biden-harris-administration-outlines-america-the-beautiful-initiative/ (accessed 8 May 2024)

Ansell, D., Macintosh, A., Butler, D. and Waschka, M. (2024), 'Developments in agri-environment schemes (AES): Australia'. In N. Reid and R. Smith (Eds), *Managing Biodiversity in Agricultural Landscapes: Conservation, Restoration and Rewilding Practices in Farming*, Burleigh Dodds Science Publishing Limited, Cambridge, UK, pp. 319–337.

Bartel, R. L. (2003), 'Compliance and complicity: an assessment of the success of land clearance legislation in New South Wales', *Environ. Planning Law J.*, 20, 116–141.

Bartel, R. L. (2004), 'Satellite imagery and land clearance legislation: a picture of regulatory efficacy?' *Australas. J. Nat. Res. Law Policy*, 9, 1–31.

Bennett, A. F., Haslem, A., Rainsford, F. W., Maisey, A. C. and Radford, J. Q. (2024), 'Key concepts for conserving biodiversity in agricultural landscapes'. In N. Reid and R. Smith (Eds), *Managing Biodiversity in Agricultural Landscapes: Conservation, Restoration and Rewilding Practices in Farming*, Burleigh Dodds Science Publishing Limited, Cambridge, UK, pp. 3–48.

Bice, C. M., Wedderburn, S. D., Hammer, M. P., Ye, Q. and Zampatti, B. P. (2018), 'Fishes of the Lower Lakes and Coorong: a summary of life history, population dynamics and management'. In L. Mosley, Q. Ye, S. Shepherd, S. Hemming and R. Fitzatrick (Eds), *Natural History of the Coorong, Lower Lakes, and Murray Mouth Region (Yarluwar-Ruwe)*, Royal Society of South Australia, Adelaide, SA, Aust, pp. 371–399.

Blanchard, G. and Munoz, F. (2023), 'Revisiting extinction debt through the lens of multitrophic networks and meta-ecosystems', *Oikos*, e09435. https://doi.org/10.1111/oik.09435

Bureau of Meteorology (2024), Previous droughts. http://www.bom.gov.au/climate/drought/knowledge-centre/previous-droughts.shtml (accessed 19 May 2024)

Published by Burleigh Dodds Science Publishing Limited, 2024.

Carroll, C. and Noss, R. F. (2022), 'How percentage-protected targets can support positive biodiversity outcomes', *Conserv. Biol.*, 36, e13869. https://doi.org/10.1111/cobi.13869

Chubb, I., Bennett, A., Gorring, A. and Hatfield-Dodds, S. (2022), *Independent review of ACCUs*, Department of Climate Change, Energy, the Environment and Water, Canberra, ACT, Aust. https://www.dcceew.gov.au/sites/default/files/documents/independent-review-accu-final-report.pdf (accessed 23 June 2024)

Convention on Biological Diversity (2022), Decision adopted by the conference of the parties to the convention on biological diversity 15/4. Kunming–Montreal global biodiversity framework. https://www.cbd.int/doc/decisions/cop-15/cop-15-dec-04-en.pdf (accessed 19 May 2024)

Cozim-Melges, F., Ripoll-Bosch, R., Veen, G. F., Oggiano, P., Bianchi, F. J. J. A., van der Putten, W. H. and van Zanten, H. H. E. (2024), 'Farming practices to enhance biodiversity across biomes: a systematic review', *NPJ Biodivers.*, 3, 1. https://doi.org/10.1038/s44185-023-00034-2

Davidson, S. (2023), *Why Australia's Food Bowl, the Murray–Darling Basin, Is Making Headlines*, National Farmers Federation, Barton, ACT, Australia. https://farmers.org.au/news/why-australias-food-bowl-the-murray-darling-basin-is-making-headlines/ (accessed 19 May 2024)

Day, A., Njeru, E. M. and Johnson, D. (2024), 'Soil health and ecological restoration'. In N. Reid and R. Smith (Eds), *Managing Biodiversity in Agricultural Landscapes: Conservation, Restoration and Rewilding Practices in Farming*, Burleigh Dodds Science Publishing Limited, Cambridge, UK, pp. 163–196.

Department of Climate Change, Energy, the Environment and Water (2022a), Australian protected areas dashboard: terrestrial areas as at June 2022. https://www.dcceew.gov.au/environment/land/nrs/science/capad/dashboard (accessed 19 May 2024)

Department of Climate Change, Energy, the Environment and Water (2022b), *Threatened Species Strategy Action Plan 2022–2032*, Department of Climate Change, Energy, the Environment and Water, Canberra, Aust. https://www.dcceew.gov.au/environment/biodiversity/threatened/publications/action-plan-2022-2032 (accessed 19 May 2024)

Department of Climate Change, Energy, the Environment and Water (2024), Australia's strategy for nature: updates to Australia's strategy for nature. https://www.dcceew.gov.au/environment/biodiversity/conservation/strategy (accessed 10 May 2024)

Deutz, A., Heal, G. M., Niu, R., Swanson, E., Townshend, T., Li, Z., Dear, A., Meghji, A., Sethi, S. A. and Tobin-de la Puente, J. (2020), *Financing Nature: Closing the Global Biodiversity Financing Gap*, The Paulson Institute, The Nature Conservancy, and the Cornell Atkinson Center for Sustainability, Chicago, IL, USA.

Dickman, C. R., Greenville, A. C. and Wardle, G. M. (2024), 'Animal rewilding in theory and practice: Australia and New Zealand'. In N. Reid and R. Smith (Eds), *Managing Biodiversity in Agricultural Landscapes: Conservation, Restoration and Rewilding Practices in Farming*, Burleigh Dodds Science Publishing Limited, Cambridge, UK, pp. 463–500.

Dietz, M. S., Belote, R. T., Gage, J. and Hahn, B. A. (2020), 'An assessment of vulnerable wildlife, their habitats, and protected areas in the contiguous United States', *Biol. Conserv.*, 248, 108646. https://doi.org/10.1016/j.biocon.2020.108646

Dodsworth, J., Lasko, R. and Little, R. (2024), 'Stick your wellies on: messy development and participation processes with England's new Environmental Land Management policy – Co-design and the Sustainable Farming Incentive in context'. In N. Reid and R. Smith (Eds), *Managing Biodiversity in Agricultural Landscapes: Conservation, Restoration and Rewilding Practices in Farming*, Burleigh Dodds Science Publishing Limited, Cambridge, UK, pp. 275–300.

Dudley, N. and Alexander, S. (2017), 'Agriculture and biodiversity: a review', *Biodiversity*, 18, 45–49. https://doi.org/10.1080/14888386.2017.1351892

European Commission (2021), *EU biodiversity strategy for 2030 – Bringing nature back into our lives*, Directorate-General for Environment, Publications Office of the European Union. https://data.europa.eu/doi/10.2779/677548

European Union (2021), European Parliament resolution of 9 June 2021 on the EU Biodiversity Strategy for 2030: bringing nature back into our lives. https://eur-lex.europa.eu/legal-content/EN/TXT/?uri=CELEX:52021IP0277 (accessed 19 May 2024)

Food and Agriculture Organization of the United Nations (2019), *The State of the World's Biodiversity for Food and Agriculture*, J. Bélanger and D. Pilling (Eds), FAO Commission on Genetic Resources for Food and Agriculture Assessments, Rome, Italy. http://www.fao.org/3/CA3129EN/CA3129EN.pdf

Ford, H. A. and Howe, R. W. (1980), 'The future of birds in the Mount Lofty Ranges', *S. Aust. Ornithol.*, 27, 88–95.

Fowler, R. (1986), 'Vegetation clearance controls in South Australia – a change of course', *Environ. Planning Law J.*, 3, 48–66.

Gann, G. D., McDonald, T., Walder, B., Aronson, J., Nelson, C. R., Jonson, J., Hallett, J. G., Eisenberg, C., Guariguata, M. R., Liu, J., Hua, F., Echeverria, C., Gonzales, E. K., Shaw, N., Decleer, K. and Dixon, K. W. (2019), *International Principles and Standards for the Practice of Ecological Restoration*, Second edition, *Restoration Ecology*, S1–S46. https://www.ser.org/page/SERStandards

Harris, C. (2024), *History of Ãgriculture in SA – Native Vegetation Heritage – South Australian Initiatives to Better Manage the State's Native Vegetation*, Govt of S. Aust., Dept Primary Industries and Regions, Adelaide, SA, Aust., https://www.pir.sa.gov.au/aghistory/natural_resources/native_vegetation_conservation/national_parks#toc5 (accessed 14 May 2024)

Hanusch, M. (2023), *A Balancing Act for Brazil's Amazonian States: An Economic Memorandum*, The World Bank, Washington, DC.

High Ambition Coalition for Nature and People (2024), The HAC for Nature & People story. https://www.hacfornatureandpeople.org/the-hac-for-nature-people-story/ (accessed 8 May 2024)

Jones, T. A. (2024), 'Rewilding grasslands and rangelands'. In N. Reid and R. Smith (Eds), *Managing Biodiversity in Agricultural Landscapes: Conservation, Restoration and Rewilding Practices in Farming*, Burleigh Dodds Science Publishing Limited, Cambridge, UK, pp. 369–418.

Karolyi, G. A. and Tobin-de la Puente, J. (2023), 'Biodiversity finance: a call for research into financing nature', *Financial Managem.*, 52, 231–251. https://doi.org/10.1111/fima.12417

Kingsford, R. T. (2000), 'Ecological impacts of dams, water diversion and river management on floodplain wetlands in Australia', *Austral Ecol.*, 25, 109–127. https://doi.org/10.1046/j.1442-9993.2000.01036x

Kingsford, R. T., Walker, K. F., Lester, R. E. et al. (2011), 'A Ramsar wetland in crisis – The Coorong, Lower Lakes and Murray Mouth, Australia', *Mar. Freshwat. Res.*, 62, 255–265. https://doi.org/10.1071/MF09315

Klebl, F., Parisi, A., Häfner, K. et al. (2024), 'How values and perceptions shape farmers' biodiversity management: insights from ten European countries,' *Biol. Conserv.*, 291, 110496. https://doi.org/10.1016/j.biocon.2024.110496

Legge, S., Lindenmayer, D. B., Robinson, N. M. et al. (2018), *Monitoring Threatened Species and Ecological Communities*, CSIRO Publishing, Clayton South, Vic., Aust.

Lewis-Reddy, L. (2024), 'Implementing sustainable land use change programmes'. In N. Reid and R. Smith (Eds), *Managing Biodiversity in Agricultural Landscapes: Conservation, Restoration and Rewilding Practices in Farming*, Burleigh Dodds Science Publishing Limited, Cambridge, UK, pp. 137–160.

Lindenmayer, D., Burns, E., Thurgate, N. and Lowe, A. (2014a), 'The value of long-term research and how to design effective ecological research and monitoring', In D. Lindenmayer, E. Burns, N. Thurgate and A. Lowe (Eds), *Biodiversity and Environmental Change: Monitoring, Challenges and Direction*, CSIRO Publishing, Collingwood, Vic., Aust., pp. 21–48.

Lindenmayer, D., Burns, E., Thurgate, N. and Lowe, A. (2014b), *Biodiversity and Environmental Change: Monitoring, Challenges and Direction*, CSIRO Publishing, Collingwood, Vic., Aust.

Lindenmayer, D. B., Lavery, T. and Scheele, B. C. (2022), 'Why we need to invest in large-scale, long-term monitoring programs in landscape ecology and conservation biology', *Current Landscape Ecol. Rep.*, 7, 137–146. https://doi.org/10.1007/s40823-022-00079

Lindenmayer, D., Smith, D., Florance, D., Crane, C., Lang, E., Siegrist, A., Young, M. and Scheele, B. C. (2024), 'Lessons from reforestation of agricultural landscapes in south-eastern Australia'. In N. Reid and R. Smith (Eds), *Managing Biodiversity in Agricultural Landscapes: Conservation, Restoration and Rewilding Practices in Farming*, Burleigh Dodds Science Publishing Limited, Cambridge, UK, pp. 419–436.

Lovett, G. M., Burns, D. A., Driscoll, C. T. et al. (2007), 'Who needs environmental monitoring?' *Front. Ecol. Environ.*, 5, 253–260.

Macdonald, R. W. and Feber, R. (2015), *Wildlife Conservation on Farmland. Volume 1. Managing for Nature on Lowland Farms*, Oxford University Press, Oxford, UK.

McCallum, K. P., Breed, M. F., Paton, D. C. and Lowe, A. J. (2019), 'Clumped planting arrangements improve seed production in a revegetated woodland', *Restor. Ecol.*, 27, 636–646. https://doi.org/10.1111/rec.12905

Mcintosh, A. and Butler, D. (2023), Chubb review of Australia's carbon credit scheme falls short – and problems will continue to fester, *The Conversation* 9 January, https://theconversation.com/chubb-review-of-australias-carbon-credit-scheme-falls-short-and-problems-will-continue-to-fester-197401 (accessed 24 June 2024)

McNeely, J. A. and Scherr, S. J. (2012), *Ecoagriculture: Strategies to Feed the World and Save Biodiversity*, Island Press, Washington, DC, USA.

Meli, P., Schweizer, D., Winowiecki, L. A., Chomba, S., Aynekulu, E. and Guariguata, M. R. (2023), 'Mapping the information landscape of the United Nations Decade on Ecosystem Restoration Strategy', *Restor. Ecol.*, 31, e13810. https://doi.org/10.1111/rec.13810

Micklin, P. (2007), 'The Aral Sea disaster', *Ann. Rev. Earth Planetary Sci.*, 35, 47–72. https://doi.org/10.1146/annurev.earth.35.031306.140120

Montgomery, I. and Reid, N. (2024), 'The impact and management of hedgerows in promoting biodiversity in agricultural landscapes'. In N. Reid and R. Smith (Eds), *Managing Biodiversity in Agricultural Landscapes: Conservation, Restoration and Rewilding Practices in Farming*, Burleigh Dodds Science Publishing Limited, Cambridge, UK, pp. 221–271.

Morrison, J. (2024), 'The impact and design of field margins in promoting biodiversity in agricultural landscapes'. In N. Reid and R. Smith (Eds), *Managing Biodiversity in Agricultural Landscapes: Conservation, Restoration and Rewilding Practices in Farming*, Burleigh Dodds Science Publishing Limited, Cambridge, UK, pp. 197–220.

Murray–Darling Basin Authority (2010), *Guide to the Proposed Basin Plan: Overview*, Murray–Darling Basin Authority, Canberra, ACT, Aust., https://cdn.environment.sa.gov.au/environment/docs/guide-to-basin-plan-vol-1-gen.pdf (accessed 21 May 2024)

Murray–Darling Basin Authority (2023), Latest Basin Plan report card – June 2023. https://www.mdba.gov.au/water-management/basin-plan/basin-plan-report-cards (accessed 15 May 2024)

Myers, M. (1987). 'The extinction spasm impending: synergisms at work', *Conserv. Biol.*, 1(1), 14–21.

Nelson, C. R., Hallett, J. G., Romero Montoya, A. E. et al. (2024), *Standards of Practice to Guide Ecosystem Restoration – A Contribution to the United Nations Decade on Ecosystem Restoration 2021-2030*, FAO, Rome, Italy, Society for Ecological Restoration, Washington, DC, USA, and IUCN CEM, Gland, Switzerland. https://doi.org/10.4060/cc9106en

New South Wales Department of Climate Change, Energy, the Environment and Water (2024), Public register of private land conservation agreements. https://www.bct.nsw.gov.au/sites/default/files/2024-05/public-register-of-conservation-agreements.xlsx (accessed 20 May 2024)

Nisbet, E. (2007), 'Cinderella science', *Nature*, 450, 789–790. https://doi.org/10.1038/450789a

Norton, D. and Reid, N. (2013), *Nature and Farming: Sustaining Native Biodiversity in Agricultural Landscapes*, CSIRO Publishing, Collingwood, Vic., Aust.

Paton, D. C. (1996), *Overview of Feral and Managed Honeybees in Australia: Distribution, Abundance, Extent of Interactions with Native Biota, Evidence of Impacts and Future Research*, Australian Nature Conservation Agency, Canberra, ACT, Aust.

Paton, D. C. (2000), 'Disruption of bird-plant pollination systems in southern Australia', *Conserv. Biol.*, 14, 1232–1234.

Paton, D. C. (2010a), 'Restoring woodlands to the Mount Lofty Ranges region'. In D. Paton and J. O'Connor (Eds), *The State of Australia's Birds 2009: Restoring Woodland Habitat for Birds, Wingspan,* 20 supplement, 22–23.

Paton, D. C. (2010b), *At the End of the River: The Coorong and Lower Lakes*, ATF Press, Adelaide, SA, Aust.

Paton, D. C., Allan, J., Rogers, D. and Northeast, P. (2010), 'Bird use of the Monarto plantations: insights from a study of home ranges'. In D. Paton and J. O'Connor (Eds), *The State of Australia's Birds 2009: Restoring Woodland Habitat for Birds, Wingspan*, 20 supplement, 14–15.

Paton, D. C., Carpenter, G. and Sinclair, R. G. (1994), 'A second bird atlas of the Adelaide region. Part 1. Changes in the distribution of birds: 1974–75 vs 1984–85', *S. Aust. Ornithol.*, 31, 151–193.

Paton, D. C. and Paton, F. L. (2022), 'Monitoring of *Ruppia tuberosa* in the southern Coorong, summer 2021–22', University of Adelaide, Adelaide, SA, Aust., www.mdba.gov.au/publications-and-data/publications/lower-lakes-coorong-and-murray-mouth-monitoring-reports (accessed 18 June 2024)

Paton, D. C., Paton, F. L., Whittaker, D. A. and Markos, D. G. (2022), *Condition Monitoring of the Coorong, Lower Lakes and Murray Mouth Icon Site: Waterbirds in the Coorong and Lower Lakes 2022*, University of Adelaide, Adelaide, SA, Aust. www.mdba.gov.au/publications-and-data/publications/lower-lakes-coorong-and-murray-mouth-monitoring-reports (accessed 18 June 2024)

Paton, D. C., Prescott, A. M., Davies, R. J.-P. and Heard, L. M. (2000), 'The distribution, status and threats to temperate woodlands in South Australia'. In R. J. Hobbs and C. J. Yates (Eds), *Temperate Eucalypt Woodlands in Australia. Biology, Conservation, Management and Restoration*, Surrey Beatty & Sons, Chipping Norton, NSW, Aust., pp. 57–85.

Paton, D. C., Rogers, D. J. and Harris, W. (2004), 'Birdscaping the environment: restoring the woodland system of the Mt Lofty region, South Australia'. In D. Lunney (Ed.), *Conservation of Australia's Forest Fauna* (2nd ed.), Roy. Zool. Soc. NSW, Mosman, pp. 331–358.

Paton, F. L., Paton, D. C. and Paton, P. A. (2018), 'Bird responses to large scale revegetation at Cygnet Park, Kangaroo Island using systematic area searches'. In R. Smith (Ed.), *Proceedings of Restore, Regenerate, Revegetate: A Conference on Restoring Ecological Processes, Ecosystems and Landscapes in a Changing World*, University of New England, Armidale, NSW, Aust., 67–68. https://rune.une.edu.au/web/bitstream/1959.11/22627/2/open/SOURCE01.pdf

Pauli, B. P., Spaul, R. J. and Heath, J. A. (2017), 'Forecasting disturbance effects on wildlife: tolerance does not mitigate effects of increased recreation on wildlands', *Anim. Conserv.*, 20, 251–260.https://doi.org/10.1111/acv.12308

Prowse, T. A. A., O'Connor, P. J., Collard, S. J., Peters, K. J. and Possingham, H. P. (2021), 'Optimising monitoring of trend detection after 16 years of woodland bird surveys', *J. Appl. Ecol.*, 58, 1090–1100. https://doi.org/10.1111/1365-2664.13860

Prowse, T. A. A., O'Connor, P. J., Collard, S. J. and Rogers, D. J. (2019), 'Eating away at protected areas: total grazing pressure is undermining public land conservation', *Global Ecol. Conserv.*, 20, e00754. https://doi.org/10.1016/j.gecco.2019.e00754

Rausser, G. and Zilberman, D. (2024), 'Developments in agri-environment schemes (AES): North America'. In N. Reid and R. Smith (Eds), *Managing Biodiversity in Agricultural Landscapes: Conservation, Restoration and Rewilding Practices in Farming*, Burleigh Dodds Science Publishing Limited, Cambridge, UK, pp. 301–318.

Schmidt, J. C., Yackulic, C. B. and Kuhn, E. (2023), 'The Colorado River water crisis: its origin and the future', *WIREs Water*, 10, e1672. https://doi.org/10.1002/wat2.1672

Smith, K. J., Gordon, I. J., Wilson, B. A. and Manning, A. D. (2024), 'The future of animal rewilding in agricultural landscapes'. In N. Reid and R. Smith (Eds), *Managing Biodiversity in Agricultural Landscapes: Conservation, Restoration and Rewilding Practices in Farming*, Burleigh Dodds Science Publishing Limited, Cambridge, UK, pp. 437–462.

Squire, G. R. and Hawes, C. (2024), 'Biodiversity for agriculture: the role of integrated farm management in supporting agricultural production through biodiversity'. In N. Reid and R. Smith (Eds), *Managing Biodiversity in Agricultural Landscapes: Conservation, Restoration and Rewilding Practices in Farming*, Burleigh Dodds Science Publishing Limited, Cambridge, UK, pp. 49–96.

Szabo, J. K., Vesk, P. A., Baxter, P. W. J. and Possingham, H. P. (2011), 'Paying the extinction debt: woodland birds in the Mount Lofty Ranges, South Australia', *Emu*, 111, 59–70.

Takeuchi, Y., Muraoka, H., Yamakita, T., et al. (2021), 'The Asia-Pacific Biodiversity Observation Network: 10-year achievements and new strategies to 2030', *Ecol. Res.*, 36, 232–257. https://doi.org/10.1111/1440-1703.12212

Taylor, G. S., Braby, M. F., Moir, M. L. et al. (2018), 'Strategic national approach for improving the conservation management of Insects and allied invertebrates in Australia', *Austral Entomol.*, 57, 124–149. https://doi.org/10.1111/aen.12343

Tilman, D., May, R. M., Lehman, C. L. and Nowak, M. A. (1994), 'Habitat destruction and the extinction debt', *Nature*, 371, 65–66.

Tussopova, K., Anchita, Hjorth, P. and Moravej, M. (2020), 'Drying lakes: a review on the applied restoration strategies and health conditions in contiguous areas', *Water*, 12, 749. https://doi.org/10.3390/w12030749

United Nations Environment Program (2021), State of Finance for Nature 2021. Nairobi, Kenya.

United Nations Environment Program (UNEP) and Food and Agricultural Organization of the United Nations (FAO) (2020), *Strategy of the United Nations Decade on Ecosystem Restoration.* https://wedocs.unep.org/bitstream/handle/20.500.11822/31813/ERDStrat.pdf?sequence=1&isAllowed=y (accessed 26 June 2024)

Vanbergen, A. J., Aizen, M. A., Cordeau, S., Garibaldi, L. A., Garratt, M. P. D., Kovács-Hostyánszki, A., Lecuyer, L., Ngo, H. T., Potts, S. G., Settele, J., Skrimizea, E. and Young, J. C. (2020), 'Transformation of agricultural landscapes in the Anthropocene: nature's contributions to people, agriculture and food security', *Adv. Ecol. Res.*, 63, 193–253. https://doi.org/10.1016/bs.aecr.2020.08.002

van Diggelen, R., Ceulemans, T., Aggenbach, C. and Emsens, W-J. (2024), 'Restoring peatlands in European landscapes'. In N. Reid and R. Smith (Eds), *Managing Biodiversity in Agricultural Landscapes: Conservation, Restoration and Rewilding Practices in Farming*, Burleigh Dodds Science Publishing Limited, Cambridge, UK, pp. 341–368.

Walker, B. (2019), *Murray-Darling Basin Royal Commission Report*. Govt S. Aust., Adelaide, SA, Aust.

Watson, J. E. M., Venegas-Li, R., Grantham, H. et al. (2023), 'Priorities for protected area expansion so nations can meet their Kunming-Montreal Global Biodiversity Framework commitments', *Integrative Conserv.*, 2, 140–155. https://doi.org/10.1002/inc3.24

Wedderburn, S. D., Bailey, C. P., Delean, S. and Paton, D. C. (2016), 'Population and osmoregulatory responses of a euryhaline fish to extreme salinity fluctuations in coastal lagoons of the Coorong, Australia', *Estuarine Coastal Shelf Sci.*, 168, 50–57.

Williams, D. R., Rondinini, C. and Tilman, D. (2022), 'Global protected areas seem insufficient to safeguard half of the world's mammals from human-induced extinction', *Proc. Nat. Acad. Sci.*, 119(24), e2200118119. https://doi.org/10.1073/pnas.2200118119

Wheeler, S. A. (2023), Water buybacks are back on the table in the Murray–Darling Basin. Here's a refresher on how they work, The Conversation 23 February, https://theconversation.com/water-buybacks-are-back-on-the-table-in-the-murray-darling-basin-heres-a-refresher-on-how-they-work-200529 (accessed 15 May 2024).

Index

www.ingramcontent.com/pod-product-compliance
Lightning Source LLC
Jackson TN
JSHW010737020125
76326JS00003B/13
* 9 7 8 1 8 0 1 4 6 4 5 4 3 *